T0418432

Advances in Experimental Medicine and Biology

Editorial Board:

NATHAN BACK, *State University of New York at Buffalo, NY, USA*
IRUN R. COHEN, *The Weizmann Institute of Science, Rehovot, Israel*
ABEL LAJTHA, *N.S. Kline Institute for Psychiatric Research, Orangeburg, NY, USA*
JOHN D. LAMBRIS, *University of Pennsylvania, Philadelphia, PA, USA*
RODOLFO PAOLETTI, *University of Milan, Milan, Italy*

For further volumes:
http://www.springer.com/series/5584

Athena P. Kourtis · Marc Bulterys
Editors

Human Immunodeficiency Virus type 1 (HIV-1) and Breastfeeding

Science, Research Advances, and Policy

Editors
Athena P. Kourtis
Division of Reproductive Health
NCCDPHP, Centers for Disease Control
and Prevention, Atlanta,
GA, USA

Marc Bulterys
Division of Global HIV/AIDS (DGHA) Center
for Global Health, Centers for Disease Control
and Prevention (CDC) Atlanta, GA, USA

CDC Global AIDS Program, Beijing, China

Adjunct Professor of Epidemiology, UCLA School
of Public Health, Los Angeles, CA, USA

ISSN 0065-2598
ISBN 978-1-4614-2250-1   e-ISBN 978-1-4614-2251-8
DOI 10.1007/978-1-4614-2251-8
Springer New York Dordrecht Heidelberg London

Library of Congress Control Number: 2012932592

© Springer Science+Business Media, LLC 2012
All rights reserved. This work may not be translated or copied in whole or in part without the written permission of the publisher (Springer Science+Business Media, LLC, 233 Spring Street, New York, NY 10013, USA), except for brief excerpts in connection with reviews or scholarly analysis. Use in connection with any form of information storage and retrieval, electronic adaptation, computer software, or by similar or dissimilar methodology now known or hereafter developed is forbidden.
The use in this publication of trade names, trademarks, service marks, and similar terms, even if they are not identified as such, is not to be taken as an expression of opinion as to whether or not they are subject to proprietary rights.
While the advice and information in this book are believed to be true and accurate at the date of going to press, neither the authors nor the editors nor the publisher can accept any legal responsibility for any errors or omissions that may be made. The publisher makes no warranty, express or implied, with respect to the material contained herein.

Printed on acid-free paper

Springer is part of Springer Science+Business Media (www.springer.com)

«Ασκέειν, ...περί τα νοσήματα, δύο, ωφελέειν ή μη βλάπτειν»

"The physician must...have two special objects in view with regard to disease, namely, to do good or to do no harm"

          Hippocrates, Epidemics, 5th century B.C.

«Ποταμοίσι τοίσιν αυτοίσιν εμβαίνουσιν, έτερα και έτερα ύδατα επιρρεί»

"Ever-newer waters flow on those who step into the same rivers"

          Heraclitus, 6th century B.C.

Foreword

This book presents the collective scholarship by well-recognized experts summarizing the current knowledge on HIV and infant feeding. Although the focus is upon HIV postnatal transmission through breast milk, the lessons learned to date have shown the importance and benefits of breastfeeding as well as the disadvantages and risks of attempting to reduce or replace this mode of infant feeding. The risks of replacing breastfeeding are becoming appreciated through well-intended, but poor, practice. The absolute necessity of breastfeeding, especially in settings with high levels of infectious diseases and poor water and sanitation, the importance of exclusive breastfeeding in the first 6 months, the continued importance of breastfeeding to the second year of life, and the inherent risks incurred by early breastfeeding cessation (what is more commonly referred to as early weaning) with its measurable adverse consequences on infant morbidity, nutrition, and survival have been well understood in the child-survival literature and now relearned by HIV researchers.

It continues to be a mystery why the majority (60–85%) of infants born to HIV-infected women are not infected with HIV without receipt of any intervention. It is also mystifying that a majority (85–90%) of all HIV-exposed babies, who consume hundreds of liters of HIV-infected breast milk with frequent daily feedings, escape infection. These facts suggest that the balance of all factors contributing to transmission of virus is tipped away from HIV transmission and we are fortunate to be able to push the balance even further from transmission by the science which has been learned. In 2009, an estimated 90% of 370,000 infants (the annual incidence of all HIV infections in children) is due to mother-to-child transmission (MTCT) of the virus. The postnatal transmission of virus through breastfeeding is estimated to be a third to a half of all transmission from mothers to their babies. At the same time, in the same context in which most of the HIV infections from MTCT occur, approximately 3.5 million children under five continue to die annually from diarrhea, pneumonia, and neonatal sepsis, all of which are significantly reduced by breastfeeding, and over one-third of the ten million childhood deaths annually are associated with undernutrition (WHO 2010). Breastfeeding has been identified as the most effective child-survival intervention, estimated to prevent 1.3 million deaths to children under 5 in the world annually (Jones, G., Steketee, R. W., Black, R. E., Bhutta, Z. A., Morris, S. & the Bellagio Child Survival Study Group (2003): How many child deaths can we prevent this year? Lancet 362: 65–71.).

We have learned a lot over the last two decades about effective interventions to lower MTCT. Research and interventions to diminish MTCT initially focused on the peripartum period, successfully using antepartum and intrapartum administration of antiretrovirals to curtail the infant acquisition of virus. In clinical trials, in the developed and the developing world and in the surveillance data of the developed world, transmission rates of 1–2% are achieved with optimum regimens initiated antepartum, and continued through labor and delivery. The postpartum transmission of HIV by ingestion of breast milk has only recently been studied and it too has shown to be substantially and significantly reduced with antiretroviral interventions. Antiretroviral drugs taken by the mother, a single drug taken by the baby for as long as 6 months or some combination of drugs to the mother and baby, successfully reduce breast milk transmission. These data are incorporated into the 2010 WHO recommendations extrapolated to continuing medications for the duration of breastfeeding.

Reviewing the lessons learned (or relearned) in HIV and breastfeeding is informative for improving maternal and child health. First, in the USA, early guidelines (1985) advised strongly against HIV-infected mothers feeding their children from the breast. This guidance was provided as it prevented infant exposure to the HIV virus after delivery. It was safe to avoid breastfeeding in the developed world with adequate nutrition provided to the babies in an environment not encumbered by multiple other infectious challenges. It is germane to point out that medical and public health authorities took responsibility for the decision making, not individual women. The decision on infant feeding was, and continues to be, established as public health policy in the USA.

In the developing world, not only is extended breastfeeding for about 2 years of life the predominant practice but it is also the medically recommended form of optimal feeding due to its many scientifically proven benefits, including reduced risk of infant mortality, immune protection which reduces risk of the incidence and severity of common childhood illnesses like diarrhea and respiratory infections, as well as the considerable nutritional and economic benefits. The medical advice to exclusively breastfeed for the first 6 months followed by continued breastfeeding through the child's second birthday or longer was backed by sound, scientific evidence and strongly promoted as optimal infant feeding by WHO, UNICEF and the public health world in general. The fear of HIV transmission by breast milk by the mid-1990s began to undermine both the practice and medical advice about the importance of breastfeeding and of specific breastfeeding practices.

Researchers in the late 1990s then showed that exclusive breastfeeding in the first 3–6 months of life was associated with significantly less HIV transmission than mixed feeding. These data were not readily accepted, to many it seemed counterintuitive to advise more exposure to breast milk virus. The lower HIV transmission rates associated with exclusive breastfeeding, however, were repeatedly confirmed and finally accepted and incorporated into WHO guidelines in 2001, although the preferred infant feeding option was still replacement feeding. This led to policies that differed between the developed and developing world—no breastfeeding for HIV-infected women in the former and exclusive breastfeeding for at least the first few months followed by abrupt weaning in the latter. Not surprisingly, health workers and policy makers were confused.

Instead of providing clear medical advice, women were individually asked to weigh rather complex scientific evidence, take into account their own circumstances, and make an informed choice to replacement feed or not, based on AFASS criteria of replacement feeding being acceptable, feasible, affordable, sustainable, and safe. Otherwise, exclusive breastfeeding with early and abrupt weaning was recommended (and clearly seen as a second best solution). There was a prevailing opinion in the international HIV community that emulation of the developed world with replacement feeding was the ideal way to proceed, the difficulties (including stigma) and mortality risks associated with replacement feeding, using bottles and early and abrupt cessation of breastfeeding tended to be underestimated, the difficulties in getting women to change to exclusive breastfeeding exaggerated, and the risk of HIV transmission associated with abrupt weaning was not yet understood.

It is interesting to note that many developing-country women and the programs that worked closely with them were somewhat quicker to seize on the findings on the protective effects of exclusive breastfeeding as compared to mixed feeding in the first 3–6 months of life than many in the international community. These stakeholders were more aware of the dangers and difficulties inherent in implementing a policy that relied on breast milk substitutes and bottle feeding, and had more confidence that mothers would be willing to alter longstanding feeding practices in the best interests of their children. The HIV epidemic has had both negative and positive unintended consequences on breastfeeding in resource poor settings—one of the positive unintended consequences is that rates of exclusive breastfeeding in the first 6 months have increased significantly and substantially in the last decade in many countries throughout the world, with some of the largest increases in countries with high HIV prevalence, attributed to an increased understanding of the protective effects of exclusive breastfeeding relative to mixed feeding to prevent transmission of HIV, in addition to its other benefits (Kothari, M and Noureddine A. 2010. *Nutrition Update 2010*. Calverton, Maryland, USA: ICF Macro).

As data documented the cumulative risk of HIV transmission with lengthening of the period of breastfeeding, it seemed logical to decrease the exposure to breast milk by shortening the duration of breastfeeding. Clinical trials in the mid 2000s, however, began documenting, to the world's consternation, that there was increased morbidity and mortality associated with early cessation of breastfeeding. In fact, when the concept of HIV-free survival was incorporated, so that death from any cause was valued equally and an added benefit was achieved if a child was HIV-free, it was shown that shortening the duration of breastfeeding led to increased mortality from causes other than HIV. Early weaning did not improve HIV-free survival. HIV-free survival is an accepted concept but should be qualified. It ignores the morbidity of hospitalizations and other nonfatal illnesses. It also ignores the longer survival of HIV-infected infants who breastfeed. The 2010 WHO guidelines endorse HIV- free survival, swing the pendulum clearly back in favor of breastfeeding in resource poor settings based on the accumulated scientific evidence, and recommend that national policy in each country decide on infant feeding practices for HIV-infected women. This recommendation removes the decision making from individual women and places it in the guidelines/policy arena, some 2½ decades after the decision was taken at the policy level in the USA.

The book thoroughly presents the science of HIV and infant feeding. The existing science clearly shows that most cases of breast milk transmission of HIV are preventable through antiretroviral drugs and modifications to infant feeding practices. The scientific world now knows how to employ antiretrovirals to prevent MTCT of HIV. The scientific and programming worlds also know, but have been less effective, in assessing and effectively incorporating the nondrug-related infant feeding interventions, like exclusive breastfeeding and later and more gradual breastfeeding cessation, that have proven effective in reducing MTCT of HIV and increasing HIV-free survival in children. We should be both encouraged and disheartened by findings like those shown in Ch 16, Table 2 that health workers' advice (in this case for early breastfeeding cessation/early weaning) led to reducing the proportion of women still breastfeeding at 9 months from 98 to 35%—encouraged, as it demonstrates that health workers' advice can be effective in changing long engrained practices; disheartened by the fact that for most of these women, we now know that the advice on early weaning likely reduced HIV-free survival in this population.

Programmatic and more limited scientific data also suggest that the most effective drug regimens (which antiretrovirals, to whom, and for how long) and the most effective modifications to infant feeding practices for a given setting are highly dependent on the underlying epidemiological, service delivery, political, and sociocultural environments.

It is also clear that while the science on these issues has rapidly advanced, PMTCT programming practice has not kept pace. Why has PMTCT been implemented so successfully in the developed world and still relatively slowly in the developing world? There is an estimated decrease of 25% in HIV transmission in 2009 as compared to 2004. The difficulties of implementation at every level of service delivery continue to plague programmatic action. These difficulties are evident in the relative failure to effectively and rapidly reduce transmission in the developing world to 1–2%.

Implementation science must more quickly uncover and address the specific barriers in each setting which continue to prevent the elimination of MTCT. It is not lack of knowledge, lack of clinical trials, lack of drugs, resistant cultural practices, or lack of interest in implementation which is responsible for the incomplete uptake of all PMTCT services, including those related to infant feeding. There is no acceptable excuse for failure to provide these services to all pregnant women who access any antenatal care and to continue to offer follow-up services to these women and their affected children for at least 2 years postpartum, if not beyond, for those who remain on antiretroviral treatment. Coverage of the population will be incomplete unless those not accessing care can also be reached.

The developing world has had particular difficulty reaching immunocompromised pregnant women with appropriate therapy. Solving the political, behavioral, and implementation problems of suboptimal uptake of PMTCT services for all HIV-infected women will prove essential if these services are to be delivered optimally to curtail postpartum transmission of HIV by breastfeeding.

The editors and authors are to be commended for compiling this information and presenting it in such a way as to provide a factual basis for essential next steps to improve worldwide programs to prevent MTCT of HIV.

Durham, NC, USA Catherine M. Wilfert, MD
Washington, DC, USA Katherine A. Krasovec, ScD

Foreword

2011 marks the 30th anniversary of the first reports of what came to be called acquired immunodeficiency syndrome (AIDS), as well as the 15th anniversary of the introduction of combination antiretroviral therapy. The early 1980s were characterized by remarkably rapid scientific advances, including recognition of HIV-1, documentation of the modes of HIV-1 transmission, proposal of means of prevention that have largely stood the test of time, and introduction of a serologic diagnostic test. This bland summary of milestones fails to capture the human and scientific drama behind these historic events, as well as the fierce debates among scientists, medical workers, and the general public about the implications of the AIDS epidemic.

As always, science alone could not address societal concerns, nor were its conclusions always readily accepted even in the face of strong evidence. Acceptance that HIV-1 was transmitted through the blood supply, for example, took time. Additional areas of initial controversy were whether the AIDS virus could be transmitted heterosexually and from mother to child. Realization that HIV-1 could also be transmitted through breastfeeding was especially difficult to accept by legions of public health workers and development specialists who had toiled for years in low-income countries to increase breastfeeding as a key strategy for improving child survival by preventing diarrheal diseases, respiratory infections, and malnutrition.

AIDS offered stark alternatives to HIV-infected women, especially in sub-Saharan Africa, the region most heavily affected by HIV/AIDS: continue to breastfeed and risk infants acquiring HIV-1; avoid breastfeeding, and risk increased infant and under-5 mortality from other causes. The emphasis on autonomy and individual rights that was part of the AIDS response paradoxically left women isolated with an unfair burden of choice, while public health officials and agencies grappled with their own conflicts and mixed opinions about the limited interventions, such as replacement feeding with formula.

Fortunately, because of widespread access to antiretroviral therapy and extensive research, options are now available for those dire choices that seemed irresolvable in the 1980s and 1990s. The evolution of our understanding of HIV-1 transmission by breastfeeding, the relevant virology and immunology, modalities for prevention, and the emergent policy and programmatic approaches are comprehensively captured in this timely book. HIV/AIDS science continues to progress with major implications for HIV prevention policies. Perhaps the greatest debate currently is around how best to use antiretroviral therapy for maximal benefit for both individual health as well as to prevent transmission to others, and the pendulum is swinging towards earlier and more widespread therapy. From the pessimism of the past, we have entered an era of possibility—even in our fiscally constrained times—so that we even dream of virtual elimination of pediatric HIV disease on a worldwide basis. If that vision seems unduly optimistic, it at least is not for lack of available tools.

The longer view in life is often necessary to fully understand an issue, and so it seems with HIV/AIDS and child survival. Despite the tragedy of almost eight million deaths annually in children under 5 across the world (the great majority in sub-Saharan Africa and South Asia), tremendous improvements in child survival have occurred and continue to do so. HIV-1 incidence is declining in many African countries and the scale-up of treatment and prevention programs that has occurred would have been unthinkable a decade ago. It is now possible to reduce mother-to-child transmission in sub-Saharan Africa to levels not very different from those achieved in Europe or North America. This represents success, and this book usefully and eloquently describes it, for which I congratulate the editors and authors.

Atlanta, GA, USA   Kevin M. De Cock, MD, FRCP (UK), DTM&H

Acknowledgments

We gratefully acknowledge the many mothers and infants who have participated selflessly in the randomized clinical trials to prevent HIV transmission through breastfeeding.

We would like to thank Christine Korhonen, MPH, and Serena Fuller, MPH, Allan Rosenfield Global Health Fellows at the CDC—Global AIDS Program in China; as well as Jeremy Sueker, medical student at Harvard University School of Medicine and intern at the Centers for Disease Control—Global AIDS program, for their helpful review and editing of many of the chapters. We also thank Dr. R.J. Simonds for insightful comments and discussion.

We dedicate this book to our parents (Dr. Kamiel Bulterys, Dr. Petros and Ms. Maria Kourtis), who provided fundamental guidance on our medical/scientific paths, and to our spouses (Ann and Alexander) and children (Philip and Michelle; Peter and Constantine) for their patience, sacrifice, and unending support, with loving gratitude.

The Editors

Contents

Part I Transmission of HIV-1 Infection and Other Viruses to the Infant Through Breastfeeding: General Issues for the Mother and Infant

1. **Breastfeeding and Transmission of HIV-1: Epidemiology and Global Magnitude** 3
   Mary Glenn Fowler, Athena P. Kourtis, Jim Aizire, Carolyne Onyango-Makumbi, and Marc Bulterys

2. **Breastfeeding and Transmission of Viruses Other than HIV-1** 27
   Claire L. Townsend, Catherine S. Peckham, and Claire Thorne

3. **Breastfeeding Among HIV-1 Infected Women: Maternal Health Outcomes and Social Repercussions** 39
   Elizabeth Stringer and Kate Shearer

4. **Early Diagnosis of HIV Infection in the Breastfed Infant** 51
   Chin-Yih Ou, Susan Fiscus, Dennis Ellenberger, Bharat Parekh, Christine Korhonen, John Nkengasong, and Marc Bulterys

Part II Mechanisms of HIV-1 Transmission Through Breast Milk: Virology

5. **Virologic Determinants of Breast Milk Transmission of HIV-1** 69
   Susan A. Fiscus and Grace M. Aldrovandi

6. **HIV-1 Resistance to Antiretroviral Agents: Relevance to Mothers and Infants in the Breastfeeding Setting** 81
   Michelle S. McConnell and Paul Palumbo

7. **Animal Models of HIV Transmission Through Breastfeeding and Pediatric HIV Infection** 89
   Koen K.A. Van Rompay and Kartika Jayashankar

8. **Antiretroviral Pharmacology in Breast Milk** 109
   Amanda H. Corbett

Part III Mechanisms of HIV-1 Transmission Through Breast Milk: Immunology

9 The Immune System of Breast Milk: Antimicrobial and Anti-inflammatory Properties 121
Philippe Lepage and Philippe Van de Perre

10 B Lymphocyte-Derived Humoral Immune Defenses in Breast Milk Transmission of the HIV-1 139
Laurent Bélec and Athena P. Kourtis

11 Cellular Immunity in Breast Milk: Implications for Postnatal Transmission of HIV-1 to the Infant 161
Steffanie Sabbaj, Chris C. Ibegbu, and Athena P. Kourtis

Part IV Prevention of Breast Milk Transmission of HIV-1

12 Antiretroviral Drugs During Breastfeeding for the Prevention of Postnatal Transmission of HIV-1 173
Athena P. Kourtis, Isabelle de Vincenzi, Denise J. Jamieson, and Marc Bulterys

13 Immune Approaches for the Prevention of Breast Milk Transmission of HIV-1 185
Barbara Lohman-Payne, Jennifer Slyker, and Sarah L. Rowland-Jones

14 Non-antiretroviral Approaches to Prevention of Breast Milk Transmission of HIV-1: Exclusive Breastfeeding, Early Weaning, Treatment of Expressed Breast Milk 197
Jennifer S. Read

15 Breast Milk Micronutrients and Mother-to-Child Transmission of HIV-1 205
Monal R. Shroff and Eduardo Villamor

Part V Research Implementation and Policy Related to Breastfeeding by HIV-1-Infected Mothers

16 Historical Perspective of African-Based Research on HIV-1 Transmission Through Breastfeeding: The Malawi Experience 217
Taha E. Taha

17 Breastfeeding and HIV Infection in China 237
Christine Korhonen, Liming Wang, Linhong Wang, Serena Fuller, Fang Wang, and Marc Bulterys

18 The Role of the President's Emergency Plan for AIDS Relief in Infant and Young Child Feeding Guideline Development and Program Implementation 247
Michelle R. Adler, Margaret Brewinski, Amie N. Heap, and Omotayo Bolu

19 HIV-1 and Breastfeeding in the United States 261
Kristen M. Little, Dale J. Hu, and Ken L. Dominguez

Part VI DEBATE: Should Women with HIV-1 Infection Breastfeed Their Infants? Balancing the Scientific Evidence, Ethical Issues and Cost-Policy Considerations

20 Pendulum Swings in HIV-1 and Infant Feeding Policies: Now Halfway Back 273
Louise Kuhn and Grace Aldrovandi

21 Should Women with HIV-1 Infection Breastfeed Their Infants? It Depends on the Setting 289
Grace John-Stewart and Ruth Nduati

Part VII Epilogue

22 The Future of Breastfeeding in the Face of HIV-1 Infection: Science and Policy 301
Marc Bulterys and Athena P. Kourtis

Index 305

Part I
Transmission of HIV-1 Infection and Other Viruses to the Infant Through Breastfeeding: General Issues for the Mother and Infant

Chapter 1
Breastfeeding and Transmission of HIV-1: Epidemiology and Global Magnitude*

Mary Glenn Fowler, Athena P. Kourtis, Jim Aizire, Carolyne Onyango-Makumbi, and Marc Bulterys

Introduction

Over the past two decades, major strides have been made in HIV-1 research and prevention. Among these advances, some of the most remarkable and sustained achievements have been in reducing the risk of transmission of HIV-1 from mothers to their infants. In resource-rich settings such as the USA and Europe, mother-to-child transmission (MTCT) of HIV-1 has successfully been reduced to less than 1–2% [1] with the goal of virtual elimination of new cases. This success in prevention of mother-to-child transmission (PMTCT) of HIV-1 has been achieved by widespread implementation of effective PMTCT antiretroviral therapy (ART) regimens, and obstetrical interventions as well as avoidance of breastfeeding through the use of infant formula.

In resource-limited international settings, progress toward reduction in PMTCT has been more modest. One major challenge has been balancing the importance of extended breastfeeding to overall infant health and survival against its known risk of HIV-1 transmission. Initial PMTCT studies in the mid to late 1990s focused on short-course peripartum regimens to reduce MTCT, targeting transmission around the time of labor/delivery. More recently, studies have focused on extended prophylactic ART regimens given to the HIV-1-infected mother during the third trimester and labor/delivery, in combination with postnatal regimens provided either to the mother or her HIV-1-exposed infant throughout breastfeeding.

---

*The findings and conclusions in this report are those of the authors and do not necessarily represent the official position of the Centers for Disease Control and Prevention.

M.G. Fowler, M.D., M.P.H. (✉)
Department of Pathology, Johns Hopkins Medical Institutes, Baltimore, MD, USA

Onsite Makerere University–Johns Hopkins University Research Collaboration, Kampala, Uganda
e-mail: mgfowler@mujhu.org

A.P. Kourtis, M.D., Ph.D., M.P.H.
Division of Reproductive Health, NCCDPHP, Centers for Disease Control and Prevention, 4770 Buford Highway, NE, MSK34, Atlanta, GA 30341, USA

J. Aizire, M.B.Ch.B., M.H.S. • C. Onyango-Makumbi, M.B.Ch.B., M.S.
Makerere University–Johns Hopkins University Research Collaboration, Upper Mulago Hill Rd, Kampala, Uganda

M. Bulterys, M.D., Ph.D.
Division of Global HIV/AIDS (DGHA), Center for Global Health, Centers for Disease Control and Prevention (CDC), 1600 Clifton Road, NE, Atlanta, GA 30333, USA

CDC Global AIDS Program, Beijing, China

Adjunct Professor of Epidemiology, UCLA School of Public Health, Los Angeles, CA, USA

This chapter provides an overview on the current state of HIV-1 transmission from mothers to infants through breastfeeding. In this context, we present current statistics on the worldwide epidemic, detail the epidemiology of MTCT including risk factors and timing of transmission, discuss possible mechanisms of transmission, and recent research findings which led to updated World Health Organization (WHO) PMTCT recommendations in 2010. We also address remaining research gaps and planned future research studies.

Epidemiology of HIV-1 Transmission: Global Trends

The HIV-1 pandemic continues to levy a heavy health and economic burden on the human race worldwide. The distribution of incident HIV-1 cases, shown in Fig. 1.1 and based on 2009 Joint United Nations Program on HIV-1/AIDS (UNAIDS) data [2], highlights that the largest numbers of infected individuals live in sub-Saharan Africa (SSA) followed by India and the Far East. According to UNAIDS data, the estimated number of people who became newly infected with HIV-1 in 2009 was 2.6 million (2.3–2.8 million) representing a 19% reduction since 1999. While significant declines of more than 25% were registered in several countries including high HIV-1 burden-settings in SSA, rates of new annual HIV-1 infections remained stable in Western, Central, and Eastern Europe; Central

GLOBAL REPORT

Regional HIV and AIDS statistics and features | 2009

	Adults and children living with HIV	Adults and children newly infected with HIV	Adult prevalence (15–49) [%]	Adult & child deaths due to AIDS
Sub-Saharan Africa	22.5 million [20.9 million – 24.2 million]	1.8 million [1.6 million – 2.0 million]	5.0% [4.7% – 5.2%]	1.3 million [1.1 million – 1.5 million]
Middle East and North Africa	460 000 [400 000 – 530 000]	75 000 [61 000 – 92 000]	0.2% [0.2% – 0.3%]	24 000 [20 000 – 27 000]
South and South-East Asia	4.1 million [3.7 million – 4.6 million]	270 000 [240 000 – 320 000]	0.3% [0.3% – 0.3%]	260 000 [230 000 – 300 000]
East Asia	770 000 [560 000 – 1.0 million]	82 000 [48 000 – 140 000]	0.1% [0.1% – 0.1%]	36 000 [25 000 – 50 000]
Central and South America	1.4 million [1.2 million – 1.6 million]	92 000 [70 000 – 120 000]	0.5% [0.4% – 0.6%]	58 000 [43 000 – 70 000]
Caribbean	240 000 [220 000 – 270 000]	17 000 [13 000 – 21 000]	1.0% [0.9% – 1.1%]	12 000 [8500 – 15 000]
Eastern Europe and Central Asia	1.4 million [1.3 million – 1.6 million]	130 000 [110 000 – 160 000]	0.8% [0.7% – 0.9%]	76 000 [60 000 – 95 000]
Western and Central Europe	820 000 [720 000 – 910 000]	31 000 [23 000 – 40 000]	0.2% [0.2% – 0.2%]	8500 [6800 – 19 000]
North America	1.5 million [1.2 million – 2.0 million]	70 000 [44 000 – 130 000]	0.5% [0.4% – 0.7%]	26 000 [22 000 – 44 000]
Oceania	57 000 [50 000 – 64 000]	4500 [3400 – 6000]	0.3% [0.2% – 0.3%]	1400 [<1000 – 2400]
TOTAL	33.3 million [31.4 million – 35.3 million]	2.6 million [2.3 million – 2.8 million]	0.8% [0.7% - 0.8%]	1.8 million [1.6 million – 2.1 million]

The ranges around the estimates in this table define the boundaries within which the actual numbers lie, based on the best available information

Fig. 1.1 Regional HIV/AIDS Statistics and Features, 2009. UNAIDS Report on the Global AIDS Epidemic 2010. Available at: http://www.unaids.org/globalreport/Epi_slides.htm. Accessed January 2011
Note: The ranges around the estimates in this table define the boundaries within which the actual numbers lie, based on the best available information

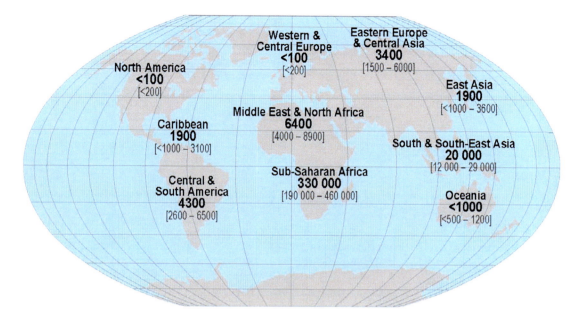

Fig. 1.2 Estimated number of children (<15 years) newly infected with HIV, 2009. (Data from World Health Organization, UNAIDS. UNAIDS report on the global AIDS epidemic, 2010.) Available at: http://www.unaids.org/documents/20101123_epislides_core_en.pdf. Accessed January 2011
Note: The ranges around the estimates in this figure define the boundaries within which the actual numbers lie, based on the best available information

Asia; and North America. New pediatric HIV-1 infections worldwide (Fig. 1.2), estimated to be 370,000 (230,000–510,000) globally in 2009, are almost all due to MTCT, occurring either during pregnancy, during labor and delivery, or through breastfeeding [3]. While still a major global health concern, this represents roughly a 25% reduction compared with 2004 figures [2].

Sharp declines in the rates of MTCT have been realized for the past 15 years in North America and Europe [1, 3, 4]. More recently, similar successes have been seen in rural Yunnan Province in China [5] with infant infection rates of less than 2%. This is a result of national implementation of effective PMTCT strategies including (1) universal, opt-out antenatal HIV-1 counseling and testing; (2) obstetrical interventions to reduce HIV-1 transmission during labor/delivery including scheduled caesarian section prior to active labor and membrane rupture; (3) combination ARVs for prophylaxis during the antepartum, intrapartum, and postpartum periods; and (4) avoidance of breastfeeding in settings where formula use is safe, available, and affordable. In the absence of effective PMTCT intervention, between 15 and 40% of infants become HIV-1 infected [3, 6–8], whereas with these interventions, infection rates of less than 1–2% are seen.

This progress notwithstanding new pediatric HIV-1 infections remain unacceptably high in resource-limited settings, particularly in SSA, a region that accounted for more than 90% of the estimated global MTCT in 2009 [1]. This high level of transmission is attributed to the high levels of endemic HIV-1 infection among pregnant women in many African countries (Table 1.1). Slow scale-up of proven PMTCT interventions such as routine opt-out counseling and testing for HIV-1; limited

Table 1.1 HIV prevalence—national adult estimates; sentinel surveillance estimates among pregnant women (15–49 years) and mother-to-child transmission risk from 2000 to 2009 (where data is available) in selected sub-Saharan African countries

	Country	National prevalence, % (year)	HIV prevalence—pregnant women aged 15–49 years, % (year)	Percentage of infants born to HIV infected mothers who are infected, % (year)
Southern Africa	Botswana[a]	17.6 (2008)	32.5 (2009)	3.8 (2009)
		17.1 (2004)	33.7 (2007)	4.8 (2007)
			32.4 (2006)	11.5 (2004)
			33.4 (2005)	20.7 (2002)
			37.4 (2003)	
			35.4 (2002)	
			36.2 (2001)	
	South Africa[b,c]	16.9 (2008)	29.3 (2008)	16 (2009)
		16.2 (2005)	29.4 (2007)	16 (2008)
			29.0 (2006–2007)	
	Namibia[d]	No population survey has been conducted	17.8 (2008)	12.7 (2009)
			19.9 (2006)	25 (2007)
			19.7 (2004)	28 (2005)
			22.0 (2002)	
			19.3 (2000)	
	Lesotho[e]	23.6 (2008)	43.3 (2007)	15 (2007)
		23.2 (2007)	19.2 (2006)	25 (2005)
			11 (2005)	
			6.3 (2004)	
East Africa	Kenya[f]	6.2 (2009)	–	27 (spectrum modeling)
		6.3 (2008–2009)		
		7.4 (2007)		
	Rwanda[g]	2.9 (2009)[c]	4.3 (2007)	4.1 (2009)
		3 (2005)	4.1 (2005)	6.9 (2008)
			5.2 (2003)	10.3 (2007)
				11 (2006)
				11.9 (2005)
				10 (2004)
				23.4 (2003)
	Tanzania[h]	5.7 (2009)	8.2 (2005–2006)	18 (2008)
		6.4 (2005–2006)	8.7 (2003–2004)	26.5 (2009)
		6.6 (2003–2004)	9.6 (2001–2002)	
		7.1 (2001–2002)		
	Uganda[i]	6.5 (2009)	5.9 (2008–2009)	9.9 (2008–2009)
		6.1 (2002)	6 (2007–2008)	
West Africa	Mali[j]	1.3 (2006)	2.2 (2009)	21 (spectrum modeling)
		1.7 (2001)		
	Cote D'Ivoire[k]	3.7 (2008)	4.5 (2008)	26.7 (2008–2009)
		4.7 (2005)		
	Nigeria[l]	4.6 (2009)	4.6 (2008)	29.1 (spectrum modeling)
		3.6 (2007)	5.8 (2001)	

[a]UNAIDS 2010, Country progress report (Botswana). http://www.unaids.org/fr/dataanalysis/monitoringcountryprogress/2010progressreportssubmittedbycountries/file,33650,fr..pdf. Accessed January 2011
[b]UNAIDS 2010, Country progress report (South Africa). http://www.unaids.org/en/dataanalysis/monitoringcountryprogress/2010progressreportssubmittedbycountries/southafrica_2010_country_progress_report_en.pdf. Accessed January 2011
[c]South Africa has the largest number of HIV infected individuals in the world (5.2–5.7 million in 2009), followed by Nigeria (2.98 million in 2009)
[d]UNAIDS 2010, Country progress report (Namibia). http://data.unaids.org/pub/Report/2010/namibia_2010_country_progress_report_en.pdf. Accessed January 2011

Table 1.1 (continued)

eUNAIDS 2010, Country progress report (Lesotho). http://www.unaids.org/en/dataanalysis/monitoringcountryprogress/2010progressreportssubmittedbycountries/lesotho_2010_country_progress_report_en.pdf. Accessed January 2011
fUNAIDS 2010, Country progress report (Kenya). Available at http://www.unaids.org/en/dataanalysis/monitoringcountryprogress/2010progressreportssubmittedbycountries/kenya_2010_country_progress_report_en.pdf. Accessed January 2011
gUNAIDS 2010, Country progress report (Rwanda). http://www.unaids.org/en/dataanalysis/monitoringcountryprogress/2010progressreportssubmittedbycountries/rwanda_2010_country_progress_report_en.pdf. Accessed January 2011
hUNAIDS 2010, Country progress report (Tanzania). http://www.unaids.org/en/dataanalysis/monitoringcountryprogress/2010progressreportssubmittedbycountries/tanzania_2010_country_progress_report_en.pdf. Accessed January 2011
iUNAIDS 2010, Country progress report (Uganda). Available at http://data.unaids.org/pub/Report/2010/uganda_2010_country_progress_report_en.pdf. Accessed January 2011
jUNAIDS 2010, Country progress report (Mali). Available at: http://data.unaids.org/pub/Report/2010/mali_2010_country_progress_report_fr.pdf. Accessed January 2011
kUNAIDS 2010, Country progress report (Cote D'Ivoire). Available at: http://data.unaids.org/pub/Report/2010/cotedivoire_2010_country_progress_report_fr.pdf. Accessed January 2011
lUNAIDS 2010, Country progress report (Nigeria). http://www.unaids.org/en/dataanalysis/monitoringcountryprogress/2010progressreportssubmittedbycountries/nigeria_2010_country_progress_report_en.pdf. Accessed January 2011

provision of ARVs for prophylaxis or maternal treatment (Fig. 1.3); and lack of acceptable, feasible, affordable, sustainable, and safe alternatives to breastfeeding all have contributed to the high MTCT rate in SSA. Extended breastfeeding, into the second year, is the cultural norm in many African countries [9] where it is crucial for infant survival. In these settings, among infants born to HIV-1-infected women who breastfeed, breast milk transmission of HIV-1 is estimated to account for about one third of all new pediatric HIV-1 infections. In total, about 10–15% of all HIV-1 exposed infants are infected through prolonged breastfeeding [10].

Risk Factors for HIV-1 Transmission During Breastfeeding

There has been substantial increase in our knowledge about risk factors and timing of breast milk transmission of HIV-1 which has helped drive the development of effective PMTCT strategies. Although the exact mechanisms of MTCT during breastfeeding have not been clearly elucidated, the most important risk factors associated with breast milk HIV-1 transmission may be summarized as follows: viral factors; maternal factors (clinical, immunological, and behavioral); factors related to patterns of breastfeeding; and maternal–infant host factors.

Viral Factors

Maternal HIV-1 RNA levels in plasma and breast milk are the strongest independent predictors of MTCT through breastfeeding [11–15], and as such, women with advanced HIV-1 disease and associated breast milk high viral load (VL) are more likely to transmit HIV-1 during breastfeeding [11–18].

Maternal primary infection during the postnatal period, which is associated with transient elevated levels of breast milk VL [19], is linked to particularly high rates of infant HIV-1 acquisition (about a 33% risk) [19]. This high transmission risk is presumably due to high levels of maternal viremia prior to emergence of maternal immunologic responses capable of containing viral replication at a lower set point. Increased MTCT risk with maternal primary HIV-1 infection while breastfeeding has been reported in several earlier studies [19, 20] and reaffirmed more recently in a Zimbabwean prospective cohort study [21] and a retrospective case-series in China among women infected by HIV-1 via

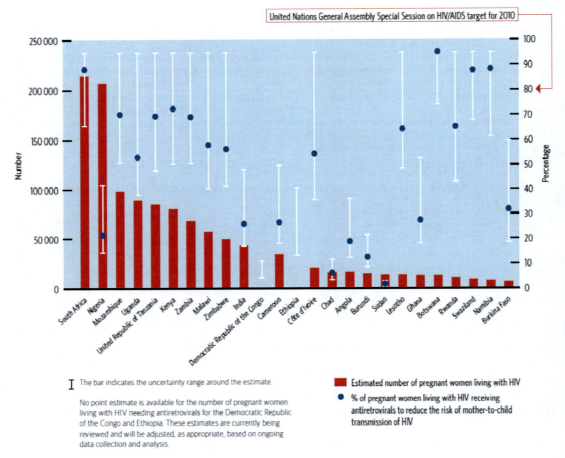

Fig. 1.3 Percentage of pregnant women living with HIV receiving antiretrovirals to prevent mother-to-child transmission of HIV in 25 countries with the highest HIV disease burden among pregnant women, in descending order, 2009. WHO, UNICEF, and UNAIDS. Towards universal access: scaling up priority HIV/AIDS intervention in the health sector. Progress report 2010. Geneva, World Health Organization, 2010. Available at: http://www.who.int/hiv/pub/2010progressreport/report/en/index.htm. Accessed November 2010

postpartum blood transfusion [22]. In the Zimbabwean study, the majority (two-thirds) of infant HIV-1 infections occurred within 90 days of maternal seroconversion. Breast milk transmission risk was nearly fourfold greater among infants who were ≤9 months old [21].

HIV-1 detected in breast milk may originate from plasma cell-free virus that crosses the mammary epithelial tight junctions into the breast milk compartment or be produced locally by replication in macrophages and epithelial cells [8]. Both cell-associated and cell-free HIV-1 virus have been reported [15, 23], although the differential roles in MTCT risk are inconclusive. Gray et al. [24] reported that viral variants may initially be seeded from plasma into breast milk but that thereafter there continued mixing of viral lineages over time between plasma and breast milk compartments.

Virologic subtypes and MTCT: HIV-1 subtypes and recombination forms exhibit functional and phenotypic variations that may influence MTCT efficiency [25–27]. Recombination, which occurs as a result of co-infection of an individual host cell with two or more distinct viruses, may lead to the emergence of HIV-1 recombinant viruses that are evasive to immune pressures or have fitness that enhances transmission [28]. For instance, in a Tanzanian study, MTCT risk was higher among women with HIV-1 subtype C compared to those with A, D, or A/D recombinant [27] while in a Kenyan study,

MTCT was more common among those with subtype D or A/D recombinant virus compared to those with subtype A [29]. Certain intersubtype recombinant viruses may be preferentially transmitted during breastfeeding [14].

Maternal Clinical and Immunologic Risk Factors

Clinical mastitis, breast abscess [30], and subclinical mastitis defined by increased sodium levels in the breast milk compartment [31] are associated with increased MTCT during breastfeeding. Severe maternal illness with malnutrition and frequent engorgement of the mammary glands from nonexclusive breastfeeding may lead to stasis of breast milk and subclinical mastitis [32, 33].

Micronutrient deficiencies have been associated with increased MTCT risk [34, 35]. Selenium deficiencies may be associated with mastitis [34] while vitamin D deficiency is associated with compromised fetal immune development, which could result in increased vulnerability to infant HIV-1 acquisition [35]. One study demonstrated a strong correlation between vitamin A deficiency and increased MTCT risk [36], although subsequent studies offering vitamin A supplementation showed no PMTCT benefit [32, 37, 38].

Several immunologic factors in breast milk, including HIV-1-specific IgG and IgA immunoglobulins, T lymphocytes, natural killer cells, interleukins, lactoferrin, lysozyme, defensins, chemokines, secretory leukocyte protease inhibitor (SLPI), and epidermal growth factors, have antimicrobial and immune-modulation properties that may influence MTCT risk [39]. For example, attachment of HIV-1 particles to dendritic cell membranes and transfer to CD4+ lymphocytes is inhibited by specific IgA secretory antibodies in breast milk [40].

Breastfeeding Patterns

Longer duration of breastfeeding is associated with increased MTCT risk due to its cumulative nature [11, 12, 14, 17]. A meta-analysis summarized late breastfeeding transmission rates as <1%/month [17]. Early weaning, prior to 6 months, on the other hand, has been associated with increased infant morbidity and mortality due to diarrheal disease, pneumonia, and other infectious diseases [41–43]. Abrupt weaning within a few hours or days has been associated with elevated breast milk sodium levels and breast engorgement, which leads to maternal illness including fever and mastitis, with resultant increase in breast milk VL [44]. Considering the potential for increased MTCT risk from abrupt weaning, WHO in 2010 issued revised detailed guidelines [45]. The WHO encouraged mothers who chose to breastfeed to do so exclusively for 6 months and if they could then safely use breast milk substitutes, wean over several weeks and then stop breastfeeding if suitable breast milk substitutes can be safely used and sustained. Mothers who cannot provide safe breast milk substitutes at 6 months should continue breastfeeding for 12 months or longer, along with providing nutritious locally available complementary foods. WHO also recommended ARV prophylaxis should be provided for the duration of breastfeeding [45].

Maternal and Infant Host Factors

Maternal host factors associated with the higher risk of MTCT, in addition to VL in breast milk, include illness severity and lower CD4 count. In addition, maternal–infant HLA concordance, as well as maternal HLA homozygosity, have been associated with increased risk of MTCT [46, 47].

Infant host factors that have been associated with increased transmission risk include increased permeability of infant gut mucosa to HIV-1 particles related to immature dendritic cells in the mucosa. These immature cells transport the virus to the lymph nodes (Peyer's patches) and then to CD4+ cells [48]. A breach in the integrity of the infant's gastro-intestinal mucosa from inflammation related to mixed feeding may also facilitate passage of HIV-1 [49]. This is postulated as the likely mechanism for increased MTCT risk with mixed infant feeding [37, 49]. Infant oral candidiasis has also been associated with MTCT during the postpartum period [30].

Timing of Transmission Among Breastfeeding Populations

MTCT of HIV-1 can occur in utero, intrapartum, and postpartum through breastfeeding. Based on fetal and infant data from several HIV-1 prevention studies, the following consensus definitions have been adopted to categorize the timing of transmission:

1. In utero infection: infants with a positive HIV-1 RNA or DNA polymerase chain reaction (PCR) result detected in a sample collected within 72 h of birth.
2. Intrapartum or early postpartum infection: infants with a negative PCR at birth (within 72 h) but a positive PCR result by 4–6 weeks of age. These infections are presumed to have occurred late in pregnancy, during labor and delivery, or very soon after delivery through oral or mucosal exposure to virus in maternal blood, secretions, or colostrum/early breast milk.
3. Late postpartum infection: infants with a negative PCR result at 4–6 weeks of age, but a detectable HIV-1 infection later. These infections are presumed to have occurred via breastfeeding [50, 51].

These definitions, however, remain arbitrary, and the relative contribution of each period to perinatal HIV-1 infections shifts with the introduction of PMTCT interventions. For example, in nonbreastfeeding populations prior to the introduction of potent combination ARVs, about 30% of MTCT was estimated to occur during pregnancy and the remainder during labor and delivery [51, 52]. With the introduction of ARV prophylaxis in the antepartum and intrapartum periods in the mid-1990s, there were major reductions in intrapartum transmission. As a result, currently, 50–67% of MTCT is estimated to occur during pregnancy among populations with high rates of intrapartum coverage of ARV prophylaxis.

Timing of Breastfeeding Transmission

There appears to be an increased risk of MTCT during the first weeks of breastfeeding. Several studies, including a randomized trial in Nairobi, Kenya [12], compared MTCT timing between breastfed and formula-fed infants. A substantial proportion (as high as 80%) of breast milk transmission of HIV-1 occurred by 1–2 months after birth [11, 12, 14, 15, 17]. The absolute breast milk transmission risk was estimated to be 2.7–4.2%/month in the first 1–2 months [12, 17]. Additional studies, including a meta-analysis by an International Collaborative Group, reported a low and constant risk from breastfeeding of about 0.6–0.9%/month through 12–18 months [11, 12, 14, 17]. An early prospective study in rural Rwanda found 0.5% HIV transmission per month after 20 months of age among children still being breastfed beyond 2 years of age, a common practice in rural Africa [18].

The increased risk of MTCT in the first weeks of breastfeeding may be related to several factors. In one study, median VL in colostrum and early breast milk was substantially higher than in breast milk collected 14 days after birth. This could lead to higher transmission rates in the first weeks of infant life [53]. Other factors could be the lack of acid secretions in the newborn gut resulting in an inability to deactivate the HIV-1 virus. The general immature immunologic responses of the newborn could also play a role [54].

Mechanisms of Breastfeeding Transmission of HIV-1 to the Infant

Transmission of HIV-1 from mother to infant through breastfeeding is influenced, as mentioned above, by a number of host (genetic, immunologic, and infectious), viral (quantity and characteristics), and behavioral/environmental (type and frequency of breastfeeding) factors of the mother–infant dyad [53, 54]. There are many gaps in our understanding of how HIV-1 is transmitted through breastfeeding and which immune or other parameters most protect infants from such transmission. As previously mentioned, HIV-1 in breast milk of infected mothers can originate either from blood cell-free virus released into breast milk through the epithelial barrier of the mammary gland or can be produced by local replication in macrophages and in ductal and alveolar mammary epithelial cells [55]. Whether mammary epithelial cells themselves support HIV-1 replication remains controversial [56, 57]. What is clear is that the mammary epithelium is quite effective in curtailing viral entry [58].

The concentration of HIV-1 in cell-free breast milk is generally lower than in plasma by about 2 logs [5, 23]. HIV-1 has been detected both in the cellular compartment of breast milk and in cell-free milk with varying frequencies. Detection ranged from 39 to 89% of breast milk samples of HIV-1-infected women in different studies [15, 23]. Viral detection is associated with lower maternal CD4+ T cell count, vitamin A deficiency, and clinical or subclinical mastitis and breast abscess [59–61]. HIV-1 load in breast milk seems to be highest just after birth [16]. Intermittent shedding and differences in VL between the two breasts have been noted in several studies [61, 62]. Highly active antiretroviral therapy (HAART) started during pregnancy or postpartum suppresses HIV-1 RNA but may not suppress DNA in breast milk [63–65]. Some researchers have found that detection of cell-associated virus in breast milk is a stronger predictor of transmission than cell-free virus [16, 28]: a twofold increased risk for every tenfold increase in cell-free HIV-1 RNA compared to a threefold increased risk for every tenfold increase in infected cells harboring HIV-1 DNA has been estimated [4, 15, 16, 28].

Entry of HIV-1 into the infant blood stream requires passage across mucosal surfaces and interaction of HIV-1-infected cells with epithelial cells in the oral, nasopharyngeal, or gastrointestinal tract. Saliva provides a hostile milieu to infection through a range of mechanisms. Salivary hypotonicity is thought to disrupt HIV-1-infected cells and to inactivate HIV-1 in leucocytes. Saliva also contains various substances with anti-HIV-1 activity including mucins, lysozyme, lactoferrin, SLPI [66], and cytokines and chemokines [67, 68].

Contact between HIV-1 infected cells and epithelial cells is postulated to occur through binding of HIV-1 envelope glycoproteins to the epithelial receptor galactosyl ceramide, with stabilization of the interaction from integrin-dependent engagement [55]. There is increasing appreciation of the fact that epithelial cells are not simply passive barriers but have reciprocal regulating interactions with innate and adaptive immune factors guarding against foreign antigens [56, 58].

Resistance to infection via breastfeeding may also involve natural killer cells and natural antibodies to HIV-1 on exposed mucosal surfaces or in the systemic compartment of the infant. Natural mucosal antibodies to the chemokine co-receptor (CCR5) and to the viral protein *tat* have been described [56, 58]. The role of such preadaptive immunity to resistance against HIV-1 transmitted through breast milk is unclear.

Breast milk itself contains many antimicrobial and immunomodulatory factors with diverse effects on HIV-1. Some factors have in vitro anti-HIV-1 activity or the capability to interfere with HIV-1 binding (including SLPI, lactoferrin, RANTES, interferon-γ, and α- and β-defensins, mucins, glycans, and polyunsaturated fatty acids) [54, 69], as well as HIV-1-specific antibodies [69]. Other factors display pro-inflammatory activity that might promote local HIV-1 replication (including interleukin-6, IL-8,

IL-7, IL-1β, and TNF-α) [69, 70]. Specific human leukocyte antigen (HLA) alleles have been associated with protection against MTCT of HIV-1; in particular, HLA B18 may protect the infant against breastfeeding transmission [71]. The immunology of breast milk in the context of mother-to-child transmission of HIV-1 is addressed in more detail in other chapters of this book.

Infant Gut Mucosal Activation Markers and Protective Effects of Exclusive Breastfeeding

Several studies have shown that the type of infant feeding in the first 3–4 months has an effect on postnatal transmission of HIV-1. Mixed feeding during the first several months of life is associated with an increased risk of transmission whereas exclusive breastfeeding appears protective [49, 72–74]. In the Zimbabwe ZVITAMBO trial, the rate of postnatal transmission of HIV-1 was 5.1, 6.7, and 10.5 infections per 100 child-years of breastfeeding for infants who were exclusively breastfed, predominantly breast-fed (breast milk plus non-milk fluids), and mixed fed, respectively [49]. This may be due to damage to the intestinal mucosa from early introduction of nonbreast milk foods leading to delayed closure of the enterocyte junctions in the intestinal mucosal barrier. Alternatively, intestinal immune activation resulting from early introduction of foreign antigens or pathogens may be involved. Both of these mechanisms can increase the risk of HIV-1 transmission to the infant [75].

Maternal Breast Factors Related to Risk of Transmission

Mammary epithelial permeability is increased during the immediate postpartum period, during weaning [74], and during periods of inflammation such as mastitis when pericellular sodium and albumin can move into breast milk resulting in elevated levels [28, 31, 76]. In addition, infrequent breast emptying, which might occur with nonexclusive breastfeeding or when weaning, can increase the risk of ductal inflammation in the breast with the production of cytokines and other inflammatory mediators leading to subclinical mastitis and increases in mammary epithelial permeability. Elevation in breast milk sodium in the absence of clinical symptoms (subclinical mastitis) and symptomatic breast inflammation (mastitis) has been associated with higher levels of HIV-1 in breast milk and increased risk of postnatal transmission [71]. However, one study indicated that laboratory markers of mastitis such as increased sodium level and cell count are poor predictors of HIV-1 RNA levels in breast milk [76]. Several mechanisms have been postulated for breast milk transmission related to clinical or subclinical mastitis including increased permeability of HIV-1 cell-free particles and white blood cells from the plasma into the lumen as a result of inflammation of the mammary epithelial tight junctions [77, 78]. Increased mobilization of inflammatory cells which up-regulate local HIV-1 expression leading to higher virus levels in breast milk may also play a role [54].

The Impact of Antiretroviral Drugs in Reducing MTCT: Clinical Trial Findings Among Breastfeeding Populations

Given the protective effects of breastfeeding on infant and child survival, but mindful of the ongoing risk of HIV-1 transmission, recent research has focused on developing and testing interventions to reduce the risk of transmission throughout the breastfeeding period (Table 1.2). Initial interventions focused on the peripartum period and the first few weeks of breastfeeding, while more recent trials have assessed the safety and efficacy of extended ART prophylaxis given to either the mother or the infant.

Table 1.2 The impact of antiretroviral drugs in reducing MTCT: clinical trial findings among breastfeeding populations

Study name	Location	Objectives	Study design	Maternal regimen	Infant regimen	Number of study participants	Cumulative rate of HIV infection
ACTG 076	USA, France	Assess the efficacy and safety of ZDV in reducing maternal–infant HIV transmission	Randomized, double-blind, placebo controlled trial	Antepartum ZDV (100 mg orally five times daily), Intrapartum ZDV (2 mg/kg IV over 1 h, then 1 mg/kg until delivery)	ZDV 2 mg/kg orally every 6 h for 6 weeks)	477 Pregnant women enrolled 409 Women delivered 415 Live births Infection status available for 363 infants: ZDV—180 Placebo—183	At 18 months: ZDV arm—8.3% Placebo arm—25.5%
CDC-Thai	Thailand (nonbreastfeeding population)	Investigate the safety and efficacy of short-course oral ZDV administered during late pregnancy and labor	Randomized, double blind, placebo controlled trial	ZDV 300 mg orally twice daily from 36 weeks gestation and every 3 h from onset of labor until delivery	N/A	397 Women randomized 393 Delivered 395 Live births	At >2 months : ZDV arm—9.4% Placebo arm—18.9%
DITRAME STUDY GROUP	Abidjan, Côte d'Ivoire and Bobo-Dioulasso, Burkina Faso	Assess the acceptability, tolerance, and 6 month efficacy of a short regimen of oral ZDV in African populations practising breastfeeding	Randomized, double blind, placebo controlled trial	ZDV from 36 to 38 weeks gestation (300 mg twice daily until labor, 600 mg at beginning of labor and 300 mg twice daily for 7 days postpartum	N/A	421 Women 400 Live born infants	At 6 months: ZDV arm—18.0% Placebo arm—27.5%
RETRO-CI	Abidjan, Côte d'Ivoire	Assess the safety and efficacy of short course perinatal ZDV among HIV-1 seropositive breastfeeding women in Abidjan, Côte d'Ivoire	Randomized, placebo controlled trial	ZDV (300 mg tablet one tablet twice daily from 36 weeks gestation until onset of labor, one tablet at onset of labor and one tablet every 3 h until delivery)	N/A	280 Women enrolled 230 Infants with HIV infection status	At 4 weeks: ZDV arm—12.2% Placebo arm—21.7% At 3 months: ZDV arm—15.7% Placebo arm—24.9%
Pooled analysis of data from DITRAME-ANRS049a and RETRO-CI	Abidjan, Côte d'Ivoire and Bobo-Dioulasso, Burkina Faso	Assess the 24 month efficacy of a maternal short course ZDV regimen to prevent mother-to-child transmission of HIV1 in a breastfeeding population in West Africa	Pooled analysis of two randomized placebo controlled clinical trials	ZDV (300 mg) one tablet twice daily from 36 to 38 weeks gestation until delivery then in DITRAME study continue ZDV for 7 days only postpartum	N/A	662 Live born children 641 With evaluable HIV infection status	At 24 months: ZDV arm—22.5% Placebo arm—30.2%

(continued)

Table 1.2 (continued)

Study name	Location	Objectives	Study design	Maternal regimen	Infant regimen	Number of study participants	Cumulative rate of HIV infection
HIVNET 012	Kampala, Uganda	Assess the safety and efficacy of short-course NVP compared with ZDV given to women during labor and to neonates during the first week of life	Originally Randomized, double blind, placebo controlled phase III trial, later changed to randomized open-label trial	NVP, 200 mg orally to the mother at onset of labor *or* ZDV 300 mg orally every 3 h from 36 weeks gestation until delivery, two 300 mg tablets at onset of labor followed by one 300 mg tablet every 3 h during labor	NVP, 2 mg/kg within 72 h *or* ZDV, 4 mg/kg twice daily for 7 days after birth	645 Women enrolled 626 Randomly assigned ZDV or NVP 616 Infants assessable	At birth: NVP arm—8.1% ZDV arm—10.3% At 6–8 weeks: NVP arm—11.8% ZDV arm—20.0% At 14–16 weeks: NVP arm—13.5% ZDV arm—22.1% At 18 months: NVP arm—15.7% ZDV arm—25.8%
PETRA STUDY	Tanzania South Africa Uganda	Assess the efficacy of short-course regimens with ZDV and 3TC in a predominantly breastfeeding population	Randomized, double blind, placebo controlled trial	Regimen A: from 36 weeks gestation ZDV (300 mg)+3TC (150 mg) twice daily until the onset of labor, intrapartum ZDV (300 mg)+3TC (150 mg) at onset of labor and ZDV (300 mg every 3 h) and 3TC (150 mg every 12 h until delivery) postpartum ZDV (300 mg)+3TC (150 mg) twice daily for 7 days Regimen B: intrapartum ZDV+3TC and postpartum ZDV+3TC as regimen A Regimen C: intrapartum ZDV+3TC only (as regimen A)	ZDV, 4 mg/kg 3TC, 2 mg/kg twice daily for 7 days	1,457 Women randomized 1,501 Infants	At 6 weeks: Regimen A—5.7% Regimen B—8.9% Regimen C—14.2% Placebo arm—15.3% At 18 months: Regimen A—15% Regimen B—18% Regimen C—20% Placebo arm—22%
SIMBA STUDY	Rwanda Uganda	Evaluate the efficacy of postnatal prophylaxis in infants with 3TC or NVP during the first 6 months of breastfeeding	Randomized open-label trial	All women received ZDV+ddI from 36 weeks gestation until 1 week after delivery	Randomized to receive either 3TC or NVP from birth until 1 month after breastfeeding cessation	413 Women enrolled 397 Live born infants randomized	After 4 weeks of age: 3TC arm—1.1% NVP arm—0.6%
MITRA STUDY	Tanzania	Investigate the possibility of reducing mother-to-child-transmission (MTCT) of HIV-1 through breastfeeding by prophylactic antiretroviral (ARV) treatment of the infant during the breastfeeding period	Open-label, nonrandomized, prospective cohort study	ZDV+3TC from week 36 to 1 week postpartum	ZDV+3TC from birth to 1 week of age	398 Infants	At 6 weeks: 3.8% At 6 months: 4.9%

PEPI STUDY	Malawi	Determine whether extended prophylaxis of infants with NVP or with NVP+ZDV until the age of 14 weeks would decrease the rate of HIV-1 infection as compared with single dose NVP combined with 1 week of ZDV	Randomized, controlled, open-label phase III clinical trial	All women received Intrapartum sdNVP	Random assignment at birth to: sdNVP combined with 1 week of daily ZDV (control group) or Control regimen followed by extended daily prophylaxis with either oral NVP (extended NVP group) or Oral NVP plus ZDV (extended dual prophylaxis group) until age 14 weeks	3,016 Infants	At 9 months: Extended NVP group—5.2% Extended dual prophylaxis group—6.4% Control group—10.6%
SWEN STUDY	Ethiopia India Uganda	Assess whether daily NVP given to breastfed infants through 6 weeks of age can decrease HIV transmission via breastfeeding	Randomized controlled trial	All women received intrapartum sdNVP	Infants received sdNVP after birth or 6 week extended dose NVP	2,024 Live born infants sdNVP arm—1,047, extended NVP arm—977	At 6 weeks: sdNVP arm—5.2% Extended NVP arm—2.5% At 6 months: sdNVP arm—9.0% Extended NVP arm—6.9%
MITRA PLUS	Tanzania	To reduce breast-milk transmission of HIV-1 by treating HIV-1 infected women with highly active antiretroviral therapy (HAART) during breastfeeding	Open-label, nonrandomized, prospective cohort study	ZDV+3TC+NVP/NLF/ from 34 weeks gestation	ZDV+3TC for 1 week after birth	441 Infants	At 6 weeks—4.1% At 6 months—5%
KIBS STUDY	Kisumu Kenya	Assessment of transmission rates in the Kisumu Breastfeeding Study (KIBS) using ZDV/3TCand NVP/NLF	Phase IIb single arm study	ZDV+3TC+NVP/NLF from 34 weeks gestation to 6 months postpartum	sdNVP at birth	497 Infants	At 0–7 days—2.4% At 6 weeks—3.9% At 3 months—4.1% At 6 months—5.0% At 12 months 5.9%
AMATA STUDY	Rwanda	Comparison of triple ART given to breastfeeding mothers with formula feeding for prevention of postnatal mother-to-child-transmission	Open label observational study	d4T+3TC+NVP (CD4 <350/mm³) or 3TC+NVP+EFV (CD4 ≥350/mm³) from 26 weeks to delivery for formula feeding arm and until 1 month after breastfeeding cessation for breastfeeding arm	N/A	572 Women enrolled 528 Children born	Overall transmission rate at 6 months—1.4%

(continued)

Table 1.2 (continued)

Study name	Location	Objectives	Study design	Maternal regimen	Infant regimen	Number of study participants	Cumulative rate of HIV infection
Kesho Bora Study	Burkina Faso Kenya South Africa	Assessment of safety and efficacy of continued maternal ARVs during breastfeeding	Randomized trial	Randomized between 28 and 36 weeks gestation to receive either Triple ARV (ZDV+3TC+LPV/r to 6.5 months postdelivery) *or* Short ARV (ZDV through delivery plus sdNVP in labor) followed by 1 week of ZDV+3TC	All received sdNVP+1 week of ZDV	824 Women 805 Infants	At birth: Triple ARV—1.8% Short ARV—2.0% At 6 weeks: Triple ARV—3.3% Short ARV—4.5% At 6 months: Triple ARV—4.9% Short ARV—8.2% At 12 months: Triple ARV—5.6% Short ARV—9.3%
Kesho Bora study	Burkina Faso Kenya South Africa	Assessment of the effectiveness and safety of antiretrovirals (ARVs) used for treatment or prophylaxis in a breastfeeding population of HIV-1 infected women	Prospective cohort studies	Cohort A—CD4 count <200/mm³ or WHO stage 4 initiated on HAART Cohort B—asymptomatic women with >500/mm³ CD4 count receiving ZDV from 34 to 36 weeks gestation until delivery with sdNVP in labor	All infants received sdNVP	248 Women enrolled Cohort A—111 infants Cohort B—125 infants	At 18 months: Cohort A—7.5% Cohort B—5.8%
Mma Bana	Botswana	Comparison of different HAART regimens used in pregnancy and during breastfeeding to determine whether the regimens differ with respect to virologic suppression during pregnancy, breastfeeding outcomes, and toxic effects in mothers and infants	Randomized trial+prospective observational cohort	Randomized to protease inhibitor (PI) based HAART regimen (ZDV+3TC+LPV/r) *or* Triple Nucleoside regimen (ZDV+3TC+ABC) Observational cohort received NVP+ZDV+3TC	All infants received sdNVP and 4 weeks of ZDV	560 Randomly assigned to PI regimen or triple nucleoside regimen 170 Women in observational group 709 Live born infants	Overall transmission rate at 6 months—1.1%
BAN STUDY	Malawi	Evaluate the efficacy of a maternal triple-drug antiretroviral regimen or infant nevirapine prophylaxis for 28 weeks during breastfeeding to reduce postnatal transmission of human immunodeficiency virus type 1 (HIV-1) in Malawi	Randomized clinical trial	All mothers received intrapartum sdNVP and 1 week of ZDV Mothers randomized to maternal triple ARV prophylaxis arm received ZDV+3TC+NVP/NLF/LPV/r twice daily to 28 weeks Control group did not receive extended prophylaxis	All infants received sdNVP and 1 week of ZDV Infants randomized to extended prophylaxis received NVP for 28 weeks Control group did not receive extended prophylaxis	2,369 Mother-infant pairs randomized	At 28 weeks postpartum: Maternal ARV regimen—2.9% Extended infant nevirapine—1.7% Control group—5.7%

Initial PMTCT Trial Results Testing Late Antepartum and Intrapartum Regimens

In 1994, the historic Pediatric AIDS Clinical Trial Group (PACTG) 076 trial results were announced which showed that maternal zidovudine (ZDV) oral prophylaxis provided from as early as 14 weeks of gestation, administered intravenously during labor/delivery, and then given to the infant for 6 weeks could reduce MTCT by 67% among nonbreastfeeding mothers with CD4 counts >200/mm^3. Transmission rates were 8.3% in the ZDV arm and 25.5% in the placebo arm at 18 months postpartum [79].

Based on this proof of concept, researchers designed simpler regimens that would be more feasible to deliver in international settings including interventions beginning in the last trimester, administered during labor, and having shorter prophylaxis duration for the infant. In 1999, a placebo-controlled trial conducted in Thailand showed that ZDV given from 36 weeks of gestation through delivery produced a 50% reduction in vertical transmission (18.9% in the placebo group compared to 9.4% in the ZDV group) [80]. However, in this Thailand trial, the mothers did not breastfeed; therefore, the efficacy of the shorter regimens on breastfeeding transmission was unknown [80].

Similar placebo-controlled trials were conducted in Côte d'Ivoire and Burkina Faso, West Africa, among breastfeeding populations [81, 82]. These trials assessed the efficacy, tolerance, and acceptability of oral ZDV regimens provided from 36 weeks and administered during labor. One of the studies also added a week of oral ZDV for both mothers and infants postdelivery [81]. A pooled analysis of these two studies reported transmission rates of 22.5% in the ZDV group and 30.2% in the placebo group, or a 26% reduction by 2 years with ongoing breastfeeding [83]. Efficacy was highest among women with CD4 counts greater than or equal to 500/mm^3 [83]. ZDV use was found to be safe and efficacious. However, the programmatic scale up of even short-term antepartum regimens has proven challenging, particularly in African settings, due to financial and logistical issues.

The HIVNET 012 Study [84, 85] announced in 1999 was unique because it targeted only the labor/delivery period. Conducted in Kampala, Uganda, it assessed the safety and efficacy of a single dose of Nevirapine (sdNVP) taken by the mother at labor onset followed by a sdNVP to the newborn within 72 h of birth. The comparison group of mothers received ZDV during labor and their newborns received a 1-week course of daily ZDV. The HIVNET 012 sdNVP regimen resulted in a 47% reduction in perinatal transmission at 14 weeks compared to short course ZDV [84]. Infants were followed through 18 months. Despite continued breastfeeding, the long-term transmission rates showed a persistent 41% reduction for infants receiving sdNVP [85]. This prophylaxis regimen, in addition to having a high efficacy rate, was found to be safe, highly cost-effective, and deliverable in high seroprevalence settings [86]. However, concerns arose regarding the emergence of resistant mutations which could negatively impact on later HIV-1 treatment success.

A number of studies evaluated the use of short-course combination ARVs for PMTCT. The PETRA trial conducted in South Africa, Tanzania, and Uganda evaluated the efficacy of short-course regimens combining ZDV and lamivudine (3TC) in predominantly breastfeeding populations. There were three regimens of ZDV + 3TC. The first targeted the antepartum, intrapartum, and postpartum periods. The second targeted the intrapartum and postpartum periods. The third targeted only the intrapartum period. All three regimens were compared to placebo. At 6 weeks, the first and second ZDV + 3TC regimens were effective in reducing HIV-1 transmission (63 and 42% reduction, respectively) with transmission rates of 5.7 and 8.9%. However, after 18 months, the transmission rate of the most intensive intervention was 14.9%, or double that observed at 6 weeks [87].

Thus, short-course perinatal ARV trials conducted in the mid to late 1990s demonstrated effectiveness in reducing HIV-1 transmission during the first months after delivery. However, except for the HIVNET 012 sdNVP trial, the efficacy observed shortly after birth decreased over time with continued breastfeeding. These studies highlight the need to supplement short-course regimens to prevent MTCT of HIV-1 with continued interventions to minimize the risk of subsequent transmission through breastfeeding.

Extended Infant Prophylaxis Regimens to Reduce Postpartum Transmission

One of the strategies proposed to reduce the risk of postpartum breast milk HIV-1 transmission is the use of infant ARV prophylaxis during the breastfeeding period. An open-label trial, SIMBA, was conducted in Uganda and Rwanda with infants of HIV-1 infected women who had received ZDV plus didanosine (ddI) from 36 weeks gestation until 1 week postpartum. Infants were randomized to receive either 3TC or NVP from birth until 1 month after breastfeeding cessation. The study provided empirical evidence for the use of infant ARV prophylaxis. Four weeks postpartum, the infant infection rates were 1.1% among infants who received 3TC and 0.6% among infants who received NVP among HIV-1 uninfected infants at birth. The overall transmission rate at 6 months was 10% [88]. Likewise, the MITRA study conducted in Tanzania was an open-label, nonrandomized, prospective cohort study investigating prophylactic ARV treatment for the infant during the breastfeeding period. The estimated risk of HIV-1 infection 6 weeks after delivery was 3.8%, whereas at 6 months it was 4.9%. Among infants uninfected at 6 weeks, the cumulative risk of HIV-1 infection was 1.2% at 6 months [89].

More recent data from three large randomized clinical trials provide stronger evidence showing that improved HIV-1-free survival is significantly associated with extended provision of prophylactic ATRs for the infant during breastfeeding. The SWEN study conducted in Uganda, Ethiopia, and India reported a 50% reduction in postnatal transmission at 6 weeks for infants who received daily NVP from birth compared to those who received sdNVP (2.5% vs. 5.3%, $p=0.009$). At 6 months there were fewer HIV-1-infected children among those who received the extended-dose NVP compared to those who received sdNVP, although this difference was not statistically different ($p=0.16$) [90].

The PEPI study conducted in Malawi reported a reduction in postnatal HIV-1 transmission at 9 months for infants who received 14 weeks of daily NVP compared to those who received sdNVP plus 1 week of daily ZDV (5.2% vs. 10.6%, $p<0.001$). Infants who received 14 weeks of daily NVP and ZDV compared to those who received sdNVP plus 1 week of ZDV also had significantly lower transmission (6.4% vs. 10.6%, $p=0.002$) [91].

HPTN 046, a phase III randomized double-blind placebo-controlled trial, compares the relative safety and efficacy of an extended regimen of infant NVP covering 6 weeks to 6 months postbirth. The trial is currently completing study participant follow-up and will compare 6 months versus 6 weeks of infant NVP on safety and cost effectiveness [92].

Likewise, the BAN study conducted in Malawi demonstrated that either 28 weeks of infant prophylaxis with NVP or maternal triple ATRs given daily during breastfeeding decreased postnatal transmission significantly compared to 1 week infant prophylaxis postdelivery [93].

Extended Maternal Prophylaxis Regimens During Breastfeeding

Another approach to the prevention of postnatal HIV-1 transmission is the use of maternal triple ARV prophylaxis during breastfeeding. Several studies suggest that it may be effective to give relatively healthy, HIV-1-infected pregnant women ARV solely for prevention of postnatal transmission. Four open-label observational trials conducted in SSA where maternal triple ARV regimens were given during the third trimester through cessation of breastfeeding at 6 months showed overall transmission rates ranging from 1.2 to 4.1% at 4–6 weeks of age, and from 1.4 to 5% at 6 months [93–98]. The MITRA PLUS study conducted in Tanzania was an open-label, nonrandomized prospective cohort study where HIV-1-infected pregnant women received ZDV, 3TC, and NVP from 34 weeks gestation through 6 months of breastfeeding. Women with CD4 counts >200 cells/mm^3 or adverse reactions to NVP received Nelfinavir (NLF). Cumulative transmission rates were 4.1% at 6 weeks and 5.0% at 6 months [94].

The Kisumu Breastfeeding Study (KiBS) study conducted in Kisumu, Kenya, was a phase IIb study using a similar regimen and duration to the MITRA PLUS study. KIBS was later modified to provide NLF for women with CD4 counts >250 cells/mm^3. Postnatal transmission rates were 4.2% at 6 weeks and 5% at 6 months [95].

The AMATA study, conducted in urban Rwanda, was an open-label, observational study comparing triple ARV therapy started at 26 weeks gestation through delivery if the infant was formula fed or through 6 months if breastfed. Women received Stavudine (d4T), 3TC, and NVP if her CD4 count was less than 350/mm^3 or 3TC, ZDV, and Efavirenz (EFV) if her CD4 count was greater than 350 cells/mm^3. Only six children were infected after birth, a 1.3% postnatal transmission rate [96].

In the Kesho Bora trial, a multisite trial in Kenya, South Africa and Burkina Faso, HIV-1-infected pregnant women with CD4 counts between 200 and 500 cells/mm^3 were randomized between 28 and 36 weeks gestation to either triple ARV [ZDV, 3TC and lopinavir enhanced with ritonavir (LPV/r)] for 6.5 months postdelivery or until breastfeeding cessation or a short ARV regimen (ZDV through delivery followed by 1 week of ZDV and 3TC). All infants received sdNVP after delivery followed by 1 week of ZDV. The study concluded that triple ARV prophylaxis given during pregnancy and continued during breastfeeding significantly reduced the risk of HIV-1 infection at 12 months of age by 40% ($p=0.05$) compared to short-course ARV prophylaxis given only during late pregnancy and peripartum [97].

At the time of the trial, there were parallel Kesho Bora observational cohort studies assessing effectiveness and safety of ARVs used for treatment or prophylaxis. Women with CD4 counts less than 200 cells/mm^3 or with WHO stage 4 disease were initiated on HAART, while women with CD4 counts greater than 500 cells/mm^3 received ARV prophylaxis consisting of ZDV from 34 to 36 weeks gestation until delivery and sdNVP in labor. All infants received sdNVP. Eighteen-month cumulative transmission rates were 7.5% for women who received HAART for treatment and 5.8% when women received ARV prophylaxis with ZDV. This study helped shed light on the importance of initiating HAART early in pregnancy in women with advanced HIV-1 disease to avert the increased risk of transmission associated with insufficient VL suppression [98].

The Mma Bana trial [99] conducted in Botswana compared the efficacy of a protease inhibitor (PI)-based ARV regimen (ZDV+3TC+LPV/r) with a triple nucleoside regimen [ZDV+3TC+abacavir (ABC)]. There was 92–95% efficacy in virologic suppression at delivery and during the breastfeeding period in both arms, with an overall MTCT rate of 1.1%.

The BAN study [93] conducted in Malawi evaluated the efficacy of either maternal triple ARV or extended infant NVP prophylaxis administered over 28 weeks of breastfeeding in reducing postpartum HIV-1 transmission. These regimens were each compared to a control group where mothers and infants did not receive extended prophylaxis after delivery. Both the maternal ARV regimen and the extended infant NVP prophylaxis were superior to the control arm. At 28 weeks postdelivery, the HIV-1 transmission rates were 1.7% for the extended NVP prophylaxis group and 2.9% for the maternal triple ARV group compared with 5.7% for the control group. This trial demonstrated that use of either maternal triple ARV or extended infant NVP prophylaxis are effective in the reduction of HIV-1 transmission during breastfeeding.

Evolving WHO Infant Feeding Guidelines for HIV-1-Exposed Infants

While the intervention trials addressing use of extended maternal and infant prophylaxis were being undertaken, other studies were demonstrating that early weaning, prior to or at 6 months of age, among HIV-1-exposed infants, as had been recommended by WHO and other groups, was associated with increased risk of infant morbidity and mortality among HIV-1-uninfected infants [41, 43, 100–102]. Based on these data, in 2006, the WHO refined their infant feeding guidelines for infants born to HIV-1-infected women. They recommended 6 months of exclusive breastfeeding if replacement feeding is not acceptable, feasible, affordable, sustainable, and safe. At 6 months, replacement feeding should be reassessed. If replacement feeding was not safe or sustainable at 6 months, then continued breastfeeding with additional complementary foods was recommended [103]. These infant feeding recommendations were updated further in 2010 [104].

WHO 2010 Advice and Implementation

Based on the above trial data demonstrating that both infant prophylaxis and maternal ARV prophylaxis were efficacious in decreasing the risk of HIV-1 transmission during the first 6 months of life among HIV-1-infected breastfeeding women who did not require treatment for their own health, the WHO in their 2010 revised PMTCT guidelines provided updated recommendations for use of either of these two approaches to reduce transmission throughout breastfeeding [45]. The other recommendations were to provide longer antepartum ARV prophylaxis to further reduce the risk of HIV-1 transmission during pregnancy with provision of ARV prophylaxis from as early as 14 weeks gestation for HIV-1-infected women with CD4 counts >350/mm^3 who did not yet require HAART for their own health. The 2010 WHO guidelines also recommended providing ARVs throughout the duration of breastfeeding, through the first year of life or longer, to either the mother or her infant.

This revised advice was strongly backed by WHO expert consultants but was based on limited clinical trial data. Data supporting the recommended PMTCT interventions applying to breastfeeding after 6 months are particularly limited. IMPAACT 1077, also known as PROMISE (promoting maternal and infant survival everywhere) is a large, multisite trial currently underway which will compare the use of maternal triple prophylaxis during pregnancy to that of maternal ZDV. In the postpartum period, the use of maternal triple ARV prophylaxis will be compared to that of infant NVP throughout breastfeeding. PROMISE will also study the impact of stopping versus continuing triple ARVs after the discontinuation of breastfeeding for women who do not yet require therapy. The study will also evaluate the effect of continuing infant cotrimoxazole prophylaxis among uninfected children after cessation of breastfeeding and whether it reduces morbidity and improves 2-year survival.

Breast Milk HIV-1 Transmission: Challenges, Gaps, and Future Research Directions

The HIV-1 pandemic is showing signs of maturing, but women of childbearing age remain a highly susceptible at-risk group. In some settings, HIV-1 prevalence is much higher among women than men, especially in the younger age groups. Despite 30 years of research, there is still no vaccine to prevent HIV-1 infection. Other interventions are urgently needed to protect women and adolescent girls from acquiring HIV-1 infection. Thus, a primary challenge, which will also reduce MTCT transmission among breastfeeding women, is to find efficacious methods to protect women from becoming HIV-1 infected.

Better understanding of innate and adaptive immunity that protects against breastfeeding transmission of HIV-1, and elucidating the role of immunogenetic factors, as well as the mechanisms of HIV-1 transfer across the gut lumen, will be important in designing future tailored preventive strategies. Other challenges and research gaps include providing clinical trial data to test the updated WHO guidelines on PMTCT among breastfeeding HIV-1 infected women as well as translating these extended and more complicated PMTCT strategies into actual practice in urban and in rural resource-limited settings. In addition, operational research to determine how best to integrate PMTCT interventions into maternal/child health services infrastructure and how to use the increased PMTCT funding to help strengthen general maternal/child health services will be important [105]. Finally, longer term safety follow-up to assess evidence of late health sequelae among HIV-1-exposed but uninfected infants who receive prolonged ARV exposure both antenatally and postnatally based on the 2009–2010 PMTCT guidelines currently being adopted will be a critical future research need.

Summary and Conclusions

Protecting the current generation of uninfected adults, especially women of childbearing age and adolescent girls, as well as the next generation of uninfected children from HIV-1 are two of the most critical global health challenges. Much progress has been made in the prevention of MTCT among nonbreastfeeding, HIV-1 infected women living in resource-rich settings. Transmission rates of less than 2% are being seen, and virtual elimination of new pediatric HIV infections is the goal. In resource-limited settings where breastfeeding is a necessity, there has also been progress in clinical trial research to reduce the risk of postnatal HIV transmission. Some of these results have been translated into reduced numbers of infected infants worldwide since 2007. However, many challenges still remain in the quest for a global HIV-1-free generation of children.

References

1. Cooper ER, Charurat M, Mofenson LM et al (2002) Combination antiretroviral strategies for the treatment of pregnant HIV-1-infected women and prevention of perinatal HIV-1 transmission. J Acquir Immune Defic Syndr Hum Retrovirol 29:484–94
2. Global report: UNAIDS report on the global AIDS epidemic 2010. http://www.unaids.org/documents/20101123_GlobalReport_em.pdf. Accessed Dec 2010
3. DeCock KM, Fowler MG, Mercier E et al (2000) Prevention of mother-to-child HIV-1 transmission in resource-poor countries: translating research into policy and practice. JAMA 283:1175–1182
4. Townsend CL, Cortina-Borja M, Peckham CS et al (2008) Low rates of MTCT of HIV-1 following effective pregnancy interventions in the United Kingdom and Ireland, 2000–2006. AIDS 22:973–81
5. Zhou Z, Meyers K, Li X et al (2010) Prevention of mother-to-child transmission of HIV-1 using highly active antiretroviral therapy in rural Yunnan, China. J Acquir Immune Defic Syndr 53:S15–22
6. Mofenson LM (2010) Prevention in neglected subpopulations: prevention of mother-to-child transmission of HIV-1 infection. Clin Infect Dis 50:130–48
7. Bulterys M, Wilfert CM (2009) HAART during pregnancy and during breast feeding among HIV-1-infected women in the developing world: has the time come? AIDS 23:2473–7
8. Kourtis AP, Duerr A (2003) Prevention of perinatal HIV-1 transmission: a review of novel strategies. Expert Opin Investig Drugs 12:1–10
9. Kuhn L, Reitz C, Abrams EJ (2009) Breastfeeding and AIDS in the developing world. Curr Opin Pediatr 21:83–93
10. Newell ML (2006) Current issues in the prevention of mother to child transmission of HIV-1 infection. Trans R Soc Trop Med Hyg 100:1–5
11. Mmiro FA, Aizire J, Mwatha AK, Eshleman SH, Donnell D, Fowler MG et al (2009) Predictors of early and late mother-to-child transmission of HIV-1 in a breastfeeding population: HIV-1 network for prevention trials 012 experience, Kampala, Uganda. J Acquir Immune Defic Syndr 52:32–9
12. Nduati R, John G, Mbori-Ngacha D et al (2000) Effect of breastfeeding and formula feeding on transmission of HIV-1: a randomized clinical trial. JAMA 283:1167–1174
13. Jamieson DJ, Sibailly TS, Sadek R et al (2003) HIV-1 viral load and other risk factors for mother-to-child transmission of HIV-1 in a breast-feeding population in Côte d'Ivoire. J Acquir Immune Defic Syndr 34:430–436
14. Koulinska IN, Villamor E, Msamanga G et al (2006) Risk of HIV-1 transmission by breastfeeding among mothers infected with recombinant and non-recombinant HIV-1 genotypes. Virus Res 120:191–8
15. Rousseau CM, Nduati RW, Richardson BA et al (2003) Longitudinal analysis of human immunodeficiency virus type 1 RNA in breast milk and of its relationship to infant infection and maternal disease. J Infect Dis 187:741–7

16. Rousseau CM, Nduati RW, Richardson BA et al (2004) Association of levels of HIV-1 infected breast milk cells and risk of mother-to-child transmission. J Infect Dis 190:1880–8
17. The Breastfeeding and HIV International Transmission Study Group (2004) Late postnatal transmission of HIV-1 in breast-fed children: an individual patient data meta-analysis. J Infect Dis 189:2154–66
18. Bulterys M, Chao A, Dushimimana A, Saah A (1995) HIV-1 seroconversion after 20 months of age in a cohort of breastfed children born to HIV-1-infected women in Rwanda. AIDS 9:93–94
19. Dunn DT, Newell M-L, Ades AE, Peckham CS (1992) Risk of human immunodeficiency virus type 1 transmission through breastfeeding. Lancet 340:585–8
20. Van de Perre P, Simonon A, Msellati P, Hitimana DG, Vaira D, Bazubagira A et al (1991) Postnatal transmission of human immunodeficiency virus type 1 from mother to infant. A prospective cohort study in Kigali, Rwanda. N Engl J Med 325:593–8
21. Humphrey JH, Marinda E, Mutasa K, Moulton LH, Iliff PJ, Ntozini R et al (2010) Mother to child transmission of HIV-1 among Zimbabwean women who seroconverted postnatally: prospective cohort study. Br Med J 341:c6580
22. Liang K, Gui X-E, Zhang YZ, Zhuang K, Meyers K, Ho DD (2009) A case series of 104 women infected by HIV-1 via blood transfusion postnatally: high rate of HIV-1 transmission to infants through breastfeeding. J Infect Dis 200:682–6
23. Lewis P, Nduati R, Kreiss JK et al (1998) Cell-free human immunodeficiency virus type 1 in breast milk. J Infect Dis 177:34–9
24. Gray RR, Salemi M, Lowe A et al (2011) Multiple independent lineages of HIV-1 persist in breast milk and plasma. AIDS 25:143–152
25. Eshleman SH, Becker-Pergola G, Deseyve M et al (2001) Impact of human immunodeficiency virus type1 (HIV-1) subtype on women receiving single-dose nevirapine prophylaxis to prevent HIV-1 vertical transmission (HIVNET 012 trial). J Infect Dis 184:914–917
26. Bjorndal A, Sonnerborg A, Tscherning C et al (1999) Phenotypic characteristics of human immunodeficiency virus type 1 subtype C isolates of Ethiopian AIDS patients. AIDS 15:647–53
27. Renjifo B, Gilbert P, Chaplin B et al (2004) Preferential in utero transmission of HIV-1 subtype C compared to subtype A or D. AIDS 18:1629–36
28. Koulinska IN, Villamor E, Chaplin B et al (2006) Transmission of cell-free and cell-associated HIV-1 through breast-feeding. J Acquir Immune Defic Syndr 41:93–9
29. Yang C, Li M, Newman RD et al (2003) Genetic diversity of HIV-1 in western Kenya: subtype-specific differences in mother-to-child transmission. AIDS 11:1667–74
30. Embree JE, Njenga S, Datta P et al (2000) Risk factors for postnatal mother-child transmission of HIV-1. AIDS 14:2535–41
31. Semba RD, Kumwenda N, Hoover DR et al (1999) Human immunodeficiency virus load in breast milk, mastitis, and mother-to-child transmission of human immunodeficiency virus type 1. J Infect Dis 180:93–8
32. Kumwenda N, Miotti PG, Taha TE et al (2002) Antenatal vitamin A supplementation increases birth weight and decreases anemia among infants born to human immunodeficiency virus-infected women in Malawi. Clin Infect Dis 35:618–24
33. Phiri W, Kasonka L, Collin S et al (2006) Factors influencing breast milk HIV-1 RNA viral load among Zambian women. AIDS Res Hum Retroviruses 22:607–14
34. Kupka R, Garland M, Msamanga G et al (2005) Selenium status, pregnancy outcomes, and mother-to-child transmission of HIV-1. J Acquir Immune Defic Syndr 29:201–10
35. Mehta S, Hunter DJ, Mugusi FM et al (2009) Perinatal outcomes, including mother-to-child transmission of HIV-1, and child mortality and their association with maternal vitamin D status in Tanzania. J Infect Dis 200: 1022–30
36. Semba RD, Miotti P, Chiphangwi J et al (1994) Maternal vitamin A deficiency and mother-to-child transmission of HIV-1. Lancet 343:1593–7
37. Coutsoudis A, Pillay K, Spooner E et al (1999) Randomized trial testing the effect of vitamin A supplementation on pregnancy outcomes and early mother-to-child transmission in Durban, South Africa. AIDS 13:1517–24
38. Fawzi W, Msamanga G, Hunter D et al (2002) Randomized trial of vitamin supplements in relation to transmission of HIV-1 through breastfeeding and early child mortality. AIDS 16:1935–44
39. Kuhn L (2010) Milk mysteries: why are women who exclusively breastfeed less likely to transmit HIV-1 during breastfeeding? Clin Infect Dis 50:770–2
40. Requena M, Bouhlal H, Nasreddine N et al (2008) Inhibition of HIV-1 transmission in trans from dendritic cells to CD4+ by natural antibodies to the CRD domain of DC-SIGN purified from breast milk and intravenous immunoglobulins. Immunology 123:508–18
41. Kafulafula G, Hoover DR, Taha TE et al (2010) Frequency of gastroenteritis and gastroenteritis-associated mortality with early weaning in HIV-1-uninfected children born to HIV-1-infected women in Malawi. J Acquir Immune Defic Syndr 53:6–13

42. Kuhn L, Sinkala M, Semrau K et al (2010) Elevations in mortality associated with weaning persist into the second year of life among uninfected children born to HIV-1 infected mothers. Clin Infect Dis 50:437–44
43. Onyango-Makumbi C, Bagenda D, Mwatha A et al (2010) Early weaning of HIV-1 exposed uninfected infants and risk of serious gastroenteritis: findings from two perinatal HIV-1 prevention trials in Kampala, Uganda. J Acquir Immune Defic Syndr 53:20–7
44. Thea DM, Aldrovandi G, Kankasa C et al (2006) Post-weaning breast milk HIV-1 viral load, blood prolactin levels and breast milk volume. AIDS 20:1539–47
45. World Health Organization (2010) Antiretroviral drugs for treating pregnant women and preventing HIV-1 infection in infants: towards universal access. Recommendations for a public health approach (2010 version). http://www.who.int/HIV-1/pub/mtct/antiretroviral2010/en/index.html. Accessed Mar 2011
46. Winchester R, Pitt J, Charurat M et al (2004) Mother- to- child transmission of HIV-1 strong association with certain maternal HLA-B alleles independent of viral load implicates innate immune mechanism. J Acquir immune Defic Syndr 36:639–70
47. Mackelprang RD, John-Stewart G, Carrington M et al (2008) Maternal HLA homozygosity and mother-child HLA concordance increase the risk of vertical transmission of HIV-1. J Infect Dis 197:1156–1161
48. Belyakov IM, Berzofsky JA (2004) Immunobiology of mucosal HIV-1 infection and the basis for development of a new generation of mucosal AIDS vaccines. Immunity 20:247–53
49. Iliff PJ, Piwoz EG, Tavengwa NV et al (2005) Early exclusive breastfeeding reduces the risk of postnatal HIV-1 transmission and increases HIV-1-free survival. AIDS 19:699–708
50. Kourtis AP, Bulterys M, Nesheim S et al (2001) Understanding the timing of HIV-1 transmission from mother to infant. JAMA 285:709–12
51. Kourtis AP, Lee FK, Jamieson DJ et al (2006) Mother-to-child transmission of HIV-1: timing and implications for prevention. Lancet Infect Dis 6:726–32
52. Bryson YY, Luzuriaga K, Sullivan JL et al (1992) Proposed definitions for in utero versus intrapartum transmission of HIV-1. N Engl J Med 327:1246–7
53. John GC, Nduati R, Mbori-Ngacha DA et al (2001) Correlates of mother-to-child human immunodeficiency virus type 1 (HIV-1) transmission: association with maternal plasma HIV-1 RNA load, genital HIV-1 DNA shedding, and breast infections. J Infect Dis 183:206–212
54. Lohman-Payne B, Slyker J, Rowland-Jones SL (2010) Immune-based approaches to the prevention of mother-to-child transmission of HIV-1: active and passive immunization. Clin Perinat 37:787–805
55. Toniolo A, Serra C, Conaldi PG, Basolo F, Falcone V, Dolei A (1995) Productive HIV-1 infection of normal human mammary epithelial cells. AIDS 9:859–866
56. Dorosko SM, Connor RI. Primary human mammary epithelial cells endocytose HIV-1 and facilitate viral infection of CD4+ T lymphocytes. J Virol 84:10533–10542
57. Lyimo MA, Howell AL, Balandya E, Eszterhas SK, Connor RI (2009) Innate factors in human breast milk inhibit cell-free HIV-1 but not cell-associated HIV-1 infection of CD4+ cells. J Acquir Immune Defic Syndr 51:117–24
58. Bulek K, Swaidani S, Aronica M, Li X (2010) Epithelium: the interplay between innate and Th2 immunity. Immunol Cell Biol 88:257–68
59. Kalish LA, Pitt J, Lewis J et al (1997) Defining the time of fetal or perinatal acquisition of HIV-1 infection on the basis of age at first positive culture. J Infect Dis 175:712–5
60. Lunney KM, Iliff P, Mutasa K et al (2010) Associations between breast milk viral load, mastitis, exclusive breastfeeding, and postnatal transmission of HIV-1. Clin Infect Dis 50:762–769
61. Willumsen JF, Filteau SM, Coutsoudis A et al (2003) Breastmilk RNA viral load in HIV-1-infected South African women: effects of subclinical mastitis and infant feeding. AIDS 17:407–414
62. Hartmann SU, Berlin CM, Howett MK (2006) Alternative modified infant-feeding practices to prevent postnatal transmission of human immunodeficiency virus type 1 through breast milk: past, present, and future. J Hum Lact 22:75–88
63. Lehman DA, Farquhar C (2007) Biological mechanisms of vertical human immunodeficiency virus (HIV-1) transmission. Rev Med Virol 17:381–403
64. Giuliano M, Guidotti G, Andreotti M et al (2007) Triple antiretroviral prophylaxis administered during pregnancy and after delivery significantly reduces breast milk viral load: a study within the Drug Resource Enhancement Against AIDS and Malnutrition Program. J Acquir Immune Defic Syndr 44:286–291
65. Shapiro RL, Ndung'u T, Lockman S, Shapiro RL et al (2005) Highly active antiretroviral therapy started during pregnancy or postpartum suppresses HIV-1 RNA, but not DNA, in breast milk. J Infect Dis 192:713–719
66. Farquhar C, VanCott TC, Mbori-Ngacha DA et al (2002) Salivary secretory leukocyte protease inhibitor is associated with reduced transmission of human immunodeficiency virus type 1 through breast milk. J Infect Dis 186:1173–1176
67. Baron S, Poast J, Richardson CJ et al (2000) Oral transmission of HIV-1 by infected seminal fluid and milk: a novel mechanism. J Infect Dis 181:498–504
68. Shugars DC (1999) Endogenous mucosal antiviral factors of the oral cavity. J Infect Dis 179:S431–5

69. Villamor E, Koulinska IN, Furtado J et al (2007) Long-chain n-6 polyunsaturated fatty acids in breast milk decrease the risk of HIV-1 transmission through breastfeeding. Am J Clin Nutr 86:682–9
70. Becquart P, Hocini H, Levy M, Sepou A, Kazatchkine MD, Belec L (2000) Secretory anti-human immunodeficiency virus (HIV-1) antibodies in colostrum and breast milk are not a major determinant of the protection of early postnatal transmission of HIV-1. J Infect Dis 181:532–539
71. Walter J, Kuhn L, Aldrovandi GM (2008) Advances in basic science understanding of mother-to-child HIV-1 transmission. Curr Opin HIV AIDS 3:146–150
72. Becquet R, Bland R, Leroy V et al (2009) Duration, pattern of breastfeeding and postnatal transmission of HIV-1: pooled analyses of individual data from West and South African cohorts. PLoS One 4:e7397
73. Coutsoudis A, Pillay K, Spooner E, Kuhn L, Coovadia HM (1999) Influence of infant-feeding patterns on early mother-to-child transmission of HIV-1 in Durban, South Africa: a prospective cohort study. South African Vitamin A Study Group. Lancet 354:471–476
74. Kuhn L, Sinkala M, Kankasa C et al (2007) High uptake of exclusive breastfeeding and reduced early post-natal HIV-1 transmission. PLoS One 2:e1363
75. Kourtis AP, Jamieson DJ, de Vincenzi I et al (2007) Prevention of human immunodeficiency virus-1 transmission to the infant through breastfeeding: new developments. Am J Obstet Gynecol 197:S113–S122
76. Gantt S, Shetty AK, Seidel KD et al (2007) Laboratory indicators of mastitis are not associated with elevated HIV-1 DNA loads or predictive of HIV-1 RNA loads in breast milk. J Infect Dis 196:570–576
77. Neville MO, Allen JC, Archer P et al (1991) Studies in human lactation: milk volume and nutrient composition during weaning and lactogenesis. Am J Clin Nutr 54:81–92
78. Lehman DA, Chung MH, John-Stewart GC et al (2008) HIV-1 persists in breast milk cells despite antiretroviral treatment to prevent mother-to-child transmission. AIDS 22:1475–1485
79. Connor EM, Sperling RS, Gelber R et al (1994) Reduction of maternal-infant transmission of human immunodeficiency virus type 1 with zidovudine treatment. Pediatric AIDS Clinical Trials Group Protocol 076 Study Group. N Engl J Med 331:1173–80
80. Shaffer N, Chuachoowong R, Mock PA et al (1999) Short-course zidovudine for perinatal HIV-1 transmission in Bangkok, Thailand: a randomised controlled trial. Bangkok Collaborative Perinatal HIV-1 Transmission Study Group. Lancet 353:773–80
81. Dabis F, Msellati P, Meda N et al (1999) 6-Month efficacy, tolerance, and acceptability of a short regimen of oral zidovudine to reduce vertical transmission of HIV-1 in breastfed children in Côte d'Ivoire and Burkina Faso: a double-blind placebo-controlled multicentre trial. DITRAME Study Group. Diminution de la Transmission Mère-Enfant. Lancet 353:786–92
82. Wiktor SZ, Ekpini E, Karon JM et al (1999) Short-course oral zidovudine for prevention of mother-to-child transmission of HIV-1 in Abidjan, Côte d'Ivoire: a randomized trial. Lancet 353:781–785
83. Leroy V, Karon J, Alioum A et al (2002) Twenty-four month efficacy of a maternal short-course zidovudine regimen to prevent mother-to-child transmission of HIV-1 in West Africa. AIDS 16:631–641
84. Guay LA, Musoke P, Fleming T et al (1999) Intrapartum and neonatal single dose nevirapine compared with zidovudine for prevention of mother-to-child transmission of HIV-1 in Kampala, Uganda: HIV-1NET 012 randomized trial. Lancet 354:795–802
85. Jackson JB, Musoke P, Fleming T et al (2003) Intrapartum and neonatal single-dose nevirapine compared with zidovudine for prevention of mother-to-child transmission of HIV-1 in Kampala, Uganda: 18-month follow-up of the HIVNET 012 randomised trial. Lancet 362:859–868
86. Marseille E, Kahn JG, Mmiro F et al (1999) Cost effectiveness of single-dose nevirapine regimen for mothers and babies to decrease vertical HIV-1 transmission in sub-Saharan Africa. Lancet 354:803–9
87. The Petra Study Team (2002) Efficacy of three short-course regimens of zidovudine and lamivudine in preventing early and late transmission of HIV-1 from mother to child in Tanzania, South Africa and Uganda [Petra Study]: a randomized, double blind, placebo-controlled trial. Lancet 359:1178–1186
88. Vyankandondera J, Lutchers S, Hassink E et al (2003) Reducing risk of HIV-1 transmission from mother to infant through breastfeeding using antiretroviral prophylaxis in infants (SIMBA). In: 2nd IAS conference on HIV-1 pathogenesis, treatment and prevention, Paris, France, 2003; Abstract LB07
89. Kilewo C, Karlsson K, Massawe A et al (2008) Prevention of mother-to-child transmission of HIV-1 through breast-feeding by treating infants prophylactically with lamivudine in Dar es Salaam, Tanzania: The Mitra Study. J Acquir Immune Defic Syndr 48:315–323
90. Six Week Extended-Dose Nevirapine (SWEN) Study Team (2008) Extended-dose nevirapine to 6 weeks of age for infants to prevent HIV-1 transmission via breastfeeding in Ethiopia, India, and Uganda: an analysis of three randomised controlled trials. Lancet 372:300–13
91. Kumwenda NI, Hoover DR, Mofenson LM et al (2008) Extended antiretroviral prophylaxis to reduce breast-milk HIV-1 transmission. N Engl J Med 359:119–29

92. Coovadia H, Brown E, Maldonado B et al (2011) HPTN 046: efficacy of extended daily infant nevirapine through age 6 months compared to 6 weeks for prevention of postnatal mother-to-child transmission (MTCT) of HIV-1 through breastfeeding (BF). In: 18th conference on retroviruses and opportunistic infections program abstract book, Boston, 2011; Abstract 123LB
93. Chasela CS, Hudgens MG, Jamieson DJ et al (2010) Maternal or infant antiretroviral drugs to reduce HIV-1 transmission. N Engl J Med 362:2271–81
94. Kilewo C, Karlsson K, Ngarina M et al (2009) Prevention of mother to child transmission of HIV-1 through breastfeeding by treating mothers with triple antiretroviral therapy in Dar es Salaam, Tanzania: the MITRA-PLUS study. J Acquir Immune Defic Syndr 52:406–16
95. Thomas T, Masaba R, Borkowf CG et al (2011) Triple antiretroviral prophylaxis to prevent mother-to-child HIV-1 transmission through breastfeeding–The Kisumu Breastfeeding Study, Kenya: a clinical trial. PLoS Med e1001015
96. Peltier CA, Ndayisaba GF, Lepage P et al (2009) Breastfeeding with maternal antiretroviral therapy or formula feeding to prevent HIV postnatal mother-to-child transmission in Rwanda. AIDS 23:2415–23
97. The Kesho Bora Study Group (2011) Triple antiretrovirals compared with zidovudine and single-dose nevirapine prophylaxis during pregnancy and breastfeeding for prevention of mother-to-child transmission of HIV-1 (Kesho Bora study): a randomised controlled trial. Lancet 3099(10):70288-7
98. The Kesho Bora Study Group (2010) Eighteen-month follow-up of HIV-1 infected mothers and their children enrolled in the Kesho Bora study observational cohorts. J Acquir Immune Defic Syndr 54:553–41
99. Shapiro RL, Hughes MD, Ogwu A et al (2010) Antiretroviral regimens in pregnancy and breastfeeding in Botswana. N Engl J Med 362:2282–94
100. Creek TL, Kim A, Lu L, Bowen A et al (2010) Hospitalization and mortality among primarily nonbreastfed children during a large outbreak of diarrhea and malnutrition in Botswana, 2006. J Acquir Immune Defic Syndr 53:14–9
101. Homsy J, Moore D, Barasa A et al (2010) Breastfeeding, mother-to-child HIV-1 transmission, and mortality among infants born to HIV-1-Infected women on highly active antiretroviral therapy in rural Uganda. J Acquir Immune Defic Syndr 53:28–35
102. Thomas T, Masaba R, van Eijk A et al (2007) Rates of diarrhea associated with early weaning among infants in Kisumu, Kenya [Abstract 774]. In: 14th conference on retroviruses and opportunistic infections, Los Angeles, CA, 2007
103. HIV-1 and infant feeding technical consultation held on behalf of the Inter-Agency Task Team (IATT) on prevention of HIV-1 infections in pregnant women, mothers and their infants. Consensus statement, Geneva, 25–27 October 2006
104. World Health Organization (2010) Guidelines on HIV-1 and infant feeding. Principles and recommendations for infant feeding in the context of HIV-1 and a summary of evidence. pp 1–49. http://whqlibdoc.who.int/publications/2010/9789241599535_eng.pdf. Accessed 16 Feb 2011
105. Stringer EM, Ekouevi DK, Coetzee D et al (2010) Coverage of nevirapine-based services to prevent mother-to-child HIV transmission in 4 African countries. JAMA 304:293–302
106. Kourtis AP, Bulterys M (2010) Mother-to-child transmission of HIV-1: pathogenesis, mechanisms and pathways. Clin Perinatol 37:721–38
107. Eslahpazir J, Jenabian MA, Bouhlal H et al (2008) Infection of macrophages and dendritic cells with primary R5-tropic HIV-1 type 1 inhibited by natural polyreactive anti-CCR5 antibodies purified from cervicovaginal secretions. Clin Vaccine Immunol 15:872–84
108. Ward S (2001) Natural anti-HIV-1 antibodies. Trends Immunol 22:544
109. de Perre V, Simonon A, Hitimana DG et al (1993) Infective and anti-infective properties of breast milk from HIV-1 infected women. Lancet 341:914–918
110. Farquhar C, Rowland-Jones S, Mbori-Ngacha D et al (2004) Human leukocyte antigen (HLA) B*18 and protection against mother-to-child HIV-1 type 1 transmission. AIDS Res Hum Retrovir 20:692–697

Chapter 2
Breastfeeding and Transmission of Viruses Other than HIV-1

Claire L. Townsend, Catherine S. Peckham, and Claire Thorne

Introduction

The risk of HIV-1 transmission from mother to infant through breastfeeding is well established, but less attention has been paid to the implications of other viruses present in breast milk. Acquisition of human immunodeficiency virus type 2 (HIV-2) through breastfeeding has been reported, but the risk of transmission appears to be lower than for HIV-1. For other viruses, such as cytomegalovirus (CMV) and human T-cell lymphotropic virus type 1 (HTLV-1), transmission through breastfeeding is common and well documented, and infection can result in short- or long-term consequences in the infant. For hepatitis B and C viruses, although mother-to-child transmission occurs, breastfeeding does not appear to be a major route of transmission. In this chapter, the implications of these viruses identified in breast milk are described.

For most other viruses that have been detected in breast milk, acquisition of infection through breastfeeding is not an important route of transmission. For example, Epstein–Barr virus (EBV) and human herpesvirus-6 (HHV-6) are shed in breast milk but are not considered an important source of infection in infancy [1, 2]. Both of these viral infections are acquired early in life, but their prevalence increases throughout childhood, with similar rates reported in breastfed and formula-fed babies, and it is therefore unlikely that breast milk is a frequent route of transmission [2]. Human herpesvirus-7 (HHV-7) and human herpesvirus-8 (HHV-8) have been identified in breast milk, albeit rarely, but are often present in saliva, a more likely route of transmission [3–5]. Herpes simplex viruses-1 and -2 are shed into breast milk [6], but reported cases of neonatal infection associated with breastfeeding are usually in mothers with herpetic lesions on the breast [7, 8]. Breastfeeding is therefore only contraindicated if lesions are apparent. The presence of EBV and CMV has been reported to be independently associated with HIV-1 RNA concentration in breast milk [9]; however, it is unknown whether infection with EBV or CMV has implications for increased shedding of other concurrent infections in breast milk.

There is a theoretical risk of acquisition of infection in breastfed infants whose mothers are vaccinated postnatally. Rubella vaccine virus can be recovered from breast milk of women receiving postpartum vaccination and may be transmitted to the breastfed neonate [10, 11]. However, this does not result in clinical disease and there is no evidence of a significant alteration in the infant's

C.L. Townsend (✉) • C.S. Peckham • C. Thorne
MRC Centre of Epidemiology for Child Health, UCL Institute of Child Health,
University College London, 30 Guilford Street, London WC1N 1EH, UK
e-mail: c.townsend@ucl.ac.uk

immune response to subsequent rubella vaccinations or of an increased risk of reactions at a later date [12]. Breastfeeding is not a contraindication for postpartum vaccination with the combined measles–mumps–rubella vaccine. Varicella immunisation is not recommended for breastfeeding mothers by the vaccine manufacturer. However, a study in the United States (US) showed no evidence of virus in breast milk or transmission to breastfeeding babies from mothers receiving a 2-dose schedule at 6 and 10 weeks after birth [13]. US guidance from the Centers for Disease Control and Prevention, and UK national guidance states that vaccination is not contraindicated in breastfeeding mothers.

The chapter focuses on the risk of transmission through breastfeeding of HIV-2, HTLV-1 and CMV, as well as hepatitis B and C viruses, and explores the implications of early acquisition of these different viruses for both the short and long term.

HIV-2

HIV-2 is found predominantly in West Africa and in countries with historical links to the region, such as France and Portugal. HIV-2 is less pathogenic, results in lower viral loads and progresses more slowly than HIV-1, but ultimately leads to immunosuppression and AIDS [14]. Reported mother-to-child transmission rates in the absence of antiretroviral therapy range from 0 to 4% [15–19]; these rates are significantly lower than for HIV-1, possibly due to the lower plasma viral loads occurring with HIV-2 infection [18].

Although vertical transmission has been studied less extensively for HIV-2 than HIV-1, postnatal HIV-2 transmission through breastfeeding has been documented. In a study in the Gambia, 8 of 202 children born to HIV-2-positive women were infected, with timing of transmission reported for 7; 3 had negative PCR tests at 2 months of age and subsequently tested positive, suggesting acquisition through breastfeeding [18]. In the French Perinatal Study, one of the two reported cases of vertically acquired HIV-2 infection probably occurred through breastfeeding [20]. Although the risk of postnatal transmission through breastfeeding is probably lower than for HIV-1, there is a lack of robust evidence, and HIV-2 is usually managed according to HIV-1 infant feeding guidelines.

Human T-Cell Lymphotropic Virus

HTLV-1 was the first human retrovirus to be discovered, in 1980 [21, 22], followed soon afterwards by the discovery of HTLV-2 in 1982 [23]. HTLV-1 infection is life-long and in most individuals remains asymptomatic. However, after a long latent period a small but significant proportion of individuals infected with HTLV-1 develop serious neurological and lymphoproliferative disorders. The association of HTLV-2 with such long-term sequelae is less certain [23].

Both HTLV-1 and HTLV-2 can be acquired by mother-to-child transmission (mainly through breastfeeding); sexual transmission, particularly from male to female; and from contaminated blood products through blood transfusion or injecting drug use [23, 24]. HTLV-1 is endemic in the Caribbean basin, Japan, South America and West and Central Africa with isolated foci in other areas [23, 24]. The natural history of HTLV-1 has mainly been described in these populations. In contrast, the prevalence of HTLV-2 is highest in the Amerindian and some central African populations, and among injecting drug users [25, 26].

Mother-to-Child Transmission of Infection

Mother-to-child transmission is an important mode of acquisition of HTLV-1 [23, 27, 28], and seems to occur mainly through breastfeeding. Based on HTLV-1 studies carried out in Japan in the 1980s, rates of mother-to-child transmission were estimated to be around 15–25%, although numbers in all studies were small [27–32]. A study in French Guyana reported a rate of 10% [33].

Infected lymphocytes have been detected in breast milk from HTLV-1 carrier mothers, and milk from infected mothers given orally to marmosets has been shown to result in infection [34–36]. However, since HTLV-1 infection also occurs in bottle-fed infants born to carrier mothers, it is likely that transmission can also occur in utero and/or during delivery, although the risk appears to be low [37]. HTLV-1 proviral DNA has been detected in the placenta and in cord blood [30], but the mechanism for transmission remains unclear. Transmission of infection due to breastfeeding has been well documented, and breastfeeding is a major risk factor for acquisition of HTLV-1 in the offspring of infected mothers, with longer duration of breastfeeding associated with increased rates of infection [37–40]. After the acquisition of infection through breastfeeding there appears to be no further transmission prior to puberty [41].

Other documented risk factors for perinatal HTLV-1 transmission include high maternal proviral load and antibody titre, increased maternal age and clinical condition and transfusion history of the mother [42–47]. In a study in Peru, the offspring of mothers who presented with HTLV-1-associated myelopathy/tropical spastic paresis (HAM/TSP) or strongyloidiasis were significantly more likely to be infected than those of asymptomatic mothers, suggesting that maternal disease is a risk factor for infection, possibly due to increased proviral load [48].

Consequences of Vertically Acquired Infection

HTLV-1 infection is life-long and most of those infected during early childhood remain asymptomatic in adulthood, although serious HTLV-1-related disease develops in a small proportion of infected individuals. Late sequelae include adult T-cell leukaemia/lymphoma (ATL) and HAM/TSP. ATL is a rare and rapidly progressive lymphoproliferative malignancy of mature T-cells which occurs in people aged 20–60 years, while TSP is a slower but incapacitating disease of the nervous system for which there is at present no treatment [24]. HTLV-1 is also associated with uveitis and other inflammatory disorders [24]. Disease typically manifests later in life, and due to the long latent period there is a paucity of information on the true risk of long-term sequelae in children with vertically acquired infection. However, it is estimated that 5–10% of vertically infected children will progress to one of the HTLV-1-related diseases after a long latent period [24]. An estimated 1–5% of infants with infection will develop ATL with an average latency of 30 years [35], while the lifetime risk of developing HAM/TSP is lower and in the order of 0.25–3% [49, 50].

Paediatric manifestations of HTLV-1 are extremely rare but cases of chronic and relapsing dermatitis of childhood have been reported, which may progress to ATL and HAM/TSP. HTLV-1-associated infective dermatitis appears to be a risk factor for disease progression, and children with this condition have a significant increase in provirus DNA load compared with asymptomatic children [52–54]. It has been suggested that among HTLV-1-infected children eczema could be a marker of risk for subsequent HTLV-associated diseases. Juvenile presentation has also been well documented but is extremely rare, and if it occurs progression to late sequelae is rapid [52].

Prevention of Transmission

Given the lack of treatment or availability of vaccines to prevent infection, many clinicians advocate prenatal screening in populations where the prevalence of infection is high, thus enabling women with infection to be advised to avoid breastfeeding, or at least reduce its duration. However, decisions about breastfeeding need to balance the potential risk of infection-related adverse sequelae against the innumerable benefits of breast milk. One study has reported that the identification of positive women caused discrimination that was worse than that for HIV and stressed the need for counselling [55]. In addition, the potential harm that could result from identifying a maternal diagnosis of HTLV-1 on the family's quality of life requires consideration. In populations where screening is carried out, uninfected women also need to be informed about other routes of acquiring the infection so that these can be avoided.

Antenatal screening for HTLV-1 was introduced in Japan in settings where the prevalence of infection was high, and women identified as seropositive were advised not to breastfeed. The response to this programme was good, and maternal transmission dropped from around 20% to 3%. In the same study, shortened duration of breastfeeding was reported to significantly reduce transmission [56]. This programme was estimated to prevent around 80% of mother-to-child transmission of HTLV-1. In Okinawa, Japan, there was also a marked reduction in the prevalence of antibody to HTLV-1 among pregnant women (from 5.6% to 3.7%) and among nursery school children (from 1.8% to 0.2%) between 1968 and 2000. This was thought to be mainly due to reduced levels and shortened duration of breastfeeding among women identified as seropositive in pregnancy. However, mother-to-child transmission rates among non-breastfed infants also decreased from 12.8% to 3.2% from 1995 to 1999, so other factors were also likely to be responsible [57].

The epidemiology of HTLV-2 has not been so well characterised. Mother-to-child transmission of HTLV-2 appears to occur in a similar pattern to HTLV-1 [51], but information on rates of transmission are sparse as only small numbers of infected women and their babies have been followed up. Similarly, information on the long-term adverse effects is limited.

Cytomegalovirus

CMV is a ubiquitous herpes virus which is commonly acquired in early life and is often asymptomatic. However, congenital CMV, acquired in utero, is symptomatic in 10–15% of newborns and has more serious long-term consequences. The incidence of congenital CMV ranges from about 0.4% of births in resource-rich countries to over 1% in resource-poor countries and in disadvantaged populations [58]. Congenital CMV can result in permanent neurological sequelae, including mental retardation, developmental and motor impairment and sensorineural hearing loss. In contrast, postnatally acquired CMV is rarely symptomatic in full-term infants, although there may be consequences for preterm and low birth weight infants.

Presence of CMV in Breast Milk

Since the 1970s human breast milk has been known to be a source of CMV infection [59–62]. The presence of CMV in breast milk was first demonstrated in 1967 [59], and in 1972, Hayes et al. isolated CMV from the breast milk of 17 of 63 (27%) seropositive mothers [60]. Subsequently, Stagno et al.

reported the isolation of CMV at least once in breast milk or colostrum from 38 of 265 (14%) seropositive women, with isolation less frequent from colostrum (8%) than breast milk (36%) [61].

Shedding of CMV in breast milk may occur following reactivation in the mammary glands, and does not necessarily correlate with shedding in urine [60, 61, 63]. Hayes et al. found that the virus was detectable in breast milk significantly more often after the first week postpartum than before [60]. In a more recent study where breast milk was tested at 2-week intervals, CMV was detectable in most samples 2 weeks after delivery, with a peak in DNA viral load at around 4–6 weeks postpartum [64].

The use of polymerase chain reaction (PCR) techniques in recent studies suggests that reactivation of CMV in breast milk occurs in the majority of seropositive mothers. A study in Tübingen, Germany, demonstrated reactivation in 96% (73 of 76) of seropositive lactating mothers [65], and other studies have found similarly high rates [66, 67]. In a systematic review of CMV transmission to preterm infants, detection of CMV in breast milk was reported to range from 67% to 97% [68]. Although the seroprevalence of CMV varies widely by age, ethnicity and region, rates in pregnant women are high in most populations, ranging from 40% in some developed countries to as high as 98% in less developed regions [69, 70]. Breastfeeding is therefore a common source of exposure to CMV and is thought to function as a form of natural immunisation.

Transmission to the Infant

Transmission of CMV to the infant was demonstrated by Stagno et al., who followed 19 breastfed infants whose mothers were shedding virus in their milk; 11 of these infants (58%) acquired CMV, despite the detection of maternal antibodies in the breast milk, but none showed any signs of disease [61]. In a population-based prospective study in the United Kingdom (UK), infants were screened for CMV at birth to exclude congenital infection, and followed up at regular intervals; 12% were excreting virus at 3 months and 20% at 1 year of age [62]. Among infants of seropositive mothers, 33% had evidence of infection at 1 year of age (compared with 4% of those born to seronegative mothers), including 76% of those who were breastfed. A recent review of CMV transmission to preterm and low birth weight infants highlighted a wide range of transmission rates from seropositive mothers, from as low as 6% to over 50%, probably due to differences in feeding practices and study inclusion criteria [68].

Consequences of Postnatal Acquisition in Preterm and Low Birth Weight Infants

Although there was no sign of disease in infants who acquired CMV in the study by Stagno et al., a subsequent study suggested that early acquisition in preterm infants could result in CMV disease, when two premature infants developed CMV pneumonitis [71]. Preterm infants may be particularly susceptible to CMV disease due to the lack of transfer of maternal antibodies. In a study of 33 infected preterm infants in Germany, clinical manifestations occurred in 16 infants (48%) and included hepatopathy, neutropenia, thrombocytopenia and sepsis syndrome [72]. Symptomatic transmission was significantly more likely among infants of lower birth weight. It may be difficult, however, to distinguish signs of CMV from complications of prematurity [73], and definitions of CMV disease often differ. Most other studies have reported lower rates of CMV disease, ranging from 0% to 34% [68]. Nevertheless, symptoms can be severe, and frequently include hepatitis, thrombocytopenia and sepsis-like syndrome.

Long-Term Outcomes

Data on long-term outcomes of postnatally acquired CMV are much more limited. Although congenital CMV is associated with neurological sequelae, including hearing loss, there have been no reports to date of severe long-term impairment in symptomatic preterm infants with postnatal infection. In the study by Stagno et al., which included psychometric, hearing and vision assessments, no abnormalities were detected in 39 term infants with asymptomatic postnatal infection at a mean age of 51 months [61]. Theoretically, sequelae would be more likely in infants who were preterm or of low birth weight and developed CMV symptoms. However, the number of such infants in any given study is usually small. In a matched follow-up study of 22 preterm infants with CMV infection (half with symptoms) and 22 uninfected controls, there was no hearing loss in either group at 2–4.5 years of age, and no evidence of a difference between groups in neurological problems, language or motor development [74]. Only one study has suggested a possible increase in neurological sequelae in preterm infants with early onset of CMV excretion who were followed up until 3 years of age [75]. In most studies, however, duration of follow up is more limited. Although there is no clear evidence that early acquisition of CMV through breastfeeding poses a significant risk of long-term sequelae for preterm and low birth weight infants, a risk cannot be excluded.

Prevention

CMV can be eliminated from breast milk by Holder pasteurisation (30 min at 62.5°C), but there are concerns that pasteurisation reduces the activity of immunologically beneficial components of breast milk, which are particularly important for preterm and low birth weight infants [76]. Freezing also reduces the infectivity of CMV in breast milk but may not completely eliminate the virus [77, 78]: there have been reports of transmission to infants who were fed frozen milk [64, 79]. Further evaluation of the relative benefits of freezing versus pasteurisation is needed to inform guidelines on early management of preterm and low birth weight infants.

Hepatitis B Virus

Discussions around the role of breastfeeding in the transmission of hepatitis B virus (HBV) from an infected mother to her infant have been ongoing for many years. This is driven by the fact that in many countries women are screened for HBV so that, if infected, their exposed baby can be protected by a full course of immunisation, the first dose being given at birth. An estimated 5% of pregnant women worldwide are chronic HBV carriers, but there is wide variation by region. A recent study reported estimated median prevalence of hepatitis B surface antigen (HBsAg) in pregnant women according to the endemicity of the region, ranging from 0.7% in low endemicity regions (i.e. the Americas and Europe) to 10% in high endemicity regions (i.e. Africa and Western Pacific) [80]. Furthermore, HBsAg carriage rates can vary substantially within populations: in a study of pregnant women in Taiwan, prevalence was 11% overall but ranged from 3.5% among women of Vietnamese ethnicity to 21% among aboriginal Taiwanese [81].

HBV in Breast Milk

Although HBsAg, hepatitis B envelope antigen (HBeAg) and HBV DNA have been isolated from breast milk samples [82–84], evidence to date suggests that breastfeeding does not increase the risk of vertical transmission of HBV. In early studies, predating the availability of hepatitis B vaccine, there was no evidence to support a substantial role of breastfeeding in HBV transmission from mother to child [85]. Beasley et al. reported a nearly identical frequency of acquisition of HBsAg and anti-HBs antibody in 92 breastfed and 55 non-breastfed infants in a study of 147 mother–infant pairs [86], while a smaller German study similarly found no association between breastfeeding mode and infant HBV infection [87].

Breast Feeding and HBV Immunisation

In the context of infant immunisation, HBV transmission does not appear to be affected by breastfeeding and the World Health Organization, the Centers for Disease Control and Prevention in the US and other international and national guidelines recommend that mothers infected with HBV should breastfeed. In a small Italian study of 85 infants born to HBsAg-positive mothers (2% were HBeAg positive), 22 were breastfed and 63 were formula fed and all received immunoglobulin and vaccination; there were no infections in either group and no significant difference between groups with respect to seroconversion (95% in the breastfed group and 97% in the formula-fed group) [88]. Wang et al. investigated whether breastfeeding impacts upon the efficacy (with respect to anti-HBs response) of immunisation of infants of HBV-infected mothers, and found no difference in incidence of immunoprophylaxis failure by breastfeeding mode [89]. In a study in the US, infections were compared in breastfed versus formula-fed infants ($n=369$) born to HBV-infected mothers, all of whom received immunoglobulin at birth and the full course of vaccination; there were no infections in the 101 breastfed infants (mean duration of breastfeeding, around 5 months) and three infections among the 268 formula-fed infants [90]. There is a theoretically increased risk of HBV transmission via breastfeeding if an infected mother has cracked and bleeding nipples, but no evidence base exists to support this hypothesis.

Hepatitis C Virus

HCV is a blood-borne virus that is transmitted most effectively by direct percutaneous exposure to blood. Infants are at risk of infection primarily as a result of transmission from their infected mothers. The risk of vertical transmission of HCV is small, occurring in 3–6% of cases, although this may be higher for some sub-groups including women co-infected with HIV.

Although excretion of HCV in colostrum has been described in some studies [91, 92], other investigators have not been able to detect HCV in breast milk, including in viremic mothers [93–95]. Furthermore, variability over time and correlates of excretion have not been evaluated. There are only a small number of observational studies carried out to date with the statistical power to address the question of whether breastfeeding increases risk of HCV vertical transmission. None of these have found any evidence to support an increased risk of HCV transmission through breastfeeding. In a cohort study of the European Paediatric HCV Network involving 1,758 mother–child pairs, the mother-to-child transmission rate was 6.2% (95% confidence interval 5.0%, 7.5%) in HCV-infected women delivering between 1998 and 2004, of whom 15% were co-infected with HIV and no protective effect of formula feeding was observed [96]. In a UK study of 441 HCV-positive mothers and their infants, 59 infants were breastfed and vertical transmission rates were 7.7% among those breastfed and 6.7% among those formula fed, a non-significant difference [97]. In a similar cohort in Italy, infants of nearly 300 HCV RNA-positive mothers were followed from birth, of whom around 30% breastfed: vertical transmission rates did not differ significantly between the breastfed and formula-fed groups at 7% and 4%, respectively [98].

A large individual patient data meta-analysis involving European data ($n=1,474$) demonstrated no effect of breastfeeding on vertical transmission among infants of women with HCV infection only; however, the study identified a significantly increased risk of transmission among the 13 breastfed infants of HCV/HIV co-infected mothers [99]. In settings where formula feeding is safe, acceptable, affordable, feasible and sustainable, HIV-infected women are advised to avoid breastfeeding due to established risks of postnatal transmission of HIV, and thus HIV/HCV co-infected women in such settings should receive the same advice.

Kumar et al. carried out a study in the United Arab Emirates involving 65 HCV seropositive mothers, to investigate the role of breastfeeding in vertical transmission of HCV [91]. Of these 65 women, all of whom breastfed their infants, five had developed symptomatic liver disease by 3 months postpartum, during the breastfeeding period; three of the five infants of these symptomatic women became infected, most likely as the result of breastfeeding. This study provides some limited evidence of an increased risk of postnatal transmission through breastfeeding among HCV-positive women with liver disease and high viral loads.

However, for asymptomatic HCV seropositive women, without HIV co-infection, current international and national guidelines recommend breastfeeding. As with HBV, a theoretical increased risk of HCV transmission where a mother is suffering from cracked and bleeding nipples exists, and it may therefore be prudent for such women to abstain from breastfeeding.

Conclusions

Breast milk provides numerous nutritional and immunological benefits, and protects the infant against viral and bacterial pathogens. For the majority of viral infections including several herpes viruses (e.g. EBV, HHV-6, HHV-7, HHV-8, HSV-1 and HSV-2), breastfeeding is not a common route of transmission even if the virus can be detected in the milk of infected mothers. There is a possibility of an increased risk of transmission of these infections in women co-infected with HIV-1, but evidence is limited. As described here, HTLV-1 and CMV are commonly transmitted through breast milk and can cause early- or late-onset disease. In areas of high HTLV-1 prevalence, antenatal screening may be warranted, to enable infected women to avoid or reduce the duration of breastfeeding, thereby reducing the risk of mother-to-child transmission and possible long-term sequelae. On the other hand, although CMV seroprevalence is high in most populations, transmission through breastfeeding does not result in clinical disease in most infants. Exposure of newborns to infections such as CMV in the presence of maternal antibodies may represent a form of natural immunisation, and breastfeeding is usually advised. However, preterm infants who lack maternal antibody may be particularly susceptible to viral infections in milk. Screening and pasteurisation of banked breast milk ensures that preterm infants receiving donor milk are protected from infection. However, for preterm infants receiving their mother's own milk, the relative benefits of freezing and pasteurisation remain a matter for debate.

References

1. Junker AK, Thomas EE, Radcliffe A, Forsyth RB, Davidson AG, Rymo L (1991) Epstein-Barr virus shedding in breast milk. Am J Med Sci 302(4):220–223
2. Kusuhara K, Takabayashi A, Ueda K et al (1997) Breast milk is not a significant source for early Epstein-Barr virus or human herpesvirus 6 infection in infants: a seroepidemiologic study in 2 endemic areas of human T-cell lymphotropic virus type I in Japan. Microbiol Immunol 41(4):309–312
3. Dedicoat M, Newton R, Alkharsah KR et al (2004) Mother-to-child transmission of human herpesvirus-8 in South Africa. J Infect Dis 190(6):1068–1075

4. Brayfield BP, Phiri S, Kankasa C et al (2003) Postnatal human herpesvirus 8 and human immunodeficiency virus type 1 infection in mothers and infants from Zambia. J Infect Dis 187(4):559–568
5. Fujisaki H, Tanaka-Taya K, Tanabe H et al (1998) Detection of human herpesvirus 7 (HHV-7) DNA in breast milk by polymerase chain reaction and prevalence of HHV-7 antibody in breast-fed and bottle-fed children. J Med Virol 56(3):275–279
6. Kotronias D, Kapranos N (1999) Detection of herpes simplex virus DNA in maternal breast milk by in situ hybridization with tyramide signal amplification. In Vivo 13(6):463–466
7. Sullivan-Bolyai J, Hull HF, Wilson C, Corey L (1983) Neonatal herpes simplex virus infection in King County, Washington. Increasing incidence and epidemiologic correlates. JAMA 250(22):3059–3062
8. Quinn PT, Lofberg JV (1978) Maternal herpetic breast infection: another hazard of neonatal herpes simplex. Med J Aust 2(9):411–412
9. Gantt S, Carlsson J, Shetty AK et al (2008) Cytomegalovirus and Epstein-Barr virus in breast milk are associated with HIV-1 shedding but not with mastitis. AIDS 22(12):1453–1460
10. Buimovici-Klein E, Hite RL, Byrne T, Cooper LZ (1977) Isolation of rubella virus in milk after postpartum immunization. J Pediatr 91(6):939–941
11. Losonsky GA, Fishaut JM, Strussenberg J, Ogra PL (1982) Effect of immunization against rubella on lactation products. II. Maternal-neonatal interactions. J Infect Dis 145(5):661–666
12. Krogh V, Duffy LC, Wong D, Rosenband M, Riddlesberger KR, Ogra PL (1989) Postpartum immunization with rubella virus vaccine and antibody response in breast-feeding infants. J Lab Clin Med 113(6):695–699
13. Bohlke K, Galil K, Jackson LA et al (2003) Postpartum varicella vaccination: is the vaccine virus excreted in breast milk? Obstet Gynecol 102(5 Pt 1):970–977
14. Matheron S, Pueyo S, Damond F et al (2003) Factors associated with clinical progression in HIV-2 infected-patients: the French ANRS cohort. AIDS 17(18):2593–2601
15. Cavaco-Silva P, Taveira NC, Lourenco MH, Santos Ferreira MO, Daniels RS (1997) Vertical transmission of HIV-2. Lancet 349(9046):177–178
16. (1994) Comparison of vertical human immunodeficiency virus type 2 and human immunodeficiency virus type 1 transmission in the French prospective cohort. The HIV Infection in Newborns French Collaborative Study Group. Pediatr Infect Dis J 13(6):502–506
17. Andreasson PA, Dias F, Naucler A, Andersson S, Biberfeld G (1993) A prospective study of vertical transmission of HIV-2 in Bissau, Guinea-Bissau. AIDS 7(7):989–993
18. O'Donovan D, Ariyoshi K, Milligan P et al (2000) Maternal plasma viral RNA levels determine marked differences in mother-to-child transmission rates of HIV-1 and HIV-2 in The Gambia. MRC/Gambia Government/University College London Medical School working group on mother-child transmission of HIV. AIDS 14(4): 441–448
19. Adjorlolo-Johnson G, De Cock KM, Ekpini E et al (1994) Prospective comparison of mother-to-child transmission of HIV-1 and HIV-2 in Abidjan, Ivory Coast. JAMA 272(6):462–466
20. Burgard M, Jasseron C, Matheron S et al (2010) Mother-to-child transmission of HIV-2 infection from 1986 to 2007 in the ANRS French Perinatal Cohort EPF-CO1. Clin Infect Dis 51(7):833–843
21. Poiesz BJ, Ruscetti FW, Gazdar AF, Bunn PA, Minna JD, Gallo RC (1980) Detection and isolation of type C retrovirus particles from fresh and cultured lymphocytes of a patient with cutaneous T-cell lymphoma. Proc Natl Acad Sci USA 77(12):7415–7419
22. Yoshida M, Miyoshi I, Hinuma Y (1982) Isolation and characterization of retrovirus from cell lines of human adult T-cell leukemia and its implication in the disease. Proc Natl Acad Sci USA 79(6):2031–2035
23. Manns A, Blattner WA (1991) The epidemiology of the human T-cell lymphotrophic virus type I and type II: etiologic role in human disease. Transfusion 31(1):67–75
24. Goncalves DU, Proietti FA, Ribas JG et al (2010) Epidemiology, treatment, and prevention of human T-cell leukemia virus type 1-associated diseases. Clin Microbiol Rev 23(3):577–589
25. Lowis GW, Sheremata WA, Minagar A (2002) Epidemiologic features of HTLV-II: serologic and molecular evidence. Ann Epidemiol 12(1):46–66
26. Mauclere P, Afonso PV, Meertens L et al (2011) HTLV-2B strains, similar to those found in several Amerindian tribes, are endemic in central African Bakola Pygmies. J Infect Dis 203(9):1316–1323
27. Nakano S, Ando Y, Saito K et al (1986) Primary infection of Japanese infants with adult T-cell leukaemia-associated retrovirus (ATLV): evidence for viral transmission from mothers to children. J Infect 12(3):205–212
28. Kajiyama W, Kashiwagi S, Ikematsu H, Hayashi J, Nomura H, Okochi K (1986) Intrafamilial transmission of adult T cell leukemia virus. J Infect Dis 154(5):851–857
29. Ureta-Vidal A, Angelin-Duclos C, Tortevoye P et al (1999) Mother-to-child transmission of human T-cell-leukemia/lymphoma virus type I: implication of high antiviral antibody titer and high proviral load in carrier mothers. Int J Cancer 82(6):832–836
30. Fujino T, Nagata Y (2000) HTLV-I transmission from mother to child. J Reprod Immunol 47(2):197–206

31. Sugiyama H, Doi H, Yamaguchi K, Tsuji Y, Miyamoto T, Hino S (1986) Significance of postnatal mother-to-child transmission of human T-lymphotropic virus type-I on the development of adult T-cell leukemia/lymphoma. J Med Virol 20(3):253–260
32. Kusuhara K, Sonoda S, Takahashi K, Tokugawa K, Fukushige J, Ueda K (1987) Mother-to-child transmission of human T-cell leukemia virus type I (HTLV-I): a fifteen-year follow-up study in Okinawa, Japan. Int J Cancer 40(6):755–757
33. Carles G, Tortevoye P, Tuppin P et al (2004) HTLV1 infection and pregnancy. J Gynecol Obstet Biol Reprod (Paris) 33(1 Pt 1):14–20
34. Ikeda S, Kinoshita K, Amagasaki T et al (1989) Mother to child transmission of human T cell leukemia virus type-1. Rinsho Ketsueki 30(10):1732–1737
35. Tsuji Y, Doi H, Yamabe T, Ishimaru T, Miyamoto T, Hino S (1990) Prevention of mother-to-child transmission of human T-lymphotropic virus type-I. Pediatrics 86(1):11–17
36. Hino S (1989) Milk-borne transmission of HTLV-I as a major route in the endemic cycle. Acta Paediatr Jpn 31(4):428–435
37. Oki T, Yoshinaga M, Otsuka H, Miyata K, Sonoda S, Nagata Y (1992) A sero-epidemiological study on mother-to-child transmission of HTLV-I in southern Kyushu, Japan. Asia Oceania J Obstet Gynaecol 18(4):371–377
38. Wiktor SZ, Pate EJ, Rosenberg PS et al (1997) Mother-to-child transmission of human T-cell lymphotropic virus type I associated with prolonged breast-feeding. J Hum Virol 1(1):37–44
39. Takezaki T, Tajima K, Ito M et al (1997) Short-term breast-feeding may reduce the risk of vertical transmission of HTLV-I. The Tsushima ATL Study Group. Leukemia 11 Suppl 3:60–62
40. Ando Y, Saito K, Nakano S et al (1989) Bottle-feeding can prevent transmission of HTLV-I from mothers to their babies. J Infect 19(1):25–29
41. Ando Y, Matsumoto Y, Nakano S et al (2003) Long-term follow up study of vertical HTLV-I infection in children breast-fed by seropositive mothers. J Infect 46(3):177–179
42. Ureta-Vidal A, Angelin-Duclos C, Tortevoye P et al (1999) Mother-to-child transmission of human T-cell-leukemia/lymphoma virus type I: implication of high antiviral antibody titer and high proviral load in carrier mothers. Int J Cancer 82(6):832–836
43. Hisada M, Maloney EM, Sawada T et al (2002) Virus markers associated with vertical transmission of human T lymphotropic virus type 1 in Jamaica. Clin Infect Dis 34(12):1551–1557
44. Wiktor SZ, Pate EJ, Murphy EL et al (1993) Mother-to-child transmission of human T-cell lymphotropic virus type I (HTLV-I) in Jamaica: association with antibodies to envelope glycoprotein (gp46) epitopes. J Acquir Immune Defic Syndr 6(10):1162–1167
45. Maehama T, Nakayama M, Nagamine M, Nakashima Y, Takei H, Nakachi H (1992) Studies on factor affecting mother-to-child HTLV-I transmission. Nippon Sanka Fujinka Gakkai Zasshi 44(2):215–222
46. Takahashi K, Takezaki T, Oki T et al (1991) Inhibitory effect of maternal antibody on mother-to-child transmission of human T-lymphotropic virus type I. The Mother-to-Child Transmission Study Group. Int J Cancer 49(5):673–677
47. Hino S, Katamine S, Miyamoto T et al (1995) Association between maternal antibodies to the external envelope glycoprotein and vertical transmission of human T-lymphotropic virus type I. Maternal anti-env antibodies correlate with protection in non-breast-fed children. J Clin Invest 95(6):2920–2925
48. Gotuzzo E, Moody J, Verdonck K et al (2007) Frequent HTLV-1 infection in the offspring of Peruvian women with HTLV-1-associated myelopathy/tropical spastic paraparesis or strongyloidiasis. Rev Panam Salud Publica 22(4):223–230
49. Kaplan JE, Osame M, Kubota H et al (1990) The risk of development of HTLV-I-associated myelopathy/tropical spastic paraparesis among persons infected with HTLV-I. J Acquir Immune Defic Syndr 3(11):1096–1101
50. Murphy EL, Fridey J, Smith JW et al (1997) HTLV-associated myelopathy in a cohort of HTLV-I and HTLV-II-infected blood donors. The REDS investigators. Neurology 48(2):315–320
51. Van Dyke RB, Heneine W, Perrin ME et al (1995) Mother-to-child transmission of human T-lymphotropic virus type II. J Pediatr 127(6):924–928
52. Bittencourt AL, Primo J, Oliveira MF (2006) Manifestations of the human T-cell lymphotropic virus type I infection in childhood and adolescence. J Pediatr (Rio J) 82(6):411–420
53. Maloney EM, Yamano Y, Vanveldhuisen PC et al (2006) Natural history of viral markers in children infected with human T lymphotropic virus type I in Jamaica. J Infect Dis 194(5):552–560
54. Mahe A, Meertens L, Ly F et al (2004) Human T-cell leukaemia/lymphoma virus type 1-associated infective dermatitis in Africa: a report of five cases from Senegal. Br J Dermatol 150(5):958–965
55. Oni T, Djossou F, Joubert M, Heraud JM (2006) Awareness of mother-to-child transmission of human T-cell lymphotropic virus (HTLV) type I through breastfeeding in a small group of HTLV-positive women in Maripasoula and Papaichton, French Guiana. Trans R Soc Trop Med Hyg 100(8):715–718
56. Hino S, Katamine S, Miyata H, Tsuji Y, Yamabe T, Miyamoto T (1996) Primary prevention of HTLV-I in Japan. J Acquir Immune Defic Syndr Hum Retrovirol 13(Suppl 1):S199–S203

57. Kashiwagi K, Furusyo N, Nakashima H et al (2004) A decrease in mother-to-child transmission of human T lymphotropic virus type I (HTLV-I) in Okinawa, Japan. Am J Trop Med Hyg 70(2):158–163
58. Dollard SC, Grosse SD, Ross DS (2007) New estimates of the prevalence of neurological and sensory sequelae and mortality associated with congenital cytomegalovirus infection. Rev Med Virol 17(5):355–363
59. Diosi P, Babusceac L, Nevinglovschi O, Kun-Stoicu G (1967) Cytomegalovirus infection associated with pregnancy. Lancet 2(7525):1063–1066
60. Hayes K, Danks DM, Gibas H, Jack I (1972) Cytomegalovirus in human milk. N Engl J Med 287(4):177–178
61. Stagno S, Reynolds DW, Pass RF, Alford CA (1980) Breast milk and the risk of cytomegalovirus infection. N Engl J Med 302(19):1073–1076
62. Peckham CS, Johnson C, Ades A, Pearl K, Chin KS (1987) Early acquisition of cytomegalovirus infection. Arch Dis Child 62(8):780–785
63. Vochem M, Hamprecht K, Jahn G, Speer CP (1998) Transmission of cytomegalovirus to preterm infants through breast milk. Pediatr Infect Dis J 17(1):53–58
64. Yasuda A, Kimura H, Hayakawa M et al (2003) Evaluation of cytomegalovirus infections transmitted via breast milk in preterm infants with a real-time polymerase chain reaction assay. Pediatrics 111(6 Pt 1):1333–1336
65. Hamprecht K, Maschmann J, Vochem M, Dietz K, Speer CP, Jahn G (2001) Epidemiology of transmission of cytomegalovirus from mother to preterm infant by breastfeeding. Lancet 357(9255):513–518
66. Jim WT, Shu CH, Chiu NC et al (2004) Transmission of cytomegalovirus from mothers to preterm infants by breast milk. Pediatr Infect Dis J 23(9):848–851
67. Croly-Labourdette S, Vallet S, Gagneur A et al (2006) Pilot epidemiologic study of transmission of cytomegalovirus from mother to preterm infant by breastfeeding. Arch Pediatr 13(7):1015–1021
68. Kurath S, Halwachs-Baumann G, Muller W, Resch B (2010) Transmission of cytomegalovirus via breast milk to the prematurely born infant: a systematic review. Clin Microbiol Infect 16(8):1172–1178
69. Kenneson A, Cannon MJ (2007) Review and meta-analysis of the epidemiology of congenital cytomegalovirus (CMV) infection. Rev Med Virol 17(4):253–276
70. Gaytant MA, Steegers EA, Semmekrot BA, Merkus HM, Galama JM (2002) Congenital cytomegalovirus infection: review of the epidemiology and outcome. Obstet Gynecol Surv 57(4):245–256
71. Dworsky M, Yow M, Stagno S, Pass RF, Alford C (1983) Cytomegalovirus infection of breast milk and transmission in infancy. Pediatrics 72(3):295–299
72. Maschmann J, Hamprecht K, Dietz K, Jahn G, Speer CP (2001) Cytomegalovirus infection of extremely low-birth weight infants via breast milk. Clin Infect Dis 33(12):1998–2003
73. Townsend CL, Peckham CS, Tookey PA (2011) Surveillance of congenital cytomegalovirus in the UK and Ireland. Arch Dis Child Fetal Neonatal Ed. doi:10.1136/adc.2010.199901
74. Vollmer B, Seibold-Weiger K, Schmitz-Salue C et al (2004) Postnatally acquired cytomegalovirus infection via breast milk: effects on hearing and development in preterm infants. Pediatr Infect Dis J 23(4):322–327
75. Paryani SG, Yeager AS, Hosford-Dunn H et al (1985) Sequelae of acquired cytomegalovirus infection in premature and sick term infants. J Pediatr 107(3):451–456
76. Schleiss MR (2006) Acquisition of human cytomegalovirus infection in infants via breast milk: natural immunization or cause for concern? Rev Med Virol 16(2):73–82
77. Friis H, Andersen HK (1982) Rate of inactivation of cytomegalovirus in raw banked milk during storage at −20 degrees C and pasteurisation. Br Med J (Clin Res Ed) 285(6355):1604–1605
78. Hamprecht K, Maschmann J, Muller D et al (2004) Cytomegalovirus (CMV) inactivation in breast milk: reassessment of pasteurization and freeze-thawing. Pediatr Res 56(4):529–535
79. Sharland M, Khare M, Bedford-Russell A (2002) Prevention of postnatal cytomegalovirus infection in preterm infants. Arch Dis Child Fetal Neonatal Ed 86(2):F140
80. Merrill RM, Hunter BD (2011) Seroprevalence of markers for hepatitis B viral infection. Int J Infect Dis 15(2):e78–e121
81. Liu CY, Chang NT, Chou P (2007) Seroprevalence of HBV in immigrant pregnant women and coverage of HBIG vaccine for neonates born to chronically infected immigrant mothers in Hsin-Chu County, Taiwan. Vaccine 25(44):7706–7710
82. Linnemann CC Jr, Goldberg S (1974) Letter: HBAg in breast milk. Lancet 2(7873):155
83. Lin HH, Hsu HY, Chang MH, Chen PJ, Chen DS (1993) Hepatitis B virus in the colostra of HBeAg-positive carrier mothers. J Pediatr Gastroenterol Nutr 17(2):207–210
84. de Oliveira PR, Yamamoto AY, de Souza CB et al (2009) Hepatitis B viral markers in banked human milk before and after Holder pasteurization. J Clin Virol 45(4):281–284
85. Shi Z, Yang Y, Wang H et al (2011) Breastfeeding of newborns by mothers carrying hepatitis B virus: a meta-analysis and systematic review. Arch Pediatr Adolesc Med 165(9):837–46
86. Beasley RP, Stevens CE, Shiao IS, Meng HC (1975) Evidence against breast-feeding as a mechanism for vertical transmission of hepatitis B. Lancet 2(7938):740–741

87. Sacher M, Eder G, Baumgarten K, Thaler H (1983) Vertical transmission of hepatitis B. Results of a prospective study 1978 to 1981. Wien Klin Wochenschr 95(13):447–451
88. de Martino M, Resti M, Appendino C, Vierucci A (1987) Different degree of antibody response to hepatitis B virus vaccine in breast- and formula-fed infants born to HBsAg-positive mothers. J Pediatr Gastroenterol Nutr 6(2):208–211
89. Wang JS, Zhu QR, Wang XH (2003) Breastfeeding does not pose any additional risk of immunoprophylaxis failure on infants of HBV carrier mothers. Int J Clin Pract 57(2):100–102
90. Hill JB, Sheffield JS, Kim MJ, Alexander JM, Sercely B, Wendel GD (2002) Risk of hepatitis B transmission in breast-fed infants of chronic hepatitis B carriers. Obstet Gynecol 99(6):1049–1052
91. Kumar RM, Shahul S (1998) Role of breast-feeding in transmission of hepatitis C virus to infants of HCV-infected mothers. J Hepatol 29(2):191–197
92. Lin HH, Kao JH, Hsu HY et al (1995) Absence of infection in breast-fed infants born to hepatitis C virus-infected mothers. J Pediatr 126(4):589–591
93. Spencer JD, Latt N, Beeby PJ et al (1997) Transmission of hepatitis C virus to infants of human immunodeficiency virus-negative intravenous drug-using mothers: rate of infection and assessment of risk factors for transmission. J Viral Hepat 4(6):395–409
94. Polywka S, Schroter M, Feucht HH, Zollner B, Laufs R (1999) Low risk of vertical transmission of hepatitis C virus by breast milk. Clin Infect Dis 29(5):1327–1329
95. Kage M, Ogasawara S, Kosai K et al (1997) Hepatitis C virus RNA present in saliva but absent in breast-milk of the hepatitis C carrier mother. J Gastroenterol Hepatol 12(7):518–521
96. Pembrey L, Newell ML, Tovo PA (2005) The management of HCV infected pregnant women and their children European paediatric HCV network. J Hepatol 43(3):515–525
97. Gibb DM, Goodall RL, Dunn DT et al (2000) Mother-to-child transmission of hepatitis C virus: evidence for preventable peripartum transmission. Lancet 356(9233):904–907
98. Resti M, Azzari C, Mannelli F et al (1998) Mother to child transmission of hepatitis C virus: prospective study of risk factors and timing of infection in children born to women seronegative for HIV-1. Tuscany Study Group on Hepatitis C Virus Infection. BMJ 317(7156):437–441
99. European paediatric Hepatitis C Network (2001) Effects of mode of delivery and infant feeding on the risk of mother-to-child transmission of hepatitis C virus. BJOG 108(4):371–377

Chapter 3
Breastfeeding Among HIV-1 Infected Women: Maternal Health Outcomes and Social Repercussions

Elizabeth Stringer and Kate Shearer

Worldwide, the majority of HIV-infected women live in resource-constrained areas and must breastfeed because replacement feeding is not a viable option for them due to its lack of feasibility, safety, and affordability [1]. The benefits of breastfeeding are many and are often overshadowed by the risk of HIV transmission in HIV-infected mother–infant pairs. Breastfeeding confers immunological benefits to infants [2], protects infants from diarrhea and pneumonia [3, 4], and may improve cognitive function, only to name a few [5]. In low-income countries, the benefits of breastfeeding are even greater than in high-resource countries. In 2000, the World Health Organization estimated that breastfeeding could prevent 1.3 million infant deaths worldwide [6, 7].

New evidence shows that HIV-infected women who take antiretrovirals during breastfeeding [8–10] have very low rates of postnatal HIV transmission. Likewise, infants who are given extended nevirapine prophylaxis during breastfeeding experience similarly low rates of transmission [9–11]. The World Health Organization and many governments now promote exclusive breastfeeding for 6–12 months, regardless of the HIV status of the mother. When antiretroviral medications are combined with exclusive breastfeeding, breastfeeding becomes much safer for HIV-exposed infants. Recently, the impact of breastfeeding on maternal health in HIV-infected women has been more closely examined. This is especially pertinent in Africa where women may not be able to meet their nutritional requirements during breastfeeding. In this chapter, we review current literature on the impact of breastfeeding on maternal health with a specific focus on morbidity, mortality, nutrition, and social issues.

Evidence of the Impact of Breastfeeding on Maternal Morbidity, Nutritional Status, and Maternal Weight

During breastfeeding, the maternal nutritional requirements are higher than during pregnancy. The average nutritional requirements during pregnancy are an additional 300 kcal/day, whereas 500 kcal/day above the normal requirements are required during breastfeeding [12, 13]. Micronutrient needs

---

E. Stringer, MD, MSc, FACOG (✉)
Department of Obstetrics and Gynecology, University of North Carolina, Chapel Hill, NC

Center for Infectious Disease Research in Zambia, Counting House, Thabo Mbeki Road, Lusaka, Zambia
e-mail: estringer@med.unc.edu

K. Shearer, MPH
Health Economics and Epidemiology Research Office, Department of Medicine, Faculty of Health Sciences, University of the Witwatersrand, Johannesburg, South Africa

Table 3.1 Recommended dietary allowance (RDA) during pregnancy and lactation for women ages 19–50 [14–18]

	Pregnancy	Lactating
kcal	Second trimester: +340 kcal	0–6 months: +500 kcal
	Third trimester: +452 kcal	6–12 months: +500 kcal
Protein	71 g	71 g
Fat[a]	20–35 g	20–35 g
Vitamin B12	2.6 µg	2.8 µg
Vitamin K[b]	90 µg	90 µg
Riboflavin	1.4 mg	1.6 mg
Zinc	11 mg	12 mg
Iodine	220 µg	290 µg
Vitamin A	770 µg	1,300 µg
Vitamin B-6	1.9 mg	2.0 mg
Vitamin C	85 mg	120 mg
Vitamin D	600 IU	600 IU
Thiamin	1.4 mg	1.4 mg
Niacin	18 mg	17 mg
Folate	600 µg	500 µg
Vitamin E	15 mg	19 mg
Selenium	60 µg	70 µg
Calcium	1,000 mg	1,000 mg
Phosphorus	700 mg	700 mg
Magnesium	350–360 mg	310–320 mg
Iron	27 mg	9 mg

[a] Acceptable macronutrient distribution range
[b] Adequate intake

also increase and the daily requirements during pregnancy and breastfeeding are outlined in Table 3.1 [13]. Although the data are scarce in this area, nutritional requirements are believed to increase in asymptomatic HIV-infected pregnant woman by as much as 10%, and with AIDS to as much as an additional 25–30% [19]. An HIV-infected pregnant woman who subsequently breastfeeds potentially has nutritional requirements that are 50% higher than a nonpregnant HIV-uninfected woman [19].

Fraser and Grimes conducted an extensive review of the literature on the impact of breastfeeding on postpartum weight loss, focusing on 28 well-conducted studies examining this issue [20]. The studies included a variety of different populations ranging from women from resource-rich countries to resource-constrained countries. In general, women who breastfed were compared to women who replacement fed, and, after a thorough review, the authors concluded that there is not enough evidence to suggest that breastfeeding has an impact on maternal postpartum weight loss [20].

In an article published in 2006, Papathakis et al. reported on 142 HIV-infected and HIV-uninfected breastfeeding women from South Africa. Their results showed that, between 8 and 24 weeks postpartum, HIV-infected women lost an average of 1.4 kg, while HIV-uninfected women gained 0.4 kg ($p=0.004$), and HIV-infected women had lower subcutaneous fat. The authors noted that fat-free mass, fat mass, and percent body fat did not differ between HIV-infected and HIV-uninfected women at any time postpartum [21].

In Zambia, Murnane and colleagues found that among HIV-infected women in Lusaka, most women actually gained weight over the course of the study period. Women were randomized into one of two groups, a long-duration breastfeeding group in which the median duration of breastfeeding was 16 months, and a short-duration breastfeeding group in which women were supposed to cease breastfeeding at 4 months. While women in the shorter duration of breastfeeding group gained 2.3 kg, women in the longer duration group gained 1.0 kg ($p=0.01$) from 4 to 24 weeks postpartum [22].

Although the majority of the literature does not suggest that HIV impacts weight loss in a postpartum breastfeeding population, recent literature suggests that breastfeeding women who do lose weight have an increased risk of mortality. Data from the ZVITAMBO study showed that breastfeeding HIV-infected women who had 10% weight loss had a sevenfold increased risk of mortality compared to HIV-infected women without weight loss [23]. These women, however, are likely to have had other causes for weight loss, such as opportunistic infections, that may have placed them at greater risk for morbidities and death. A Kenyan prospective study supports this. Of 535 in the cohort, the incidence of upper respiratory infections was 161 per 100-person year and diarrhea was 63 per 100-person years [24].

Surprisingly, breastfeeding was associated with diarrhea and the authors postulated that formula feeding mothers may have adhered to better hygiene practices.

Maternal Mortality and Morbidity in HIV-Infected Breastfeeding Women

In 2001, Nduati and colleagues published the results of a secondary analysis of a randomized trial, examining maternal mortality among HIV-infected mothers [25]. 425 HIV-infected pregnant women were enrolled between 1992 and 1997 and randomized to either breastfeeding or replacement feeding. Women were followed monthly for the first year and quarterly for the second year. The authors found that women who were randomized to breastfeeding were three times as likely to die compared to women who were randomized to formula feeding (RR: 3.2; 95% CI 1.3–8.1; $p=0.01$). The authors also showed a positive relationship between postpartum weight loss and maternal mortality for every kg lost per month (RR of 3.4; 95% CI: 2.0–5.8).

The effects of breastfeeding seemed to persist even when entry viral load and CD4 counts were controlled for. One concerning finding in the study was that almost half of the women who died had a CD4 count >350. However, this study had several weaknesses. Reporting was an issue and mainly pertained to the quality and quantification of breast or formula feeding. As is common in many areas in Africa, infants were not exclusively breastfed or formula fed, however, the extent of this was not described in this chapter.

Since the publication of this study, a number of other studies and secondary analyses have been conducted. Otieno and colleagues conducted a prospective cohort study which enrolled pregnant HIV-infected women <28 weeks gestation between 2000 and 2005 [26]. All women were given short course AZT regardless of CD4 count. Women were also provided free formula and counseled on options for breastfeeding. Follow-up was monthly in the first year and quarterly in the second year. In the analysis, feeding options were dichotomized into breastfed, even if a woman breastfed for a very limited time, and never breastfed. Women who breastfed had a faster rate of CD4 decline (−7.7 vs. −4.4 CD4 cells/μL/month; $p=0.014$), however, there were not significant differences in viral load changes. The decline in CD4 cell counts was most evident for those who breastfed for longer periods of time. There was also a greater decline in BMI at 25 months postpartum, although mortality rates did not differ between the two groups, nor did time to CD4 cell count falling below 200 or time to start ART.

A number of other studies have been performed that do not show a significant impact of breastfeeding on maternal morbidity or mortality in HIV-infected women (Table 3.2). Coutsoudis and colleagues conducted a secondary analysis of breastfeeders and never breastfeeders. The mean time of follow-up for breastfeeding mothers was about 10.5 months, a shorter amount of follow-up time compared to the Nduati and Otieno studies. In this cohort, there were 2/410 maternal deaths (0.49%) in the breastfeeding group and 3/156 (1.92%) maternal deaths in the never breastfeeding group. There was no difference between the two groups with regard to mortality, new co-morbidities, CD4 cell count changes, or change in hemoglobin [32].

Sedgh and colleagues also examined the relationship of breastfeeding with time to death, CD4 cell decline, anemia, and weight loss in a prospective cohort of HIV-infected women who were enrolled

Table 3.2 Review of literature on maternal morbidity and mortality as it relates 0 to infant feeding method in HIV-infected women

Author	Year	Country	Population	Design	Results
Nduati [25]	2001	Kenya	397 HIV-infected pregnant women Women followed until death, for 2 years after delivery, or until end of study	RCT of breastfeeding versus formula feeding	Median duration of breastfeeding = 17 months 18/197 deaths in the breastfeeding group versus 6/200 in the formula feeding group ($p=0.009$)
Coutsoudis [27]	2001	South Africa	566 HIV-infected women followed postpartum (410 breastfed; 156 never breastfed)	Vitamin A intervention study No infant feeding randomization No ARVs were available	Mean time of follow-up breastfed = 10.4 months Never breastfed = 10.6 months 2/410 (0.49%) deaths among breastfeeding mothers 3/156 (1.92%) deaths among formula feeding mothers $p=0.10$
Kuhn [28]	2005	Zambia	653 HIV-infected women Early weaning vs. prolonged BF Follow up for 2 years	Secondary analysis of a RCT—abrupt weaning at 4 months vs. prolonged breastfeeding	Median duration of breastfeeding in the prolonged group = 15 months No difference in mortality up to 24 months after delivery ($p=0.38$)
Taha [29]	2006	Malawi	2,000 HIV-infected women 2 MTCT trials Followed up for 2 years	Two RCT	Median duration of BF = 18 months Over 24 months, 44 (2.2%) women died No increased risk of maternal morbidity or mortality in breastfeeding women
Otieno [26]	2007	Kenya	296 women (98 formula fed; 198 breastfed) Followed for 2 years	Prospective cohort study	Current BF had a higher rate of CD4 cell decline 12/166 deaths among breastfeeders 4/88 deaths among never breastfeeders (no significant difference in time to death)
Walson [24]	2007	Kenya	489 HIV-infected women Followed up for 12–24 months	Prospective cohort study	All cause mortality at 12 months (1.9%) 24 months mortality (4.8%) among all women
Sedgh [30]	2004	Tanzania	698 HIV-infected women All women breastfed	Observational cohort from RCT from the Trial of Vitamin Study (TOV) 2 year outcomes	Mean duration of breastfeeding was 18.2 months 109 (15.6%) deaths No association between breastfeeding and risk of death (adjusted RR = 0.73, 95% CI: 0.29–1.83), anemia, weight loss, or CD4 cell count decline
Breastfeeding and HIV International Transmission Study Group [31]	2005	Meta analysis	4,237 HIV-infected women 3,717 (87.7%) breastfed 520 (12.3%) never breastfed	Meta analysis on HIV-infected women from eligible clinical trials	Median duration of breastfeeding: 10.0 months 162 (3.8%) died within 18 months of delivery. No association between infant feeding status and probability of death ($p=0.16$)

into the Trial of Vitamin Study in Tanzania, a randomized controlled trial enrolling between 1995 and 1997 [30]. Women were seen monthly for up to 80 months. This group tried to very accurately describe duration and type of breastfeeding. While the mean duration of breastfeeding in this study was 18.2 months, the mean duration of exclusive breastfeeding was just 2.9 months. The authors examined the impact of breastfeeding on maternal status in a variety of ways and ultimately concluded that breastfeeding did not have a negative effect on mortality, CD4+ decline below 200, anemia, or weight loss. The authors also did not note a "dose response" effect; women who breastfed longer did not have a higher risk of death or other morbidities. These results are encouraging despite the fact that the authors noted that their power to detect a difference was limited by the sample size [30].

Using data from two randomized trials in Malawi, Taha and colleagues examined the impact of breastfeeding on the health of HIV-infected mothers [29]. All mothers received single dose nevirapine (sdNVP) and the infants were randomized to receive either sdNVP or sdNVP+zidovudine (ZDV). Mother–infant pairs were followed every 3 months for up to 24 months. Outcome was maternal morbidity and mortality and infant and child mortality at 2 years of age. 2,000 women were enrolled into both studies. Cumulative maternal mortality was 12/1,000 and 32/1,000 at 12 and 24 months, respectively. 44/2,000 died (2.2%) overall. The mean duration of breastfeeding was 15.4 months (median 18 months, IQR 9–22). The authors compared outcomes of breastfeeding women at 3, 6, and 12 months to those of women who were not breastfeeding at those time points. There were no differences between the two groups with regard to maternal viral load, BMI, or hemoglobin. Breastfeeding was also not associated with the various morbidities that were described but was associated with decreased risk of child mortality (HR 0.44; 95% CI: 0.28–0.70) [29].

In 2005, Kuhn and colleagues published a secondary analysis of a randomized trial which compared rapid weaning to extended breastfeeding in pregnant HIV-infected women [28]. They found no differences in mortality between the two groups of women even when the first 4 months postdelivery were excluded.

The Breastfeeding and HIV International Transmission Study Group aggregated data from nine randomized controlled trials [31]. 4,237 HIV-infected women (3,717 women who breastfed and 520 who never breastfed) were included from a variety of prevention of mother-to-child transmission (PMTCT) trials. Of all women, 162 (3.8%) died during the 18-month period of follow-up. The median duration of breastfeeding was 10 months. The probability of death did not differ by feeding mode. Women who were breastfeeding at the time of maternal death were at a lower risk of death compared to those who were not breastfeeding at the time of death (HR=0.03, 95% CI: 0.02–0.05).

Tuberculosis and Breastfeeding

There have been a number of articles written on tuberculosis and breastfeeding, as tuberculosis is the most common opportunistic infection in HIV-infected women [33]. While tuberculosis bacilli are not found in breast milk, newborns can contract tuberculosis through close contact with their mothers, especially if the mother is breastfeeding. The Global Programme for Vaccines and Immunization recommends different treatments for breastfeeding women who are smear negative and smear positive. If a mother is diagnosed with tuberculosis prior to delivery, but is smear negative she can breastfeed normally and have her infant immunized with BCG as soon as possible after birth. If a mother is smear positive during pregnancy, she is advised to breastfeed normally and give isoniazid prophylaxis to the infant for 6 months (5 mg/kg/day). The infant can be immunized with BCG after stopping isoniazid and if the infant is HIV infected it should not receive BCG. All first line tuberculosis medications for nonpregnant adults are suitable for breastfeeding mothers, because they are minimally secreted into the breast milk [34, 35].

Vitamin Supplementation, Breastfeeding, and HIV Disease

Papathakis and colleagues in South Africa examined micronutrient status and serum protein levels in HIV-infected and HIV-uninfected breastfeeding women. Ninety-two HIV-infected and 52 HIV-uninfected women were enrolled into the study and followed for up to 6 months postpartum. HIV-infected women were more likely to have lower concentrations of folate, vitamin A, serum albumin, and pre-albumin. There were no differences between the two groups with respect to vitamin B-12, vitamin E, ferritin, zinc, selenium, or copper [21].

Multivitamins

Evidence for multivitamin use in HIV-infected individuals dates back to the 1990s when an observational study in the USA showed that men infected with HIV who took multivitamins had a reduction in progression to AIDS [36]. In 1998, Fawzi and team published the first results of the Trial of Vitamins (TOV) study [37]. The TOV study was a study comparing placebo, multivitamins with vitamin A+betacarotene, multivitamins without vitamin A, and vitamin A+betacarotene. One thousand seventy-five HIV-infected pregnant women were enrolled in Tanzania between the gestational ages of 12 and 27 weeks and followed for up to 80 months postpartum; the mean follow-up time was 72 months. One of the most important results of this study was the observed decrease in HIV disease progression among women randomized to the multivitamin (MVI) arm. Women in the MVI arm were less likely to progress to stage 4 disease or AIDS (RR 0.71; 0.51–0.98) and were more likely to have sustained increases in CD4 cell counts and CD4 cell percentage postpartum. Although all women were enrolled during pregnancy, the majority of follow-up time was during breastfeeding. MVI seemed to decrease HIV transmission via breastfeeding in women with the lowest CD4 cell counts (RR 0.37; 95% CI 0.16–0.85; $p=0.02$). Multivitamins also had other positive effects on maternal reduction in thrush, gingival erythema, angular chelitis, oral ulcers, fatigue, and rash. Additional effects were a 44% reduction in low birth weight, and a 39% reduction in severe prematurity [38].

The TOV study provides convincing evidence that HIV-infected women who took multivitamins had improved outcomes which ranged from a decrease in HIV disease progression to other HIV morbidities, and is viewed as the definitive trial on this issue.

The role of vitamin A and other micronutrients in breastfeeding and HIV infection is discussed in detail in Chapter 15.

Social Issues and Stigma

HIV infection places a heavy emotional and mental burden on infected mothers to protect their infants from acquiring infection, especially after birth through feeding [39–41]. In many well-resourced countries, HIV-infected mothers replacement feed, a widely available and acceptable infant feeding method. However, in many low-resource countries, especially among those found in sub-Saharan Africa, replacement feeding is neither easily accessible nor widely acceptable [42–44]. Breastfeeding is a long-standing cultural norm, often accompanied by complementary food and drinks, and replacement feeding is often perceived as suspicious [39, 45, 46]. HIV-infected women who deviate from cultural norms by using formula or practicing only exclusive breastfeeding often face heavy criticism from family and neighbors and the risk that their HIV status could be unwittingly exposed. Furthermore, infant feeding decisions are often not the sole responsibility of the mother. Husbands and grandmothers often play an integral role in decision-making, complicating matters even further for many infected women, especially those that have yet to disclose their HIV status to family members or friends.

Breastfeeding and Mixed Feeding as a Cultural Norm

Exclusive breastfeeding for the first 6 months of life is recommended by the World Health Organization for HIV-infected mothers [47]. However, in many countries where breastfeeding is normative, mixed feeding, which includes providing the infant with food and drinks in addition to breast milk, tends to predominate. While it has been shown that mixed feeding confers a risk of transmission four times greater than exclusive breastfeeding, in some communities it is mistakenly believed that mixed feeding helps reduce the risk of HIV transmission through breastfeeding as the infant consumes a smaller amount of the HIV-infected mother's breast milk [48, 49]. Several studies conducted throughout sub-Saharan Africa have found high rates of mixed feeding among participants, ranging from 12.4% in a study in South Africa to 100% in a small study from Tanzania [49–54].

Supplementing breastmilk with additional food and drinks is common in many countries. In Burkina Faso, for example, it is routine to provide the infant with fluids, including herbal teas, in addition to breast milk, from the first day of life. It is also traditional in many communities to bathe the infant in water combined with "medicinal herbs," part of which is then given to the infant to drink [44]. When mixed feeding is culturally normative, it can be extremely difficult for an HIV-infected mother to practice exclusive breastfeeding, even when the mother is aware of the risks associated with mixed feeding, due to societal pressures to supplement breast milk with complementary food and drinks from friends, family, and other members of her surrounding community. Not conforming to such societal pressures may make the mother a target for suspicion and criticism from the people around her.

Replacement Feeding Creates Suspicion

For HIV-infected women who choose to formula-feed their infants, the obstacles they face are often as great or greater than those faced by HIV-infected women attempting to exclusively breastfeed in communities in which mixed feeding is normative. In many countries where replacement feeding is uncommon for the vast majority of the population, usually at least in part due to its prohibitive cost, the use of infant formula is often perceived as suspicious and may be considered to be an indicator of a positive HIV status [40, 44, 45, 49]. In order to deal with these suspicions, many HIV-infected women who choose to formula feed, either because they have the means to do so or because they are enrolled in a program which provides free infant formula, turn to deception in order to avoid questions and protect their HIV status from becoming public knowledge. For example, one focus group participant in South Africa noted, "…there's a lot of thinking within the community that the babies that are fed on formula are those who are HIV-positive so sometimes you'll find that the mother will go to the clinic to collect the milk and then tear off the cover which shows what kind of milk it is because she doesn't want to be asked questions" [46].

In order to cope with the stigma surrounding formula feeding, some women report conjuring up lies and stories to mask the true reason for their choice. Claiming to have breast problems has been reported by women as an excuse for using formula and, in some cases where the woman has disclosed her status to her husband, her husband is complicit in supporting these lies in order to protect his family from backlash associated with being infected with HIV. As one focus group participant in Burkina Faso stated, "I'm so happy because my husband told them that I had some tests and that my breasts are not good" [39]. Some women will also claim that health care providers or their method of delivery are the reason for their decision to use formula. In Tanzania, an HIV-infected woman who had delivered by caesarean section reported telling others that she "was advised by a doctor to only breastfeed my baby at night because I don't have enough breast milk after the operation" [49].

The Role of Relatives in Infant-Feeding Decision-Making

In many communities, women do not have much control over infant-feeding decision-making. Often, the mother-in-law and husband hold a significant amount of power in decisions over the care of infants. In South Africa, 13% of mothers reported that their infant feeding decision was influenced by either their mothers or other female relatives [45]. This can create great conflict for HIV-infected women who want to choose the best infant feeding method for themselves and their infant without having to disclose their status to family members. As one woman in Burkina Faso stated, "I breastfeed because of my mother-in-law, but deep down, I'm not calm, I am afraid" [39]. In many cases, women are forced to succumb to the wishes of their mother-in-law in order to maintain peace in the household and not raise suspicions about their HIV status, which in turn puts their infant at greater risk of acquiring HIV infection through infant feeding.

Husbands can also have a strong hand in infant-feeding decision making. In Burkina Faso, for example, fathers control the family's finances and must give the wife permission to be able to purchase infant formula [44]. In Kenya, mothers reported that their husbands held great influence in their infant feeding decisions and would push the mothers to obey their decisions [55]. While husbands exerting influence in infant-feeding decision-making can cause distress for HIV-infected women, husbands can also be a great asset in protecting the infant feeding decision from influence from the mother-in-law and other family members, primarily in situations where the mother has disclosed her status to her husband. The father may be the only person in the household with enough societal power to override the decision of the mother-in-law, making him a very important ally to HIV-infected mothers [39, 44].

Stigma in the Community

Stigma and social pressure around infant feeding decisions are not limited to family members. HIV-infected women also face similar pressures from members of their surrounding communities and, in some cases, from health care workers. Some HIV-infected women have reported fear of discrimination and stigma at health facilities, especially those in which baby friendly initiatives heavily promote breastfeeding. Mothers in South Africa reported feeling pressured to explain why they would not breastfeed and instead of facing that discussion, they chose to breastfeed so as to avoid having to disclose their HIV status [40]. Even in situations where HIV-infected mothers are appropriately counseled and understand the risks associated with breastfeeding, especially nonexclusive breastfeeding, many will still choose to do so due to social pressure, regardless of whether or not they are in a position to afford to safely formula-feed their infants. As one mother from South Africa put it, "Babies feeding from the breast grow well and look big and healthy…but really, nobody asks funny questions when you breastfeed" [43].

Mothers Feelings of Guilt Over Breastfeeding

Stigma and pressure to make appropriate choices in regards to infant feeding is not only external for many HIV-infected mothers. Many women report feeling anxiety and fear over breastfeeding and the potential that they may infect their infants with HIV [39, 41]. Self-stigma and guilt weigh heavily on the minds of many mothers struggling to ensure HIV-free survival for their children. In Ethiopia, mothers compared themselves to criminals for breastfeeding their children. One mother stated, "You feel like you are a criminal doing something bad to your own child." This sentiment was echoed by another mother who stated, "I cannot believe I was breastfeeding my child knowing that I have the virus in my breast milk. If you were a judge, you might have sent me to jail. I almost killed him" [41].

Conclusion

In most places in the world, HIV-infected women must breastfeed their infants because of the numerous constraints surrounding replacement feeding. With the discovery that maternal ART or infant prophylaxis throughout the breastfeeding period can prevent infant HIV transmission, the impact of breastfeeding on maternal HIV disease has taken on a greater focus. Breastfeeding is associated with increased caloric demands, but if those nutritional demands are met, breastfeeding does not affect postpartum weight loss, even in HIV-infected women. Despite two studies which suggested that breastfeeding in HIV-infected women may increase their morbidity and mortality, the majority of studies do not suggest a deleterious effect. Multivitamins provide many benefits to HIV-infected breastfeeding women and should be routinely incorporated into postnatal care. Women also face a lot of pressure to breastfeed in areas where breastfeeding is culturally the norm, but the societies do not always support safe exclusive breastfeeding. The overwhelming evidence shows that breastfeeding in HIV-infected women is not detrimental to their health and that HIV-infected women can now safely breastfeed with low rates of HIV transmission. HIV-infected women also require a lot of support whether they choose to breastfeed or formula feed.

References

1. WHO. New data on the prevention of mother-to-child transmission of HIV and their policy implications. Conclusions and recommendations. Geneva, 11–13 Oct 2000
2. Pabst H, Spady D (1990) Effect of breast-feeding on antibody response to conjugate vaccine. Lancet 336(8710): 269–270
3. Dujits L, Jaddoe V, Hofman A, Moll H (2010) Prolonged and exclusive breastfeeding reduces the risk of infectious diseases in infancy. Pediatrics 126(1):e18–e25
4. Wobudeya E, Bachou H, Karamagi C, Kalyango J, Mutebi E, Wamani H (2011) Breastfeeding and the risk of rotavirus diarrhea in hospitalized infants in Uganda: a matched case control study. BMC Pediatr 11:17
5. Lucas A, Morley R (1996) Breastfeeding, dummy use, and adult intelligence. Lancet 347(9017):1765
6. Jones G, Steketee R, Black R, Bhutta Z, Morris S (2003) Bellagio Child Survival Study Group. How many child deaths can we prevent this year? Lancet 362(9377):65–71
7. (2000) WHO Collaborative Study Team on the Role of Breastfeeding on the Prevention of Infant Mortality. Effect of breastfeeding on infant and child mortality due to infectious diseases in less developed countries: a pooled analysis. WHO. Lancet 355(9202):451–455
8. Thomas T, Masaba R, Ndivo R, Zeh C, Borkowf C, Thigpen M et al (2008) Prevention of mother-to-child transmission of HIV-1 among breastfeeding mothers using HAART: the Kisumu Breastfeeding Study, Kisumu, Kenya, 2003–2007. Abstract 45aLB. 15th conference on retroviruses and opportunistic infections, Boston, MA
9. Palombi L, Marazzi MC, Voetburg A, Magid NA (2007) Treatment acceleration program and the experience of the DREAM program in prevention of mother-to-child transmission of HIV. AIDS 21(4):S65–S71

10. Chasela CS, Hudgens MG, Jamieson DJ, Kayira D, Hosseinipour MC, Kourtis AP et al (2010) Maternal or infant antiretroviral drugs to reduce HIV-1 transmission. N Engl J Med 362(24):2271–2281
11. The Kesho Bora Study Group (2011) Triple antiretroviral compared with zidovudine and single-dose nevirapine prophylaxis during pregnancy and breastfeeding for prevention of mother-to-child transmission of HIV-1 (Kesho Bora study): a randomised controlled trial. Lancet Infect Dis 11(3):171–180
12. Mehta SH (2008) Nutrition and pregnancy. Clin Obstet Gynecol 51(2):409–418
13. Picciano MF (2003) Pregnancy and lactation: physiological adjustments, nutritional requirements and the role of dietary supplements. J Nutr 133:1997S–2002S
14. Institute of Medicine (2010) Dietary reference intakes for calcium and vitamin D. Washington, DC
15. Institute of Medicine (2002) Dietary reference intakes for energy, carbohydrate, fiber, fat, fatty acids, cholesterol, protein, and amino acids. Washington, DC
16. Institute of Medicine (2001) Dietary reference intakes for vitamin A, vitamin K, arsenic, boron, chromium, copper, iodine, iron, manganese, molybdenum, nickel, silicon, vanadium, and zinc. Washington, DC
17. Institute of Medicine (2009) Weight gain during pregnancy: reexamining the guidelines. Washington, DC
18. Institute of Medicine (1991) Nutrition during lactation. Washington, DC
19. (2003) Nutrient requirements for people living with HIV/AIDS: report of a technical consultation. World Health Organization, Geneva
20. Fraser AB, Grimes DA (2003) Effect of lactation on maternal body weight: a systematic review. Obstet Gynecol Surv 58(4):265–269
21. Papathakis PC, Van Loan MD, Rollins NC, Chantry CJ, Bennish ML, Brown KH (2006) Body composition changes during lactation in HIV-infected and HIV-uninfected South African women. J Acquir Immune Defic Syndr 43(4): 467–474
22. Murnane PM, Arpadi SM, Sinkala M, Kankasa C, Mwiya M, Kasonde P et al (2010) Lactation-associated postpartum weight changes among HIV-infected women in Zambia. Int J Epidemiol 39:1299–1310
23. Koyanagi A, Humphrey JH, Moulton LH, Ntozini R, Mutasa K, Iliff P et al (2011) Predictive value of weight loss on mortality of HIV-positive mothers in a prolonged breastfeeding setting. AIDS Res Hum Retroviruses 27(11): 1141–1148
24. Walson JL, Brown ER, Otieno PA, Mbori-Ngacha DA, Wariua G, Obimbo EM et al (2007) Morbidity among HIV-1-infected mothers in Kenya: prevalence and correlates of illness during 2-year postpartum follow-up. J Acquir Immune Defic Syndr 46(2):208–215
25. Nduati R, Richardson BA, John G, Mbori-Ngacha D, Mwatha A, Ndinya-Achola J et al (2001) Effect of breastfeeding on mortality among HIV-1 infected women: a randomised trial. Lancet 357(9269):1651–1655
26. Otieno PA, Brown ER, Mbori-Ngacha DA, Nduati RW, Farquhar C, Obimbo EM et al (2007) HIV-1 disease progression in breast-feeding and formula-feeding mothers: a prospective 2-year comparison of T cell subsets, HIV-1 RNA levels, and mortality. J Infect Dis 195(2):220–229
27. Coutsoudis A, Pillay K, Kuhn L, Spooner E, Tsai WY, Coovadia HM (2001) Method of feeding and transmission of HIV-1 from mothers to children by 15 months of age: prospective cohort study from Durban, South Africa. AIDS 15(3):379–387
28. Kuhn L, Kasonde P, Sinkala M, Kankasa C, Semrau K, Vwalika C et al (2005) Prolonged breast-feeding and mortality up to two years post-partum among HIV-positive women in Zambia. AIDS 19(15):1677–1681
29. Taha TE, Kumwenda NI, Hoover DR, Kafulafula G, Fiscus SA, Nkhoma C et al (2006) The impact of breastfeeding on the health of HIV-positive mothers and their children in sub-Saharan Africa. Bull World Health Organ 84(7):546–554
30. Sedgh G, Spiegelman D, Larsen U, Msamanga G, Fawzi WW (2004) Breastfeeding and maternal HIV-1 disease progression and mortality. AIDS 18(7):1043–1049
31. Breastfeeding and HIV International Study Group (2005) Mortality among HIV-1-infected women according to children's feeding modality: an individual patient data meta-analysis. J Acquir Immune Defic Syndr 39(4):430–438
32. Coutsoudis A, Coovadia H, Pillay K, Kuhn L (2001) Are HIV-infected women who breastfeed at increased risk of mortality? AIDS 15(5):653–655
33. Nhan-Chang C-L, Jones TB (2010) Tuberculosis in pregnancy. Clin Obstet Gynecol 53(2):311–321
34. Aquilina S, Winkelman T (2008) Tuberculosis: a breast-feeding challenge. J Perinat Neonatal Nurs 22(3):205–213
35. American Academy of Pediatrics Committee on Drugs (2001) Transfer of drugs and other chemicals into human milk. Pediatrics 108(3):776–789
36. Abrams B, Duncan D, Hertz-Picciotto I (1993) A prospective study of dietary intake and acquired immune deficiency syndrome in HIV-seropositive homosexual men. J Acquir Immune Defic Syndr 6:949–958
37. Fawzi WW, Msamanga GI, Spiegelman D, Urassa EJ, McGrath N, Mwakagile D et al (1998) Randomised trial of effects of vitamin supplements on pregnancy outcomes and T cell counts in HIV-1-infected women in Tanzania. Lancet 351(9114):1477–1482
38. Fawzi WW, Msamanga GI, Hunter D, Renjifo B, Antelman G, Bang H et al (2002) Randomized trial of vitamin supplements in relation to transmission of HIV-1 through breastfeeding and early child mortality. AIDS 16:1935–1944

39. Cames C, Saher A, Ayassou KA, Cournil A, Meda N, Simondon KB (2010) Acceptability and feasibility of infant-feeding options: experiences of HIV-infected mothers in the World Health Organization Kesho Bora mother-to-child transmission prevention (PMTCT) trial in Burkina Faso. Matern Child Nutr 6:253–265
40. Thairu LN, Pelto GH, Rollins NC, Bland RM, Ntshangase N (2005) Sociocultural influences on infant feeding decisions among HIV-infected women in rural Kwa-Zulu Natal, South Africa. Matern Child Nutr 1:2–10
41. Koricho A, Moland KM, Blystad A (2010) Poisonous milk and sinful mothers: the changing meaning of breastfeeding in the wake of the HIV epidemic in Addis Ababa, Ethiopia. Int Breastfeed J 5:12
42. Israel-Ballard KA, Maternowska MC, Abrams BF, Morrison P, Chitibura L, Chipato T et al (2006) Acceptability of heat treating breast milk to prevent mother-to-child transmission of human immunodeficiency virus in Zimbabwe: a qualitative study. J Hum Lact 22(1):48–60
43. Sibeko L, Coutsoudis A, Nzuza S, Gray-Donald K (2009) Mothers' infant feeding experiences: constraints and supports for optimal feeding in an HIV-impacted urban community in South Africa. Public Health Nutr 12(11):1983–1990
44. Desclaux A, Alfieri C (2009) Counseling and choosing between infant-feeding options: overall limits and local interpretations by health care providers and women living with HIV in resource-poor countries (Burkina Faso, Cambodia, Cameroon). Soc Sci Med 69:821–829
45. Swarts S, Kruger HS, Dolman RC (2010) Factors affecting mothers' choice of breastfeeding vs. formula feeding in the lower Umfolozi district war memorial hospital, KwaZulu-Natal. Health SA Gesondheid 15:1
46. Doherty T, Chopra M, Nkonki L, Jackson D, Greiner T (2006) Effect of the HIV epidemic on infant feeding in South Africa: "When they see me coming with the tins they laugh at me". Bull World Health Organ 84:90–96
47. WHO Library Cataloguing-in-Publication Data (2010) Guidelines on HIV and infant feeding. 2010. Principles and recommendations for infant feeding in the context of HIV and a summary of evidence
48. Piwoz EG, Iliff PJ, Tavengwa N, Gavin L, Marinda E, Lunney K et al (2005) An education and counseling program for preventing breast-feeding-associated HIV transmission in Zimbabwe: design and impact on maternal knowledge and behavior. J Nutr 135:950–955
49. Leshabari S, Blystad A, Moland K (2007) Difficult Choices: infant feeding experiences of HIV-positive mothers in northern Tanzania. SAHARA J 4(1):544–555
50. Laar AK, Ampofo W, Tuakli JM, Quakyi IA (2009) Infant feeding choices and experiences of HIV-positive mothers from two Ghanaian districts. JAHR 1(2):23–33
51. Becquet R, Ekouevi DK, Viho I, Sakarovitch C, Toure H, Castetbon K et al (2005) Acceptability of exclusive breast-feeding with early cessation to prevent HIV transmission through breast milk, ANRS 1201/1202 Ditrame Plus, Abidjan, Cote d'Ivoire. J Acquir Immune Defic Syndr 40(5):600–608
52. Kiarie JN, Richardson BA, Mbori-Ngacha D, Nduati RW, John-Stewart GC (2004) Infant feeding practices of women in a perinatal HIV-1 prevention study in Nairobi, Kenya. J Acquir Immune Defic Syndr 35(1):75–81
53. Ladzani R, Peltzer K, Mlambo MG, Phaweni K (2010) Infant-feeding practices and associated factors of HIV-positive mothers at Gert Sibande, South Africa. Acta Paediatr
54. Maru S, Datong P, Selleng D, Mang E, Inyang B, Ajene A et al (2009) Social determinants of mixed feeding behavior among HIV-infected mothers in Jos, Nigeria. AIDS Care 21(9):1114–1123
55. Morgan MC, Masaba RO, Nyikuri M, Thomas TK (2010) Factors affecting breastfeeding cessation after discontinuation of antiretroviral therapy to prevent mother-to-child transmission of HIV. AIDS Care 22(7):866–873

Chapter 4
Early Diagnosis of HIV Infection in the Breastfed Infant*

Chin-Yih Ou, Susan Fiscus, Dennis Ellenberger, Bharat Parekh, Christine Korhonen, John Nkengasong, and Marc Bulterys

Introduction

More than 90% of the 370,000 pediatric human immunodeficiency virus type 1 (HIV-1) infections globally in 2009 were acquired through mother-to-child transmission (MTCT) [1], and most of these transmissions occurred in sub-Saharan Africa. MTCT of HIV-1 occurs either during late pregnancy, the intrapartum period, or breastfeeding [2, 3]. With the application of prophylactic antiretroviral (ARV) therapy and breastfeeding avoidance, MTCT is now observed in only 1–2% of at-risk infants in developed countries [4, 5]. The majority of pregnant women residing in high HIV-burden, resource-limited countries (RLCs) are still not aware of their infection status and do not receive timely intervention measures to prevent vertical transmission [6–11]. Untreated infected infants have high HIV-related morbidity and mortality. Approximately 33% of the untreated infected infants in RLCs die during their first year of life, and >50% die within their first 2 years [12]. Treating infants early greatly reduces mortality and morbidity [13]. Recognition of the importance of reducing infant HIV mortality has facilitated the development of methods to bring appropriate testing closer to pregnant and lactating mothers, to identify HIV-infected infants earlier, and to provide timely access to life-saving ARV treatment and care. New and accurate diagnostic methods have emerged in the last few years, and many of these methods have been field-validated. This diagnostic service should not comprise a stand-alone program but must be integrated into the overall mother and child health programs to achieve the goal of prevention of mother-to-child transmission (PMTCT) [14, 15]. In this chapter, we review currently available diagnostic methodologies, including their advantages and disadvantages, their testing

---

*Conflict of interest statement for individual authors
None of the authors report conflict of interest.

C.-Y. Ou, Ph.D. (✉) • C. Korhonen • M. Bulterys
Division of Global HIV/AIDS (DGHA), Center for Global Health, Centers for Disease Control and Prevention (CDC), 1600 Clifton Road, NE, Atlanta, GA 30333, USA

CDC Global AIDS Program, Beijing, China

Adjunct Professor of Epidemiology, UCLA School of Public Health, Los Angeles, CA, USA
e-mail: cho2@cdc.gov

S. Fiscus
Department of Microbiology and Immunology, University of North Carolina, Chapel Hill, NC 27599, USA

D. Ellenberger • B. Parekh • J. Nkengasong
Global AIDS Program, U.S. Centers for Disease Control and Prevention, Atlanta, GA, USA

algorithms, and their quality assurance requirements, with a particular focus on early HIV diagnosis in the breastfed infant. Further, we discuss efforts toward the development of simple, accurate, and rapid diagnostic applications.

General Perspectives and the Roles of Laboratory Testing in Early Infant Diagnosis

First recognized among homosexual men in 1981, acquired immunodeficiency syndrome (AIDS) was soon found among injection drug users, recipients of blood or blood products, and infants born to women with AIDS [16]. In 1985, a case of infant infection via breastfeeding from a mother who acquired HIV from the transfusion of HIV-contaminated blood after cesarean section was recognized [17].

As shown in Table 4.1, two distinct types of serological and virological assays are useful in the fields of PMTCT and early infant diagnosis (EID). Serological assays are highly sensitive and specific in detecting HIV antibodies in infected adults but are ineffective in determining the infection status in the first few months of life in exposed infants because of the presence and slow decay of cross-placental maternal IgG antibodies. Maternal IgG has a half-life of approximately 25 days [18, 19], and takes 12–13 months (range 10–16 months) to disappear from the circulation of an uninfected infant [20–24]. Prior to the availability of virological assays, the presence of HIV antibodies in exposed children older than 18 months was used as the sole marker for HIV infection. Although there had been numerous attempts to improve test performance, the sensitivity and specificity of serological assays for infants less than 6–9 months of age remained low and thus were not useful for early diagnosis [25–37]. With the advent of the polymerase chain reaction (PCR) in the mid-1980s [38], PCR was used to overcome the problems of sensitivity and specificity of serological assays [39–41]. Over the last two decades, numerous PCR-based and other nucleic acid techniques (NATs) were developed for EID. The 2010 WHO recommendations suggest the use of virological tests with a sensitivity of >95 % and specificity of >98 % for infant and child diagnosis of HIV infection [42]. The detection of viral nucleic acid is considered a virological assay and is currently the method of choice for EID. A second type of virological assay is based on the detection of viral structural p24 antigens (Table 4.1).

Table 4.1 Current virological and serological assays useful in early infant HIV diagnosis and PMTCT settings

Assays	Utility and operational challenges
Virological assays	
Nucleic acid	Methods of choice for definitive early diagnosis
DNA	Manual Roche Amplicor DNA version 1.5 is the most used assay in RLCs. Can be used with blood and DBS; discontinuation of manufacturing and replacement by real-time assays are anticipated
RNA or TNA[a]	Several quantitative and qualitative methods are available and found to be highly sensitive and specific (Table 4.2); POC methods are being developed (Table 4.4)
P24 antigen	Has been greatly improved and can be used with both plasma and DBS specimens (Table 4.3). There is a great need for a commercial kit that contains all required testing components. POC methods are being developed (Table 4.4).
Reverse transcriptase	Cavidi assay detects reverse transcriptase enzyme activity of retroviruses, not widely used because it requires fresh specimen
Serological assays	Not useful for early identification of infected infants
Rapid tests	Useful in rapid determination of the serostatus of pregnant women in labor room or exposure status of infants
Maternal antibody decay	The majority of uninfected infants lose maternal antibodies by 9–12 months of age and thus serologic assay can be used to rule out infection. Exposed infants having antibody at 18 months or older are judged as infected

[a]TNA = total nucleic acid

A third type of virological assay, the Cavidi ExaVir assay, detects viral reverse transcriptase enzymatic activity [43]. To maintain the enzymatic activity, blood specimens have to be kept frozen and require a cold chain for transportation prior to testing in a laboratory. This method is thus not as useful as the other two virological assays in RLCs.

Breastfeeding and HIV Diagnosis

In RLCs, breast milk remains an important source of nutrients but also of HIV transmission in infants. In the most recent WHO guidelines for breastfeeding [44], it is recommended that HIV-infected mothers receive ARV when breastfeeding. If ARV drugs are not available, it is recommended that mothers be counseled to exclusively breastfeed for the first 6 months and continue thereafter unless replacement feeding is available [44]. Since many HIV-infected pregnant women may not receive ARV for various reasons, the potential of HIV transmission via breastfeeding in RLCs remains.

When testing exposed infants who are receiving breast milk, the timing of the test and the timing of breastfeeding cessation have to be taken into account in deciding the infection status of the infants [42]. Exposed infants may be presented to the testing facility regardless of their feeding status. Two test results are considered. First, a positive virological test irrespective of breastfeeding status is indicative of HIV transmission. In this case, a second specimen should be taken as soon as possible and tested to confirm transmission. In the meantime, ARV treatment (ART) should be started without delay. Second, a negative virological test indicates the absence of infection. However, if the infant is being breastfed or was breastfed until recently, this infant should be tested again at least 6 weeks after complete weaning to rule out HIV transmission via breast milk. Breastfeeding should not be discontinued while diagnostic tests are being performed or while waiting for results [42, 44].

HIV-1 DNA Assays

Most of the vertical HIV transmission burden has fallen on RLCs. Given the limitations that such settings can place on the application of novel technologies, any EID diagnostic service considered for RLC-deployment should meet the following three criteria. First, the technique and related technologies should not require an overwhelming degree of laboratory capacity building and training. Second, the method should be able to use dried blood spot (DBS) samples in order to circumvent the many challenges associated with venipuncture of infants and specimen transportation to provide wider access to exposed infants. Third, and most important of all, the method should detect HIV infection early. Depending on the methodologies, NAT can be used to detect either proviral DNA, viral RNA, or both (total nucleic acid). These assays can amplify HIV DNA or RNA up to 10^{10}–10^{12}-fold and thus without proper built-in anticontamination measures, false-positive results may occur [45, 46]. It is thus recommended that high-quality commercial assays with built-in anti-contamination controls are used in dedicated PCR laboratories for EID following good laboratory practices [47, 48]. The manual Roche Amplicor HIV-1 DNA PCR version 1.5 assay was the best method in the last decade that fulfilled these requirements and was quickly adopted in many RLCs [48, 49]. This assay qualitatively amplifies and detects the HIV-1 proviral DNA integrated in the mononuclear cells of an infected person. It uses relatively inexpensive thermocyclers to conduct PCR amplification of HIV-1 DNA. It also employs the same equipment commonly present in HIV serological laboratories for enzyme-linked immunosorbent assays (ELISA) to colorimetrically measure amplified DNA. This assay was originally designed for whole blood but was soon adopted to use with DBS [50, 51]. Furthermore, it allows the collection of specimens to occur at healthcare facilities close to the residences of exposed infants and thus significantly increases infant access and EID service coverage [48, 49]. The performance of the Roche DNA assay has been extensively investigated in many countries and was recently reviewed [14]. The sensitivity of detection on specimens taken at birth has been shown to be between 10% and 40%, but sensitivity increases rapidly as the infants reach 4–6 weeks of age [52–58]. Sherman et al. demonstrated that performance using whole blood and DBS was essentially identical, with the sensitivity and specificity of detection in 6-week-old infants at 100% and 99.6%, respectively [59]. Using DBS and this assay, successful EID programs have been established in Botswana [10], Uganda [60], South Africa [51, 61], Kenya [62, 63], Zambia [64], Tanzania [65, 66], Malawi [67], and other high burden countries [49, 68–71]. Although there is a trend toward the use of automated, high-volume assay technologies [48], this manual assay remains the primary technology used in most RLCs [72].

Qualitative and Quantitative HIV-1 RNA Assays

The infant immune system at the time of birth is not fully developed and cannot control viral replication effectively. Further, it contains an abundance of CD4 cells for the vertically transmitted HIV virus to infect and propagate. When an infant is infected in utero, HIV viral load is usually in the range of 10^3–10^5 copies per ml of plasma at birth. Viral load quickly increases to 10^5–10^6 copies per ml of plasma within 1–2 weeks [73–77]. Current commercial viral load assays that are used to monitor the effectiveness of ARV treatment typically have a detection limit ranging from 50–100 copies per ml of plasma. The rise of high viral load content in the infants soon after birth can thus be readily detected. Five "real-time" viral load assays are listed in Table 4.2. They are named "real-time" because these assays employ fluorescence-labeled HIV-1-specific oligonucleotide probes that can emit fluorescence during the process of amplification/detection cycles. Three of the listed assays use an initial reverse transcription step to convert HIV-1 RNA to DNA prior to DNA PCR (Roche TaqMan, Abbott m2000, Biometric Generic Viral Load). Thus, if the initial specimens are whole blood or DBS and the nucleic extraction procedure does not discriminate between RNA and DNA, cellular DNA would also be measured. For EID, there is no need to differentiate RNA and DNA.

Using the bioMerieux NucliSens EasyQ assay, Lilian et al. reported 100% sensitivity with DBS specimens from infants less than 3 months of age [63]. Using Roche TaqMan real-time assay, Stevens reported sensitivity of 99.7% and specificity of 100% in infants from South Africa [78]. Three study groups reported on the use of the qualitative GenProbe APTIMA assay [79, 80]. One of these studies [80] used DBS specimens from eight countries with several HIV-1 subtypes (A, B, C, D, and AE) and obtained 99.2% sensitivity and 100% specificity. To increase the detection throughput, Stevens and

Table 4.2 Performance characteristics of major quantitative and qualitative real-time assays

Instrument	First author (references)	Country	Specimen	Target	DBS processing automation	Infant age	Sensitivity	Specificity
bioMerieux EasyQ	Lilian [119]	S Africa	DBS	RNA	No	<3 months	22/22 (100)	151/157 (96.3)
			DBS	RNA	No	3–12 months	30/30 (100)	59/59 (100)
Roche TaqMan v2.0[b]	Stevens [48]	S. Africa	DBS	TNA	Yes	NA[a]	303/304 (99.7)	496/496 (100)
			Blood	TNA	Yes	NA	303/304 (99.7)	691/691 (100)
Abbott m2000	Lofgren [66]	Tanzania	DBS	RNA	Yes	<18 months	35/36 (97)*	140/140 (100)*
Gen-Probe APTIMA	Sullivan [79]	USA	DBS	RNA	Yes	NA	68/68 (100)	28/29 (96.6)
	Kerr [80]	USA	DBS	RNA	No	NA	128/129 (99.2)	162/162 (100)
	Stevens [81]	S. Africa	DBS	RNA	Yes	6 weeks	151/151 (100)	333/343 (97)
Biocentric Generic Viral load	Viljoen [82]	Burkina Fasso	DBS	RNA	No	6 weeks	20/20 (100)	94/94 (100)
			DBS	RNA	No	3–6 months	34/34 (100)	4/4 (100)
			DBS	RNA	No	9–18 months	52/52 (100)	7/7 (100)

[a]NA = not available in the original publication
[b]Roche TaqMan v2.0 assay could also detect DNA when whole blood or DBS samples are used
*Used a viral load cut off of 1,000 copies/ml

colleagues extracted RNA from DBS using GenProbe SB100 extractor and documented excellent performance by GenProbe APTIMA with sensitivity and specificity of 100% and 97%, respectively [81]. They estimated that a centralized laboratory could process as many as 1,900 DBS specimens in a 24-hour period [81]. In Tanzania, RNA extracted from two full circles of DBS (each 16 mm in diameter) was successfully detected by the Abbott m2000 instrument [66]. Successful employment of the Biocentric assay was reported to yield 100% sensitivity and 100% specificity with DBS samples obtained from 6-week-old infants from Burkina Faso [82].

NAT Testing Times and Testing Algorithm in Developed Countries

The NAT testing methods, specimen collection time, and testing algorithms vary significantly across countries. The basic operational principle is to (1) begin testing exposed infants as early as possible, (2) if the first sample yields a positive result, confirm with a second sample as early as possible, and (3) breastfed infants should be tested 6 weeks after the cessation of breastfeeding. Although the sensitivity of detection in the first few days of life is low, in the USA and other developed countries, many healthcare facilities collect specimens for HIV testing in the first 48 hours (h) of life in order to find the infected infant early [83]. The American Academy of Pediatrics (AAP) currently recommends commercial NAT testing (qualitative or quantitative) be performed on HIV-exposed infants within the first 14 days of life, at 4 weeks, and at 4 months. If the NAT is positive, it is recommended to promptly obtain a second specimen for confirmation with a second quantitative NAT and then begin ARV treatment if positive [83]. It is important to note that none of the quantitative viral load or qualitative assays is currently FDA-approved for infant diagnosis, and when used injudiciously they may occasionally yield false-positive results as recently reported by Patel et al. [84]. When EID testing is performed in a research laboratory or a facility conducting a high volume of viral load determinations, extra safeguards against contamination are advised. False-positive results could come from cross contamination between samples, specimen mix-ups, the presence of amplified HIV DNA, or specimen labeling errors. It is important to use a second sample to confirm positivity. The AAP also recommends the use of a threshold of 10,000 copies/ml for diagnosis of pediatric HIV infection [14]. This recommendation may need to be carefully verified. Weak signals around 100 copies/ml or lower are most likely due to contaminations [84].

For exposed children with negative virological tests, many experts confirm the absence of HIV antibody using a serologic assay between 12 and 18 months. In exposed infants younger than 18 months who have not been breastfed, a presumptive exclusion of HIV infection can be made based on several combinations of negative results (1) three NAT tests: two at ≥2 weeks and one at ≥4 weeks; (2) one NAT test at ≥2 months; or (3) one antibody test at ≥6 months [15, 83, 85].

A definitive exclusion of HIV infection in a nonbreastfed infant with no positive HIV-1 virological test result can be made based on serial negative results from either (1) two NAT tests: one at ≥1 month and one at ≥4 months or, (2) two negative ELISA antibody tests at ≥6 months and no other clinical or laboratory evidence of HIV-1 infection.

HIV p24 Antigen Assay

An antigen assay could be a cost-effective or complementary alternative to NAT for infant diagnosis. The most evaluated and validated p24 assay is the ultrasensitive p24 (Up24) assay originally developed by Schupbach and others [86]. The Up24 method uses (1) heat denaturation (100°C for 5 min) to disrupt antibody–antigen immune complexes, (2) biotinyl tyramide ELAST ELISA signal amplification system to increase detection sensitivity, and (3) plasma or DBS specimens. The Up24 assay has

Table 4.3 HIV-1 p24-based infant diagnostic studies

First author (references)	Sample	Countries	Subtype	Confirmation of infection	Infant age	Sensitivity	Specificity
Nsojo [72]	Plasma or serum	Tanzania	A, D	DNA PCR	1–8 weeks	18/18 (100)	106/106 (100)[a]
					9–26 weeks	35/36 (97.2)	
					27–52 weeks	40/40 (100)	
					>52 weeks	30/31 (96.8)	
Nadal [120]	Plasma	Switzerland	B	DNA PCR	<10 days	6/12 (50)	146/148 (98.6)
					11 days to 3 months	10/10 (100)	114/116 (98.3)
					3–6 months	19/19 (100)	68/68 (100)
					>6 months	191/191 (100)	310/311 (99.7)
Sutthent [121]	Plasma	Thailand	AE	Viral load	1–2 months	21/21 (100)	100/100 (100)
					4–6 months	21/21 (100)	100/100 (100)
Sherman [122]	Plasma	S. Africa	C	DNA PCR	6 weeks	22/23 (95.7)	62/62 (100)
					3 months	20/20 (100)	62/62 (100)
					4 months	2/2 (100)	3/3 (100)
					7 months	7/7 (100)	22/24 (93.6)
Zijenah [94]	Plasma	Zimbabwe	C	DNA PCR	<6 months	NA (98.1)	NA (96.9)
					7–18 months	NA (89.5)	NA (91.5)
De Baets [123]	DBS	DRC[b]	C	DNA PCR	<18 months	5/5 (100)	82/82 (100)
Nouhin [124]	DBS	Cambodia	AE	DNA PCR & culture	3.7–24 months	42/46 (91.3)	123/123 (100)
Patton [90]	DBS	S. Africa	C	DNA PCR	1 month to 12 years	82/83 (98.8)	59/59 (100)
George [88]	Plasma	Haiti	B	Viral load	0.2–3.6 months	40/43 (93)	156/157 (99)
Fiscus [92]	DBS	USA	B	DNA PCR and culture	<7 days	0/5 (0)	108/109 (99.1)
					8–30 days	24/29 (90)	148/151 (97.4)
					31–90 days	88/94 (93.3)	268/273 (98.2)
					91–180 days	44/45 (94.1)	95/96 (99)
Patton [91]	DBS	S. Africa	C	DNA PCR, Viral load	6 weeks	72/72 (100)	NA
					6 weeks to 6 years	48/49 (98)	NA
Cachafeiro [89]	DBS	DR[c], Malawi, S. Africa, USA, Vietnam	AE, B, C	DNA PCR, RNA PCR	<7 days	2/4 (50)	80/80 (100)
					1–6 weeks	4/5 (80)	95/95 (100)
					6–26 weeks	21/22 (95)	192/192 (100)
					26–72 weeks	24/24 (100)	63/63 (100)
Wittawatmongkol [93]	DBS	Thailand	AE, B	DNA PCR antibody at >18 months	1–2 months	1/5 (20)	116/118 (98.3)
					4–6 months	5/5 (100)	116/118 (98.3)

NA not available
[a]Age not specified
[b]Democratic Republic of the Congo
[c]Dominican Republic

a sensitivity of >98% and works with either plasma or DBS specimens and on diverse HIV-1 subtypes (A, B, C, E, and AE) (Table 4.3). Similar to the NAT assays, many versions of p24 antigen assays with modification of p24 extraction solutions exist. It was also recently reported that a simplified assay without the use of signal amplification had a sensitivity of 84% and specificity of 98% in Malawi [87] and a sensitivity of 91% and specificity of 97% in Haiti [88]. The usefulness of the simplified p24 assay remains to be determined.

Assays using p24 antigen, despite having excellent performance characteristics and appearing to detect most HIV-1 subtypes, have yet to be widely used. One of the reasons for this is the lack of a complete commercial kit that allows a standardized extraction of blood or DBS p24 [89] and subsequent

Table 4.4 Innovative point-of-care technologies potentially useful for EID

First authors (references)	Target	Methodology	Starting material	Testing time
Parpia [112]	P24	Dip-stick, lateral flow	25 µl plasma	45 min
Tang [125]	P24	Nanoparticle-based immunoassay	Not mentioned	3 h
Jangam [126]	DNA	Use of FINA* disk to extract blood DNA followed by Abbott real-time PCR	100 µl whole blood	1–2 h
Lee [114]	RNA	Dip-stick in cartridge, NAT real-time machine	240 µl plasma	2-h
Steinmetzer [115]	RNA, total nucleic acid	Cartridge-closed system, NAT real-time machine	10 µl whole blood	1–2 h
Tanriverdi [116]	RNA	Cartridge-closed system, NAT real-time machine	200 µl plasma	1.5 h

*FINA=filtration isolation of nucleic acids

detection. Excellent performance of p24 assays have been recently shown in large reference laboratories in South Africa [90, 91], the USA [89, 92], and Thailand [93]. The cost of a noncommercial p24 assay was reported to be fivefold lower than that of current commercial NAT assays [94]. However, when fully commercialized, the true cost difference may not be as large. Nevertheless, the successful demonstration of p24 provides an important tool, in addition to NAT, to independently verify vertical transmission. Point-of-care p24 assays are being vigorously explored (Table 4.4).

Rapid HIV Testing for Late-Presenting Pregnant Women and for Infants of Unknown Serostatus

Laboratory-based EID activity is an integrated part of a PMTCT program. The coverage of PMTCT services varies greatly from country to country and facility to facility but is generally suboptimal in RLCs, thus adversely affecting the overall performance of EID efforts [95]. A successful EID program requires early testing of pregnant women and their infants, quick transfer of samples to the laboratory for testing, and delivery of results to original clinics [49, 96, 97]. Rapid serological tests can yield results in 30 minutes (min) or less. Since 2002, the U.S. Food and Drug Administration (FDA) has approved seven rapid tests (RTs) (OraQuick Advance Rapid HIV 1/2 Antibody Test, Reveal G3 Rapid HIV-1 Antibody Test, Uni-Gold Recombigen HIV Test, Multispot HIV1/2 Rapid Test, Clearview HIV1/2 Stat-PAK, Clearview Complete HIV1/2, and INSTI Test) [98]. The most recent FDA-approved INSTI test yields a result in 1 minute. There are many commercial rapid tests used outside of the USA that have received USAID/CDC evaluation [99] and WHO prequalification [100]. Rapid tests are extremely valuable in labor rooms to identify late-presenting HIV-infected women for timely prophylaxis to prevent vertical transmission [101, 102]. Maternal acquisition of HIV during pregnancy is also known to be a strong risk for vertical HIV transmission [103]. The AAP recommends RT be performed on the mother and the result reported to healthcare professional within 12 hours for ARV prophylaxis for the infant to commence [104]. RT was successfully used for pregnant women who were unaware of their serostatus in the labor room [101, 105–107] or at outpatient obstetric–gynecologic settings [108].

Many young children seen in pediatric healthcare facilities are HIV-infected in high burden countries. If the maternal status is not documented, RT can be used to identify the HIV-exposure status of the infant in a single clinical visit and result in immediate provision of co-trimoxazole prophylaxis and collection of blood for virological testing. It was found in a recent study conducted in a PMTCT clinic in Johannesburg, South Africa, that many RTs could be used to identify exposed infants at <3 months of age with a sensitivity of 100% [48]. An oral test (OraQuick) only identified

85% of the exposed infants younger than 6 months in a recent study [109]. Interestingly, while the 1-minute INSTI test detected 99.4% of the exposed infants younger than 3 months, it is negative in 90% and 98% of uninfected infants at 7 months and 12 months, respectively [59]. Thus INSTI may be used as a test to exclude infection at 7 months or later [59]. Homsy et al. also previously reported that blood-based rapid testing could identify uninfected infants at an age of 3 months or older [110], and it was determined that using such a strategy to reduce the number of infants requiring PCR is cost-effective [60].

Point-of-Care and Innovative Technologies

Two conspicuous obstacles are present in many RLCs, (1) low EID coverage and therefore late infant testing and (2) high loss to follow-up after EID testing. Although infants could be tested at the age of 4–6 weeks, in a recent survey, Tripathi et al. reported that the average age of infants tested by PCR in Senegal, Namibia, and Uganda was 4.0, 4.4, and 7.2 months, respectively [111]. Obtaining a second specimen for PCR confirmation would further lengthen the time to diagnosis and increase the loss to follow-up. It was found that 50% of guardians in Malawi did not come back for results and two-thirds of the identified infected infants were not referred to care [67]. One way to overcome these challenges is to use simple and rapid point-of-care (POC) methodologies to bring testing services closer to the affected families and their exposed infants (Table 4.4).

Parpia et al. [112] developed a rapid dip-stick p24 antigen assay and evaluated it with fresh plasma samples from 24 HIV infected (19 younger than 6 months) and 365 uninfected infants 1 year of age or younger in South Africa. This 45-min test required the use of plasma and yielded sensitivity and specificity of 95.8% and 99.4%, respectively. The same research group also designed a low cost vertical flow device to separate plasma from whole blood [113] to further simplify the test for field use. In a different dip-stick assay based on NAT testing, Lee et al. qualitatively measured HIV-1 RNA using 240 µl of plasma [114]. Plasma RNA was isolated using nonorganic solvents and then amplified in an enclosed cartridge to eliminate contamination. The assay has a detection sensitivity of 200 copies per ml of plasma and could detect major HIV-1 subtypes (A, B, C, D, AE, AG, G, H, J, and K) and groups N and O. Steinmetzer et al., showed that HIV-1 total nucleic acid from 10 µl of finger-prick blood can be detected in an hour using a small credit-card sized cartridge and a light-weight, portable thermocycler with comparable or better detection sensitivity than commercial HIV real-time NAT [115]. This assay may prove to be useful for EID. Another quantitative RNA assay, the Liat assay with 200-µl plasma input, was shown [116, 117] to have a detection sensitivity limit of 58 copies per ml of plasma and could also detect most of the HIV-1 group M subtypes. Most of these novel assays are still in the developmental stage and require vigorous field evaluation outside of a reference laboratory setting to truly document their field utility. Nevertheless, it is encouraging that many of these assays are being developed by commercial entities or researchers with the commercialization goal to determine viral load for ART monitoring. These well-standardized and sustainable viral load technologies are applicable to EID diagnosis.

Quality Assurance and Global Proficiency Testing Programs for NAT Testing

As discussed earlier, NAT-based assays based on exponential amplification of viral nucleic acid tend to yield false-positive results if a nonstandardized commercial assay is used [46, 118]. The World Health Organization (WHO), the U.S. Department of Health and Human Services (HHS), and other health agencies recommend the use of two virological tests using two different specimens to confirm a positive diagnosis. The surge in funding from the President's Emergency Plan for AIDS Relief (PEPFAR), the Global Fund, and other major donors and initiatives has allowed for the rapid expansion of EID testing in many RLCs. To ensure the quality of test performance, the Global AIDS Program, U.S. Centers for Disease Control and Prevention (CDC), began to provide an external quality assurance (EQA) program with DBS specimens in 2006. This program consists of two components: (1) provision of known positive ($n=2$) and negative ($n=1$) DBS specimens and (2) provision of a proficiency testing (PT) panel. The PT panel consists of blinded, ten-member positive and negative DBS samples. Participating laboratories are expected to return the results within 4 weeks mimicking primary testing turnaround time requirements and expectations. Laboratories are tested three times a year. This program has increased from 17 EID testing facilities in 11 countries in 2006 to over 100 laboratories in 35 countries by the end of 2010. The global distribution of participating facilities is as follows: 79 laboratories in 21 sub-Saharan African countries, 13 in 5 Asian countries, 5 in the Caribbean region, and 2 each in North and South America. The increase in the number of laboratories and countries with EID capacity reaffirms the strong need for a standardized EQA Program. Laboratories enrolled in the program have shown consistent improvement in their mean scores over time ($R=0.71$). Most laboratories used the Roche DNA assay until recently. In 2010, 15 laboratories transitioned to automated nucleic extraction platforms and real-time PCR to accommodate their higher demand for infant testing.

With the anticipated use of p24 tests, similar quality assurance/quality control (QA/QC) programs for PT and self-assisted capacity building are also needed.

Conclusion

Since the first recognition of mother-to-child HIV transmission in 1983 and breast milk-acquired transmission in 1985, great advances in scientific knowledge, programmatic implementation, and diagnostic technologies have been made. Two independent virological methods to detect HIV nucleic acid and the structural protein p24 have been shown to possess both high sensitivity and specificity of detection needed for early diagnosis. All exposed infants are recommended to receive virological testing 4–6 weeks after birth. In many RLCs, exclusive breastfeeding is recommended by WHO to provide the needed infant nutrition and to reduce mortality associated with mixed feeding. Transmission of HIV can occur throughout the breastfeeding period and thus a negative virological test result during the breastfeeding period does not rule out infection. Exposed infants who continue breastfeeding should be tested 6 weeks after weaning. Although nucleic acid testing is the predominant method currently used, the p24 antigen assay, if provided in a cost-effective manner, may become the preferred method because of its simpler laboratory equipment and training requirements. The use of DBS to collect blood specimens enables a wider reach of exposed infants in hard-to-reach areas. The impact of automated nucleic acid extraction and fluorescent real-time technology in a centralized laboratory has been shown to produce accurate and high-throughput results in South Africa. Up to this date, due to the inadequate healthcare system, human resources, and laboratory capacity in many RLCs, reaching exposed infants at an early age and providing timely results are still very challenging. The overall number of exposed infants, as compared to the number of persons on ART who receive viral load determination, is too small to elicit the rapid manufacture of affordable, high quality, and standardized commercial testing kits and equipment. It is fortunate that many POC technologies that are currently under development to determine viral load for ART monitoring can also be used for EID.

Disclaimer The findings and conclusions in this chapter are those of the authors and do not necessarily represent the views of the U.S. Centers for Disease Control and Prevention. The use of trade names is for identification purposes only and does not constitute endorsement by the U.S. Centers for Disease Control and Prevention or the Department of Health and Human Services.

References

1. UNAIDS (2010) Report on the global AIDS epidemic. http://www.unaids.org/globalreport/Global_report.htm. Accessed 21 May 2011
2. De Cock KM, Fowler MG, Mercier E, de Vincenzi I, Saba J, Hoff E et al (2000) Prevention of mother-to-child HIV transmission in resource-poor countries: translating research into policy and practice. JAMA 283:1175–1182
3. Kourtis AP, Bulterys M, Nesheim SR, Lee FK (2001) Understanding the timing of HIV transmission from mother to infant. JAMA 285:709–712
4. Dorenbaum A, Cunningham CK, Gelber RD, Culnane M, Mofenson L, Britto P et al (2002) Two-dose intrapartum/newborn nevirapine and standard antiretroviral therapy to reduce perinatal HIV transmission: a randomized trial. JAMA 288:189–198
5. Cooper ER, Charurat M, Mofenson L, Hanson IC, Pitt J, Diaz C et al (2002) Combination antiretroviral strategies for the treatment of pregnant HIV-1-infected women and prevention of perinatal HIV-1 transmission. J Acquir Immune Defic Syndr 29:484–494
6. Anderson JE, Ebrahim SH, Sansom S (2004) Women's knowledge about treatment to prevent mother-to-child human immunodeficiency virus transmission. Obstet Gynecol 103:165–168
7. Chou R, Smits AK, Huffman LH, Korthuis PT (2005) Screening for human immunodeficiency virus in pregnant women: evidence synthesis. Rockville (MD): Agency for Healthcare Research and Quality (US). http://www.ncbi.nlm.nih.gov/books/NBK33383/
8. Larsson EC, Thorson A, Pariyo G, Conrad P, Arinaitwe M, Kemigisa M et al (2011) Opt-out HIV testing during antenatal care: experiences of pregnant women in rural Uganda. Health Policy Plan
9. Larsson EC, Waiswa P, Thorson A, Tomson G, Peterson S, Pariyo G et al (2009) Low uptake of HIV testing during antenatal care: a population-based study from eastern Uganda. AIDS 23:1924–1926
10. Creek TL, Ntumy R, Seipone K, Smith M, Mogodi M, Smit M et al (2007) Successful introduction of routine opt-out HIV testing in antenatal care in Botswana. J Acquir Immune Defic Syndr 45:102–107
11. Kharsany AB, Karim QA, Karim SS (2010) Uptake of provider-initiated HIV testing and counseling among women attending an urban sexually transmitted disease clinic in South Africa - missed opportunities for early diagnosis of HIV infection. AIDS Care 22:533–537
12. Newell ML, Coovadia H, Cortina-Borja M, Rollins N, Gaillard P, Dabis F (2004) Mortality of infected and uninfected infants born to HIV-infected mothers in Africa: a pooled analysis. Lancet 364:1236–1243
13. Violari A, Cotton MF, Gibb DM, Babiker AG, Steyn J, Madhi SA et al (2008) Early antiretroviral therapy and mortality among HIV-infected infants. N Engl J Med 359:2233–2244
14. Read JS (2007) Diagnosis of HIV-1 infection in children younger than 18 months in the United States. Pediatrics 120:e1547–e1562
15. Schneider E, Whitmore S, Glynn KM, Dominguez K, Mitsch A, McKenna MT (2008) Revised surveillance case definitions for HIV infection among adults, adolescents, and children aged <18 months and for HIV infection and AIDS among children aged 18 months to <13 years–United States, 2008. MMWR Recomm Rep 57:1–12
16. De Cock KM, Jaffe HW, Curran JW (2011) Refections on 30 years of AIDS. Emerg Infect Dis 17:1044–1048
17. Ziegler JB, Cooper DA, Johnson RO, Gold J (1985) Postnatal transmission of AIDS-associated retrovirus from mother to infant. Lancet 1:896–898
18. Palasanthiran P, Robertson P, Ziegler JB, Graham GG (1994) Decay of transplacental human immunodeficiency virus type 1 antibodies in neonates and infants. J Infect Dis 170:1593–1596
19. Parekh BS, Shaffer N, Coughlin R, Hung CH, Krasinski K, Abrams E et al (1993) Dynamics of maternal IgG antibody decay and HIV-specific antibody synthesis in infants born to seropositive mothers. The New York City Perinatal HIV Transmission Study Group. AIDS Res Hum Retroviruses 9:907–912
20. Mok JQ, Giaquinto C, De Rossi A, Grosch-Worner I, Ades AE, Peckham CS (1987) Infants born to mothers seropositive for human immunodeficiency virus: Preliminary findings from a multicentre European study. Lancet 1:1164–1168
21. Pyun KH, Ochs HD, Dufford MT, Wedgwood RJ (1987) Perinatal infection with human immunodeficiency virus. Specific antibody responses by the neonate. N Engl J Med 317:611–614
22. Mother-to-child transmission of HIV infection (1988) The European Collaborative Study. Lancet 2:1039–1043

23. Nielsen K, Bryson YJ (2000) Diagnosis of HIV infection in children. Pediatr Clin North Am 47:39–63
24. Oxelius VA (1979) IgG subclass levels in infancy and childhood. Acta Paediatr Scand 68:23–27
25. Weiblen BJ, Lee FK, Cooper ER, Landesman SH, McIntosh K, Harris JA et al (1990) Early diagnosis of HIV infection in infants by detection of IgA HIV antibodies. Lancet 335:988–990
26. Landesman S, Weiblen B, Mendez H, Willoughby A, Goedert JJ, Rubinstein A et al (1991) Clinical utility of HIV-IgA immunoblot assay in the early diagnosis of perinatal HIV infection. JAMA 266:3443–3446
27. Quinn TC, Kline RL, Halsey N, Hutton N, Ruff A, Butz A et al (1991) Early diagnosis of perinatal HIV infection by detection of viral-specific IgA antibodies. JAMA 266:3439–3442
28. Shaffer N, Ou CY, Abrams EJ, Krasinski K, Parekh B, Thomas P et al (1992) PCR and HIV-IGA for early infant diagnosis of perinatal HIV infection. The New York City Perinatal HIV Transmission Collaborative Study. International Conference on AIDS; Amsterdam, The Netherlands
29. Mokili JL, Connell JA, Parry JV, Green SD, Davies AG, Cutting WA (1996) How valuable are IgA and IgM anti-HIV tests for the diagnosis of mother-child transmission of HIV in an African setting? Clin Diagn Virol 5:3–12
30. Martin NL, Rautonen J, Crombleholme W, Rautonen N, Wara DW (1992) A screening test for the detection of anti-HIV-1 IgA in young infants. Immunol Invest 21:65–70
31. Basualdo Mdel C, Moran K, Alcantara P, Gonzalez E, Puentes E, Soler C (2004) IgA antibody detection and PCR as first options in the diagnosis of perinatal HIV-1 infection. Salud Publica Mex 46:49–55
32. Liberatore D, Avila MM, Calarota S, Libonatti O (1996) Martinez Peralta L. Diagnosis of perinatally acquired HIV-1 infection using an IgA ELISA test. Pediatr AIDS HIV Infect 7:164–167
33. Archibald DW, Hebert CA, Sun D, Tacket CO (1990) Salivary antibodies to human immunodeficiency virus type 1 in a phase I AIDS vaccine trial. J Acquir Immune Defic Syndr 3:954–958
34. Amadori A, de Rossi A, Giaquinto C, Faulkner-Valle G, Zacchello F, Chieco-Bianchi L (1988) In-vitro production of HIV-specific antibody in children at risk of AIDS. Lancet 1:852–854
35. Arico M, Caselli D, Marconi M, Avanzini MA, Colombo A, Pasinetti G et al (1991) Immunoglobulin G3-specific antibodies as a marker for early diagnosis of HIV infection in children. AIDS 5:1315–1318
36. Moodley D, Bobat RA, Coutsoudis A, Coovadia HM (1995) Predicting perinatal human immunodeficiency virus infection by antibody patterns. Pediatr Infect Dis J 14:850–852
37. Madurai S, Moodley D, Coovadia HM, Bobat RA, Gopaul W, Smith AN et al (1996) Use of HIV-1 specific immunoglobulin G3 as a serological marker of vertical transmission. J Trop Pediatr 42:359–361
38. Saiki RK, Scharf S, Faloona F, Mullis KB, Horn GT, Erlich HA et al (1985) Enzymatic amplification of beta-globin genomic sequences and restriction site analysis for diagnosis of sickle cell anemia. Science 230:1350–1354
39. Laure F, Courgnaud V, Rouzioux C, Blanche S, Veber F, Burgard M et al (1988) Detection of HIV1 DNA in infants and children by means of the polymerase chain reaction. Lancet 2:538–541
40. Edwards JR, Ulrich PP, Weintrub PS, Cowan MJ, Levy JA, Wara DW et al (1989) Polymerase chain reaction compared with concurrent viral cultures for rapid identification of human immunodeficiency virus infection among high-risk infants and children. J Pediatr 115:200–203
41. Rogers MF, Ou CY, Rayfield M, Thomas PA, Schoenbaum EE, Abrams E et al (1989) Use of the polymerase chain reaction for early detection of the proviral sequences of human immunodeficiency virus in infants born to seropositive mothers. New York City Collaborative Study of Maternal HIV Transmission and Montefiore Medical Center HIV Perinatal Transmission Study Group. N Engl J Med 320:1649–1654
42. WHO (2010) Recommendations on the diagnosis of HIV infection in infants and children. Geneva. http://whqlibdoc.who.int/publications/2010/9789241599085_eng.pdf. Accessed 5 May 2011
43. Sivapalasingam S, Patel U, Itri V, Laverty M, Mandaliya K, Valentine F et al (2007) A reverse transcriptase assay for early diagnosis of infant HIV infection in resource-limited settings. J Trop Pediatr 53:355–358
44. WHO (2010) Guidelines on HIV and infant feeding. http://www.who.int/child_adolescent_health/documents/9789241599535/en/. Accessed 20 May 2011
45. Kwok S (1990) Procedures to minimize PCR-product carry-over. In: Innis MA, Gelfand DH, Sninski JJ, White TJ (eds) PCR Protocols, A Guide to Methods and Applications. Academic, San Diego, pp 142–145
46. Jackson JB, Drew J, Lin HJ, Otto P, Bremer JW, Hollinger FB et al (1993) Establishment of a quality assurance program for human immunodeficiency virus type 1 DNA polymerase chain reaction assays by the AIDS Clinical Trials Group. ACTG PCR Working Group, and the ACTG PCR Virology Laboratories. J Clin Microbiol 31:3123–3128
47. Parry JV, Mortimer PP, Perry KR, Pillay D, Zuckerman M (2003) Towards error-free HIV diagnosis: guidelines on laboratory practice. Commun Dis Public Health 6:334–350
48. Stevens W, Sherman G, Downing R, Parsons LM, Ou CY, Crowley S et al (2008) Role of the laboratory in ensuring global access to ARV treatment for HIV-infected children: consensus statement on the performance of laboratory assays for early infant diagnosis. Open AIDS J 2:17–25
49. Creek TL, Sherman GG, Nkengasong J, Lu L, Finkbeiner T, Fowler MG et al (2007) Infant human immunodeficiency virus diagnosis in resource-limited settings: issues, technologies, and country experiences. Am J Obstet Gynecol 197:S64–S71

50. Cassol S, Butcher A, Kinard S, Spadoro J, Sy T, Lapointe N et al (1994) Rapid screening for early detection of mother-to-child transmission of human immunodeficiency virus type 1. J Clin Microbiol 32:2641–2645
51. Biggar RJ, Miley W, Miotti P, Taha TE, Butcher A, Spadoro J et al (1997) Blood collection on filter paper: a practical approach to sample collection for studies of perinatal HIV transmission. J Acquir Immune Defic Syndr Hum Retrovirol 14:368–373
52. Kline MW, Lewis DE, Hollinger FB, Reuben JM, Hanson LC, Kozinetz CA et al (1994) A comparative study of human immunodeficiency virus culture, polymerase chain reaction and anti-human immunodeficiency virus immunoglobulin A antibody detection in the diagnosis during early infancy of vertically acquired human immunodeficiency virus infection. Pediatr Infect Dis J 13:90–94
53. Bremer JW, Lew JF, Cooper E, Hillyer GV, Pitt J, Handelsman E et al (1996) Diagnosis of infection with human immunodeficiency virus type 1 by a DNA polymerase chain reaction assay among infants enrolled in the Women and Infants' Transmission Study. J Pediatr 129:198–207
54. Cunningham CK, Charbonneau TT, Song K, Patterson D, Sullivan T, Cummins T et al (1999) Comparison of human immunodeficiency virus 1 DNA polymerase chain reaction and qualitative and quantitative RNA polymerase chain reaction in human immunodeficiency virus 1-exposed infants. Pediatr Infect Dis J 18:30–35
55. Dunn DT, Brandt CD, Krivine A, Cassol SA, Roques P, Borkowsky W et al (1995) The sensitivity of HIV-1 DNA polymerase chain reaction in the neonatal period and the relative contributions of intra-uterine and intra-partum transmission. AIDS 9:F7–F11
56. Owens DK, Holodniy M, McDonald TW, Scott J, Sonnad S (1996) A meta-analytic evaluation of the polymerase chain reaction for the diagnosis of HIV infection in infants. JAMA 275:1342–1348
57. Sherman GG, Matsebula TC, Jones SA (2005) Is early HIV testing of infants in poorly resourced prevention of mother to child transmission programmes unaffordable? Trop Med Int Health 10:1108–1113
58. Lambert JS, Harris DR, Stiehm ER, Moye J Jr, Fowler MG, Meyer WA 3rd et al (2003) Performance characteristics of HIV-1 culture and HIV-1 DNA and RNA amplification assays for early diagnosis of perinatal HIV-1 infection. J Acquir Immune Defic Syndr 34:512–519
59. Sherman GG, Driver GA, Coovadia AH (2008) Evaluation of seven rapid HIV tests to detect HIV-exposure and seroreversion during infancy. J Clin Virol 43:313–316
60. Menzies NA, Homsy J, Chang pitter JY, Pitter C, Mermin J, Downing R et al (2009) Cost-effectiveness of routine rapid human immunodeficiency virus antibody testing before DNA-PCR testing for early diagnosis of infants in resource-limited settings. Pediatr Infect Dis J 28:819–825
61. Sherman GG, Stevens G, Jones SA, Horsfield P, Stevens WS (2005) Dried blood spots improve access to HIV diagnosis and care for infants in low-resource settings. J Acquir Immune Defic Syndr 38:615–617
62. Cherutich P, Inwani I, Nduati R, Mbori-Ngacha D (2008) Optimizing paediatric HIV care in Kenya: challenges in early infant diagnosis. Bull World Health Organ 86:155–160
63. Khamadi S, Okoth V, Lihana R, Nabwera J, Hungu J, Okoth F et al (2008) Rapid identification of infants for antiretroviral therapy in a resource poor setting: the Kenya experience. J Trop Pediatr 54:370–374
64. Kankasa C, Carter RJ, Briggs N, Bulterys M, Chama E, Cooper ER et al (2009) Routine offering of HIV testing to hospitalized pediatric patients at university teaching hospital, Lusaka, Zambia: acceptability and feasibility. J Acquir Immune Defic Syndr 51:202–208
65. Nuwagaba-Biribonwoha H, Werq-Semo B, Abdallah A, Cunningham A, Gamaliel JG, Mtunga S et al (2010) Introducing a multi-site program for early diagnosis of HIV infection among HIV-exposed infants in Tanzania. BMC Pediatr 10:44
66. Lofgren SM, Morrissey AB, Chevallier CC, Malabeja AI, Edmonds S, Amos B et al (2009) Evaluation of a dried blood spot HIV-1 RNA program for early infant diagnosis and viral load monitoring at rural and remote healthcare facilities. AIDS 23:2459–2466
67. Braun M, Kabue MM, McCollum ED, Ahmed S, Kim M, Aertker L et al (2011) Inadequate coordination of maternal and infant HIV services detrimentally affects early infant diagnosis outcomes in Lilongwe, Malawi. J Acquir Immune Defic Syndr 56:e122–e129
68. Stevens W, Sherman G, Cotton M, Grentholtz L, Webber L (2006) Revised guidelines for diagnosis of perinatal HIV-1 infection in South Africa. S Afr J HIV Med 7:24–28
69. Kekitiinwa A, Kelly S, Sengendo H, Kline M, Namale A, Serukka D et al Scaling up of early infant diagnosis of HIV through LInkages of PMTCT to pediatric care services in Uganda. PEPFAR HIV/AIDS Implementers' Meeting, Kigali, Rwanda, 904
70. Mudany A. Overview of the scale up of early infant diagnosis in Kenya from January to December 2006. PEPFAR HIV/AIDS Implementers' Meeting, Kigali, Rwanda, 478
71. Cook RE, Ciampa PJ, Sidat M, Blevins M, Burlison J, Davidson MA et al (2011) Predictors of successful early infant diagnosis of HIV in a rural district hospital in Zambezia, Mozambique. J Acquir Immune Defic Syndr 56:e104–e109
72. Nsojo A, Aboud S, Lyamuya E (2010) Comparative evaluation of Amplicor HIV-1 DNA test, version 1.5, by manual and automated DNA extraction methods using venous blood and dried blood spots for HIV-1 DNA PCR testing. Tanzan J Health Res. http://ajol.info/index.php/thrb/article/view/58621

73. De Rossi A, Masiero S, Giaquinto C, Ruga E, Comar M, Giacca M et al (1996) Dynamics of viral replication in infants with vertically acquired human immunodeficiency virus type 1 infection. J Clin Invest 97:323–330
74. Dickover RE, Dillon M, Leung KM, Krogstad P, Plaeger S, Kwok S et al (1998) Early prognostic indicators in primary perinatal human immunodeficiency virus type 1 infection: importance of viral RNA and the timing of transmission on long-term outcome. J Infect Dis 178:375–387
75. Shearer WT, Quinn TC, LaRussa P, Lew JF, Mofenson L, Almy S et al (1997) Viral load and disease progression in infants infected with human immunodeficiency virus type 1. Women and Infants Transmission Study Group. N Engl J Med 336:1337–1342
76. Young NL, Shaffer N, Chaowanachan T, Chotpitayasunondh T, Vanparapar N, Mock PA et al (2000) Early diagnosis of HIV-1-infected infants in Thailand using RNA and DNA PCR assays sensitive to non-B subtypes. J Acquir Immune Defic Syndr 24:401–407
77. Rouet F, Sakarovitch C, Msellati P, Elenga N, Montcho C, Viho I et al (2003) Pediatric viral human immunodeficiency virus type 1 RNA levels, timing of infection, and disease progression in African HIV-1-infected children. Pediatrics 112:e289
78. Stevens W, Erasmus L, Moloi M, Taleng T, Sarang S (2008) Performance of a novel human immunodeficiency virus (HIV) type 1 total nucleic acid-based real-time PCR assay using whole blood and dried blood spots for diagnosis of HIV in infants. J Clin Microbiol 46:3941–3945
79. Sullivan TJ, Miller TT, Warren B, Parker MM. Evaluation of an FDA-approved qualitative RNA detection assay for diagnosis of HIV-1 infection in perinatally exposed infants. HIV diagnostics conference, Orlando, FL2010
80. Kerr RJ, Player G, Fiscus SA, Nelson JA (2009) Qualitative human immunodeficiency virus RNA analysis of dried blood spots for diagnosis of infections in infants. J Clin Microbiol 47:220–222
81. Stevens WS, Noble L, Berrie L, Sarang S, Scott LE (2009) Ultra-high-throughput, automated nucleic acid detection of human immunodeficiency virus (HIV) for infant infection diagnosis using the Gen-Probe Aptima HIV-1 screening assay. J Clin Microbiol 47:2465–2469
82. Viljoen J, Gampini S, Danaviah S, Valea D, Pillay S, Kania D et al (2010) Dried blood spot HIV-1 RNA quantification using open real-time systems in South Africa and Burkina Faso. J Acquir Immune Defic Syndr 55: 290–298
83. Havens PL, Mofenson LM (2009) Evaluation and management of the infant exposed to HIV-1 in the United States. Pediatrics 123:175–187
84. Patel JA, Anderson E, Dong J (2009) False positive ultrasensitive HIV bDNA viral load results in diagnosis of perinatal HIV-infection in the era of low transmission. LabMedicine 40:611–614. 5 May 2011
85. Benjamin DK Jr, Miller WC, Fiscus SA, Benjamin DK, Morse M, Valentine M et al (2001) Rational testing of the HIV-exposed infant. Pediatrics 108:E3
86. Schupbach J, Tomasik Z, Knuchel M, Opravil M, Gunthard HF, Nadal D et al (2006) Optimized virus disruption improves detection of HIV-1 p24 in particles and uncovers a p24 reactivity in patients with undetectable HIV-1 RNA under long-term HAART. J Med Virol 78:1003–1010
87. Mwapasa V, Cachafeiro A, Makuta Y, Beckstead DJ, Pennell ML, Chilima B et al (2010) Using a simplified human immunodeficiency virus type 1 p24 antigen assay to diagnose pediatric HIV-infection in Malawi. J Clin Virol 49:299–302
88. George E, Beauharnais CA, Brignoli E, Noel F, Bois G (2007) De Matteis Rouzier P, et al. Potential of a simplified p24 assay for early diagnosis of infant human immunodeficiency virus type 1 infection in Haiti. J Clin Microbiol 45:3416–3418
89. Cachafeiro A, Sherman GG, Sohn AH, Beck-Sague C, Fiscus SA (2009) Diagnosis of human immunodeficiency virus type 1 infection in infants by use of dried blood spots and an ultrasensitive p24 antigen assay. J Clin Microbiol 47:459–462
90. Patton JC, Sherman GG, Coovadia AH, Stevens WS, Meyers TM (2006) Ultrasensitive human immunodeficiency virus type 1 p24 antigen assay modified for use on dried whole-blood spots as a reliable, affordable test for infant diagnosis. Clin Vaccine Immunol 13:152–155
91. Patton JC, Coovadia AH, Meyers TM, Sherman GG (2008) Evaluation of the ultrasensitive human immunodeficiency virus type 1 (HIV-1) p24 antigen assay performed on dried blood spots for diagnosis of HIV-1 infection in infants. Clin Vaccine Immunol 15:388–391
92. Fiscus SA, Wiener J, Abrams EJ, Bulterys M, Cachafeiro A, Respess RA (2007) Ultrasensitive p24 antigen assay for diagnosis of perinatal human immunodeficiency virus type 1 infection. J Clin Microbiol 45:2274–2277
93. Wittawatmongkol O, Vanprapar N, Chearskul P, Phongsamart W, Prasitsuebsai W, Sutthent R et al (2010) Boosted p24 antigen assay for early diagnosis of perinatal HIV infection. J Med Assoc Thai 93:187–190
94. Zijenah LS, Tobaiwa O, Rusakaniko S, Nathoo KJ, Nhembe M, Matibe P et al (2005) Signal-boosted qualitative ultrasensitive p24 antigen assay for diagnosis of subtype C HIV-1 infection in infants under the age of 2 years. J Acquir Immune Defic Syndr 39:391–394
95. UNICEF (2008) Countdown to 2015 MNCH: the 2008 report: tracking progress in maternal, newborn and child survival. New York

96. Read JS, Mwatha A, Richardson B, Valentine M, Emel L, Manji K et al (2009) Primary HIV-1 infection among infants in sub-Saharan Africa: HPTN 024. J Acquir Immune Defic Syndr 51:317–322
97. Doherty T, Chopra M, Nsibande D, Mngoma D (2009) Improving the coverage of the PMTCT programme through a participatory quality improvement intervention in South Africa. BMC Public Health 9:406
98. Delaney KP, Branson BM, Uniyal A, Phillips S, Candal D, Owen SM et al (2011) Evaluation of the performance characteristics of 6 rapid HIV antibody tests. Clin Infect Dis 52:257–263
99. USAID (2001) List of approved HIV/AIDS rapid test kits – 04/15/11. http://www.usaid.gov/our_work/global_health/aids/TechAreas/treatment/scms.html. Accessed 6 May 2011
100. WHO (2011) Status of applications to the prequalification of diagnostics programme. http://www.who.int/diagnostics_laboratory/pq_status/en/index.html
101. Bulterys M, Fowler MG, Van Rompay KK, Kourtis AP (2004) Prevention of mother-to-child transmission of HIV-1 through breast-feeding: past, present, and future. J Infect Dis 189:2149–2153
102. Levison J, Williams LT, Moore A, McFarlane J, Davila JA (2010) Increasing use of rapid HIV testing in labor and delivery among women with no prenatal care: a local initiative. Matern Child Health J. http://www.springerlink.com/content/w8ww43496370149t/
103. Birkhead GS, Pulver WP, Warren BL, Hackel S, Rodriguez D, Smith L (2010) Acquiring human immunodeficiency virus during pregnancy and mother-to-child transmission in New York: 2002–2006. Obstet Gynecol 115:1247–1255
104. (2008) HIV testing and prophylaxis to prevent mother-to-child transmission in the United States. Pediatrics 122:1127–1134
105. Pai NP, Barick R, Tulsky JP, Shivkumar PV, Cohan D, Kalantri S et al (2008) Impact of round-the-clock, rapid oral fluid HIV testing of women in labor in rural India. PLoS Med 5:e92
106. (2004) Introduction of routine HIV testing in prenatal care – Botswana, 2004. MMWR Morb Mortal Wkly Rep 53:1083–1086
107. Bulterys M, Jamieson DJ, O'Sullivan MJ, Cohen MH, Maupin R, Nesheim S et al (2004) Rapid HIV-1 testing during labor: a multicenter study. JAMA 292:219–223
108. Tepper NK, Farr SL, Danner SP, Maupin R, Nesheim SR, Cohen MH et al (2009) Rapid human immunodeficiency virus testing in obstetric outpatient settings: the MIRIAD study. Am J Obstet Gynecol 201:31.e1–31.e6
109. Sherman GG, Lilian RR, Coovadia AH (2010) Oral fluid tests for screening of human immunodeficiency virus-exposed infants. Pediatr Infect Dis J 29:169–172
110. Homsy J, Kalamya JN, Obonyo J, Ojwang J, Mugumya R, Opio C et al (2006) Routine intrapartum HIV counseling and testing for prevention of mother-to-child transmission of HIV in a rural Ugandan hospital. J Acquir Immune Defic Syndr 42:149–154
111. Tripathi S, Kiyaga C, Nghatanga M, Chhi Vun M, Wade AS, Gass R et al (2010) Increasing uptake of HIV early infant diagnosis (EID) services in four countries (Cambodia, Namibia, Senegal and Uganda). International AIDS conference, July 18–23, Vienna, Austria
112. Parpia ZA, Elghanian R, Nabatiyan A, Hardie DR, Kelso DM (2010) p24 antigen rapid test for diagnosis of acute pediatric HIV infection. J Acquir Immune Defic Syndr 55:413–419
113. Nabatiyan A, Parpia ZA, Elghanian R, Kelso DM (2011) Membrane-based plasma collection device for point-of-care diagnosis of HIV. J Virol Methods 173:37–42
114. Lee HH, Dineva MA, Chua YL, Ritchie AV, Ushiro-Lumb I, Wisniewski CA (2010) Simple amplification-based assay: a nucleic acid-based point-of-care platform for HIV-1 testing. J Infect Dis 201(suppl 1):S65–S72
115. Steinmetzer K, Seidel T, Stallmach A, Ermantraut E (2010) HIV load testing with small samples of whole blood. J Clin Microbiol 48:2786–2792
116. Tanriverdi S, Chen L, Chen S (2010) A rapid and automated sample-to-result HIV load test for near-patient application. J Infect Dis 201(suppl 1):S52–S58
117. Coombs R, Dragavon J, Harb S (2011) Validation of a novel lab-in-a-tube analyzer and single-tube system for simple/rapid HIV-1 RNA quantification. 18th conference on retroviruses and opportunistic infections
118. Kwok S, Higuchi R (1989) Avoiding false positives with PCR. Nature 339:237–238
119. Lilian RR, Bhowan K, Sherman GG (2010) Early diagnosis of human immunodeficiency virus-1 infection in infants with the NucliSens EasyQ assay on dried blood spots. J Clin Virol 48:40–43
120. Nadal D, Boni J, Kind C, Varnier OE, Steiner F, Tomasik Z et al (1999) Prospective evaluation of amplification-boosted ELISA for heat-denatured p24 antigen for diagnosis and monitoring of pediatric human immunodeficiency virus type 1 infection. J Infect Dis 180:1089–1095
121. Sutthent R, Gaudart N, Chokpaibulkit K, Tanliang N, Kanoksinsombath C, Chaisilwatana P (2003) p24 Antigen detection assay modified with a booster step for diagnosis and monitoring of human immunodeficiency virus type 1 infection. J Clin Microbiol 41:1016–1022
122. Sherman GG, Stevens G, Stevens WS (2004) Affordable diagnosis of human immunodeficiency virus infection in infants by p24 antigen detection. Pediatr Infect Dis J 23:173–176

123. De Baets AJ, Edidi BS, Kasali MJ, Beelaert G, Schrooten W, Litzroth A et al (2005) Pediatric human immunodeficiency virus screening in an African district hospital. Clin Diagn Lab Immunol 12:86–92
124. Nouhin J, Nguyen M (2006) Evaluation of a boosted-p24 antigen assay for the early diagnosis of pediatric HIV-1 infection in Cambodia. Am J Trop Med Hyg 75:1103–1105
125. Tang S, Hewlett I (2010) Nanoparticle-based immunoassays for sensitive and early detection of HIV-1 capsid (p24) antigen. J Infect Dis 201(suppl 1):S59–S64
126. Jangam SR, Yamada DH, McFall SM, Kelso DM (2009) Rapid, point-of-care extraction of human immunodeficiency virus type 1 proviral DNA from whole blood for detection by real-time PCR. J Clin Microbiol 47: 2363–2368

Part II
Mechanisms of HIV-1 Transmission Through Breast Milk: Virology

Chapter 5
Virologic Determinants of Breast Milk Transmission of HIV-1

Susan A. Fiscus and Grace M. Aldrovandi

Introduction

In 1985, the first report of presumed HIV breast milk transmission was described in an Australian infant whose mother received a postpartum transfusion from an apparently healthy male homosexual donor, who subsequently developed Kaposi's sarcoma and *Pneumocystis* pneumonia [1]. Later that year, HIV was cultured from the cell-free fraction of breast milk of 3 HIV-infected women [2]. On the basis of these reports, guidelines in the USA were changed advising HIV-infected women not to breastfeed, but for the majority of HIV-infected women worldwide this was not and is not an option. In the intervening 25 years, much has been learned about breast milk transmission, including ways to reduce its occurrence. Some of this knowledge may appear "obvious"; e.g., avoidance of breast feeding prevents breast milk transmission [3]. Less obvious is the fact that breastfeeding behavior, i.e., exclusive breastfeeding, as opposed to mixed feeding, substantially reduces the risk of transmission [4]. However, despite these advances, there are still many questions regarding key pathogenic mechanisms of HIV breast milk transmission. Among these are whether cell-free or cell-associated virus is responsible for transmission; where and when transmission occurs; which virologic determinants predispose mother–infant pairs to transmission; and whether the breast is a separate virologic compartment? This chapter reviews the virologic factors that have been associated with breast milk transmission of HIV.

What Is Transmitted: Cell-Free or Cell-Associated Virus?

One of the most remarkable and consistent features of HIV breast milk transmission is its inefficiency. Despite immunologic immaturity and ingestion of hundreds of liters of breast milk, only 15–20% of infants will become infected [5, 6]. This observation has raised questions about the nature and

S.A. Fiscus, Ph.D. (✉)
Departments of Microbiology & Immunology and Pathology & Laboratory Medicine,
University of North Carolina at Chapel Hill, CB# 7290, Chapel Hill, NC 27599-7290, USA
e-mail: fiscussa@med.unc.edu

G.M. Aldrovandi, MD, CM
Departments of Pediatrics and Pathology and Laboratory Medicine, Children's Hospital of Los Angeles,
Los Angeles, CA, USA

"infectiousness" of milk-borne virus. Thiry et al. [2] were able to isolate infectious virus from filtered skim milk, but attempts to isolate HIV from breast milk cells failed due to bacterial contamination. However, successful viral isolation of HIV from the cellular fraction of colostrum has been reported [7, 8] and in vitro, breast milk cells and breast epithelium are known to be susceptible to HIV infection [9, 10]. Milk contains many inhibitors of cell-free HIV and likely inactivates many of the potentially infectious virions [11–17]. The paucity of studies successfully culturing virus from this secretion has been cited as evidence of decreased infectiousness. However, culturing HIV from bodily fluids, including plasma, semen, and cervicovaginal fluids is much more difficult than obtaining it from cells. Advances in molecular biological techniques allowed investigators first to determine that the HIV genome could be detected in breast milk cells [18–29]. Later, HIV RNA assays were used to detect and quantitate viral load in cell-free lactoserum [23–27, 30–38].

Cell-Associated HIV DNA

HIV-infected cells can be detected in breast milk [9, 18–29, 39]. Detection of proviral DNA ranges from 21 to 88% depending on the study, the time of sampling, and the sensitivity of the assay used [18–29]. Confounding all these studies is the marked natural variation in breast milk cell numbers—which decrease from about 10^6/ml in colostrum to 10^2–10^3/ml in mature milk—as well as cellular composition. Several studies have assessed the role of proviral HIV DNA in breast milk in transmission to the infant. One study found an association between detection of breast milk HIV DNA at 15 days after birth and transmission—63% of transmitting mothers had detectable HIV DNA compared to only 39% of nontransmitters [22]. However, this study did not consider the timing of infection of the baby and so could not differentiate between in utero, peripartum, and breast milk transmission. Koulinska et al. [24] found that breast milk cell-associated HIV was predictive of transmission both before and after 9 months postpartum, while cell-free virions were only associated with transmission after 9 months. Rousseau et al. [25] found that each $\log_{10}$ increase in the number of proviral HIV DNA copies/million cell equivalents was associated with an increase in the risk of vertical transmission, and this association held after adjustment for plasma and breast milk HIV RNA levels. However, when the analysis was limited to the four infants known to be infected via breast milk, the results were no longer significant. In addition, two other reports found no association in the detection of proviral DNA and breast milk transmission. Guay et al. [19] reported that 80% of transmitters and 72% of nontransmitters had detectable proviral DNA in breast milk cells; for mothers of infants who were negative by p24 antigen or HIV DNA at birth, the transmission rate of those with detectable HIV DNA in breast milk was 13% compared to 7% in those with no detectable breast milk HIV DNA ($p=0.48$). Likewise, John et al. found overall detection rates of breast milk DNA of 79–82%, and no correlation with late postnatal infection (32 months of age) with an odds ratio (95% CI) of 0.9 (0.4–2.0) [29].

Cell-Free HIV RNA

Breast milk HIV RNA levels are typically two or more logs lower than matched blood plasma viral loads [23, 30, 32–34, 37, 40], but the two are significantly correlated [23, 32–34, 40, 41]. Detectable HIV RNA was found in cell-free milk of 37–89% of untreated women [23, 24, 30–32, 34, 38] and is more likely in women with advanced HIV disease. Viral loads in these studies ranged from undetectable to 5,000,000 copies/ml, although the mean or median viral loads were typically less than 1,000 copies/ml in all studies. The highest viral loads were observed in women with breast inflammation, subclinical mastitis, or cracked nipples [23, 30, 34–38, 40]. Intermittent shedding and discordant shedding of HIV RNA were observed in studies that collected samples at more than one time or from both breasts at the same time [30, 33, 38, 42], and there is disagreement as to how breast milk HIV levels vary during lactation in the absence of antiretrovirals (ARVs). Two studies found no change in the frequency of detection or mean/median viral load when samples were collected within 1 week of birth up to 3 months [32, 35]. One found that colostrum/early milk (0–10 days) had significantly higher viral loads compared to samples collected later [33]. A fourth study determined that the prevalence of breast milk HIV RNA was higher in mature milk (greater than 7 days) than in colostrum [31]. Abrupt cessation of breastfeeding has been associated with significant increases in HIV RNA, as well as clinical mastitis [43], underscoring how normal physiologic processes influence HIV entry into milk.

At least eight studies have evaluated the role of breast milk HIV RNA in transmission, with six finding a significant association with some caveats [24–26, 32–35, 40]. In one study, breast milk RNA levels were predictive only after 9 months [24], and in another only a trend towards an association with late postpartum transmission was found [35]. Cumulative exposure to cell-free HIV in breast milk, rather than the mode of breast feeding (exclusive vs. mixed) was associated with transmission in another study [40]. However, since cell-associated HIV was not evaluated in this study, its possible contribution remains unclear. Rousseau et al. [25] assessed both breast milk HIV RNA and DNA and found that the concentration of infected breast milk cells was marginally associated with breast milk transmission after adjustments for plasma and breast milk HIV RNA levels. However, this analysis only included four infected infants. Lastly, in a case–control study mastitis was positively related to both cell-free and cell-associated shedding, but only HIV DNA remained statistically significant in multivariate analyses [26].

Factors That Change HIV RNA and DNA Concentrations in Breast Milk

Effect of Mastitis

Mastitis has been associated with increased risk of transmission in several studies [23, 26, 34–38]. All studies that only evaluated the association between cell-free viral loads in breast milk and mastitis found significant associations [34–38]. Although an elevated Na/K ratio (>1) was positively correlated with both cell-free and cell-associated HIV shedding in breast milk, only cell-associated HIV DNA remained statistically significant in the multivariate analysis in one study [26]. In contrast, Gantt et al. [23] found that indicators of mastitis (Na>12 mmol/L, Na/K ratio>1, or total leukocyte count>10^6 cells/ml) were associated with increased breast milk HIV RNA load ($p<0.05$), but not with HIV DNA load. However, none of these mastitis surrogates were predictive of HIV RNA levels in breast milk. A third study found that mastitis was associated with postnatal transmission only when plasma, not breast milk, viral load was elevated [36].

Effect of Antiretrovirals on Breast Milk HIV RNA and DNA Levels

Of ten ARVs which have been evaluated to date (zidovudine, lamivudine, nevirapine, lopinavir, ritonavir, efavirenz, nelfinavir, indinavir, tenofovir, and emtricitabine) [44–51], only lamivudine (3TC) and zidovudine (ZDV) concentrate to any extent in breast milk, with breast milk levels two to eight times higher than concurrent plasma levels [46, 49–51]. Frequency of detection and levels of breast milk HIV RNA were lower in women who had received ZDV during gestation compared to those who received placebo [48]. Between days 3 and 21 postpartum, nevirapine (NVP) was associated with significantly greater suppression of breast milk HIV RNA compared to ZDV [52, 53] and this was associated with a significantly lower transmission rate at 6 weeks [53]. Lehman et al. [54] noted lower breast milk HIV RNA over 4 months postpartum in women receiving ZDV/3TC/NVP compared to those receiving ZDV alone, single-dose NVP alone, or single-dose NVP with ZDV. Breast milk HIV

RNA levels were significantly lower at delivery and day 7 in women who received ZDV/3TC/NVP antenatally compared to women who were untreated [55]. Increases in breast milk viral load have been observed after ARV discontinuation [48] and during weaning [43].

Lehman et al. [54] collected breast milk at several time points up to 4–6 weeks after delivery in 18 women taking HAART (ZDV/3TC/NVP). They demonstrated that HIV persists in breast milk cells despite HAART, while plasma and breast milk HIV RNA levels were suppressed. Similar results were reported by Shapiro et al. [56] in a study of 26 women on HAART who provided milk samples at 2 and 5 months postpartum.

The Debate: Cell Free Versus Cell Associated

It is clear that with the sensitive assays now available, both cell-free HIV RNA and cell-associated HIV DNA can be found in the breast milk of most HIV-infected women who are not taking ARVs. Depending on the study, either cell-free or cell-associated HIV can be correlated with breast milk transmission. Compounding the confusion is the fact that breast milk RNA and DNA levels are significantly correlated, with correlation coefficients ranging from 0.28 to 0.46 [23–25]. In vitro experiments [9–11, 57–66] and animal models [67–78] have produced conflicting data. Typically, cell-free virions suspended in PBS or culture fluid, not breast milk, have been used in in vivo models, but in vitro studies have indicated that cell-free HIV is inactivated in the presence of breast milk [11–17, 79], while infected breast milk cells demonstrated prolonged survival in breast milk [80]. Differences in animal models, study populations, time points, and assays also make direct comparisons among reports difficult. There have been only six reports where both HIV DNA and HIV RNA were measured in the same study [23–26, 46, 54]. One found that HIV DNA loads were not increased during mastitis, while HIV RNA levels were [23]. Two analyzed data from the same case-cohort study finding that only cell-associated breast milk viral load was significantly associated with mastitis and with transmission throughout the breastfeeding period [24, 26]. Similarly, in a large study conducted in Nairobi, each $\log_{10}$ increase in infected breast milk cells was associated with a threefold increase in the risk of transmission ($p=0.002$) [25]. However, when the analysis was limited to the four infants known to be infected via breastfeeding (i.e., they were known to be uninfected at or after 1 month of age), this finding lost significance. Two studies reported decreases in breast milk HIV RNA, but not cell-associated DNA, during antiretroviral treatment, suggesting that cell-free virus might be more important in transmission [24, 26]. However, the identification of proviral DNA in breast milk cells may not be indicative of ongoing replication, while the presence of HIV RNA is. The pool of long-lived infected macrophages and resting CD4 cells would probably persist despite the presence of ARVs.

However, it is possible that either cell-free or cell-associated virus may be responsible depending on factors such as the age of the infant at the time of transmission, breast feeding practices, mastitis or other inflammatory processes, innate immune factors in the milk, and other unknowns [81]. Differences observed among both the cell-free and cell-associated studies described above are undoubtedly due in large part to differences in the HIV assay used (lower limit of detection, variability, etc.), timing of collection of specimens, exclusive versus mixed breast feeding practices, sample processing, and possibly HIV subtypes.

When and Where Does Transmission Occur?

It is difficult to discriminate between perinatal and early breast milk HIV transmission. Although a meta-analysis suggested that the risk of breast milk transmission was relatively constant at about 0.9% per month after the first month of life [82], several individual studies have demonstrated that the risk of breast milk transmission is higher during early lactation [3, 31, 33, 83, 84]. For instance, Miotti et al. found the risk of breast milk transmission per month was 0.7% during age 1–5 months, 0.6% during age 6–11 months, and 0.3% during age 12–17 months [83]. As described above, whether HIV RNA loads are higher in early lactation or not is unclear [31–33, 35]. It will be difficult to assess this question in the future since guidelines now recommend ARV therapy for all pregnant women which reduces not only the plasma viral load, but also the colostrum/early milk HIV RNA levels. Two studies have both found that proviral HIV DNA is higher in colostrum and gradually decreases with time [21, 25]. Despite this, there was no significant difference in the concentration of infected breast milk cells among transmitting and nontransmitting mothers, although the number of infants known to have been infected via breastfeeding was very small [25].

There have been few investigations addressing where transmission occurs in the infant—oral mucosa, esophagus, or intestine. HIV RNA has been detected in gastric aspirates of neonates as a result of swallowing maternal blood and cervicovaginal secretions during delivery [85]. Gastric pH is neutral at birth which would reduce the potential inactivation of virions or infected cells. Moreover, the tremendous buffering capacity of breast milk seems to allow absorption of cytokines and even cells. Oral epithelial cells have been shown to be permissive to both cell-free and cell-associated HIV in vitro but the in vivo relevance is uncertain [66]. These findings may be relevant to early postnatal transmission via breastfeeding. The observation that mixed feeding more than doubles the risk of infection compared to exclusive breast feeding [4, 36, 86, 87] suggests that the gut may be the site of transmission [81]. The humanized mouse model may be able to shed light on this issue [88].

Why Does Transmission Happen? Viral Determinants of Breast Milk Transmission

Most of the early investigations of viral factors associated with vertical transmission of HIV were conducted in the USA and Europe where formula feeding was recommended for HIV-infected women. Later studies in Africa frequently did not discriminate in the analysis between infants infected in utero, peripartum, or via breastfeeding. As a consequence there are very few reports that address viral determinants of breast milk transmission.

Tropism

Single viral variants establish infection in the recipient when transmission of HIV occurs across a mucosal barrier (sexual, perinatal, or via breast feeding) [89–94], and almost all newly transmitted viruses have been shown to use the CCR5 co-receptor for entry into cells [93, 95, 96]. Few studies, however, have assessed these parameters for breast milk transmission [97, 98]. In a very small study, four infants infected via breast milk possessed HIV envelopes that used CCR5 as the co-receptor [97]. In vitro studies implicate CCR5-expressing memory CD4 T cells, breast-milk macrophages, or mammary epithelial cells as a potential source of CCR5 using HIV [10, 39, 99].

Envelope Sequence

To date there has been no evidence for signature sequence motifs that might explain why some virus variants are transmitted via breast feeding over others, and few differences between plasma or milk viral sequences in terms of potential N-linked glycosylation sites, shorter variable loops, or V regions lengths [100, 101], although others have described fewer potential glycosylation sites [94]. A study of genetic diversity in the reverse transcriptase genes of breast milk virus found more diversity in women treated with ARVs compared to untreated women, suggesting viral evolution in the breast under selective drug pressure [102].

Subtype

There is some indication that HIV subtype C might be preferentially transmitted in utero [103], although earlier studies had failed to find differences [104, 105]. A larger study found that subtype D was more commonly transmitted compared to subtype A [106], while another found that recombinant viruses, subtype C and subtype A were more commonly transmitted than subtype D when HIV *gag* (p24-p7), *env* (C2-C5), and LTR were used to classify subtypes [107, 108]. However, none of these reports differentiated between in utero, peripartum, and breastfeeding transmissions. More recently, it has been found that intersubtype recombinant genomes, especially recombination within the LTR, might render HIV-1 more fit for transmission via breast milk in comparison with nonrecombinant subtypes A, C, and D [109].

Replicative Capacity

Accumulated mutations in the RT and PR genes due to selective pressure from ARVs may lead to decreased viral replication capacity. Eshleman et al. [110] found that the average replication capacity of maternal blood isolates was higher in transmitters than in nontransmitters after adjustment for maternal HIV-1 load and other factors.

Compartmentalization

Breast milk and plasma HIV RNA viral loads have been shown to be correlated, although the amount in breast milk, in the absence of inflammation, is typically 100 times less, [23, 29, 30, 32] suggesting that there is limited mixing between these two anatomically distinct compartments. Studies have yielded conflicting results, probably due to differences in the assays used, sampling times, and/or HIV subtypes [102, 111–113]. More recent studies, several of which used single genome amplification techniques, which reduce the possibility of resampling and PCR-associated error and recombination, have concluded that the breast is not a separate compartment [100, 101, 114, 115]. In maximum likelihood analyses, all of these studies found that milk and plasma sequences were interspersed in most women with no evidence of compartmentalization in samples collected contemporaneously. In addition, monotypic virus variants were significantly more frequent in milk than in plasma suggesting local production of HIV in the breast from infected cells [100, 101, 114]. However, if breast milk and plasma samples collected 10 or more days apart were assessed, compartmentalization was usually observed [100]. Phylogenetic comparison of milk virus collected 8 weeks apart revealed synchronous viral evolution and new clonal expansion suggesting continuing seeding of the breast by blood variants [101]. Another study concluded that blood plasma was the unlikely source of the most recent common ancestor in half of the women tested [115]. These results suggest that there may be two distinct mechanisms for HIV to enter and populate the lactating breast [101]. First, there may be continual trafficking of either cell-free or cell-associated virus from the blood to the breast. Second, there appears to be transient local production of virus [100, 101] which may undergo independent evolution due to selective immune or drug pressures [116–121]. One caveat for all of these studies is that women with detectable breast milk viral loads were selected, and compartmentalization might be more evident in women with lower viral loads [100, 101, 114].

Resistance

Development and transmission of drug resistance in the genital tract have been observed for many years [122, 123]. Distinct patterns of drug resistance may arise in the genital tract since there may be subtherapeutic concentrations of ARVs in the genital compartment [124]. Similar events may occur in the breast, especially with the use of single-dose NVP, without the use of a week or two of combination ART to prevent the development of resistance [117].

Several studies have demonstrated the development of drug resistance in breast milk [116–121, 125]. NVP has a very long half life [126] and a low threshold to resistance [127] and thus it is not surprising that resistance to it develops readily [128]. This is especially true for women and their infants given single-dose NVP to prevent mother-to-child transmission of HIV [129, 130]. A meta-analysis estimated that 36% (19–76%) of women and 53% (36–87%) of infants had detectable resistance in plasma following single-dose NVP using population sequencing techniques [130]. Even more resistance in both women and infants is observed when more sensitive assays are employed [117, 131–133]. An additional risk is the development of NVP-resistant mutations in breast milk [117–121] which can then be transmitted to the infants [117, 125]. The specific resistance mutations observed in mother–infant pairs are frequently different [117, 125]. The prevalence of resistance mutations in breast milk following single-dose NVP ranges from 37 to 65% 4–12 weeks after delivery [118, 120, 121] and fades with time [121]. Women receiving HAART or a week of ZDV/3TC appear to develop less drug resistance in breast milk [116, 117]. The drug-associated mutations in breast milk can differ between right and left breasts and between plasma and breast milk [116–121].

Conclusions

Despite more than 25 years of studying the pathogenesis of breast milk transmission of HIV, many questions remain. Even in the absence of preventive measures, why do most breastfed infants escape infection? It is possible that either cell-free or cell-associated virus might be responsible for breast milk transmission, depending on factors such as timing of infection, inflammation, breast feeding practice, innate inhibitory properties of milk, and other unknown factors. Like other modes of HIV infection which involve crossing a mucosal barrier, most infections are established by a single CCR-5 using virion. Recent data suggest that there is little or no compartmentalization of breast milk variants, but do indicate local evolution of the virus within the breast. It is still unknown where in the infant transmission occurs—oral cavity, esophagus, or the intestinal tract. Understanding the factors that prevent vulnerable infants from acquiring HIV may provide valuable insights into vaccine and other preventive strategies.

References

1. Ziegler JB, Cooper DA, Johnson RO, Gold J (1985) Postnatal transmission of AIDS-associated retrovirus from mother to infant. Lancet 1:896–898
2. Thiry L, Sprecher-Goldberger S, Jonckheer T et al (1985) Isolation of AIDS virus from cell-free breast milk of three healthy virus carriers. Lancet 2:891–892

3. Nduati R, John G, Mbori-Ngacha D et al (2000) Effect of breastfeeding and formula feeding on transmission of HIV-1: a randomized clinical trial. JAMA 283:1167–1174
4. Kuhn L (2010) Milk mysteries: why are women who exclusively breast-feed less likely to transmit HIV during breast-feeding? Clin Infect Dis 50:770–772
5. Dunn DT, Newell ML, Ades AE, Peckham CS (1992) Risk of human immunodeficiency virus type 1 transmission through breastfeeding. Lancet 340:585–588
6. Richardson BA, John-Stewart GC, Hughes JP et al (2003) Breast-milk infectivity in human immunodeficiency virus type 1-infected mothers. J Infect Dis 187:736–740
7. Gray L, Fiscus S, Shugars D (2007) HIV-1 variants from a perinatal transmission pair demonstrate similar genetic and replicative properties in tonsillar tissues and peripheral blood mononuclear cells. AIDS Res Hum Retroviruses 23:1095–1104
8. Vogt MW, Witt DJ, Craven DE et al (1986) Isolation of HTLV-III/LAV from cervical secretions of women at risk for AIDS. Lancet 1:525–527
9. Southern SO (1998) Milk-borne transmission of HIV. Characterization of productively infected cells in breast milk and interactions between milk and saliva. J Hum Virol 1:328–337
10. Toniolo A, Serra C, Conaldi PG, Basolo F, Falcone V, Dolei A (1995) Productive HIV-1 infection of normal human mammary epithelial cells. AIDS 9:859–866
11. Lyimo MA, Howell AL, Balandya E, Eszterhas SK, Connor RI (2009) Innate factors in human breast milk inhibit cell-free HIV-1 but not cell-associated HIV-1 infection of CD4+ cells. J Acquir Immune Defic Syndr 51:117–124
12. Habte HH, de Beer C, Lotz ZE, Tyler MG, Kahn D, Mall AS (2008) Inhibition of human immunodeficiency virus type 1 activity by purified human breast milk mucin (MUC1) in an inhibition assay. Neonatology 93:162–170
13. Habte HH, Kotwal GJ, Lotz ZE et al (2007) Antiviral activity of purified human breast milk mucin. Neonatology 92:96–104
14. Villamor E, Koulinska IN, Furtado J et al (2007) Long-chain n-6 polyunsaturated fatty acids in breast milk decrease the risk of HIV transmission through breastfeeding. Am J Clin Nutr 86:682–689
15. Viveros-Rogel M, Soto-Ramirez L, Chaturvedi P, Newburg DS, Ruiz-Palacios GM (2004) Inhibition of HIV-1 infection in vitro by human milk sulfated glycolipids and glycosaminoglycans. Adv Exp Med Biol 554:481–487
16. Baron S, Poast J, Richardson CJ, Nguyen D, Cloyd M (2000) Oral transmission of human immunodeficiency virus by infected seminal fluid and milk: a novel mechanism. J Infect Dis 181:498–504
17. Martin V, Maldonado A, Fernandez L, Rodriguez JM, Connor RI (2010) Inhibition of human immunodeficiency virus type 1 by lactic acid bacteria from human breastmilk. Breastfeed Med 5:153–158
18. Buransin P, Kunakorn M, Petchlai B et al (1993) Detection of human immunodeficiency virus type 1 (HIV-1) proviral DNA in breast milk and colostrum of seropositive mothers. J Med Assoc Thai 76:41–44
19. Guay LA, Hom DL, Mmiro F et al (1996) Detection of human immunodeficiency virus type 1 (HIV-1) DNA and p24 antigen in breast milk of HIV-1-infected Ugandan women and vertical transmission. Pediatrics 98:438–444
20. Nduati RW, John GC, Richardson BA et al (1995) Human immunodeficiency virus type 1-infected cells in breast milk: association with immunosuppression and vitamin A deficiency. J Infect Dis 172:1461–1468
21. Ruff AJ, Coberly J, Halsey NA et al (1994) Prevalence of HIV-1 DNA and p24 antigen in breast milk and correlation with maternal factors. J Acquir Immune Defic Syndr 7:68–73
22. Van de Perre P, Simonon A, Hitimana DG et al (1993) Infective and anti-infective properties of breastmilk from HIV-1-infected women. Lancet 341:914–918
23. Gantt S, Shetty AK, Seidel KD et al (2007) Laboratory indicators of mastitis are not associated with elevated HIV-1 DNA loads or predictive of HIV-1 RNA loads in breast milk. J Infect Dis 196:570–576
24. Koulinska IN, Villamor E, Chaplin B et al (2006) Transmission of cell-free and cell-associated HIV-1 through breast-feeding. J Acquir Immune Defic Syndr 41:93–99
25. Rousseau CM, Nduati RW, Richardson BA et al (2004) Association of levels of HIV-1-infected breast milk cells and risk of mother-to-child transmission. J Infect Dis 190:1880–1888
26. Kantarci S, Koulinska IN, Aboud S, Fawzi WW, Villamor E (2007) Subclinical mastitis, cell-associated HIV-1 shedding in breast milk, and breast-feeding transmission of HIV-1. J Acquir Immune Defic Syndr 46:651–654
27. Villamor E, Koulinska IN, Aboud S et al (2010) Effect of vitamin supplements on HIV shedding in breast milk. Am J Clin Nutr 92:881–886
28. Vonesch N, Sturchio E, Humani AC, Fei PC, Coszena D, Caprilli F, Pezzella M (1992) Detection of HIV-1 genome in leukocytes of human colostrum from anti-HIV-1 seropositive mothers. AIDS Res Hum Retroviruses 8:1283–1287
29. John GC, Nduati RW, Mbori-Ngacha DA et al (2001) Correlates of mother-to-child human immunodeficiency virus type 1 (HIV-1) transmission: association with maternal plasma HIV-1 RNA load, genital HIV-1 DNA shedding, and breast infections. J Infect Dis 183:206–212

30. Hoffman IF, Martinson FE, Stewart PW et al (2003) Human immunodeficiency virus type 1 RNA in breast-milk components. J Infect Dis 188:1209–1212
31. Lewis P, Nduati R, Kreiss JK et al (1998) Cell-free human immunodeficiency virus type 1 in breast milk. J Infect Dis 177:34–39
32. Pillay K, Coutsoudis A, York D, Kuhn L, Coovadia HM (2000) Cell-free virus in breast milk of HIV-1-seropositive women. J Acquir Immune Defic Syndr 24:330–336
33. Rousseau CM, Nduati RW, Richardson BA et al (2003) Longitudinal analysis of human immunodeficiency virus type 1 RNA in breast milk and of its relationship to infant infection and maternal disease. J Infect Dis 187:741–747
34. Semba RD, Kumwenda N, Hoover DR et al (1999) Human immunodeficiency virus load in breast milk, mastitis, and mother-to-child transmission of human immunodeficiency virus type 1. J Infect Dis 180:93–98
35. Willumsen JF, Filteau SM, Coutsoudis A et al (2003) Breastmilk RNA viral load in HIV-infected South African women: effects of subclinical mastitis and infant feeding. AIDS 17:407–414
36. Lunney KM, Iliff P, Mutasa K et al (2010) Associations between breast milk viral load, mastitis, exclusive breast-feeding, and postnatal transmission of HIV. Clin Infect Dis 50:762–769
37. Phiri W, Kasonka L, Collin S et al (2006) Factors influencing breast milk HIV RNA viral load among Zambian women. AIDS Res Hum Retroviruses 22:607–614
38. Semrau K, Ghosh M, Kankasa C et al (2008) Temporal and lateral dynamics of HIV shedding and elevated sodium in breast milk among HIV-positive mothers during the first 4 months of breast-feeding. J Acquir Immune Defic Syndr 47:320–328
39. Petitjean G, Becquart P, Tuaillon E et al (2007) Isolation and characterization of HIV-1-infected resting CD4+ T lymphocytes in breast milk. J Clin Virol 39:1–8
40. Neveu D, Viljoen J, Bland RM et al (2011) Cumulative exposure to cell-free HIV in breast milk, rather than feeding pattern per se, identifies postnatally infected infants. Clin Infect Dis 52:819–825
41. Chung MH, Kiarie JN, Richardson BA et al (2007) Independent effects of nevirapine prophylaxis and HIV-1 RNA suppression in breast milk on early perinatal HIV-1 transmission. J Acquir Immune Defic Syndr 46:472–478
42. Willumsen JF, Newell ML, Filteau SM et al (2001) Variation in breastmilk HIV-1 viral load in left and right breasts during the first 3 months of lactation. AIDS 15:1896–1898
43. Thea DM, Aldrovandi G, Kankasa C et al (2006) Post-weaning breast milk HIV-1 viral load, blood prolactin levels and breast milk volume. AIDS 20:1539–1547
44. Benaboud S, Pruvost A, Coffie PA et al (2011) Concentrations of tenofovir and emtricitabine in breast milk of HIV-1-infected women in Abidjan, Cote d'Ivoire, in the ANRS 12109 TEmAA Study, step 2. Antimicrob Agents Chemother 55:1315–1317
45. Schneider S, Peltier A, Gras A et al (2008) Efavirenz in human breast milk, mothers', and newborns' plasma. J Acquir Immune Defic Syndr 48:450–454
46. Shapiro RL, Holland DT, Capparelli E et al (2005) Antiretroviral concentrations in breast-feeding infants of women in Botswana receiving antiretroviral treatment. J Infect Dis 192:720–727
47. Colebunders R, Hodossy B, Burger D et al (2005) The effect of highly active antiretroviral treatment on viral load and antiretroviral drug levels in breast milk. AIDS 19:1912–1915
48. Manigart O, Crepin M, Leroy V et al (2004) Effect of perinatal zidovudine prophylaxis on the evolution of cell-free HIV-1 RNA in breast milk and on postnatal transmission. J Infect Dis 190:1422–1428
49. Mirochnick M, Thomas T, Capparelli E et al (2009) Antiretroviral concentrations in breast-feeding infants of mothers receiving highly active antiretroviral therapy. Antimicrob Agents Chemother 53:1170–1176
50. Corbett A, Martinson F, Rezk N, Kashuba A, Jamieson D, Chasela C, Kayira D, Tegha G, Kamwendo D, van der Horst C, BAN Study Team. (2009) Lopinavir/ritonavir concentrations in breast milk and breast-feeding infants 16th CROI. Montreal, Canada
51. Corbett A, Kashuba A, Rezk N, Jamieson D, Chasela C, Hyde L, Tegha G, Joaki G, Kamwendo D, van der Horst C, BAN Study Team (2008) Antiretroviral drug concentrations in breast milk and breastfeeding infants. 15th CROI. Boston, MA
52. Rossenkhan R, Ndung'u T, Sebunya TK et al (2009) Temporal reduction of HIV type 1 viral load in breast milk by single-dose nevirapine during prevention of MTCT. AIDS Res Hum Retroviruses 25:1261–1264
53. Chung MH, Kiarie JN, Richardson BA, Lehman DA, Overbaugh J, John-Stewart GC (2005) Breast milk HIV-1 suppression and decreased transmission: a randomized trial comparing HIVNET 012 nevirapine versus short-course zidovudine. AIDS 19:1415–1422
54. Lehman DA, Chung MH, John-Stewart GC et al (2008) HIV-1 persists in breast milk cells despite antiretroviral treatment to prevent mother-to-child transmission. AIDS 22:1475–1485
55. Giuliano M, Guidotti G, Andreotti M et al (2007) Triple antiretroviral prophylaxis administered during pregnancy and after delivery significantly reduces breast milk viral load: a study within the Drug Resource Enhancement Against AIDS and Malnutrition Program. J Acquir Immune Defic Syndr 44:286–291

56. Shapiro RL, Ndung'u T, Lockman S et al (2005) Highly active antiretroviral therapy started during pregnancy or postpartum suppresses HIV-1 RNA, but not DNA, in breast milk. J Infect Dis 192:713–719
57. Alfsen A, Yu H, Magerus-Chatinet A, Schmitt A, Bomsel M (2005) HIV-1-infected blood mononuclear cells form an integrin- and agrin-dependent viral synapse to induce efficient HIV-1 transcytosis across epithelial cell monolayer. Mol Biol Cell 16:4267–4279
58. Lagaye S, Derrien M, Menu E et al (2001) Cell-to-cell contact results in a selective translocation of maternal human immunodeficiency virus type 1 quasispecies across a trophoblastic barrier by both transcytosis and infection. J Virol 75:4780–4791
59. Lehman DA, Farquhar C (2007) Biological mechanisms of vertical human immunodeficiency virus (HIV-1) transmission. Rev Med Virol 17:381–403
60. Phillips DM (1994) The role of cell-to-cell transmission in HIV infection. AIDS 8:719–731
61. Becquart P, Hocini H, Levy M, Sepou A, Kazatchkine MD, Belec L (2000) Secretory anti-human immunodeficiency virus (HIV) antibodies in colostrum and breast milk are not a major determinant of the protection of early postnatal transmission of HIV. J Infect Dis 181:532–539
62. Bomsel M, Alfsen A (2003) Entry of viruses through the epithelial barrier: pathogenic trickery. Nat Rev Mol Cell Biol 4:57–68
63. Bosire R, John-Stewart GC, Mabuka JM et al (2007) Breast milk alpha-defensins are associated with HIV type 1 RNA and CC chemokines in breast milk but not vertical HIV type 1 transmission. AIDS Res Hum Retroviruses 23:198–203
64. Farquhar C, Mbori-Ngacha DA, Redman MW et al (2005) CC and CXC chemokines in breastmilk are associated with mother-to-child HIV-1 transmission. Curr HIV Res 3:361–369
65. Kuhn L, Trabattoni D, Kankasa C et al (2006) HIV-specific secretory IgA in breast milk of HIV-positive mothers is not associated with protection against HIV transmission among breast-fed infants. J Pediatr 149:611–616
66. Moore JS, Rahemtulla F, Kent LW et al (2003) Oral epithelial cells are susceptible to cell-free and cell-associated HIV-1 infection in vitro. Virology 313:343–353
67. Burkhard MJ, Dean GA (2003) Transmission and immunopathogenesis of FIV in cats as a model for HIV. Curr HIV Res 1:15–29
68. Ruprecht RM, Baba TW, Liska V et al (1999) Oral transmission of primate lentiviruses. J Infect Dis 179(Suppl 3):S408–S412
69. Otsyula MG, Miller CJ, Tarantal AF et al (1996) Fetal or neonatal infection with attenuated simian immunodeficiency virus results in protective immunity against oral challenge with pathogenic SIVmac251. Virology 222:275–278
70. Burkhard MJ, Obert LA, O'Neil LL, Diehl LJ, Hoover EA (1997) Mucosal transmission of cell-associated and cell-free feline immunodeficiency virus. AIDS Res Hum Retroviruses 13:347–355
71. O'Neil LL, Burkhard MJ, Hoover EA (1996) Frequent perinatal transmission of feline immunodeficiency virus by chronically infected cats. J Virol 70:2894–2901
72. Obert LA, Hoover EA (2000) Feline immunodeficiency virus clade C mucosal transmission and disease courses. AIDS Res Hum Retroviruses 16:677–688
73. Howard KE, Burkhard MJ (2007) Mucosal challenge with cell-associated or cell-free feline immunodeficiency virus induces rapid and distinctly different patterns of phenotypic change in the mucosal and systemic immune systems. Immunology 122:571–583
74. Anderson DJ (2010) Finally, a macaque model for cell-associated SIV/HIV vaginal transmission. J Infect Dis 202:333–336
75. Sellon RK, Jordan HL, Kennedy-Stoskopf S, Tompkins MB, Tompkins WA (1994) Feline immunodeficiency virus can be experimentally transmitted via milk during acute maternal infection. J Virol 68:3380–3385
76. Van Rompay KK, Abel K, Lawson JR et al (2005) Attenuated poxvirus-based simian immunodeficiency virus (SIV) vaccines given in infancy partially protect infant and juvenile macaques against repeated oral challenge with virulent SIV. J Acquir Immune Defic Syndr 38:124–134
77. Van Rompay KK, Greenier JL, Cole KS et al (2003) Immunization of newborn rhesus macaques with simian immunodeficiency virus (SIV) vaccines prolongs survival after oral challenge with virulent SIVmac251. J Virol 77:179–190
78. Van Rompay KK, Schmidt KA, Lawson JR, Singh R, Bischofberger N, Marthas ML (2002) Topical administration of low-dose tenofovir disoproxil fumarate to protect infant macaques against multiple oral exposures of low doses of simian immunodeficiency virus. J Infect Dis 186:1508–1513
79. Kourtis AP, Bulterys M (2010) Mother-to-child transmission of HIV: pathogenesis, mechanisms and pathways. Clin Perinatol 37:721–737, vii
80. Yamaguchi K, Sugiyama T, Takizawa M, Yamamoto N, Honda M, Natori M (2007) Viability of infectious viral particles of HIV and BMCs in breast milk. J Clin Virol 39:222–225
81. Walter J, Kuhn L, Aldrovandi GM (2008) Advances in basic science understanding of mother-to-child HIV-1 transmission. Curr Opin HIV AIDS 3:146–150

82. Coutsoudis A, Dabis F, Fawzi W et al (2004) Late postnatal transmission of HIV-1 in breast-fed children: an individual patient data meta-analysis. J Infect Dis 189:2154–2166
83. Miotti PG, Taha TE, Kumwenda NI et al (1999) HIV transmission through breastfeeding: a study in Malawi. JAMA 282:744–749
84. Moodley D, Moodley J, Coovadia H et al (2003) A multicenter randomized controlled trial of nevirapine versus a combination of zidovudine and lamivudine to reduce intrapartum and early postpartum mother-to-child transmission of human immunodeficiency virus type 1. J Infect Dis 187:725–735
85. Mandelbrot L, Burgard M, Teglas JP et al (1999) Frequent detection of HIV-1 in the gastric aspirates of neonates born to HIV-infected mothers. AIDS 13:2143–2149
86. Coovadia HM, Rollins NC, Bland RM et al (2007) Mother-to-child transmission of HIV-1 infection during exclusive breastfeeding in the first 6 months of life: an intervention cohort study. Lancet 369:1107–1116
87. Kuhn L, Sinkala M, Kankasa C et al (2007) High uptake of exclusive breastfeeding and reduced early post-natal HIV transmission. PLoS One 2:e1363
88. Olesen R, Wahl A, Denton PW, Garcia JV (2011) Immune reconstitution of the female reproductive tract of humanized BLT mice and their susceptibility to human immunodeficiency virus infection. J Reprod Immunol 88:195–203
89. Zhu T, Mo H, Wang N et al (1993) Genotypic and phenotypic characterization of HIV-1 patients with primary infection. Science 261:1179–1181
90. Zhang LQ, MacKenzie P, Cleland A, Holmes EC, Brown AJ, Simmonds P (1993) Selection for specific sequences in the external envelope protein of human immunodeficiency virus type 1 upon primary infection. J Virol 67: 3345–3356
91. Wolinsky SM, Wike CM, Korber BT et al (1992) Selective transmission of human immunodeficiency virus type-1 variants from mothers to infants. Science 255:1134–1137
92. Derdeyn CA, Decker JM, Bibollet-Ruche F et al (2004) Envelope-constrained neutralization-sensitive HIV-1 after heterosexual transmission. Science 303:2019–2022
93. Keele BF, Giorgi EE, Salazar-Gonzalez JF et al (2008) Identification and characterization of transmitted and early founder virus envelopes in primary HIV-1 infection. Proc Natl Acad Sci USA 105:7552–7557
94. Zhang H, Tully DC, Hoffmann FG, He J, Kankasa C, Wood C (2010) Restricted genetic diversity of HIV-1 subtype C envelope glycoprotein from perinatally infected Zambian infants. PLoS One 5:e9294
95. Margolis L, Shattock R (2006) Selective transmission of CCR5-utilizing HIV-1: the 'gatekeeper' problem resolved? Nat Rev Microbiol 4:312–317
96. Salvatori F, Scarlatti G (2001) HIV type 1 chemokine receptor usage in mother-to-child transmission. AIDS Res Hum Retroviruses 17:925–935
97. Rainwater SM, Wu X, Nduati R et al (2007) Cloning and characterization of functional subtype A HIV-1 envelope variants transmitted through breastfeeding. Curr HIV Res 5:189–197
98. Mulder-Kampinga GA, Simonon A, Kuiken CL et al (1995) Similarity in env and gag genes between genomic RNAs of human immunodeficiency virus type 1 (HIV-1) from mother and infant is unrelated to time of HIV-1 RNA positivity in the child. J Virol 69:2285–2296
99. Satomi M, Shimizu M, Shinya E et al (2005) Transmission of macrophage-tropic HIV-1 by breast-milk macrophages via DC-SIGN. J Infect Dis 191:174–181
100. Heath L, Conway S, Jones L et al (2010) Restriction of HIV-1 genotypes in breast milk does not account for the population transmission genetic bottleneck that occurs following transmission. PLoS One 5:e10213
101. Salazar-Gonzalez JF, Salazar MG, Learn GH et al (2011) Origin and evolution of HIV-1 in breast milk determined by single-genome amplification and sequencing. J Virol 85:2751–2763
102. Andreotti M, Galluzzo CM, Guidotti G et al (2009) Comparison of HIV type 1 sequences from plasma, cell-free breast milk, and cell-associated breast milk viral populations in treated and untreated women in Mozambique. AIDS Res Hum Retroviruses 25:707–711
103. Renjifo B, Gilbert P, Chaplin B et al (2004) Preferential in-utero transmission of HIV-1 subtype C as compared to HIV-1 subtype A or D. AIDS 18:1629–1636
104. Tapia N, Franco S, Puig-Basagoiti F et al (2003) Influence of human immunodeficiency virus type 1 subtype on mother-to-child transmission. J Gen Virol 84:607–613
105. Murray MC, Embree JE, Ramdahin SG, Anzala AO, Njenga S, Plummer FA (2000) Effect of human immunodeficiency virus (HIV) type 1 viral genotype on mother-to-child transmission of HIV-1. J Infect Dis 181: 746–749
106. Yang C, Li M, Newman RD et al (2003) Genetic diversity of HIV-1 in western Kenya: subtype-specific differences in mother-to-child transmission. AIDS 17:1667–1674
107. Renjifo B, Fawzi W, Mwakagile D et al (2001) Differences in perinatal transmission among human immunodeficiency virus type 1 genotypes. J Hum Virol 4:16–25
108. Blackard JT, Renjifo B, Fawzi W et al (2001) HIV-1 LTR subtype and perinatal transmission. Virology 287:261–265

109. Koulinska IN, Villamor E, Msamanga G et al (2006) Risk of HIV-1 transmission by breastfeeding among mothers infected with recombinant and non-recombinant HIV-1 genotypes. Virus Res 120:191–198
110. Eshleman SH, Lie Y, Hoover DR et al (2006) Association between the replication capacity and mother-to-child transmission of HIV-1, in antiretroviral drug-naive Malawian women. J Infect Dis 193:1512–1515
111. Becquart P, Chomont N, Roques P et al (2002) Compartmentalization of HIV-1 between breast milk and blood of HIV-infected mothers. Virology 300:109–117
112. Becquart P, Courgnaud V, Willumsen J, Van de Perre P (2007) Diversity of HIV-1 RNA and DNA in breast milk from HIV-1-infected mothers. Virology 363:256–260
113. Henderson GJ, Hoffman NG, Ping LH et al (2004) HIV-1 populations in blood and breast milk are similar. Virology 330:295–303
114. Gantt S, Carlsson J, Heath L et al (2010) Genetic analyses of HIV-1 env sequences demonstrate limited compartmentalization in breast milk and suggest viral replication within the breast that increases with mastitis. J Virol 84:10812–10819
115. Gray RR, Salemi M, Lowe A et al (2011) Multiple independent lineages of HIV-1 persist in breast milk and plasma. AIDS 25:143–152
116. Andreotti M, Guidotti G, Galluzzo CM et al (2007) Resistance mutation patterns in plasma and breast milk of HIV-infected women receiving highly-active antiretroviral therapy for mother-to-child transmission prevention. AIDS 21:2360–2362
117. Farr SL, Nelson JA, Ng'ombe TJ et al (2010) Addition of 7 days of zidovudine plus lamivudine to peripartum single-dose nevirapine effectively reduces nevirapine resistance postpartum in HIV-infected mothers in Malawi. J Acquir Immune Defic Syndr 54:515–523
118. Hudelson SE, McConnell MS, Bagenda D et al (2010) Emergence and persistence of nevirapine resistance in breast milk after single-dose nevirapine administration. AIDS 24:557–561
119. Kassaye S, Lee E, Kantor R et al (2007) Drug resistance in plasma and breast milk after single-dose nevirapine in subtype C HIV type 1: population and clonal sequence analysis. AIDS Res Hum Retroviruses 23:1055–1061
120. Lee EJ, Kantor R, Zijenah L et al (2005) Breast-milk shedding of drug-resistant HIV-1 subtype C in women exposed to single-dose nevirapine. J Infect Dis 192:1260–1264
121. Pilger D, Hauser A, Kuecherer C et al (2011) Minor drug-resistant HIV type-1 variants in breast milk and plasma of HIV type-1-infected Ugandan women after nevirapine single-dose prophylaxis. Antivir Ther 16:109–113
122. Kroodsma KL, Kozal MJ, Hamed KA, Winters MA, Merigan TC (1994) Detection of drug resistance mutations in the human immunodeficiency virus type 1 (HIV-1) pol gene: differences in semen and blood HIV-1 RNA and proviral DNA. J Infect Dis 170:1292–1295
123. Eron JJ, Vernazza PL, Johnston DM et al (1998) Resistance of HIV-1 to antiretroviral agents in blood and seminal plasma: implications for transmission. AIDS 12:F181–F189
124. Kashuba AD, Dyer JR, Kramer LM, Raasch RH, Eron JJ, Cohen MS (1999) Antiretroviral-drug concentrations in semen: implications for sexual transmission of human immunodeficiency virus type 1. Antimicrob Agents Chemother 43:1817–1826
125. Moorthy A, Gupta A, Bhosale R et al (2009) Nevirapine resistance and breast-milk HIV transmission: effects of single and extended-dose nevirapine prophylaxis in subtype C HIV-infected infants. PLoS One 4:e4096
126. Shetty AK, Coovadia HM, Mirochnick MM et al (2003) Safety and trough concentrations of nevirapine prophylaxis given daily, twice weekly, or weekly in breast-feeding infants from birth to 6 months. J Acquir Immune Defic Syndr 34:482–490
127. Richman D, Shih CK, Lowy I et al (1991) Human immunodeficiency virus type 1 mutants resistant to nonnucleoside inhibitors of reverse transcriptase arise in tissue culture. Proc Natl Acad Sci USA 88:11241–11245
128. Richman DD, Havlir D, Corbeil J et al (1994) Nevirapine resistance mutations of human immunodeficiency virus type 1 selected during therapy. J Virol 68:1660–1666
129. Eshleman SH, Mracna M, Guay LA et al (2001) Selection and fading of resistance mutations in women and infants receiving nevirapine to prevent HIV-1 vertical transmission (HIVNET 012). AIDS 15:1951–1957
130. Arrive E, Newell ML, Ekouevi DK et al (2007) Prevalence of resistance to nevirapine in mothers and children after single-dose exposure to prevent vertical transmission of HIV-1: a meta-analysis. Int J Epidemiol 36:1009–1021
131. Flys T, Nissley DV, Claasen CW et al (2005) Sensitive drug-resistance assays reveal long-term persistence of HIV-1 variants with the K103N nevirapine (NVP) resistance mutation in some women and infants after the administration of single-dose NVP: HIVNET 012. J Infect Dis 192:24–29
132. Church JD, Huang W, Parkin N et al (2009) Comparison of laboratory methods for analysis of non-nucleoside reverse transcriptase inhibitor resistance in Ugandan infants. AIDS Res Hum Retroviruses 25:657–663
133. Hammer SM (2005) Single-dose nevirapine and drug resistance: the more you look, the more you find. J Infect Dis 192:1–3

Chapter 6
HIV-1 Resistance to Antiretroviral Agents: Relevance to Mothers and Infants in the Breastfeeding Setting

Michelle S. McConnell and Paul Palumbo

HIV is a retrovirus whose genetic material is comprised of RNA. It is a decidedly adaptable virus with properties of high replication rates in association with an RNA copying enzyme—reverse transcriptase—which possesses a relatively high copying error rate, in the order of 1 error or mutation for every 10,000 nucleotides copied. This translates to a single mutation for every viral replication event on average in the setting of a billion viral copies produced daily in an infected individual. As such, the viral quasispecies (the pool of viral variants present at a given time) can and does adapt rapidly to environmental pressures such as the immune response or antiretroviral agents. It should not be surprising, therefore, that as antiretroviral agents have been developed and deployed, mutations in HIV genes associated with ARV resistance have rapidly been detected, underscoring the strategy of multiple agent and multiple class ARV therapy known as highly active antiretroviral therapy (HAART).

Prevention of mother-to-child HIV transmission (PMTCT) has been one of the remarkable success stories in the era of HIV infection beginning with the ACTG 076 trial which features use of AZT [1], to the HIVNET 012 trial featuring single-dose nevirapine [2], to the use of maternal HAART. Current PMTCT standards result in approximately 1% transmission rates between an HIV-infected pregnant woman and her newborn in resource-rich settings [3–5]. PMTCT successes have also been recognized in resource-limited settings although on a more modest scale. Challenges in the form of cost, limited healthcare infrastructure, and the need to continue breastfeeding practices despite HIV transmission risk have been difficult to overcome. This chapter focuses on the unique aspects of viral resistance in the complex setting of breastfeeding and the use of ARVs for both prevention and treatment of HIV infection.

Detection of Resistance: Population Sequencing Versus Ultrasensitive Assays

The technology for the detection of HIV resistance to antiretroviral agents in human specimens, primarily blood plasma, was established and validated in the 1990s and, to some extent, has changed little to the present day [6]. The dominant, commercially available assays (genotyping) employ a technique

M.S. McConnell, MD
HCMC Office, US Centers for Disease Control and Prevention, Ho Chi Minh City, Vietnam

P. Palumbo, MD (✉)
Section of Infectious Diseases and International Health, Dartmouth Medical School,
1 Medical Center Drive, Lebanon, NH 03756, USA
e-mail: paul.palumbo@dartmouth.edu

referred to as "population sequencing" in which viral RNA in plasma samples is reverse transcribed to viral DNA, massively PCR-amplified, and then undergoes DNA sequencing. Mutations in the viral reverse transcriptase and protease genes, which are associated with resistance to antiretroviral agents of the nucleoside reverse transcriptase inhibitor (NRTI), nonnucleoside reverse transcriptase inhibitor (NNRTI), and protease inhibitor (PI) classes are then scored and packaged into a resistance profile report. Population sequencing exhibits sensitivity limits in that it detects the "dominant" sequence within the viral quasispecies pool present in a blood sample, that is, a mutation must be present in at least 20% of the virions in a sample for it to be detected. Minor viral variants which are present at frequencies less than 20% will not be detected yet may contribute to clinical resistance once exposed to antiretroviral agents. Extensions of the base assay have been developed for the detection of resistance to the newer classes of ARVs—integrase inhibitors and cell binding/fusion inhibitors.

It has recently been reported that minority resistance variants do indeed contribute to adverse clinical outcomes although lower thresholds of clinical importance have not been established [7, 8]. Multiple assays have surfaced in the research setting with enhanced, or "ultrasensitive," resistance detection capabilities. In addition, resistance testing (basic or ultrasensitive) is being applied to breast milk samples targeting either cell-free viral RNA or cell-associated proviral DNA.

Relationship of HIV in Blood and Breast Milk: Open, Communicating Systems Versus Distinct Compartments

Clinician investigators have long been concerned that regional compartments within an HIV-infected individual may possess characteristics which allow for the independent evolution of the viral quasispecies when compared with the systemic circulation. Examples of such possible compartments include the central nervous system (CNS), genital tract, and breast. What properties might comprise such a compartment? Classically, a distinct compartment would offer limited or negligible penetration by ARVs and might possess depressed or absent elements of the immune system. The CNS is the best example of a compartment with many ARVs demonstrating altered penetration and pharmacokinetic properties. Locally replicating virions clearly evolve with respect to a different set of selective pressures when compared with those in the systemic circulation.

Extensive studies of the "compartment" properties of the mammary gland and breast milk have been conducted and reported in recent years. ARV drug levels in maternal plasma, breast milk, and in infants via breast milk exposure demonstrate a complex pharmacokinetic profile reflective of plasma and breast milk half-life characteristics and sampling time point variability. The very short half-life of AZT in adults results in delayed and low concentrations achieved in breast milk [9, 10], although median breast milk/plasma ratios range from 0.5 to >1.0. This results in very low AZT exposure in breastfeeding infants, estimated at more than 1,000 times lower than the recommended infant dose for PMTCT [9]. The longer plasma half-life of lamivudine results in breast milk/plasma ratios ranging from 1.25 to >5 [9–11] and in detectable levels which approach the HIV inhibitory concentration (IC_{50}) in the breastfeeding infant.

Nevirapine has the longest half-life of the commonly used ARVs, a critical property for its effectiveness as a PMTCT agent and also a liability regarding emergence of resistance. Breast milk/maternal plasma ratios have been reported at about 0.7 [9, 11, 12] and infant NVP concentrations resulting from breast milk exposure at a range of 200–900 ng/mL which is well above the viral IC_{50}. The detectable infant levels of lamivudine and NVP acquired from breastfeeding exposure represent a risk for selection of resistance should the infant be or become HIV-infected. The protease inhibitors, such as atazanavir, lopinavir/ritonavir, and nelfinavir have also been evaluated and appear to have breast milk/maternal plasma concentration ratios of 0.1 with minimal to no detection in the breastfeeding infant [10, 11, 13].

Recent analyses of comparative viral genetic evolution in maternal plasma versus breast milk have resulted in considerable insight as well as debate. Clonal analyses of viral quasispecies within a sample can be used to construct phylogenetic (relatedness) trees. Results from such analyses have been used to assess whether "compartments" are evolving independently or whether maternal plasma and breast milk represent an intermixing continuum. Some investigators have generated data which support the compartment model with phylogenetic trees demonstrating distinct lineages evolving in maternal plasma versus breast milk [14–17]. A recent report from Heath and Aldrovandi emphasized the importance of joint sample timing in that breast milk, and plasma samples obtained at the same time were highly related but, if they were obtained ≥10 days apart, compartmentalization was often inferred [18]. This underscores the rapid evolution of HIV genetic sequences and the care with which genotypic analyses must be undertaken. Despite supporting general intermixing and lack of compartmentalization, three investigator groups recently reported evidence of local clonal expansion and the possibility of the development of discrete lineages within the breast [18–20]. We will have to stay tuned as the debate and investigations continue—it may, in fact, not be an "all-or-none" phenomenon.

Settings Which May Predispose to the Emergence of Resistant Virus in Blood and Breast Milk

In the USA, Europe, and most resource-rich settings, recommendations for pregnant HIV-infected women are to avoid breastfeeding completely. However, in many resource-limited settings, due to unavailability of formula, limited or no access to clean water, stigma associated with formula use, and high infant morbidity and mortality rates due to diarrheal diseases and pneumonia, continued breastfeeding by HIV-infected mothers is recommended. To reduce the risk of HIV transmission through breastfeeding, a number of antiretroviral interventions have been identified for PMTCT during the breastfeeding period.

Until the most recent update to the WHO recommendations [21], many women in resource-limited settings who did not require antiretroviral treatment for their own health received either single-dose nevirapine alone, or combination prophylaxis with zidovudine and lamivudine plus single-dose nevirapine in labor. In a study that randomized women to single-dose nevirapine or short-course zidovudine, it was reported that breast milk HIV-1 RNA levels and nevirapine use were independently associated with reductions in transmission of HIV-1 [22]. This suggests that the efficacy of single-dose nevirapine for reduction of HIV transmission is, in part, by way of reducing breast milk RNA levels, in addition to nevirapine in plasma. Other studies have corroborated these findings but also reported that nevirapine levels in plasma and breast milk following single-dose nevirapine persist beyond 2 weeks following delivery [12, 23, 24], thereby potentially leading to emergence of resistant virus with continued breastfeeding. In short, the presence of ante- and intrapartum ARVs in breast milk, while helping to reduce transmission of HIV through breast milk, may also lead to the development of resistant virus.

More recently, maternal ARVs given during the breastfeeding period have shown to be efficacious in terms of reduction of transmission, and the importance of these regimens in reducing postnatal HIV transmission cannot be understated. However, ARVs delivered during the breastfeeding period also pose a risk of development of mutant virus. The various clinical trials and regimens for prevention of breastfeeding HIV transmission, using a combination of maternal ARV initiated in the third trimester and continuing postpartum, all reported significant reductions in postnatal HIV transmission as compared to standard antepartum and intrapartum regimens [25–30].

The other principal approach to prevention of breastfeeding HIV transmission has been infant prophylaxis and studies have similarly promising findings. The Six Week Extended Dose Nevirapine (SWEN) study reported a reduction in infant HIV infection rates from 9% to 6.9% at 6 months among infants receiving 6 weeks of nevirapine [31], and the Post-Exposure Prophylaxis of Infants (PEPI) study reported a 66% reduction in HIV infection among infants who received 14 weeks of nevirapine or nevirapine plus zidovudine [32]. The Mitra study provided daily lamivudine to breastfeeding infants, and reported a postnatal HIV transmission rate between 6 weeks and 6 months of 1.1% [33]. Finally, the BAN study provided daily NVP to infants during 28 weeks of breastfeeding and demonstrated a postnatal 28-week HIV transmission rate of 1.7%, compared with 5.7% in those who did not received extended prophylaxis [22].

In summary, antiretroviral regimens for either mothers or infants for the prevention of breastfeeding HIV transmission are effective and have become part of WHO guidelines [21], but these regimens may also lead to development of resistant virus in breast milk, particularly as drug levels start to fall and viral load rebounds.

Other factors contributing to the emergence of resistant virus may relate to differential penetration of ARVs into breast milk [34] and the impact of those ARV concentrations on HIV RNA and DNA levels in breast milk. Prophylaxis regimens, as described above, impact breast milk RNA levels differentially. In a study in Kenya, comparison was made of RNA and DNA levels in breast milk among women receiving HAART, single-dose nevirapine, single-dose nevirapine plus zidovudine, or zidovudine alone. There were significantly greater reductions in cell-free RNA with administration of HAART than with zidovudine ($p=0.0001$), and a slightly less significant difference was noted between HAART and single-dose nevirapine plus zidovudine arms ($p=0.04$). There was no difference in any of the arms in viral DNA levels [35]. Similarly, in a Botswana study, mothers who received single-dose nevirapine plus zidovudine were more likely to have undetectable HIV RNA levels in breast milk at 2 weeks postpartum, as compared to women who received zidovudine alone (80% vs. 39%, respectively). By 2 months there was no difference in the HIV RNA levels in breast milk by study arm [36]. This again highlights both the success of ARVs, and nevirapine in particular, in reducing viral load, not just in plasma but also in breast milk, and also the resultant viral load rebound which occurs without ARVs. Suboptimal levels of drug in breast milk can lead to resistant virus, although not all studies have reported associations between ARV levels in breast milk and resistance rates [37, 38].

A number of studies have suggested that ARVs in breast milk may preferentially depress HIV RNA levels, cell-free more than cell-associated, and RNA levels more than HIV DNA levels [39–41]. Overall, cell-free and some cell-associated RNA levels decrease with ARVs, but DNA levels were not reduced [42]. Additionally, there is some evidence that cell-associated virus in the form of infected cells is a stronger predictor of transmission than cell-free virus [43]. The key effect of maternal ARV therapy during breastfeeding may be the marked reduction of viral gene expression and replication, represented by decreases in cell-free and cell-associated viral RNA, despite limited effect on cell-associated proviral DNA.

Reports of HIV Drug Resistance in Breast Milk and Infected Infants

Postpartum ARV prophylaxis for prevention of postnatal HIV transmission has clear benefits not only in terms of reducing viral load levels in breast milk itself, but also in terms of providing infants with ARVs administered through breast milk. However, infants infected with HIV despite infant prophylaxis may develop or be infected with resistant virus, either transmitted from the mother via breast milk or selected for following transmission of HIV-1 from the mother. Factors such as the frequency of infant feedings and reduced drug clearance times in young infants may result in selective drug pressure and predispose infants to the development of viral mutations in infants HIV-infected through breastfeeding. We summarize here reports of resistant HIV-1 virus in breast milk following single-dose nevirapine prophylaxis and of resistance associated with maternal or infant ARV prophylaxis during the breastfeeding period.

Receipt of single-dose nevirapine prophylaxis in labor has been reported to be associated with the development of resistant HIV in breast milk. Among women in Zimbabwe who received single-dose nevirapine, 65% of breast milk and 50% of plasma samples at 8 weeks had detectable NNRTI mutations [44]. In Uganda, nevirapine resistance was detected in 40% of 30 breast milk samples at 4 weeks following receipt of single-dose nevirapine [45].

Among women receiving HAART postnatally, resistance in breast milk has been reported from a number of studies. A study in Mozambique looked at resistance mutation patterns in breast milk HIV and reported that the genotypic resistance patterns differed between plasma and cell-free breast milk viruses in 38% of women and between cell-free and cell-associated viruses in 50% of women (see above) [46]. These data lent further evidence to the theory that breast milk and plasma virus and associated viral mutations originate in separate compartments. Another small sample of women assessed resistance rates in plasma and breast milk following receipt of maternal zidovudine, lamivudine, and nevirapine from 28 weeks gestation until 1 month postpartum. NNRTI mutations were found in 13% of plasma and breast milk samples, and lamivudine mutations were found in 9%, including both cell-free and cell-associated viral mutations. While the rates of resistance were similar in plasma and breast milk, there were differences in the resistance patterns in breast milk and plasma, suggesting that resistance patterns in plasma among women receiving HAART may not predict what patterns will emerge in breast milk [47].

Among HIV-infected infants, there have also been reports of resistance associated with infant ARV prophylaxis for prevention of breastfeeding HIV transmission. A comparison of HIV-infected infants in the SWEN study at 6 weeks and 6 months found that infants who received extended dose nevirapine for 6 weeks had higher rates of resistance than infants who received single-dose nevirapine (62% vs. 18%, respectively, at 6 months) [48, 49].

Finally, there have been a few reports of HIV drug resistance in infected infants whose mothers received HAART postnatally. A small analysis of infected infants from the SWEN study in Uganda found that all of seven HIV-infected breastfeeding infants whose mothers started HAART postnatally had nevirapine drug resistance and six of seven also had NRTI resistance [50]. However, the authors concluded that because both mothers and infants received nevirapine prophylaxis before the infant's diagnosis, it could not be determined whether resistance was acquired through nevirapine prophylaxis or nevirapine in the mothers' breast milk. NRTI resistance could also have been acquired through resistant breast milk virus or through exposure to nonsuppressive levels of NRTIs in breast milk. In the KiBs study where mothers received either nevirapine or nelfinavir-based HAART from 34 weeks gestation through 6 months postpartum, 67% of HIV-infected infants had detectable HIV drug resistance [51, 52]. Genotypic resistance mutations to NRTIs were M184V ($n=13$), K65R ($n=4$), D67N ($n=2$), and T215Y ($n=2$), and mutations to NNRTI were Y181C ($n=3$), K103N ($n=2$), G190A ($n=2$), and K101E ($n=1$). In the same study, among infants exposed to maternal nevirapine, four (67%) of six infants with resistance had an NRTI mutation and all six infants had an NNRTI mutation. Among infants exposed to maternal nelfinavir, all ten (100%) infected infants had an NRTI mutation, but none had a major protease inhibitor mutation. Notably, many of these infants did not have resistance on the first positive specimen but did on subsequent specimens at 14 and/or 24 weeks, suggesting that the virus mutated after infection.

Policy Implications and Status

Current WHO recommendations for prevention of breastfeeding HIV transmission are to initiate HAART for eligible pregnant women as soon as possible in pregnancy and continue through the breastfeeding period and thereafter. Infants should be given daily nevirapine or twice-daily zidovudine for 4–6 weeks. For women who do not require treatment for their own health, prophylaxis may include (1) twice daily zidovudine, plus single-dose nevirapine in labor, plus 7 days of zidovudine plus lamivudine immediately postpartum for the mother, and daily nevirapine for 4–6 weeks or until 1 week following cessation of all breastfeeding for the infant; or (2) triple daily ARV prophylaxis starting as early as 14 weeks gestation and continuing until 1 week after cessation of all breastfeeding for mothers, and daily nevirapine or twice daily zidovudine for 4–6 weeks for infants [21]. WHO recommendations on infant feeding are that women should (1) breastfeed and either they or their infants receive ARVs, as noted above, or (2) avoid all breastfeeding [53].

While current recommendations for PMTCT attempt to minimize HIV transmission and the development of drug resistance, many of these ARV interventions still pose a risk for development or transmission of resistant HIV. However, it should be noted that resistance to zidovudine requires multiple sequential mutations and is associated with advanced disease stage and low CD4 counts. As a result, it is unlikely that zidovudine prophylaxis for women who do not require treatment for their own health will result in significant resistance. Lamivudine resistance, on the other hand, emerges with a single mutation and has been associated with mono-therapy or when combined with zidovudine for extended periods. Similarly, a single mutation can result in resistance to both nevirapine and efavirenz.

While it is important to note the potential risk of the development of resistant HIV during the administration of maternal or infant ARV prophylaxis for the prevention of breastfeeding HIV transmission, this must be weighed against the benefits of both breastfeeding and of reductions in HIV transmission. The combination of maternal and infant prophylaxis may yield the greatest benefit in terms of reductions in transmission, health benefits to the mother, and minimizing development of resistant HIV virus in infected infants. There is evidence to suggest that early initiation of ARVs in the mother antepartum and continuance of the ARVs through the breastfeeding period leads to some of the greatest reductions in HIV transmission, particularly for women with CD4 counts <350 cells/mm. However, longer periods of prophylaxis may also lead to higher rates of resistance, and for women with higher CD4 counts, there may be comparable efficacy of providing ARVs to women and to breastfeeding infants. Nevertheless, caution should be taken in making direct comparisons between these studies of breastfeeding HIV transmission due to cohort and contextual differences among the different studies [54]. In summary, while the findings summarized here are illustrative of the various types of resistance that can and do emerge in both breast milk of infected mothers and in HIV-infected infants, further studies are needed to determine how best to prevent transmission and minimize the development of resistance.

References

1. Connor EM, Sperling RS, Gelber R et al (2004) Reduction of maternal-infant transmission of human immunodeficiency virus type 1 with zidovudine treatment. Pediatric AIDS Clinical Trials Group Protocol 076 Study Group. N Engl J Med 331(18):1173–1180
2. Guay LA, Musoke P, Fleming T et al (1999) Intrapartum and neonatal single-dose nevirapine compared with zidovudine for prevention of mother-to-child transmission of HIV-1 in Kampala, Uganda: HIVNET 012 randomised trial. Lancet 354:795–802
3. Dorenbaum A, Cunningham CK, Gelber RD et al (2003) Two-dose nevirapine and standard antiretroviral therapy to reduce perinatal HIV transmission: a randomized trial. JAMA 288:189–198
4. Cooper ER, Charurat M, Mofenson LM et al (2002) Combination antiretroviral strategies for the treatment of HIV-1 infected pregnant women and prevention of perinatal HIV-1 transmission. J Acquir Immune Defic Syndr 29:484–494
5. Chou R, Smits AK, Huffman LH et al (2005) Prenatal screening for HIV: a review of the evidence for the US Preventive Services Task Force. Ann Intern Med 143:38–54
6. Huang DD, Bremer JW, Brambilla DJ et al (2005) Model for assessment of proficiency of human immunodeficiency virus type 1 sequencing-based genotypic antiretroviral assays. J Clin Microbiol 43:3963–3970
7. Halvas EK, Wiegand A, Boltz VF et al (2010) Low frequency nonnucleoside reverse-transcriptase inhibitor-resistant variants contribute to failure of efavirenz-containing regimens in treatment-experienced patients. J Infect Dis 201(5):672–680

8. Paredes R, Lalama CM, Ribaudo HJ et al (2010) Pre-existing minority drug-resistant HIV-1 variants, adherence, and risk of antiretroviral treatment failure. J Infect Dis 201(5):662–671
9. Mirochnick M, Thomas T, Capparelli E et al (2009) Antiretroviral concentrations in breast-feeding infants of mothers receiving highly active antiretroviral therapy. Antimicrob Agents Chemother 53(3):1170–1176
10. Spencer L, Neely M, Mordwinkin N et al Intensive PK of zidovudine (AZT), lamivudine (3TC), and atazanavir (ATV) and HIV-1 viral load in breast milk and plasma in HIV+ women receiving HAART Therapy. 16th CROI 2009; Abstract #942
11. Corbett A, Martinson F, Rezk N et al Antiretroviral drug concentrations in breast milk and breastfeeding infants. 15th CROI 2008; Abstract #648
12. Kunz A, Frank M, Mugenyi K et al (2009) Persistence of nevirapine in breast milk and plasma of mothers and their children after single-dose administration. J Antimicrob Chemother 63(1):170–177
13. Corbett A, Martinson F, Rezk N et al Lopinavir/ritonavir concentrations in breast milk and breast-feeding infants. 16th CROI 2009; Abstract #947
14. Andreotti M, Galluzzo CM, Guidotti G, Germano P, Doro Altan A, Pirillo MF (2009) Comparison of HIV type 1 sequences from plasma, cell-free breast milk, and cell-associated breast milk viral populations in treated and untreated women in Mozambique. AIDS Research and Human Retroviruses, 25
15. Andreotti M, Guidotti G, Galluzzo CM et al (2007) Resistance mutation patterns in plasma and breast milk of HIV-infected women receiving highly-active antiretroviral therapy for mother-to-child transmission prevention. AIDS 21(17):2360–2362
16. Salazar-Gonzalez JF, Salazar MG et al (2011) Origin and evolution of HIV-1 in breast milk determined by single genome amplification and sequencing. J Virol 85(6):2751–63, Mar 2011
17. Neveu D, Viljoen J, Bland RM et al (2011) Cumulative exposure to cell-free HIV in breast milk, rather than feeding pattern per se, identifies postnatally infected infants. Clin Infect Dis 52(6):819–825, 15 Mar 2011
18. Heath L, Conway S, Jones L et al (2010) Restriction of HIV-1 genotypes in breast milk does not account for the population transmission genetic bottleneck that occurs following transmission. PLoS One 5:1–11
19. Gray RR, Salemi M, Lowe A et al (2011) Multiple independent lineages of HIV-1 persist in breast milk and plasma. AIDS 25(2):143–152
20. Gantt S, Carlsson J, Heath L et al (2010) Genetic analyses of HIV-1 env sequences demonstrate limited compartmentalization in breast milk and suggest viral replication within the breast that increases with mastitis. J Virol 84:10812–10819
21. WHO. Antiretroviral drugs for treating pregnant women and preventing HIV infection in infants. http://www.who.int/hiv/pub/mtct/antiretroviral/en/index.html. Accessed 17 Feb 2011
22. Chung MH, Kiarie JN, Richardson BA et al (2007) Independent effects of nevirapine prophylaxis and HIV-1 RNA suppression in breast milk on early perinatal HIV-1 transmission. J Acquir Immune Defic Syndr 46:472–478
23. Aizire J, Mudiope P, Matovu F et al (2010) Impact of systemic and mucosal nevirapine levels on serial HIV RNA levels in maternal plasma and breast milk after perinatal single dose nevirapine. 17th Conference on Retroviruses and Opportunistic Infections. http://www.retroconference.org/2010/Abstracts/37126.htm. Accessed 17 Feb 2011
24. Lee EJ, Kantor R, Zijenah L et al (2005) Breast milk shedding of drug resistant HIV-1 subtype C in women exposed to single-dose nevirapine. JID 192:1260–1264
25. Chasela CS, Hudgens MG, Jamieson DJ et al (2010) Maternal or infant antiretroviral drugs to reduce HIV-1 transmission. N Engl J Med 362(24):2271–2281
26. Thomas TK, Masaba R, Borkowf CB et al and KiBS Study Team (2011) Triple-antiretroviral prophylaxis to prevent mother-to-child HIV transmission through breastfeeding–the Kisumu Breastfeeding Study, Kenya: a clinical trial. PLoS Med 8(3):e1001015. Epub 2011 Mar 29
27. Kilewo C, Karlsson K, Ngarina M et al (2009) Prevention of mother to child transmission of HIV-1 through breastfeeding by treating mothers with triple antiretroviral therapy in Dar es Salaam, Tanzania: The Mitra Plus Study. J Acquir Immune Defic Syndr 52:406–416
28. Palombi L, Marazzi MC, Voetberg A, Magid NA, The DREAM Program prevention of mother to child transmission team (2007) Treatment acceleration program and the experience of the DREAM program in prevention of mother to child transmission of HIV. AIDS 21(Suppl 4):S65–S71
29. Shapiro RL, Hughes MD, Ogwu A et al (2010) Antiretroviral regimens in pregnancy and breast-feeding in Botswana. N Engl J Med 362(24):2282–2294, 17 June 2010
30. Kesho Bora Study Group (2011) Triple antiretroviral compared with zidovudine and single-dose nevirapine prophylaxis during pregnancy and breastfeeding for prevention of mother-to-child transmission of HIV-1 (Kesho Bora study): a randomised controlled trial. Lancet Infect Dis. published online 14 Jan 2011, doi:10.1016/S1473-3099(10)70288-7
31. Bedri A, Gudetta B, Isehak A et al (2008) Extended-dose nevirapine to 6 weeks of age for infants to prevent HIV transmission via breastfeeding in Ethiopia, India, and Uganda: an analysis of three randomized controlled trials. Lancet 372:300–313
32. Kumwenda NI, Hoover DR, Jamieson D et al (2008) Extended antiretroviral prophylaxis to reduce breast milk HIV-1 transmission. N Engl J Med 359:119–129

33. Kilewo C, Karlsson K, Massawe A et al (2008) Prevention of mother-to-child transmission of HIV-1 through breast-feeding by treating infants prophylactically with lamivudine in Dar es Salaam, Tanzania: the Mitra study. J Acquir Immune Defic Syndr 48:315–323
34. Mirochnick M, Thomas T, Capparelli E et al (2009) Antiretroviral concentrations in breast-feeding infants of mothers receiving highly active antiretroviral therapy. J Antimicrob Chemother 53:1170–1176
35. Lehman DA, Chung MH, John-Stewart GC et al (2008) HIV-1 persists in breast milk cells despite antiretroviral treatment to prevent mother-to-child transmission. AIDS 22:1475–1485
36. Rossenkhan R, Ndung'u T, Sebunya TK et al (2009) Temporal reduction of HIV type 1 viral load in breast milk by single dose nevirapine during prevention of MTCT. AIDS Res Hum Retroviruses 25:1261–1264
37. Hudelson SE, McConnell MS, Bagenda D et al (2010) Emergence and persistence of nevirapine resistance in breast milk after single-dose nevirapine administration. AIDS 24:557–561
38. Andreotti M, Guidotti G, Galluzzo CM et al (2007) Resistance mutation patterns in plasma and breast milk of HIV-infected women receiving highly-active antiretroviral therapy for mother-to-child transmission prevention. AIDS 21(17):2360–2362, 12 Nov 2007
39. Andreotti M, Galluzzo CM, Guidotti G et al (2009) Comparison of HIV type 1 sequences from plasma, cell-free breast milk, and cell-associated breast milk viral populations in treated and untreated women in Mozambique. AIDS Res Hum Retroviruses 25(7):707–711, Jul 2009
40. Shapiro RL, Ndung'u T, Lockman S et al (2005) Highly active antiretroviral therapy started during pregnancy or postpartum suppresses HIV-1 RNA, but not DNA, in breast milk. J Infect Dis 192:713–719
41. Bulterys M, Weidle PJ, Abrams EJ, Fowler MG (2005) Combination antiretroviral therapy in African nursing mothers and drug exposure in their infants: new pharmokinetic and virologic findings. J Infect Dis 192:709–712
42. Lehman DA, Chung MH, John-Stewart GC et al (2008) HIV-1 persists in breast milk cells despite antiretroviral treatment to prevent mother-to-child transmission. AIDS 22:1475–1485
43. Rousseau CM, Nduati RW, Richardson BA et al (2004) Association of levels of HIV-1-infected breast milk cells and risk of mother-to-child transmission. J Infect Dis 190:1880–1888
44. Lee EJ, Kantor R, Zijenah L et al (2005) Breast milk shedding of drug resistant HIV-1 subtype C in women exposed to single-dose nevirapine. JID 192:1260–1264
45. Hudelson SE, McConnell MS, Bagenda D et al (2010) Emergence and persistence of nevirapine resistance in breast milk after single-dose nevirapine administration. AIDS 24:557–561
46. Giuliano M, Guidotti G, Andreotti M et al (2007) Resistance mutation patterns in plasma and breast milk of HIV-infected women receiving HAART for PMTCT: a study within the DREAM program. Abstract 135; 14th Conference on Retroviruses and Opportunistic Infections, 2007
47. Andreotti M, Guidotti G, Galluzzo CM et al (2007) Resistance mutation patterns in plasma and breast milk of HIV-infected women receiving highly-active antiretroviral therapy for mother-to-child transmission prevention. AIDS 21:2360–2361
48. Persaud D, Bedri A, Ziemniak C et al and the Ethiopian SWEN Study Team (2011) Slower clearance of nevirapine resistant virus in infants failing extended nevirapine prophylaxis for prevention of mother-to-child HIV transmission. AIDS Res Hum Retroviruses 27(8):823–829. Epub 2011 Feb 25, Aug 2011
49. Moorthy A, Gupta A, Bhosale R et al (2009) Nevirapine resistance and breast milk HIV transmission: effects of single and extended-dose nevirapine prophylaxis in subtype C HIV-infected infants. PLoS One 4:e4096
50. Lidstrom J, Guay L, Musoke P et al (2010) Multi-class drug resistance arises frequently in HIV-infected breastfeeding infants whose mothers initiate HAART post-partum. Abstract 920; 17th Conference on Retroviruses and Opportunistic Infections, 2010
51. Thomas R, Masaba R, Ndivo R, et al. Kisumu breastfeeding study team (2008) Prevention of mother-to-child transmission of HIV-1 among breastfeeding mothers using HAART: the Kisumu breastfeeding study, Kisumu, Kenya 2003–2007 (45aLB). 15th Conference on Retroviruses and Opportunistic Infections, 2008, Boston, MA
52. Zeh C, Weidle PJ, Nafisa L et al (2011) HIV-1 drug resistance emergence among breastfeeding infants born to HIV-infected mothers during a single-arm trial of triple-antiretroviral prophylaxis for prevention of mother-to-child transmission: a secondary analysis. PLoS Med. 8(3):e1000430. Epub 2011 Mar 29, Mar 2011
53. WHO. Guidelines on HIV and infant feeding 2010. http://www.who.int/child_adolescent_health/documents/9789241599535/en/index.html. Accessed 17 Feb 2011
54. Mofenson L (2009) Prevention of breast milk transmission of HIV: the time is now. J Acquir Immune Defic Syndr 52:305–308

Chapter 7
Animal Models of HIV Transmission Through Breastfeeding and Pediatric HIV Infection

Koen K.A. Van Rompay and Kartika Jayashankar

Introduction

In 2009, an estimated 370,000 children contracted HIV from their mothers during the perinatal and breastfeeding period [1]. Major progress has been made in reducing intrapartum transmission of HIV-1 using prepartum antiretroviral regimens [2–4]. However, in low-resource countries where it has been difficult to make replacement feeding AFASS (affordable, feasible, acceptable, sustainable, and safe), breastfeeding continues to be a considerable risk factor for postnatal mother-to-child transmission of HIV [5–7]. Breastfeeding is a big dilemma for many HIV-infected women in such areas: even though it can transmit HIV, breast milk remains the best resource to provide the nursing infant with much-needed nutrition and protection against other serious infectious diseases [8, 9].

Whereas exclusive breastfeeding carries a lower risk of HIV-1 transmission than mixed feeding [10, 11], considerable research efforts are focused on exploring additional intervention strategies, such as heat treatment of expressed breast milk [12, 13]. Prolonged administration of antiviral drugs to mothers and their nursing infants can reduce HIV transmission [14–16], but their cost, risk of toxicity, and need for regular administration are limiting factors in resource-poor areas. Ideally, a vaccine regimen should be developed that, when administered to the infant shortly after birth, could protect against HIV transmission during prolonged breastfeeding (see review in ref. [17]).

Progress is hampered by our incomplete understanding of the many viral and host factors that are involved with breast milk transmission of HIV. In addition, because clinical trials aimed at reducing transmission rates are very tedious, time-consuming, and expensive, only relatively few intervention strategies can be explored. Accordingly, there is an important role for animal models. Appropriate animal models can be useful to gain insights into the biology of transmission and disease pathogenesis, and they can be used to screen intervention strategies in a relatively short time so that the most promising ones can enter clinical trials first.

There is currently no perfect animal model of HIV infection. A number of animal models that use different, related viruses are available, each with their advantages and limitations. Although murine and feline models of perinatal transmission have been developed and can be useful for initial screening of interventions, limitations of these models remain the significant differences in the virus, host physiology (e.g., placentation with transplacental antibody transfer, pharmacokinetics, and fetal/neonatal immune development), and disease pathogenesis (reviewed in ref. [18]).

K.K.A. Van Rompay (✉) • K. Jayashankar
California National Primate Research Center, University of California, Davis, CA 95616, USA
e-mail: kkvanrompay@ucdavis.edu

Simian immunodeficiency virus (SIV) infection of non-human primates is generally considered to be the best animal model because it resembles more closely HIV infection of humans and therefore allows a more reliable extrapolation of the results of antiviral strategies. Accordingly, non-human primate models are the focus of the remainder of this review.

Overview of Non-human Primate Models for HIV Infection

Non-human primates are phylogenetically the closest to humans, and they have very similar physiology, including immunology and pharmacology. Many non-human primate species in Africa have a long history of natural infection with different SIV strains [19, 20]. A common theme among the different non-human primate models is that natural infection rarely results in overt disease; in contrast, a disease that resembles AIDS in humans is most consistently seen during infection of non-natural hosts.

Chimpanzees in the wild are the source of SIVcpz, the immediate precursor of HIV-1. Although SIVcpz infection of chimpanzees in the wild is associated with an increased mortality rate [21], very few animals that have been experimentally infected with SIVcpz or HIV-1 in captivity have developed disease [22–24]. In addition, this animal model is not practical due to the low availability, its high price, and ethical issues such as their high intelligence and endangered status. HIV-1 infection could be induced in young pigtailed macaques, but virus replication was not sustained and no disease was observed [25]. HIV-2 infection models have been developed with hamadryas baboons (*Papio hamadryas*) and several macaque species; depending on the HIV-2 isolates, the outcome varied from an AIDS-like disease with CD4+ T cell decline to no disease [26–28].

Many other non-human primate species in Africa are naturally infected with SIV strains; examples are African green monkeys (SIVagm) and sooty mangabeys (SIVsm) [19]. These viruses are more closely related to HIV-2 than to HIV-1. Despite persistent high-level virus replication, these natural hosts rarely develop disease. Accordingly, these natural, nonprogressive SIV infections represent an evolutionary adaptation that allows a peaceful coexistence of primate lentiviruses and the host immune system; research indicates that this adaptation involves phenotypic changes to CD4+ T cell subsets, limited immune activation, and preserved mucosal immunity, all of which contribute to the avoidance of disease progression [29–33]. In contrast, due to co-housing different primate species [34], it was discovered coincidentally in the mid-1980s that SIV infection of non-natural hosts such as Asian macaques results in a disease that resembles human AIDS in many aspects [35]. Limitations of the SIV-macaque models remain their high cost, relative availability, and the subtle differences between HIV and SIV (i.e., SIV resembles more HIV-2 than HIV-1). However, the many similarities in virus, host, and disease pathogenesis have made them currently the premier animal model in HIV research (reviewed in refs. [36–40]).

Main Characteristics and Development of SIV-Macaque Models

The most commonly used species are rhesus macaques (*Macaca mulatta*), cynomolgus macaques (*M. Fascicularis*), and pigtailed macaques (*M. nemestrina*). The SIV isolates generally belong to a few groups, in particular SIVmac, SIVsm, and SIVmne. While CD4 is their main cell receptor for viral entry, most SIV isolates use CCR5 as a co-receptor. To mimic the common routes of HIV transmission in humans, macaque models have been developed for different routes of virus inoculation (i.e., intravenous, intravaginal, intrarectal, or oral) [41–43]. As discussed further, models for pediatric HIV infection have been developed by using pregnant and infant macaques [44–47].

Infection of macaques with virulent SIV isolates such as SIVmac251 results, after a variable asymptomatic period, in a disease which resembles human AIDS in many aspects, including cell tropism, generalized immune activation, CD4+ T cell depletion (especially from mucosal sites), opportunistic infections, weight loss, and wasting [35, 48]. Very importantly, the same laboratory

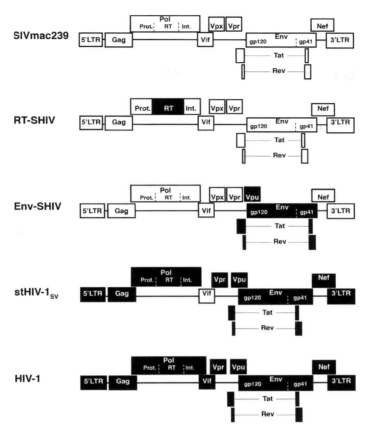

Fig. 7.1 Schematic representation of simian immunodeficiency virus (SIV), HIV-1, and chimeric constructs that are used in non-human primate studies. Although HIV-1 can replicate in chimpanzees and to a very limited extent, in pigtailed macaques [21, 25], SIV infection of macaques is a much more practical animal model to study disease pathogenesis and test intervention strategies. Because there are important genetic differences between SIV and HIV-1 that include the targets of some antiviral strategies, chimeric viruses have been constructed that, depending on the construct and further in vivo passaging, can induce transient to persistent viremia and sometimes disease. In this schematic representation, *white* and *black boxes* refer to SIV and HIV sequences, respectively. RT-SHIVs contain HIV-1 reverse transcriptase (RT) in a SIVmac239 background [50] or SIVmne background [145]. Env-SHIVs consist of SIV background containing HIV-1 env, tat, rev, and generally (to have better replication in macaques) also vpu [146]. Simian-tropic (st)HIV-1 strains differ from HIV-1 only in the vif gene [59]

markers such as viral RNA levels in plasma and CD4+ T lymphocyte counts that are used to monitor HIV-infected people are also predictive of infection and disease progression in SIV-infected macaques. Accordingly, these markers are also extremely useful to evaluate the efficacy of antiviral strategies.

It is important to note that SIV infection of macaques is not necessarily fatal. Depending on the selection of the host and the SIV isolate or clone, a broad spectrum of viremia and clinical outcomes is observed. At one end of the spectrum are models in which highly virulent SIV isolates induce persistently high viremia and rapid disease progression within ~3 months; at the other end are avirulent isolates such as SIVmac1A11, which induce transient or low-level viremia and no disease at all, even in newborn macaques [46]. Accordingly, investigators can select the virus with a particular virulence that is most appropriate to address a certain research question. In addition, this spectrum of infection outcomes makes this model also suitable to assess how genetic changes in the virus (e.g., mutations that confer drug resistance or immune escape) affect viral virulence.

Although SIV is related to HIV-1, some genetic differences remain (Fig. 7.1). Preclinical testing of antiviral strategies in the SIV-macaque model is most relevant if the viral target and its susceptibility to inhibition naturally resemble, or can be engineered to resemble, that of HIV-1. Most SIV isolates are susceptible to many anti-HIV drugs [36, 49]. In contrast, non-nucleoside RT inhibitors (NNRTI) such as nevirapine and efavirenz are active only against HIV-1 and not against HIV-2 or SIV [49]. Accordingly, the construction of infectious SIV/HIV-1 chimeric viruses, in which the reverse transcriptase (RT) gene of SIV was replaced by its counterpart of HIV-1 (so-called RT-SHIVs), has allowed evaluation of NNRTI in primate models [50] (Fig. 6.1). Similarly, because there are significant differences between the SIV and HIV envelope proteins, env-SHIVs that contain the HIV-1 envelope region have been constructed to allow direct testing of strategies that target the HIV-1 envelope region (Fig. 7.1). Many env-SHIVs are attenuated but some virulent isolates have been derived through serial passage. Most virulent env-SHIVs such as SHIV-89.6P, while useful to address specific questions, have the limitation that their disease pathogenesis (including CXCR4 co-receptor usage and very rapid CD4+ cell depletion) differs from the typical course seen with HIV and SIV infection, and some evidence suggests that they are not reliable predictors of efficacy in humans [51–55]. The few currently available CCR5-using env-SHIVs (such as SHIV-SF162P3 and the clade C env-SHIV-1157i) have the limitation that, after the initial peak of viremia, a significant portion of untreated animals suppressed viremia to low or undetectable levels and did not develop disease [56–58].

In ongoing attempts to use a virus that resembles HIV-1 as much as possible, investigators constructed simian-tropic (st)HIV-1 strains that differ from HIV-1 only in the vif gene; these viruses caused persistent viremia in pigtailed macaques for several months after which viremia was controlled by the immune system [59] (Fig. 7.1). Accordingly, although these different CCR5-tropic env-SHIVs and stHIV-1 isolates are useful to test prophylactic or early postinfection interventions, their large variability in chronic viremia and disease outcome makes them less suitable to test the efficacy of antiviral strategies during established infection, especially because animal numbers are usually limited. However, it is likely that further development of these SHIV and stHIV-1 models in the coming years will lead to a more persistent viremia.

Macaque Models of Perinatal Transmission

Whereas numerous SIV and SHIV experiments have been performed in juvenile and adult macaques, relatively few research groups have focused their attention on developing models to study mother-to-child transmission and pediatric infection. Two main categories of macaque models have been explored, depending on whether the mother is inoculated with subsequent monitoring of transmission to the offspring, or whether the newborn or infant macaque is directly inoculated with virus (Fig. 7.2). Both have their advantages and disadvantages, as neither one is perfect in mimicking every aspect of natural transmission plus being suitable to rapidly screen intervention strategies. Nonetheless, the studies that have been performed in macaques so far have provided important information on the many host and viral factors that determine transmission and the disease course in the infants. In addition, they have guided the clinical development of strategies aimed at preventing or treating infection.

Maternal Transmission Models

In the maternal transmission models, which are aimed at mimicking natural mother-to-child transmission as much as possible, the female macaques are infected with SIV either during pregnancy or after delivery. Following birth, the infants are generally allowed to suckle. The offspring is monitored regularly to determine the timing of transmission and the disease course.

To investigate in utero transmission in detail, one research group developed a chronic fetal catheterization model in which pigtailed macaques were infected intravenously with HIV-2 and fetal samples

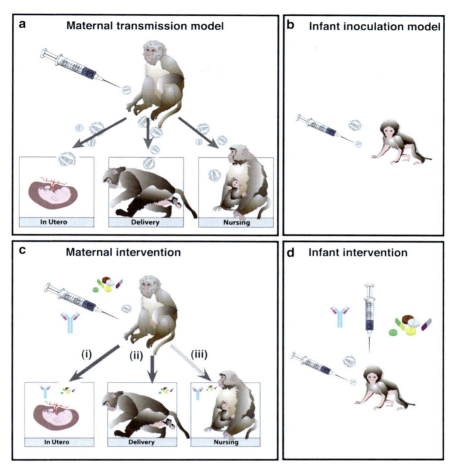

Fig. 7.2 Summary of non-human primate models to study pathogenesis and intervention strategies for mother-to-infant transmission and pediatric HIV infection. Two main categories of non-human primate models have been developed, each with their advantages and drawbacks. (**a**) In the maternal transmission model, the adult female macaque is inoculated with simian immunodeficiency virus (SIV) or SHIV either before pregnancy, during pregnancy, or after delivery; the offspring is monitored for evidence of spontaneous transmission that can occur in utero, during delivery, or through breastfeeding. (**b**) In the infant inoculation model, the newborn or infant macaque is directly inoculated with virus, which provides more control over the timing of infection and the rate of infection to address certain research questions. Both models can be used to study interventions aimed at preventing infection or disease progression. (**c**) In the maternal transmission model, antiviral strategies (e.g., active and passive immunization; antiretroviral drugs) can be administered to the mother: (i) maternal antibodies and many antiretroviral drugs can cross the placenta and reach the fetus; (ii) antiretroviral drugs given shortly before delivery (including caesarian section) can still reach the newborn in sufficient amounts; (iii) antibodies present in breast milk (including IgG and IgA), although not bioavailable to reach the systemic circulation, can still provide some local immunity to the nursing infant. Although antiretroviral drugs can be detected in breast milk of macaques and humans after dosing [73, 147], the transfer rate differs for each drug, and the amount of drug ingested by the nursing infant is generally too low or suboptimal to provide the infant with systemic prophylactic or therapeutic benefits. (**d**) Antiviral strategies (passive and active immunization; antiretroviral drugs) can also be given directly to the infants. Some studies used a combination of these models; for example, an immune- or drug-based intervention can be administered to the uninfected mother, and after birth, the infant is inoculated directly with SIV [72, 76]

could be collected regularly [60]. However, one can doubt the relevance of such a fetal catheterization model to study natural transmission, because the 100% fetal infection rates suggest that the invasive procedure may promote transmission at the site of the catheter. Such a model may therefore resemble to some extent the model of ultrasound-guided direct SIV inoculation into the fetal amniotic fluid [61].

In some of the earliest studies that used natural transmission, pregnant rhesus or pigtailed macaques were inoculated intravenously with SIV/DeltaB670 or SIVsmm during various stages of gestation, but depending on the study, few to none of the infants were infected at birth, indicating a low rate of in utero transmission [62–66]. When the infants were allowed to suckle, 3 of 12 infants became infected at 9–15 months of age, indicating late transmission through breastfeeding [63].

In follow-up studies to the more natural transmission studies, four rhesus macaques were inoculated intravenously during midgestation with SIV/DeltaB670, with the aim that infants would be delivered when females were in the chronic phase of infection. The four infants were SIV-negative at birth and for the first 6 weeks of age, indicating that in utero or peripartum transmission did not occur. However, three of the four infants became infected by breastfeeding between 2 and 4 months of birth [67].

When nine pregnant pigtailed macaques were inoculated intravenously during midgestation with SHIV-SF162P3, four of the nine infants became infected. The one infant that was infected in utero developed high viremia and CD4+ T cell depletion. The other three infants, which acquired infection peri- or postpartum, highly controlled virus replication but showed de novo antiviral antibody responses [44].

The experimental design of another set of studies mimics the situation in which women become HIV-infected shortly after delivery, so that the infants have no transplacentally acquired antiviral antibodies while breastfeeding. In these studies, 14 lactating macaques were inoculated intravenously with SIV/DeltaB670 between 15 and 45 days after vaginal delivery; 10 of the 14 nursing infants became infected through breastfeeding but the timing of infection ranged from 14–21 days (two "early transmitters") to 88–360 days (eight "late transmitters") after maternal SIV inoculation [68]. The relatively high (71%) rate of breast milk transmission in this macaque model in comparison to observations in breastfeeding HIV-infected women can be explained because of the higher virus levels and accelerated disease course in this animal model. In this study, a number of viral and host parameters were measured to identify possible correlates of transmission. In the early transmitters, peak viral loads in milk and plasma were similar to those of the other animals; however, the early transmitters subsequently displayed a rapid progressor phenotype (i.e., high virus replication, accelerated CD4+ T cell decline, and poor SIV-specific antibody responses). This suggests that host factors that are associated with failure to control the initial viremia may also play a role in promoting viral transmission in breast milk to the nursing infant [68]. Comparison of the late transmitters versus nontransmitters did not reveal any differences in plasma SIV levels (which were all similarly high), CD4+ T cell counts, total immunoglobulin levels, or SIV-specific antibody titers in plasma and milk; however, late transmitters had higher levels and more persistent detection of SIV in milk in comparison to nontransmitters [68].

Genotypic analysis that was performed in several of these SIV transmission studies revealed that the infant macaques were infected with homogenous virus populations in comparison to the more diverse viral populations in maternal samples. This indicates that viral phenotypes were selectively transmitted across the placenta and through breastfeeding [62, 67].

Altogether, the main advantage of the different maternal infection models is that they can address certain questions about mother-to-child transmission in their most natural context of maternal infection, including the presence of maternal immune responses.

Although these models have provided important insights, some limitations and weaknesses remain. The combined data indicate that most SIV transmission from infected female macaques to their offspring occurs during prolonged breastfeeding, which differs from observations in people where most HIV-infected infants acquire their infection around the time of delivery. In addition, because of the variable timing and rates of transmission and the limited animal numbers, these macaque models of maternal transmission are often not sufficiently powered (1) to identify viral or host determinants of disease progression in the infants and (2) to screen novel intervention strategies for efficacy in reducing transmission or delaying disease progression in those infants that became infected.

Although uninfected pregnant or nursing female macaques have been used to study the pharmacokinetics of antiretroviral drugs across the placenta or into breast milk [69–73], the maternal transmission model has to our knowledge not been used to study whether such interventions reduce the transmission rate of SIV to their infants.

Accordingly, to solve some of these problems and complement the maternal transmission model, direct infant inoculation models have been developed.

Infant Inoculation Models

Development of High- and Low-Dose Oral SIV Inoculation Models

To develop a model of pediatric HIV infection and AIDS, the first studies in infant macaques used the intravenous route of virus inoculation [45, 74, 75]. In subsequent models, SIV was administered orally to newborn or infant macaques [43, 76]. Depending on the age of the animals, these oral inoculation models simulate oral exposure of human newborns and infants to HIV through ingestion of maternal genital fluids or blood during delivery, or of breast milk after birth.

The advantages of these infant inoculation models are that investigators have more control over the timing of infection, the dose, and the composition of the viral inoculum to target specific research questions. A dose of virus can be selected for oral administration that can assure nearly 100% infection rates of the infant macaques. Although such high infection rates do not mimic natural transmission rates, their advantage is that they allow easier testing of the efficacy of intervention strategies with smaller group sizes. To achieve high infection rates following oral virus inoculation, two main models have been developed, namely high-dose inoculation and repeated low-dose inoculation.

In the high-dose inoculation model that was developed first, a high dose of SIV, SHIV, or HIV-2 is given once or twice to the infant macaques [43, 77–80]. Whereas a high-dose oral inoculation model is relatively easy to perform, a potential disadvantage is that the amount of virus that is administered to achieve nearly 100% infection rates is much higher than the one that human infants are exposed to. Such a model with an overwhelmingly high virus inoculum can probably identify highly potent antiviral intervention strategies, but may miss or underestimate the efficacy of prophylactic strategies with low to moderate efficacy.

To mimic the frequent exposure to lower amounts of virus that occurs during prolonged breastfeeding, repeated low-dose exposure models have been developed. In one model of early transmission, newborn or 4-week-old infant macaques are hand-held, without anesthesia, and are bottle-fed low amounts of SIV three times daily for 5 consecutive days [81, 82]. Infection rates of ~87–94% were observed. Similarly, a model of late transmission was developed, in which low doses of virulent SIVmac251 were given orally to juvenile macaques at weekly intervals [81]. An advantage of such more extended inoculation schedule is that (1) regular inoculations can be continued until virtually all control animals are infected and (2) the number of inoculations needed to induce infection provides an additional measure to determine the relative efficacy of prophylactic intervention strategies [81]. Similar repeated low-dose virus inoculation models have been developed for intravaginal and intrarectal inoculation routes of juvenile or adult macaques [56, 83–85].

Lessons Learned on Transmission and Pathogenesis

Studies in infant and adult macaques suggest that after oral inoculation, the tonsil is the most likely site of viral entry and early replication [86, 87]. Oral inoculation of infant macaques with virulent SIVmac251 resulted in rapid systemic dissemination within the first week, and the induction of strong innate immune responses in local and systemic tissues [86]. The early response at the sites of mucosal entry was dominated by the induction of pro-inflammatory cytokines, which are likely to favor

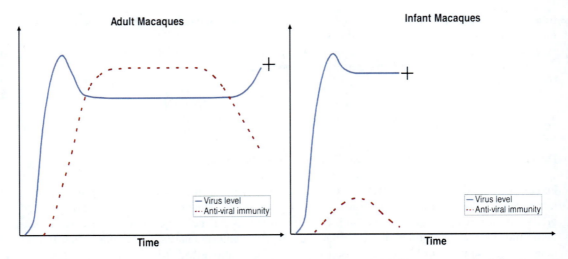

Fig. 7.3 Pathogenesis of virulent SIV infection in adult versus infant macaques. After the primary peak of viremia, most SIV-infected adult macaques mount strong antiviral immune responses which reduce virus levels in peripheral blood to a lower set-point; after months to years, these animals develop a gradually progressive immunodeficiency. In contrast, inoculation of infant macaques within the first month of life with highly virulent isolates like SIVmac251 results in a more fulminant disease course characterized by persistently high viremia, weak and/or transient antiviral immune responses, and life-threatening immunodeficiency (indicated by +) generally within 2–4 months of infection [46, 75]

immune activation, virus replication, and systemic dissemination. In contrast, cytokines with direct antiviral activity were only minimally induced at the mucosal sites of entry [86].

Several studies have demonstrated that the outcome of SIV infection is not necessarily fatal but depends highly on the virulence of the virus inoculum. Oral or intravenous inoculation of newborn macaques within the first month of life with highly virulent isolates like SIVmac251 results in persistently high viremia and, in comparison to older animals, an accelerated disease course as most infants develop life-threatening immunodeficiency within 2–4 months of infection [46, 88, 89] (Fig. 7.3). This rapid disease course characterized by persistently high viremia and poor antiviral immune responses is similar to what is observed in a minority of SIV-infected adult macaques that become rapid progressors [90]. These observations are also consistent with the accelerated disease course in many HIV-infected human infants in comparison to adults [91]. Similar to observations in older macaques, virulent SIV infection of infant macaques also results in rapid depletion of CD4+ T lymphocytes in the gut [92]. Infection of infant macaques with other virulent isolates such as the molecular clone SIVmac239 results in slightly lower virus levels, and delayed disease course (animals developed disease between 6 and 36 months). Partially attenuated isolates, such as the multiply deleted SIVmac239Δ3, induce disease more slowly or in only a fraction of the infected animals over time [93]. At the far end of the spectrum of virulence, infection of infant macaques with avirulent isolates such as SIVmac1A11 results in transient low-level viremia and no disease even after many years of follow-up [46].

The selection of a virus strain with a particular virulence requires careful consideration in the experimental design of intervention studies. The available evidence suggests that prophylactic and therapeutic efficacy is more difficult to achieve against highly virulent uncloned isolates than against molecularly cloned or attenuated virus isolates. This also means that, if an intervention strategy gives even partial efficacy in a highly virulent model (e.g., high-dose SIVmac251 inoculation of newborn macaques), its efficacy may be even more pronounced for human infants and adults [75, 94].

Interventions to Prevent or Treat Infection

The infant inoculation models have been used to explore different immune- or drug-based intervention strategies aimed at preventing infection or delaying disease progression (Fig. 7.2).

Passive Immunization

Because of the proven success of passive immunization in preventing and treating many other infectious diseases in humans (reviewed in ref. [95]), there has been interest to explore if a similar approach can be effective against HIV. A number of different studies have demonstrated conclusively that passively acquired antibodies can protect infant macaques against oral SIV infection.

In an early study, vaccination of pregnant animals against SIV resulted in transplacental transfer of anti-SIV antibody that protected two of three newborn macaques against oral–conjunctival SIVmac251 inoculation at birth [76]. In a subsequent study, SIV hyperimmune serum, administered subcutaneously prior to oral inoculation with a high dose of virulent SIVmac251, protected six of six newborn macaques against infection [79]. The antibodies in this hyperimmune serum had no neutralizing activity in vitro against the challenge virus, but had high activity in an antibody-dependent cell-mediated virus inhibition (ADCVI) assay [79, 96]. However, when this SIV hyperimmune serum was given to three newborns 3 weeks after oral SIV inoculation to test its efficacy against established infection, viremia was not reduced and all three infants died within 3 months due to AIDS and immune-complex disease.

Another research group demonstrated that pre- and/or early postexposure administration of a high dose of monoclonal antibodies with broad in vitro neutralizing activity including against the challenge virus protected newborn macaques against oral infection with SHIV-vpu+ or SHIV89.6P [77, 97–99]. The success of passive immunization depended on the time window because a delay of the administration of monoclonal antibodies from 1 to 12 or 24 h after oral SHIV89.6P inoculation reduced the prophylactic efficacy, even though infected infant macaques had reduced viremia and slower disease progression than untreated animals [100]. Despite these promising data, however, it was subsequently found that two of the monoclonal antibodies used in these studies exhibit auto-reactivity, which limits the use of these broadly neutralizing antibodies in humans [101].

In another study, neutralizing IgG antibodies were administered subcutaneously to newborn macaques 1 day before oral inoculation with a high dose of SHIV-SF162P3. The amount of antibody that was given was low, so that it resembled physiologic levels in SHIV-infected pregnant macaques that could be transferred transplacentally. Although such low amounts were insufficient to prevent infection, they conferred therapeutic benefits, because treated animals had reduced plasma virus levels and enhanced de novo generation of antiviral antibodies [102].

In summary, these passive immunization studies provide the important proof-of-concept of the passive immunization approach. They also support the continued search for potent antibodies with broad antiviral activity in vitro but lacking auto-reactivity that can then be evaluated in vivo; examples include the recently discovered PG9/PG16 and VCR01/VCR02 antibodies [103, 104].

Active Immunization

Ideally, an active immunization strategy can be developed that, when administered to the infant shortly after birth, can induce long-lasting immunity that protects against HIV transmission during prolonged breastfeeding (reviewed in ref. [17]). The development of a pediatric HIV vaccine faces many of the same problems as those of adult vaccine development, as discussed below. However, a pediatric vaccine has the additional challenges that (1) the newborn immune system is less mature and has some differences compared to that of older animals and humans, (2) the vaccine needs to be able to induce immune responses in the presence of passively acquired maternal antibodies, and (3) because breastfeeding starts early after birth, rapid induction of protective immune responses is important.

Before discussing active immunization studies in infant macaques, a brief review of the lessons learned from vaccine studies in adult macaques and human adults is warranted.

As reviewed above, passive immunization studies in infant macaques have provided the important proof-of-concept of the prophylactic potential of antibodies. However, macaque studies that used heterologous viral challenges and human clinical trials have demonstrated that it has been very difficult to induce antibodies with broad antiviral activity through active immunization [39, 105, 106]. Because many studies, including some that used in vivo CD8+ cell depletion, have found an important role of CD8+ T cell responses in determining peak viremia and viral setpoint after infection [107, 108], much attention has focused on exploring vaccines that induce also cell-mediated immune responses in non-human primate models. In general, live-attenuated vaccines have been most successful, probably because they generate persistent and broad antiviral immune responses [109–111]; but due to safety concerns, this approach is currently not being explored in humans. Many other vaccine approaches—including DNA, replicating and non-replicating vectors, and subunit proteins—have been tested in macaque models. Although many vaccines have shown various levels of efficacy in macaque models, their mechanisms and location of action are still poorly understood, because no single in vitro immune effector function or combination of markers has been consistently associated with efficacy [111, 112].

In recent years, the pursuit for an effective HIV vaccine and the quest to identify immune correlates of protection have been stirred up significantly by the results of two-phase III clinical trials that were performed in adults. The STEP study failed to show protection of an adenovirus type 5 (Ad5) vector-based vaccine, despite good induction of cell-mediated immune responses [55]. In contrast, the Thai ALVAC/AIDSVAX trial found a 30% reduction in infection rate despite—by conventional assays—a relatively low immunogenicity [113, 114]. Although the outcomes of all these vaccine studies confirm our poor understanding of in vivo antiviral immunity and the limitations of the in vitro and ex vivo assays, they promote some new trends (reviewed in ref. [115]). One of them is that "the more, the better" may not apply to the immune responses needed for an effective HIV vaccine. Instead, the available data suggest the need for an intricate balance between quantity, quality, timing, and location of the different antiviral immune responses, as too much immune activation and inflammation can be harmful by promoting virus replication and dissemination [116]. Thus, there is the need to focus future research efforts on (1) the further development and optimization of in vitro assays and markers to better capture the variety of beneficial and harmful immunological responses that occur in vivo at mucosal sites and systemically, and (2) the more effective use of non-human primate models in HIV vaccine development [40, 115, 117].

A number of immunization studies have been performed in infant macaques. In one initial study, newborn macaques received multiple DNA prime immunizations followed by HIV-1 gp160 protein booster immunizations [118]. When animals were challenged intravenously at ~26 months of age with homologous SHIV-vpu+, partial protection was observed, as some animals did not become infected and others partially controlled virus replication. An intravenous rechallenge of some animals with the heterologous SHIV89.6P revealed partial control of viremia.

In another early study, a more accelerated vaccine regimen with oral SIV challenge was explored. Groups of infant macaques were immunized at birth and 3 weeks of age with either modified vaccinia virus Ankara (MVA) expressing SIV gag, pol, and env (MVA-SIV), or live-attenuated SIVmac1A11. One MVA-SIV-immunized group had maternally derived anti-SIV antibodies prior to immunization. At 4 weeks of age, the animals were challenged orally twice (24 h apart) with a high dose of a highly virulent and genetically diverse SIVmac251 stock. Although all animals became infected, the immunized animals mounted better antiviral antibody responses, controlled virus levels more effectively, and had a longer disease-free survival than the unvaccinated monkeys. Maternal antibodies did not significantly reduce the efficacy of the MVA-SIV vaccine [89]. Plasma samples of this study were analyzed for SIV envelope variants by heteroduplex mobility assay. Early after infection, plasma of the infant macaques showed heterogeneous SIV *env* populations including one or both of the most common *env* variants in the virus inoculum, and no consistent differences in transmission patterns of envelope variants were found between vaccinated and unvaccinated infants. However, envelope variants diverged in most vaccinated animals 3–5 months after infection, in association with the development of neutralizing antibodies [119]. These patterns of viral envelope diversity, immune responses, and disease course are similar to observations in HIV-infected children, and underscore the relevance of this pediatric animal model.

In a subsequent study, infant macaques were given multiple intramuscular immunizations during the first 3 weeks of life with recombinant poxvirus vaccines expressing SIV structural proteins, namely either ALVAC-SIV or MVA-SIV [81]. This time, however, starting at 4 weeks of age, the animals were inoculated orally according to a repeated low-dose SIV challenge model (three times per day for 5 days in a row). Although ALVAC-SIV was relatively poorly immunogenic according to conventional assays, significantly fewer ALVAC-SIV-immunized infants were infected compared with unimmunized infants (relative risk of infection: 0.43). These early results of partial prophylactic efficacy of ALVAC-SIV in infant macaques despite poor immunogenicity were consistent with the later results of the Thai ALVAC/AIDSVAX trial in adults (see above; ref. [113]). In these infant macaque studies, animals not infected after oral challenge in infancy were rechallenged at a later age. When a subgroup was inoculated orally with a high dose of virus, all animals became immediately infected. In contrast, when the remaining animals were rechallenged by repeated weekly exposures to low amounts of SIV, more exposures were required to infect animals vaccinated with ALVAC-SIV or MVA-SIV than to infect unimmunized animals [81]. These observations demonstrate the role of the virus inoculation regimen in determining our ability to detect a protective effect. Once infected, the vaccinated animals also had reduced viremia compared with unimmunized animals.

In later studies, an attenuated recombinant vesicular stomatits virus expressing SIVmac239 gag, pol, and env (VSV-SIVgpe) was tested in infant macaques. Oral administration of VSV-SIVgpe to infant macaques shortly after birth and followed by an intramuscular booster immunization with MVA-SIV 2 weeks later induced local and systemic antiviral immune responses by 4 weeks of age [120]. However, this regimen did not protect any infants against infection following repeated low-dose SIVmac251 inoculation at 4 weeks of age; in addition, the effect of the vaccine regimen on peak viremia and viral setpoint was small [121]. Altogether, this VSV-SIVgpe/MVA-SIV as a prime-boost vaccine regimen was less effective than the poxvirus-based prime-boost regimens described above [81].

Finally, in another recent study, infant rhesus macaques were immunized first intradermally with a novel recombinant BCG construct expressing HIV immunogens, followed 11 and 14 weeks later by intramuscular injections of a MVA-HIV construct. The HIV-specific responses induced in infants were lower compared to historical data in adult animals. Because HIV-1 does not replicate in rhesus macaques, the infants were not challenged [122].

Overall, the results of the passive and active immunization studies in infant macaques are promising and support the concept that neonatal immunization with HIV vaccines might reduce infection rates from breastfeeding or modulate disease progression in infants that became infected. However, further research is warranted to improve our arsenal of vaccine regimens, adjuvants, and in vitro assays aimed at identifying the immune correlates of protection so that the different vaccine strategies can be screened more efficiently in animal models and the most promising ones can enter the pipeline of clinical trials first.

Antiretroviral Drugs

Since the early days of the HIV epidemic, the potential of antiretroviral drugs to reduce mother-to-infant transmission of HIV has been explored for two main reasons. First of all, administration of antiretroviral drugs to the mother can, depending on the duration and composition of the drug regimen, lead to reduced virus levels in blood, secretions, and breast-milk, which is expected to reduce the exposure of the infant to the virus. Secondly, the presence of antiretroviral drugs in the uninfected infant around the time of viral exposure may also prevent infection (chemoprophylaxis) (Fig. 7.2c, d).

The infant macaque models have been used extensively to show proof-of-concept of chemoprophylaxis. Initial studies with infant macaques demonstrated that a 6-week oral zidovudine regimen starting shortly before intravenous low-dose virus inoculation could protect some infant macaques against SIV infection [74]. This study provided an incentive for the PACTG 076 Study Group clinical trial which demonstrated that zidovudine administration to HIV-infected women during pregnancy and delivery and to their newborn for 6 weeks after birth was found to reduce the mother-to-child transmission rates of HIV by approximately two thirds [123].

Later studies with the newer and more potent RT inhibitor tenofovir (PMPA) used the oral route of SIV inoculation. Pre- and postexposure regimens of tenofovir were found to be quite effective in protecting infant macaques in both the high-dose and repeated low-dose oral SIVmac251 inoculation models, even when short regimens (such as one or two doses) were used that targeted the time of virus exposure [82]. Efficacy was dependent on having sufficient drug levels, as determined by several factors of the dosage regimen including drug amount, timing of initiation (i.e., pre-exposure is more effective then postexposure), and the duration of administration [82].

Shortly after the prophylactic efficacy of a 2-dose tenofovir was demonstrated in infant macaques, the HIVNET 012 trial showed that a 2-dose intrapartum and neonatal nevirapine regimen also reduced the rate of mother-to-infant transmission of HIV [4, 80]. Because this short nevirapine regimen frequently induces drug-resistant mutants in the mother [124], the efficacy and safety of tenofovir in the infant macaque model provided an incentive for the more recent clinical trials which demonstrated that the addition of a short regimen of tenofovir and emtricitabine reduced the risk of nevirapine resistance [125, 126].

The success of these short-term antiretroviral regimens in reducing in utero, intrapartum, and early postpartum transmission of HIV provided the impetus to extend the administration of antiretroviral drugs to lactating mother and/or to the breastfeeding infant throughout the breastfeeding period. Several studies have demonstrated that these strategies are highly effective [14–16, 127, 128].

Despite their prophylactic success, the cost of long-term administration of antiretroviral drugs to lactating mothers and their infants throughout the period of breastfeeding is still considerable. In an attempt to find cheaper alternatives, it would be ideal if topical administration of antiviral compounds at the site of viral exposure could prevent oral infection. Similar to the development of vaginal microbicides, such compounds would be given in amounts sufficient for potent antiviral activity directly at the mucosal site, but too low to give sufficient systemic antiviral drug levels. In addition to a lower cost, another advantage of such topical approach is that also compounds that are too toxic or have unfavorable pharmacokinetics for systemic use may be suitable candidates. Because the regular ingestion of considerable amounts of breast milk is likely to wash away any loose molecules from the mucosal surface, a compound that can adhere to or penetrate in the mucosal cells is more likely to provide longlasting antiviral activity after infrequent administration. So far, however, only very little research has been done in this area. In two published studies, topical oral administration of two prodrugs of tenofovir, developed to give high intracellular penetration and accumulation of the active moiety tenofovir diphosphate, was not effective in protecting infant macaques in repeated low-dose oral exposure models [82, 129]. These results are in stark contrast to the demonstrated efficacy of tenofovir as topical microbicide against vaginal and rectal SIV or SHIV infection in macaque studies and against vaginal HIV infection in the recently completed CAPRISA 004 clinical trial among women in South Africa [130–132]. Possible reasons for this discrepancy in efficacy include (1) differences in contact time to allow drug absorption by mucosal cells and (2) biologic differences in viral transmission events across different mucosal surfaces, and have been described in detail elsewhere [129].

The infant models have also been used to investigate the therapeutic efficacy of antiretroviral regimens for infant macaques that became infected following intravenous or oral SIV inoculation. Initial studies demonstrated that early zidovudine treatment provided significant benefits because treated animals had lower virus levels, enhanced antiviral immune responses, and delayed disease progression [75]. These results were at that time very important as proof-of-concept of the benefits of early antiretroviral therapy. However, the long-term benefits of zidovudine monotherapy were moderate and also limited by the emergence of zidovudine-resistant viral mutants [133].

Studies with tenofovir in SIV-infected infant macaques were the first ones to demonstrate the high therapeutic efficacy of this compound in the animal model prior to the first clinical trials in HIV-infected humans. While short-term tenofovir therapy during acute infection already provided some clinical benefits, prolonged tenofovir therapy initiated early during infection had stronger and much more durable antiviral effects than late therapy [88, 134, 135]. The emergence of tenofovir-resistant viral mutants (which is associated with a K65R mutation in RT) partially reduced but did not completely abrogate the benefits of tenofovir therapy, as some animals infected with K65R SIV mutants but maintained on long-term tenofovir therapy were able to sustain low or undetectable viremia and no disease progression for currently more than 14 years; both tenofovir and CD8+ cell-mediated immune responses were required to maintain this suppressed viremia [136–138]. While these studies with tenofovir in SIV-infected infant macaques have contributed to its clinical development and have been predictive of its efficacy in HIV-infected people [139–141], they also provide the hope that many HIV-infected children may be able to lead long and healthy lives, particularly if started early on antiretroviral therapy.

Future Directions

Studies in the pediatric macaque model have provided proof-of-concept of several immune- and drug-based intervention strategies that are effective in preventing or treating infection; these data have been useful to guide clinical trials. Because mother-to-child transmission of HIV has, in the absence of any interventions, a relatively high transmission rate during a defined window period in comparison to sexual transmission, it offers unique opportunities to evaluate prophylactic intervention strategies. Accordingly, some of the strategies, particularly antiretroviral drug prophylaxis, have been translated from preclinical research in animal models into clinical applications for prevention of mother-to-child transmission quite rapidly, well before the development of such strategies to prevent sexual transmission in adults [142]. In return, the information obtained in human clinical trials needs to be fed back into the animal models, because both positive and negative outcomes help to validate the predictability of the different macaque models so that they can be improved further.

Because of the major success in using antiretroviral drugs to reduce mother-to-child transmission of HIV, the door of opportunity for immune-based strategies is slowly closing, and it remains to be determined how we can best integrate any of the immune-based strategies into current clinical research efforts. Yet, a valid scientific incentive for immune-based strategies is that they could reduce the need for antiretroviral drugs that, despite high efficacy, still have issues of affordability, practicality, and the risk of drug-resistant mutants. For example, an attractive hypothetical strategy (similar to the perinatal hepatitis B model) would be the combination of active and passive immunoprophylaxis shortly after birth, and potentially overlapping with neonatal chemoprophylaxis. In such a strategy, passively administered antibodies with high antiviral activity and/or short-term antiretroviral drug prophylaxis might effectively cover the early breastfeeding period while active immunity is still being developed [17].

However, the development of pediatric HIV vaccines is met by a number of hurdles. Historically, many antiviral vaccines (e.g., polio) were first tested and found to be effective in pediatric populations [143, 144]. For HIV, however, the success of antiretroviral drugs in preventing pediatric infection reduces the incentive to invest in research for a pediatric HIV vaccine. In the current research climate, the majority of resources are spent on adult SIV and HIV vaccine research efforts. Accordingly, the development of pediatric HIV vaccines is becoming increasingly dependent on the results of vaccine trials in adult populations. However, because so far no highly effective HIV vaccine for adults has been identified, the immune correlates of protection are unclear, and evidence indicates a significant difference in infant and adult vaccine responses, it is currently unclear how and if the results of vaccine studies in adults that are aimed against sexual or intravenous transmission can be extrapolated to infants. The macaque models may prove to be useful for pediatric vaccine development by allowing comparative studies on the safety, immunogenicity, and efficacy in adult and infant macaques.

Considering the bleak prognosis for HIV-infected children and adults during the early years of the epidemic, our current ability to manage HIV infection with highly active antiretroviral therapy (ART) represents a major triumph of modern medicine. However, as even the newer drugs are still relatively expensive and carry some risk for toxicity (which may be cumulative after many years), the ultimate goal remains to have a strategy that would allow permanent withdrawal of ART, either by gaining long-term immunological control or by purging out viral reservoirs and totally curing infection. Many potential strategies to reactivate latent virus are likely to carry the risk of adverse effects (such as toxicity); in case of incomplete immunological control or eradication, the withdrawal of the antiretroviral drugs may lead to a viral rebound that may affect future treatment options. Without proof-of-concept, only few people who fare well on a stable ART regimen may be interested to enroll in such clinical trials. Accordingly, as reviewed elsewhere [36], macaque models can play an important role in better understanding viral persistence and testing novel concepts aimed at permanent control or elimination.

Conclusions

The studies that have been performed in macaque models so far have improved our understanding of the many host and viral determinants of transmission and pathogenesis of pediatric infection, such as the timing of infection and the dose and virulence of the virus. Studies in the pediatric macaque model have also provided proof-of-concept of several immune- and drug-based intervention strategies. The ongoing comparison of data obtained in these animal models with results from clinical studies in humans will provide further validation and improvement of these animal models so that they can continue to advance our scientific knowledge and guide future clinical trials with increasing predictability and reliability.

References

1. UNAIDS. Global report: UNAIDS report on the global AIDS epidemic 2010. City is Geneva. This publication has as URL: http://www.unaids.org/globalreport/Global_report.htm
2. Dorenbaum A, Cunningham CK, Gelber RD, Culnane M, Mofenson L, Britto P et al (2002) Two-dose intrapartum/newborn nevirapine and standard antiretroviral therapy to reduce perinatal HIV transmission. A randomized trial. JAMA 288:189–98
3. Cooper ER, Charurat M, Mofenson L, Hanson IC, Pitt J, Diaz C et al (2002) Combination antiretroviral strategies for the treatment of pregnant HIV-1-infected women and prevention of perinatal HIV-1 transmission. J Acquir Immune Defic Syndr 29(5):484–94
4. Guay LA, Musoke P, Fleming T, Bagenda D, Allen M, Nakabiito C et al (1999) Intrapartum and neonatal single-dose nevirapine compared with zidovudine for prevention of mother-to-child transmission of HIV-1 in Kampala, Uganda: HIVNET 012 randomized trial. Lancet 354(9181):795–802
5. John-Stewart G, Mbori-Ngacha D, Ekpini R, Janoff EN, Nkengasong J, Read JS et al (2004) Breast-feeding and transmission of HIV-1. J Acquir Immune Defic Syndr 35(2):196–202

6. Jackson JB, Musoke P, Fleming T, Guay LA, Bagenda D, Allen M et al (2003) Intrapartum and neonatal single-dose nevirapine compared with zidovudine for prevention of mother-to-child transmission of HIV-1 in Kampala, Uganda: 18-month follow-up of the HIVNET 012 randomised trial. Lancet 362(9387):859–68
7. The Breastfeeding and HIV International Transmission Study (BHITS) Group (2004) Late postnatal transmission of HIV-1 in breast-fed children: an individual patient data meta-analysis. J Infect Dis 189:2154–66
8. Kuhn L, Aldrovandi GM, Sinkala M, Kankasa C, Semrau K, Mwiya M et al (2008) Effects of early, abrupt weaning on HIV-free survival of children in Zambia. N Engl J Med 359(2):130–41
9. Coutsoudis A, Coovadia HM, Wilfert CM (2008) HIV, infant feeding and more perils for poor people: new WHO guidelines encourage review of formula milk policies. Bull World Health Organ 86(3):210–4
10. Coutsoudis A, Pillay K, Spooner E, Kuhn L, Coovadia HM (1999) Influence of infant-feeding patterns on early mother-to-child transmission of HIV-1 in Durban, South Africa: a prospective study. South African Vitamin A Study Group. Lancet 354:471–6
11. Becquet R, Bland R, Leroy V, Rollins NC, Ekouevi DK, Coutsoudis A et al (2009) Duration, pattern of breastfeeding and postnatal transmission of HIV: pooled analysis of individual data from West and South African cohorts. PLoS One 4(10):e7397
12. Mbuya MN, Humphrey JH, Majo F, Chasekwa B, Jenkins A, Israel-Ballard K et al (2010) Heat treatment of expressed breast milk is a feasible option for feeding HIV-exposed, uninfected children after 6 months of age in rural Zimbabwe. J Nutr 140(8):1481–8
13. Chantry CJ, Israel-Ballard K, Moldoveanu Z, Peerson J, Coutsoudis A, Sibeko L et al (2009) Effect of flash-heat treatment on immunoglobulins in breast milk. J Acquir Immune Defic Syndr 51(3):264–7
14. Kilewo C, Karlsson K, Ngarina M, Massawe A, Lyamuya E, Swai A et al (2009) Prevention of mother-to-child transmission of HIV-1 through breastfeeding by treating mothers with triple antiretroviral therapy in Dar es Salaam, Tanzania: the Mitra Plus study. J Acquir Immune Defic Syndr 52(3):406–16
15. Kilewo C, Karlsson K, Massawe A, Lyamuya E, Swai A, Mhalu F et al (2008) Prevention of mother-to-child transmission of HIV-1 through breast-feeding by treating infants prophylactically with lamivudine in Dar es Salaam, Tanzania: the Mitra Study. J Acquir Immune Defic Syndr 48(3):315–23
16. Chasela CS, Hudgens MG, Jamieson DJ, Kayira D, Hosseinipour MC, Kourtis AP et al (2010) Maternal or infant antiretroviral drugs to reduce HIV-1 transmission. N Engl J Med 362(24):2271–81
17. Safrit JT, Ruprecht R, Ferrantelli F, Xu W, Kitabwalla M, Van Rompay K et al (2004) Immunoprophylaxis to prevent mother to child transmission of HIV-1. J Acquir Immune Defic Syndr 35(2):169–77
18. Jayaraman P, Haigwood NL (2006) Animal models for perinatal transmission of HIV-1. Front Biosci 11:2828–44
19. Apetrei C, Robertson DL, Marx PA (2004) The history of SIVS and AIDS: epidemiology, phylogeny and biology of isolates from naturally SIV infected non-human primates (NHP) in Africa. Front Biosci 9:225–54
20. Worobey M, Telfer P, Souquiere S, Hunter M, Coleman CA, Metzger MJ et al (2010) Island biogeography reveals the deep history of SIV. Science 329(5998):1487
21. Keele BF, Jones JH, Terio KA, Estes JD, Rudicell RS, Wilson ML et al (2009) Increased mortality and AIDS-like immunopathology in wild chimpanzees infected with SIVcpz. Nature 460(7254):515–9
22. Nath BM, Schumann KE, Boyer JD (2000) The chimpanzee and other non-human-primate models in HIV-1 vaccine research. Trends Microbiol 8(9):426–31
23. Novembre FJ, Saucier M, Anderson DC, Klumpp SA, O'Neill SP, Brown CR II et al (1997) Development of AIDS in a chimpanzee infected with human immunodeficiency virus. J Virol 71(5):4086–91
24. Novembre FJ, de Rosayro J, Nidtha S, O'Neil SP, Gibson TR, Evans-Strickfaden T et al (2001) Rapid CD4(+) T-cell loss induced by human immunodeficiency virus type 1(NC) in uninfected and previously infected chimpanzees. J Virol 75(3):1533–9
25. Frumkin LR, Agy MB, Coombs RW, Panther L, Morton WR, Koehler J et al (1993) Acute infection of *Macaca nemestrina* by human immunodeficiency virus type 1. Virology 195(2):422–31
26. Locher CP, Witt SA, Herndier BG, Tenner-Racz K, Racz P, Levy JA (2001) Baboons as an animal model for human immunodeficiency virus pathogenesis and vaccine development. Immunol Rev 183:127–40
27. Livartowski J, Dormont D, Boussin F, Chamaret S, Guetard D, Vazeux R et al (1992) Clinical and virological aspects of HIV2 infection in rhesus monkeys. Cancer Detect Prev 16(5–6):341–5
28. Putkonen P, Bottiger B, Warstedt K, Thorstensson R, Albert J, Biberfeld G (1989) Experimental infection of cynomolgus monkeys (*Macaca fascicularis*) with HIV-2. J Acquir Immune Defic Syndr 2(4):366–73
29. Sodora DL, Allan JS, Apetrei C, Brenchley JM, Douek DC, Else JG et al (2009) Toward an AIDS vaccine: lessons from natural simian immunodeficiency virus infections of African nonhuman primate hosts. Nat Med 15(8):861–5
30. Silvestri G, Sodora DL, Koup RA, Paiardini M, O'Neil SP, McClure HM et al (2003) Nonpathogenic SIV infection of Sooty Mangabeys is characterized by limited bystander immunopathology despite chronic high-level viremia. Immunity 18:1–20
31. Silvestri G, Fedanov A, Germon S, Kozyr N, Kaiser WJ, Garber DA et al (2005) Divergent host responses during primary simian immunodeficiency virus SIVsm infection of natural sooty mangabey and nonnatural rhesus macaque hosts. J Virol 79(7):4043–54

32. Pandrea I, Silvestri G, Apetrei C (2009) AIDS in African nonhuman primate hosts of SIVs: a new paradigm of SIV infection. Curr HIV Res 7(1):57–72
33. Silvestri G, Paiardini M, Pandrea I, Lederman MM, Sodora DL (2007) Understanding the benign nature of SIV infection in natural hosts. J Clin Invest 117(11):3148–54
34. Gardner MB (2003) Simian AIDS: an historical perspective. J Med Primatol 32:180–6
35. Daniel MD, Letvin NL, King NW, Kannagi M, Sehgal PK, Hunt RD et al (1985) Isolation of T-cell tropic HTLV-III-like retrovirus from macaques. Science 228(4704):1201–4
36. Van Rompay KK (2010) Evaluation of antiretrovirals in animal models of HIV infection. Antiviral Res 85(1):159–75
37. Haigwood NL (2009) Update on animal models for HIV research. Eur J Immunol 39(8):1994–9
38. Staprans SI, Feinberg MB (2004) The roles of nonhuman primates in the preclinical evaluation of candidate AIDS vaccines. Expert Rev Vaccines 3(4 Suppl):S5–32
39. Haigwood NL (2004) Predictive value of primate models for AIDS. AIDS Rev 6(4):187–98
40. Morgan C, Marthas M, Miller C, Duerr A, Cheng-Mayer C, Desrosiers R et al (2008) The use of nonhuman primate models in HIV vaccine development. PLoS Med 5(8):e173
41. Miller CJ, Alexander NJ, Sutjipto S, Lackner AA, Gettie A, Hendrickx AG et al (1989) Genital mucosal transmission of simian immunodeficiency virus: animal model for heterosexual transmission of human immunodeficiency virus. J Virol 63(10):4277–84
42. Pauza CD, Emau P, Salvato MS, Trivedi P, MacKenzie D, Malkovsky M et al (1993) Pathogenesis of SIVmac251 after atraumatic inoculation of the rectal mucosa in rhesus monkeys. J Med Primatol 22(2–3):154–61
43. Baba TW, Koch J, Mittler ES, Greene M, Wyand M, Penninck D et al (1994) Mucosal infection of neonatal rhesus monkeys with cell-free SIV. AIDS Res Hum Retroviruses 10(4):351–7
44. Jayaraman P, Zhu T, Misher L, Mohan D, Kuller L, Polacino P et al (2007) Evidence for persistent, occult infection in neonatal macaques following perinatal transmission of simian-human immunodeficiency virus SF162P3. J Virol 81(2):822–34
45. Bohm RP, Martin LN, Davison-Fairburn B, Baskin GB, Murphey-Corb M (1993) Neonatal disease induced by SIV infection of the rhesus monkey (*Macaca mulatta*). AIDS Res Hum Retroviruses 9(11):1131–7
46. Marthas ML, Van Rompay KKA, Otsyula M, Miller CJ, Canfield D, Pedersen NC et al (1995) Viral factors determine progression to AIDS in simian immunodeficiency virus-infected newborn rhesus macaques. J Virol 69(7):4198–205
47. Abel K (2009) The rhesus macaque pediatric SIV infection model – a valuable tool in understanding infant HIV-1 pathogenesis and for designing pediatric HIV-1 prevention strategies. Curr HIV Res 7(1):2–11
48. Veazey RS, DeMaria M, Chalifoux LV, Shvetz DE, Pauley DR, Knight HL et al (1998) Gastrointestinal tract as a major site of CD4+ T cell depletion and viral replication in SIV infection. Science 280:427–31
49. Witvrouw M, Pannecouque C, Switzer WM, Folks TM, De Clercq E, Heneine W (2004) Susceptibility of HIV-2, SIV and SHIV to various anti-HIV-1 compounds: implications for treatment and postexposure prophylaxis. Antivir Ther 9(1):57–65
50. Uberla K, Stahl-Hennig C, Böttiger D, Mätz-Rensing K, Kaup FJ, Li J et al (1995) Animal model for the therapy of acquired immunodefiency syndrome with reverse transcriptase inhibitors. Proc Natl Acad Sci U S A 92:8210–4
51. Nishimura Y, Igarashi T, Donau OK, Buckler-White A, Buckler C, Lafont BA et al (2004) Highly pathogenic SHIVs and SIVs target different CD4+ T cell subsets in rhesus monkeys, explaining their divergent clinical courses. Proc Natl Acad Sci U S A 101(33):12324–9
52. Feinberg MB, Moore JP (2002) AIDS vaccine models: challenging challenge viruses. Nat Med 8(3):207–10
53. Lifson JD, Martin MA (2002) One step forwards, one step back. Nature 415:272–3
54. Watkins DI, Burton DR, Kallas EG, Moore JP, Koff WC (2008) Nonhuman primate models and the failure of the Merck HIV-1 vaccine in humans. Nat Med 14(6):617–21
55. Buchbinder SP, Mehrotra DV, Duerr A, Fitzgerald DW, Mogg R, Li D et al (2008) Efficacy assessment of a cell-mediated immunity HIV-1 vaccine (the step study): a double-blind, randomised, placebo-controlled, test-of-concept trial. Lancet 372(9653):1881–93
56. Subbarao S, Otten RA, Ramos A, Kim C, Jackson E, Monsour M et al (2006) Chemoprophylaxis with tenofovir disoproxil fumarate provided partial protection against infection with simian human immunodeficiency virus in macaques given multiple virus challenges. J Infect Dis 194(7):904–11
57. Harouse JM, Gettie A, Tan RCH, Blanchard J, Cheng-Mayer C (1999) Distinct pathogenic sequela in rhesus macaques infected with CCR5 or CXCR4 utilizing SHIVs. Science 284:816–9
58. Humbert M, Rasmussen RA, Song R, Ong H, Sharma P, Chenine AL et al (2008) SHIV-1157i and passaged progeny viruses encoding R5 HIV-1 clade C env cause AIDS in rhesus monkeys. Retrovirology 5:94
59. Hatziioannou T, Ambrose Z, Chung NP, Piatak M Jr, Yuan F, Trubey CM et al (2009) A macaque model of HIV-1 infection. Proc Natl Acad Sci U S A 106(11):4425–4429
60. Ho RJ, Larsen K, Kinman L, Sherbert C, Wang XY, Finn E et al (2001) Characterization of a maternal-fetal HIV transmission model using pregnant macaques infected with HIV-2(287). J Med Primatol 30(3):131–40
61. Tarantal AF, Marthas ML, Gargosky SE, Otsyula M, McChesney MB, Miller CJ et al (1995) Effects of viral virulence on intrauterine growth in SIV-infected fetal rhesus macaques *Macaca mulatta*. J Acquir Immune Defic Syndr Human Retrovirol 10:129–38

62. Amedee AM, Lacour N, Gierman JL, Martin LN, Clements JE, Bohm RJ et al (1995) Genotypic selection of simian immunodeficiency virus in macaque infants infected transplacentally. J Virol 69(12):7982–90
63. McClure HM, Anderson DC, Fultz PN, Ansari AA, Jehuda-Cohen T, Villinger F et al (1991) Maternal transmission of SIVsmm in rhesus macaques. J Med Primatol 20(4):182–7
64. Davison-Fairburn B, Blanchard J, Hu FS, Martin L, Harrison R, Ratterree M et al (1990) Experimental infection of timed-pregnant rhesus monkeys with simian immunodeficiency virus (SIV) during early, middle, and late gestation. J Med Primatol 19(3–4):381–93
65. McClure HM, Anderson DC, Ansari AA, Klumpp SA (1992) The simian immunodeficiency virus infected macaque: a model for pediatric AIDS. Pathol Biol 40(7):694–700
66. Ochs HD, Morton WR, Tsai CC, Thouless ME, Zhu Q, Kuller LD et al (1991) Maternal-fetal transmission of SIV in macaques: disseminated adenovirus infection in an offspring with congenital SIV infection. J Med Primatol 20(4):193–200
67. Amedee AM, Lacour N, Ratterree M (2003) Mother-to-infant transmission of SIV via breast-feeding in rhesus macaques. J Med Primatol 32:187–93
68. Amedee AM, Rychert J, Lacour N, Fresh L, Ratterree M (2004) Viral and immunological factors associated with breast milk transmission of SIV in rhesus macaques. Retrovirology 1(1):17
69. Pereira CM, Nosbisch C, Winter HR, Baughman WL, Unadkat JD (1994) Transplacental pharmacokinetics of dideoxyinosine in pigtailed macaques. Antimicrob Agents Chemother 38:781–6
70. Lopez-Anaya A, Unadkat JD, Schumann LA, Smith AL (1990) Pharmacokinetics of zidovudine (azidothymidine).I. Transplacental transfer. J Acquir Immune Defic Syndr 3:959–64
71. Tuntland T, Nosbisch C, Baughman W, Massarella J, Unadkat J (1996) Mechanism and rate of placental transfer of zalcitabine (2′,3′-dideoxycytidine) in *Macaca nemestrina*. Am J Obstet Gynecol 174(3):856–63
72. Van Rompay KKA, Marthas ML, Lifson JD, Berardi CJ, Vasquez GM, Agatep E et al (1998) Administration of 9-[2-(phosphonomethoxy)propyl]adenine (PMPA) for prevention of perinatal simian immunodeficiency virus infection in rhesus macaques. AIDS Res Hum Retroviruses 14(9):761–73
73. Van Rompay KKA, Hamilton M, Kearney B, Bischofberger N (2005) Pharmacokinetics of tenofovir in breast milk of lactating rhesus macaques. Antimicrob Agents Chemother 49:2093–4
74. Van Rompay KKA, Marthas ML, Ramos RA, Mandell CP, McGowan EK, Joye SM et al (1992) Simian immunodeficiency virus (SIV) infection of infant rhesus macaques as a model to test antiretroviral drug prophylaxis and therapy: oral 3′-azido-3′-deoxythymidine prevents SIV infection. Antimicrob Agents Chemother 36(11):2381–6
75. Van Rompay KKA, Otsyula MG, Marthas ML, Miller CJ, McChesney MB, Pedersen NC (1995) Immediate zidovudine treatment protects simian immunodeficiency virus-infected newborn macaques against rapid onset of AIDS. Antimicrob Agents Chemother 39(1):125–31
76. Van Rompay KKA, Otsyula MG, Tarara RP, Canfield DR, Berardi CJ, McChesney MB et al (1996) Vaccination of pregnant macaques protects newborns against mucosal simian immunodeficiency virus infection. J Infect Dis 173:1327–35
77. Baba TW, Liska V, Hoffman-Lehmann R, Vlasak J, Xu W, Ayehunie S et al (2000) Human neutralizing monoclonal antibodies of the IgG1 subtype protect against mucosal simian-human immunodeficiency virus infection. Nat Med 6(2):200–6
78. Herz AM, Robertson MN, Lynch JB, Schmidt A, Rabin M, Sherbert C et al (2002) Viral dynamics of early HIV infection in neonatal macaques after oral exposure to HIV-2287: an animal model with implications for maternal-neonatal HIV transmission. J Med Primatol 31(1):29–39
79. Van Rompay KKA, Berardi CJ, Dillard-Telm S, Tarara RP, Canfield DR, Valverde CR et al (1998) Passive immunization of newborn rhesus macaques prevents oral simian immunodeficiency virus infection. J Infect Dis 177(5): 1247–59
80. Van Rompay KKA, Berardi CJ, Aguirre NL, Bischofberger N, Lietman PS, Pedersen NC et al (1998) Two doses of PMPA protect newborn macaques against oral simian immunodeficiency virus infection. AIDS 12:F79–F83
81. Van Rompay KKA, Abel K, Lawson JR, Singh RP, Schmidt KA, Evans T et al (2005) Attenuated poxvirus-based SIV vaccines given in infancy partially protect infant and juvenile macaques against repeated oral challenge with virulent SIV. J Acquir Immune Defic Syndr 38(2):124–34
82. Van Rompay KKA, Kearney BP, Sexton JJ, Colón R, Lawson JR, Blackwood EJ et al (2006) Evaluation of oral tenofovir disoproxil fumarate and topical tenofovir GS-7340 to protect infant macaques against repeated oral challenges with virulent simian immunodeficiency virus. J Acquir Immune Defic Syndr 43:6–14
83. Ma ZM, Abel K, Rourke T, Wang Y, Miller CJ (2004) A period of transient viremia and occult infection precedes persistent viremia and antiviral immune responses during multiple low-dose intravaginal simian immunodeficiency virus inoculations. J Virol 78(24):14048–52
84. Regoes RR, Longini IM, Feinberg MB, Staprans SI (2005) Preclinical assessment of HIV vaccines and microbicides by repeated low-dose virus challenges. PLoS Med 2(8):e249
85. Otten RA, Adams DR, Kim CN, Jackson E, Pullium JK, Lee K et al (2005) Multiple vaginal exposures to low doses of R5 simian-human immunodeficiency virus: strategy to study HIV preclinical interventions in nonhuman primates. J Infect Dis 191(2):164–73

86. Abel K, Pahar B, Van Rompay KK, Fritts L, Sin C, Schmidt K et al (2006) Rapid virus dissemination in infant macaques after oral simian immunodeficiency virus exposure in the presence of local innate immune responses. J Virol 80(13):6357–67
87. Stahl-Hennig C, Steinman RM, Tenner-Racz K, Pope M, Stolte N, Mätz-Rensing K et al (1999) Rapid infection of oral mucosal-associated lymphoid tissue with simian immunodeficiency virus. Science 285:1261–5
88. Van Rompay KKA, Cherrington JM, Marthas ML, Berardi CJ, Mulato AS, Spinner A et al (1996) 9-[2-(Phosphonomethoxy)propyl]adenine therapy of established simian immunodeficiency virus infection in infant rhesus macaques. Antimicrob Agents Chemother 40(11):2586–91
89. Van Rompay KKA, Greenier JL, Cole KS, Earl P, Moss B, Steckbeck JD et al (2003) Immunization of newborn rhesus macaques with simian immunodeficiency virus (SIV) vaccines prolongs survival after oral challenge with virulent SIVmac251. J Virol 77:179–90
90. Dykhuizen M, Mitchen JL, Montefiori DC, Thomson J, Acker L, Lardy H et al (1998) Determinants of disease in the simian immunodeficiency virus-infected rhesus macaque: characterizing animals with low antibody responses and rapid progression. J Gen Virol 79(Pt 10):2461–7
91. Wilfert CM, Wilson C, Luzuriaga K, Epstein L (1994) Pathogenesis of pediatric human immunodeficiency virus type 1 infection. J Infect Dis 170:286–92
92. Veazey RS, Lifson JD, Pandrea I, Purcell J, Piatak M Jr, Lackner AA (2003) Simian immunodeficiency virus infection in neonatal macaques. J Virol 77(16):8783–92
93. Baba TW, Liska V, Khimani AH, Ray NB, Dailey PJ, Penninck D et al (1999) Live attenuated, multiply deleted simian immunodeficiency virus causes AIDS in infant and adult macaques. Nat Med 5(2):194–203
94. Van Rompay KKA, McChesney MB, Aguirre NL, Schmidt KA, Bischofberger N, Marthas ML (2001) Two low doses of tenofovir protect newborn macaques against oral simian immunodeficiency virus infection. J Infect Dis 184(4):429–38
95. Keller MA, Stiehm ER (2000) Passive immunity in prevention and treatment of infectious diseases. Clin Microbiol Rev 13(4):602–14
96. Forthal DN, Landucci G, Cole KS, Marthas M, Becerra JC, Van Rompay K (2006) Rhesus macaque polyclonal and monoclonal antibodies inhibit simian immunodeficiency virus in the presence of human or autologous rhesus effector cells. J Virol 80(18):9217–25
97. Hofmann-Lehmann R, Vlasak J, Rasmussen RA, Smith BA, Baba TW, Liska V et al (2001) Postnatal passive immunization of neonatal macaques with a triple combination of human monoclonal antibodies against oral simian-human immunodeficiency virus challenge. J Virol 75:7470–80
98. Ferrantelli F, Rasmussen RA, Buckley KA, Li PL, Wang T, Montefiori DC et al (2004) Complete protection of neonatal rhesus macaques against oral exposure to pathogenic simian-human immunodeficiency virus by human anti-HIV monoclonal antibodies. J Infect Dis 189(12):2167–73
99. Ferrantelli F, Hofmann-Lehmann R, Rasmussen RA, Wang T, Xu W, Li P-L et al (2003) Post-exposure prophylaxis with human monoclonal antibodies prevented SHIV89.6P infection or disease in neonatal macaques. AIDS 17:301–9
100. Ferrantelli F, Buckley KA, Rasmussen RA, Chalmers A, Wang T, Li PL et al (2007) Time dependence of protective post-exposure prophylaxis with human monoclonal antibodies against pathogenic SHIV challenge in newborn macaques. Virology 358(1):69–78
101. Haynes BF, Fleming J, St Clair EW, Katinger H, Stiegler G, Kunert R et al (2005) Cardiolipin polyspecific autoreactivity in two broadly neutralizing HIV-1 antibodies. Science 308(5730):1906–8
102. Ng CT, Jaworski JP, Jayaraman P, Sutton WF, Delio P, Kuller L et al (2010) Passive neutralizing antibody controls SHIV viremia and enhances B cell responses in infant macaques. Nat Med 16(10):1117–9
103. Walker LM, Phogat SK, Chan-Hui PY, Wagner D, Phung P, Goss JL et al (2009) Broad and potent neutralizing antibodies from an African donor reveal a new HIV-1 vaccine target. Science 326(5950):285–9
104. Wu X, Yang ZY, Li Y, Hogerkorp CM, Schief WR, Seaman MS et al (2010) Rational design of envelope identifies broadly neutralizing human monoclonal antibodies to HIV-1. Science 329(5993):856–61
105. Burton DR, Desrosiers RC, Doms RW, Koff WC, Kwong PD, Moore JP et al (2004) HIV vaccine design and the neutralizing antibody problem. Nat Immunol 5(3):233–6
106. Cohen J (2003) Public health. AIDS vaccine trial produces disappointment and confusion. Science 299(5611):1290–1
107. Jin X, Bauer DE, Tuttleton SE, Lewin S, Gettie A, Blanchard J et al (1999) Dramatic rise in plasma viremia after CD8+ T cell depletion in the simian immunodefiency virus-infected macaques. J Exp Med 189(6):991–8
108. Schmitz JE, Kuroda MJ, Santra S, Sasseville VG, Simon MA, Lifton MA et al (1999) Control of viremia in simian immunodeficiency virus infection by CD8+ T lymphocytes. Science 283:857–60
109. Marthas ML, Sutjipto S, Higgins J, Lohman B, Torten J, Luciw PA et al (1990) Immunization with a live, attenuated simian immunodeficiency virus (SIV) prevents early disease but not infection in rhesus macaques challenged with pathogenic SIV. J Virol 64(8):3694–700

110. Daniel MD, Kirchhoff K, Czajak SC, Sehgal PK, Desrosiers RC (1992) Protective effects of a live attenuated SIV vaccine with a deletion in the nef gene. Science 258:1938–41
111. Abel K, Compton L, Rourke T, Montefiori D, Lu D, Rothaeusler K et al (2003) Simian-human immunodeficiency virus SHIV89.6-induced protection against intravaginal challenge with pathogenic SIVmac239 is independent of the route of immunization and is associated with a combination of cytotoxic T-lymphocyte and alpha interferon responses. J Virol 77(5):3099–118
112. Demberg T, Robert-Guroff M (2009) Mucosal immunity and protection against HIV/SIV infection: strategies and challenges for vaccine design. Int Rev Immunol 28(1):20–48
113. Rerks-Ngarm S, Pitisuttithum P, Nitayaphan S, Kaewkungwal J, Chiu J, Paris R et al (2009) Vaccination with ALVAC and AIDSVAX to prevent HIV-1 infection in Thailand. N Engl J Med 361(23):2209–20
114. Cohen J (2009) HIV/AIDS research. Beyond Thailand: making sense of a qualified AIDS vaccine "success". Science 326(5953):652–3
115. Virgin HW, Walker BD (2010) Immunology and the elusive AIDS vaccine. Nature 464(7286):224–31
116. Genesca M, McChesney MB, Miller CJ (2009) Antiviral CD8+ T cells in the genital tract control viral replication and delay progression to AIDS after vaginal SIV challenge in rhesus macaques immunized with virulence attenuated SHIV 89.6. J Intern Med 265(1):67–77
117. Haase AT (2010) Targeting early infection to prevent HIV-1 mucosal transmission. Nature 464(7286):217–23
118. Rasmussen RA, Hofmann-Lehman R, Montefiori DC, Li PL, Liska V, Vlasak J et al (2002) DNA prime/protein boost vaccine strategy in neonatal macaques against simian human immunodeficiency virus. J Med Primatol 31(1):40–60
119. Greenier JL, Van Rompay KKA, Montefiori D, Earl P, Moss B, Marthas ML (2005) Simian immunodeficiency virus (SIV) envelope quasispecies transmission and evolution in infant rhesus macaques after oral challenge with uncloned SIVmac251: increased diversity is associated with neutralizing antibodies and improved survival in previously immunized animals. Virol J 2:11
120. Van Rompay KK, Abel K, Earl P, Kozlowski PA, Easlick J, Moore J et al (2010) Immunogenicity of viral vector, prime-boost SIV vaccine regimens in infant rhesus macaques: attenuated vesicular stomatitis virus (VSV) and modified vaccinia Ankara (MVA) recombinant SIV vaccines compared to live-attenuated SIV. Vaccine 28(6):1481–92
121. Marthas ML, Van Rompay KK, Abbott Z, Earl P, Buonocore-Buzzelli L, Moss B et al (2011) Partial efficacy of a VSV-SIV/MVA-SIV vaccine regimen against oral SIV challenge in infant macaques. Vaccine 29(17):3124–37
122. Rosario M, Fulkerson J, Soneji S, Parker J, Im EJ, Borthwick N et al (2010) Safety and immunogenicity of novel recombinant BCG and modified vaccinia virus Ankara vaccines in neonate rhesus macaques. J Virol 84(15): 7815–21
123. Connor EM, Sperling RS, Gelber R, Kiselev P, Scott G, O'Sullivan MJ et al (1994) Reduction of maternal-infant transmission of human immunodeficiency virus type 1 with zidovudine treatment. N Engl J Med 331:1173–80
124. Eshleman SH, Mracna M, Guay LA, Deseye M, Cunningham S, Mirochnick M et al (2001) Selection and fading of resistance mutations in women and infants receiving nevirapine to prevent HIV-1 vertical transmission (HIVNET012). AIDS 15:1951–7
125. Arrive E, Chaix ML, Nerrienet E, Blanche S, Rouzioux C, Coffie PA et al (2009) Tolerance and viral resistance after single-dose nevirapine with tenofovir and emtricitabine to prevent vertical transmission of HIV-1. AIDS 23(7):825–33
126. Chi BH, Sinkala M, Mbewe F, Cantrell RA, Kruse G, Chintu N et al (2007) Single-dose tenofovir and emtricitabine for reduction of viral resistance to non-nucleoside reverse transcriptase inhibitor drugs in women given intrapartum nevirapine for perinatal HIV prevention: an open-label randomised trial. Lancet 370(9600):1698–705
127. Shapiro RL, Hughes MD, Ogwu A, Kitch D, Lockman S, Moffat C et al (2010) Antiretroviral regimens in pregnancy and breast-feeding in Botswana. N Engl J Med 362(24):2282–94
128. Mofenson LM (2010) Prevention in neglected subpopulations: prevention of mother-to-child transmission of HIV infection. Clin Infect Dis 50(Suppl 3):S130–48
129. Van Rompay KKA, Schmidt KA, Lawson JR, Singh R, Bischofberger N, Marthas ML (2002) Topical administration of low-dose tenofovir disoproxil fumarate to protect infant macaques against multiple oral exposures of low doses of simian immunodeficiency virus. J Infect Dis 186:1508–13
130. Abdool-Karim Q, Abdool-Karim SS, Frohlich JA, Grobler AC, Baxter C, Mansoor LE et al (2010) Effectiveness and safety of tenofovir gel, an antiretroviral microbicide, for the prevention of HIV infection in women. Science 329(5996):1168–74
131. Parikh UM, Dobard C, Sharma S, Cong ME, Jia H, Martin A et al (2009) Complete protection from repeated vaginal SHIV exposures in macaques by a topical gel containing tenofovir alone or with emtricitabine. J Virol 83(20):10358–65
132. Cranage M, Sharpe S, Herrera C, Cope A, Dennis M, Berry N et al (2008) Prevention of SIV rectal transmission and priming of T cell responses in macaques after local pre-exposure application of tenofovir gel. PLoS Med 5(8):e157
133. Van Rompay KKA, Greenier JL, Marthas ML, Otsyula MG, Tarara RP, Miller CJ et al (1997) A zidovudine-resistant simian immunodeficiency virus mutant with a Q151M mutation in reverse transcriptase causes AIDS in newborn macaques. Antimicrob Agents Chemother 41:278–83

134. Van Rompay KKA, Brignolo LL, Meyer DJ, Jerome C, Tarara R, Spinner A et al (2004) Biological effects of short-term and prolonged administration of 9-[2-(phosphonomethoxy)propyl]adenine (PMPA; tenofovir) to newborn and infant rhesus macaques. Antimicrob Agents Chemother 48:1469–87
135. Van Rompay KKA, Dailey PJ, Tarara RP, Canfield DR, Aguirre NL, Cherrington JM et al (1999) Early short-term 9-[2-(phosphonomethoxy)propyl]adenine (PMPA) treatment favorably alters subsequent disease course in simian immunodeficiency virus-infected newborn rhesus macaques. J Virol 73(4):2947–55
136. Van Rompay KKA, Cherrington JM, Marthas ML, Lamy PD, Dailey PJ, Canfield DR et al (1999) 9-[2-(Phosphonomethoxy)propyl]adenine (PMPA) therapy prolongs survival of infant macaques inoculated with simian immunodeficiency virus with reduced susceptibility to PMPA. Antimicrob Agents Chemother 43(4):802–12
137. Van Rompay KK, Durand-Gasselin L, Brignolo LL, Ray AS, Abel K, Cihlar T et al (2008) Chronic administration of tenofovir to rhesus macaques from infancy through adulthood and pregnancy: summary of pharmacokinetics and biological and virological effects. Antimicrob Agents Chemother 52(9):3144–60
138. Van Rompay KKA, Singh RP, Pahar B, Sodora DL, Wingfield C, Lawson JR et al (2004) CD8+ cell-mediated suppression of virulent simian immunodeficiency virus during tenofovir treatment. J Virol 78:5324–37
139. Barditch-Crovo P, Deeks SG, Collier A, Safrin S, Coakley DF, Miller M et al (2001) Phase I/II trial of the pharmacokinetics, safety, and antiretroviral activity of tenofovir disoproxil fumarate in HIV-1 infected adults. Antimicrob Agents Chemother 45(10):2733–9
140. Deeks SG, Barditch-Crovo P, Lietman PS, Hwang F, Cundy KC, Rooney JF et al (1998) Safety, pharmacokinetics and antiretroviral activity of intravenous 9-[2-(R)-(phosphonomethoxy)propyl]adenine, a novel anti-human immunodeficiency virus (HIV) therapy, in HIV-infected adults. Antimicrob Agents Chemother 42(9):2380–4
141. Schooley RT, Ruane P, Myers RA, Beall G, Lampiris H, Berger D et al (2002) Tenofovir DF in antiretroviral-experienced patients: results from a 48-week, randomized, double-blind study. AIDS 16:1257–63
142. Grant RM, Lama JR, Anderson PL, McMahan V, Liu AY, Vargas L et al (2010) Preexposure chemoprophylaxis for HIV prevention in men who have sex with men. N Engl J Med 363(27):2587–99
143. Juskewitch JE, Tapia CJ, Windebank AJ (2010) Lessons from the Salk polio vaccine: methods for and risks of rapid translation. Clin Transl Sci 3(4):182–5
144. Datta SK, Bhatla N, Burgess MA, Lehtinen M, Bock HL (2009) Women and vaccinations: from smallpox to the future, a tribute to a partnership benefiting humanity for over 200 years. Hum Vaccin 5(7):450–4
145. Ambrose Z, Boltz V, Palmer S, Coffin JM, Hughes SH, Kewalramani VN (2004) In vitro characterization of a simian immunodeficiency virus-human immunodeficiency virus (HIV) chimera expressing HIV type 1 reverse transcriptase to study antiviral resistance in pigtail macaques. J Virol 78(24):13553–61
146. Li JT, Halloran M, Lord CI, Watson A, Ranchalis J, Fung M et al (1995) Persistent infection of macaques with simian-human immunodeficiency viruses. J Virol 69(11):7061–71
147. Mirochnick M, Thomas T, Capparelli E, Zeh C, Holland D, Masaba R et al (2009) Antiretroviral concentrations in breast-feeding infants of mothers receiving highly active antiretroviral therapy. Antimicrob Agents Chemother 53(3):1170–6

Chapter 8
Antiretroviral Pharmacology in Breast Milk

Amanda H. Corbett

Introduction

Despite the risks of HIV transmission via breastfeeding, it is utilized worldwide for infant nutrition, especially in resource poor countries [1]. Due to limited resources, many mothers are not receiving antiretroviral therapy for their own disease, hence, the need for interventions during the peri- and postpartum period to prevent mother-to-child transmission. One strategy to limit transmission is providing antiretroviral agents to mothers of breastfeeding infants or antiretrovirals directly to the infants during the postpartum period while breastfeeding [1]. Both strategies have been shown to be effective at decreasing transmission rates [1–14]. This chapter focuses on the pharmacology of antiretrovirals in breast milk, while being administered to mothers, and the resultant exposure to the breastfeeding infant.

Pharmacology of Medications Used During Breastfeeding

There are several factors that affect the exposure of drugs to breastfeeding infants: (1) the concentration of drug in the mother's plasma, (2) the amount of drug excreted into the breast milk, and (3) the amount of milk the infant ingests daily [15].

1. *The concentration of drug in the mother's plasma*
   Mother's plasma concentrations in the early postpartum period can be quite different than in pregnant and nonpregnant women. A multitude of physiologic changes take place during pregnancy which affect drug disposition [16]. The period of time postpartum for the reversion of these changes to baseline is not fully known. Therefore, drug disposition in the postpartum period can be unpredictable. For example, during pregnancy the amount of albumin and alpha-1-acid glycoprotein declines over time with approximately 70–80% of prepreganancy values at delivery [17]. This has

---

A.H. Corbett, Pharm.D., B.C.P.S., F.C.C.P., AAHIVE (✉)
The University of North Carolina Eshelman School of Pharmacy,
3202 Kerr Hall; CB# 7569, Chapel Hill, NC 27599, USA

The University of North Carolina Center for AIDS Research,
3202 Kerr Hall; CB# 7569, Chapel Hill, NC 27599, USA
e-mail: amanda_corbett@unc.edu

been shown to be clinically significant for anticonvulsants such as phenytoin used during pregnancy [18, 19]. When protein binding is decreased, total drug concentration of phenytoin will decrease much more than unbound concentration; therefore, adjusting doses based on unbound concentration would be critical to avoid overdosing [18, 19].

Investigations with antiretrovirals have aimed at evaluating dosing during pregnancy and during the postpartum period [21, 22]. Two of these evaluated lopinavir/ritonavir concentrations during pregnancy (with adjusted doses of drug during that time) and at various times postpartum [21, 22]. Even at 2 weeks postpartum it seems that mother's plasma concentrations of lopinavir/ritonavir are higher than that in the third trimester, suggesting very early postpartum changes in drug exposures [21, 22]. This would suggest that changes in pregnancy-related drug disposition return to baseline in the early postpartum period for this medication.

2. *The amount of drug excreted into the breast milk*
Multiple drug characteristics contribute to the degree of exposure in breast milk including the degree of protein binding, lipo- or hydrophilicity, and ionization [17]. Several in vitro and in vivo models have attempted to predict the exposure of drugs into breast milk based on pKa (the acid dissociation constant), lipophilicity, protein binding, and fat content of breast milk [23–25]. All available data have concluded that large intra- and intersubject variability exists and that the gold standard for predicting breast milk exposure is the concentration seen in the breastfeeding infant's plasma [17]. The most predictive drug characteristic of breast milk:mother plasma ratio has been protein binding [17]. In a comprehensive literature review of exposure of drugs to breastfeeding infants, no infant exposures existed for drugs with >85% protein binding. An investigation conducted in the mid-1980s evaluating beta-blocker transfer into breast milk found that regardless of lipophilicity or whole-milk-binding properties, plasma protein binding was the most predictive of breast milk exposures [26].

3. *The amount of milk the infant ingests daily*
All infants do not ingest the same amount of milk daily. Milk ingestion can vary based on infant age, sex, and water intake [27]. Also, medications and hormone production can alter the amount of milk production that is available to the infant [16, 28]. These are important factors included in the calculation for infant "doses" of drugs via breastfeeding

$$\text{Dose}_{\text{absolute}} = C_{\text{milk}} \times V_{\text{milk}},$$

where $\text{Dose}_{\text{absolute}}$ = absolute infant dose, C_{milk} = drug concentration in milk, and V_{milk} = volume of milk ingested [16].

Both of these factors must be taken into account when predicting the exposure of drugs into breast milk.

Assays for Measuring Antiretroviral Drug Concentrations in Breast Milk

As with any matrix, it can be difficult to develop an accurate and precise drug assay for measuring drug concentrations in breast milk. Milk has a combination of protein, fat, and carbohydrates in variable amounts [29]. The largest component of milk is water, accounting for 65–90% in mammalian species, with the remainder consisting of dissolved and suspended solids and various nutrients [29]. Many factors can contribute to the relative amounts of each nutrient in milk, including diet and the numbers of days since onset of lactation. In addition, drug penetration into breast milk can be affected by multiple factors including protein binding, ion trapping, and lipophilicity. Most drugs have been reported to passively diffuse into breast milk from mother's plasma. Drugs with higher protein binding, such as some antiretrovirals, would have limited penetration, as the protein-bound drugs would be too large to passively diffuse. Once these drugs are in the breast milk there is much less protein binding relative to plasma (20–60% of plasma). Secondly, weakly basic drugs tend to concentrate in milk as they become ionized in this matrix due to the lower pH of milk compared to blood. Finally, lipid-soluble drugs will concentrate in the fat component of milk which prevents return to plasma but will be ingested by the infant.

Multiple techniques have been evaluated to provide high efficiency extraction of drugs in milk [29]. Liquid–liquid and solid-phase extraction techniques for analysis with high-performance liquid chromatography (HPLC) have shown to be the most efficient. One published assay for detection of multiple antiretrovirals in one assay utilized protein precipitation prior to solid-phase extraction in order to isolate the liquid layer of milk with a resultant 92.6% extraction efficiency [30]. Subsequent HPLC/MS/MS (MS = mass spectrometry) detection accurately and precisely detected seven antiretrovirals in breast milk with >99% intra- and interday accuracy and 5.0% and 7.8% intra- and interday precision, respectively.

Antiretrovirals Used During Breastfeeding for Prevention of Mother-to-Child Transmission of HIV

Many observational studies and clinical trials have evaluated the use of antiretrovirals for prevention of mother-to-child transmission [3, 4, 6–15]. Eight of these interventions administered antiretroviral combination therapy to the mother during the postpartum breastfeeding period [3, 4, 8–11, 14]. Some studies included both formula and breastfed infants, while others enrolled only mothers willing to breastfeed for the duration of the antiretroviral intervention. It is difficult to directly compare the differences between maternal antiretroviral therapy and infant prophylaxis between studies; however, overall it is thought that the interventions are comparable [2]. The percent of HIV infections in the infants that were exposed to combination antiretroviral therapy via breastfeeding was very low with a range of 0.4–3.0% at the end of therapy and breastfeeding [3, 4, 6, 8–11, 14]. All but one regimen consisted of two nucleoside reverse transcriptase inhibitor (NRTI) combinations (mostly zidovudine + lamivudine) and one of the following: nevirapine, efavirenz, nelfinavir, or lopinavir/ritonavir. One study administered a triple NRTI therapy of zidovudine + lamivudine + abacavir. Despite very low levels of transmission, 100% protection was not achieved in any of the studies. This is likely due to multiple factors including adherence, resistance, antiretroviral drug exposure in the breast milk, and subsequent effects on infant exposures or a combination of all of these factors.

Exposure of Antiretrovirals into Breast Milk

Pharmacokinetic data from these investigations and others have demonstrated differential penetration of antiretrovirals into breast milk and subsequently to infants [31–39]. It is critical to know the extent of drug exposures in breast milk in order to provide the best protective effect to infants without a risk of toxicity or antiretroviral resistance if in fact the infant is exposed to HIV.

To identify the best prevention strategies, further analysis of current clinical data combining viral suppression, drug concentrations, and transmission is warranted. A recent investigation determined that there are two potential mechanisms for viral populations in breast milk [40]. One is by trafficking of virus from blood into breast milk, which is thought to constitute the largest proportion. The second is transient breast milk viral replication from recently trafficking virus from the blood [40]. Therefore, treatment strategies should target both these mechanisms of viral proliferation in breast milk.

Table 8.1 Summary of antiretroviral concentrations in mother's plasma, breast milk, and infant's plasma based on single concentration time points

References	Antiretroviral	BM/MP ratio	IP/BM ratio	IP/MP ratio
NRTIs				
Corbett et al. [34]	Zidovudine	1.86	0	0
Mirochnick et al. [36]	Zidovudine	0.44	0	0
Shapiro et al. [38]	Zidovudine	3.21	NR	2.5[a]
Corbett et al. [34]	Lamivudine	5.57	0.003	0.01
Corbett et al. [33]	Lamivudine	2.6	0.01	0.06
Mirochnick et al. [36]	Lamivudine	2.56	0.02[b]	0.04[b]
Shapiro et al. [38]	Lamivudine	3.34	NR	0.04
NNRTIs				
Colebunders et al. [32]	Nevirapine	0.68–0.9	NR	NR
Corbett et al. [33]	Nevirapine	0.7	0.17	0.12
Mirochnick et al. [36]	Nevirapine	0.75	0.17[b]	0.13[b]
Shapiro et al. [38]	Nevirapine	0.67	NR	0–0.20
Schneider et al. [37]	Efavirenz	0.54	0.25	0.13
PIs				
Colebunders et al. [32]	Nelfinavir	0.06–0.24	NR	NR
Corbett et al. [33]	Nelfinavir	0.08	0	0
Colebunders et al. [32]	Indinavir	0.9–5.4	NR	NR
Corbett et al. [34]	Lopinavir	0.11	0	0
	Ritonavir	0.11	0	0

MP mother's plasma, *BM* breast milk, *IP* infant's plasma, *NR* not reported
[a]Infants were also directly receiving drug
[b]Estimates based on data in manuscript

Table 8.2 Summary of antiretroviral concentrations in mother's plasma, breast milk, and infant's plasma based on multiple sampling over the dosing interval

References	Antiretroviral	BM/MP AUC ratio	IP/BM AUC ratio	IP/MP AUC ratio
NRTIs				
Corbett et al. [35]	Zidovudine	1.4	0	0
Spencer et al. [39]	Zidovudine	0.22	NR	NR
Corbett et al. [35]	Lamivudine	1.2	0.02	0.03
Spencer et al. [39]	Lamivudine	0.89	NR	NR
PIs				
Spencer et al. [39]	Atazanavir	0.09	NR	NR
Corbett et al. [35]	Lopinavir	0.21	0	0
	Ritonavir	0.21	0	0

MP mother's plasma, *BM* breast milk, *IP* infant's plasma, *NR* not reported, *AUC* area under the concentration time curve (h×ng/mL)

Nucleoside Reverse Transcriptase Inhibitors

Both zidovudine and lamivudine concentrations have been evaluated in breast milk [34–36, 38, 39].

Zidovudine

There is somewhat conflicting data on the exposures of zidovudine in breast milk (see Tables 8.1 and 8.2). Both single concentration time point and concentrations over the dosing interval have been evaluated for zidovudine. Three investigations showed 1.4–3.21-fold higher exposures of zidovudine in breast milk compared to mother's plasma [34, 38], while two others report a 66–78% lower exposure in breast milk [36, 39]. None of these investigations had seen changes in concentrations over 2–24 weeks postpartum, indicating a constant exposure over the postpartum breastfeeding period. Some explanations for the differences include sampling time after the previous dose, drug concentration assays, and inclusion or exclusion of zero values. The study with the highest exposure of zidovudine (3.21-fold) also had very high mother's plasma concentrations compared to the other studies. This could be due to the median sampling time after the dose of 4.0 h where there would be an expected higher exposure of zidovudine [38]. In addition, two studies with higher zidovudine exposures in breast milk also reported a very high rate of undetectable concentrations at the predose time point and these zero values were excluded in the analysis for BM/MP ratios [34, 35]. This could have provided a falsely elevated BM/MP ratio.

Differences also exist when looking at single concentration time points compared to concentrations over the dosing interval. Lower overall zidovudine exposures in breast milk compared to mother's plasma were seen in a 12-h dosing interval sampling as compared to single concentration time points (0.22- and 1.4-fold) [35, 39]. This discrepancy could be due to the differences in time postpartum of the evaluations: 14 days postpartum (0.22 BM/MP) versus 6–24 weeks postpartum (1.4 BM/MP). Additionally, nearly one-third of the predose zidovudine concentrations ($T=0$ h) were undetectable for the study with the higher BM/MP ratio and were excluded in the ratio analysis. In addition, zidovudine does not appear to concentrate in infant's plasma after ingestion of the breast milk. Zidovudine has only been detected in infant's plasma in breastfeeding infants who were also receiving their own daily prophylaxis dose of the drug [38]. The limitations of all these investigations include the following: intracellular zidovudine was not evaluated, sophisticated pharmacokinetic/pharmacodynamic evaluations were not reported, timing and volume of breastfeeding was not controlled, and whether or not the last dose prior to PK sampling was observed is not known for all studies.

Lamivudine

Lamivudine has higher exposures in breast milk compared to mother's plasma with low but detectable concentrations in breastfeeding infant's plasma (see Tables 8.1 and 8.2). Six investigations have evaluated the exposure of lamivudine in breast milk [33–36, 38, 39]. Five of the six showed a higher concentration of lamivudine in the breast milk compared to mother's plasma [1.2–5.6-fold]. The higher concentrations in breast milk will likely suppress viral replication in this compartment thereby preventing infection of the infant. Also, lamivudine concentrations are detectable in breastfeeding infant's plasma at 0.3–2% of that found in breast milk and 1–6% of that found in mother's plasma. The majority of the concentrations detected in the infant was not above the inhibitory concentration$_{50}$ (IC_{50}) for wild-type virus; therefore, would not lead to viral suppression in infants exposed to HIV. However, intracellular drug concentrations have not been evaluated. The implication of this exposure in infants is likely a better HIV protective effect as compared to antiretrovirals that are not absorbed by breastfeeding infants. Lastly, the low-level exposures in infant plasma may lead to lamivudine resistance if in fact the infant is exposed to virus; therefore, complete viral suppression in the breast milk is critical to prevent exposure of the infant.

Nonnucleoside Reverse Transcriptase Inhibitors

Both nevirapine and efavirenz have been studied in breastfeeding infants and their mothers and shown to have intermediate exposures in breast milk and infant's plasma compared to NRTIs and protease inhibitors (PIs). This is likely due to the degree of lipophilicity and protein binding relative to the other classes of antiretrovirals [32, 33, 36–38].

Nevirapine

Five studies have evaluated concentrations of nevirapine in breast milk of breastfeeding mother–infant pairs (Table 8.1) [32, 33, 36, 38, 41]. Overall exposures of nevirapine in breast milk are 67–90% those of mother's plasma with 17–25% in infant plasma relative to breast milk and 12–20% in infant plasma relative to mother's plasma. This relative exposure in infants is higher than the IC_{50} for wild-type virus in most cases, hence should provide a protective effect to infants exposed to virus through breastfeeding. For infants that have lower yet detectable concentrations in plasma, the potential for nevirapine resistance may exist; however, due to very low concentrations toxicity is less likely to occur. Resistance has been reported in breast milk with single-dose nevirapine at labor; therefore, this strategy should be avoided [42].

Additionally, nevirapine may be detectable in breast milk over an extended period of time following discontinuation. A case study evaluating nevirapine in breast milk detected concentrations up to 17 days after discontinuation [41]. In addition, concentrations were greater than the IC_{90} for wild-type virus for 6 days after discontinuation in breast milk. This prolonged exposure without combination antiretroviral therapy could likely lead to nevirapine resistance in breast milk and subsequently to the infant. Use of an overlap in therapy with two additional NRTIs, for example, would be warranted in a patient that discontinues nevirapine therapy.

Efavirenz

Similar to nevirapine, efavirenz penetrates into breast milk with subsequent exposures in infants (see Table 8.1) [37]. One study of 13 breastfeeding mother–infant pairs evaluated efavirenz exposures over 6 weeks to 6 months postpartum [37]. Breast milk concentrations were 54% those of mother's plasma with 25% and 13% of infant plasma concentrations relative to breast milk and mother's plasma, respectively. The degree of exposure in the majority of these infants was below the therapeutic concentration required for viral load suppression in adults; however, there is no documented concentration for protective or prophylactic effects of an infant exposed to HIV. The low-level exposure in both breast milk and infants could, however, lead to viral resistance in either breast milk or infant's plasma.

Protease Inhibitors

Atazanavir, indinavir, nelfinavir, lopinavir, and ritonavir concentrations in breast milk and infant's plasma via breastfeeding have been investigated and shown overall to have lower exposures in these compartments compared to NRTIs and NNRTIs (see Tables 8.1 and 8.2). This is most likely due to their high lipophilicity and high protein binding [32–35, 39].

Atazanavir

Concentrations of atazanavir in breast milk have been shown to be minimal (9% that of mother's plasma) (see Table 8.2) [39]. This is based on a PK study in seven mothers at 5 and 14 days postpartum.

Breast milk and plasma samples were obtained over a 24-h period postdose; however, infant plasma was not obtained as these mothers were not breastfeeding, simply lactating. Despite low levels of exposure, all atazanavir breast milk concentrations were above the IC_{50} for wild-type virus; however, to varying degrees. It is not known what degree of elevated concentrations above the IC_{50} in breast milk is required for viral suppression. In addition, there were no differences in exposures between the colostrum (day 5) and mature milk (day 14), suggesting that atazanavir concentrations are present in breast milk in the early postpartum period. Since infant plasma concentrations were not evaluated, it is not known whether a direct protective effect is possible for infants exposed to HIV via breastfeeding. However, based on breast milk concentrations above the HIV inhibitory concentration, breast milk viral load is likely to be suppressed.

Indinavir

Data on indinavir exposures in breast milk are sparce but, of that reported, 90–540% of indinaivir concentrations are present in breast milk relative to mother's plasma (see Table 8.1) [32]. This is based on a single mother who donated breast milk samples over 5 days postpartum. The plasma and breast milk viral load for this patient was <50 copies/mL. This minimal data would suggest that indinavir may be a reasonable option for breastfeeding mothers.

Lopinavir

Lopinavir is detectable in breast milk but undetectable in plasma of breastfeeding infants (see Tables 8.1 and 8.2) [34, 35]. In breast milk, 11–21% of lopinavir mother's plasma concentrations are present. Despite low exposures in breast milk, the majority of concentrations reported were >IC_{50} for wild-type virus. Nearly double the degree of exposures is seen when evaluating lopinavir over the dosing interval compared to single concentration time points; however, this exposure is still only 21% of mother's plasma. These differences are likely due to limitations such as timing of sampling after the dosing interval; however, it does represent the true exposure of lopinavir over time in breast milk. There were no detectable concentrations in infant's plasma. There is potential for breast milk viral load suppression; a direct protective effect in infants would be unlikely.

Nelfinavir

Minimal data are available for the exposure of nelfinavir in breast milk (see Table 8.1) [32, 33]. An estimated 6–24% of mother's plasma concentrations are present in breast milk with no detectable concentrations in plasma of breastfeeding infants. This data is based on single concentration time points over 0–17 h after the previous dose of nelfinavir. The majority of (but not all) concentrations in breast milk were above the IC_{50} for wild-type virus hence nelfinavir may not provide complete viral suppression in the breast milk. Since no concentration was detectable in infant plasma, a direct protective effect may not exist either.

Ritonavir

Ritonavir breast milk concentrations are 11–21% those of mother's plasma, with no detectable concentrations in infant's plasma (see Tables 8.1 and 8.2) [34, 35]. Similar to lopinavir, breast milk exposures for ritonavir were twice as high when evaluating concentrations over the dosing interval as compared to single concentration time points. Nearly all the concentrations in breast milk were above the IC_{50} for wild-type virus suggesting the potential for breast milk viral load suppression. Due to no detectable concentrations in infant plasma, a direct protective effect is unlikely.

Single Concentration Time Points Versus Extensive Sampling

Zidovudine, lamivudine, lopinavir, and ritonavir breast milk and infant plasma concentrations have been evaluated for both single concentration time points and more intensive sampling over the dosing interval as described previously [34, 35]. Zidovudine exposures were similar for both of these approaches. Lamivudine levels were much higher when looking at single concentrations only (5.57-fold BM/MP) versus over the dosing interval (1.2-fold BM/MP). The opposite was true for lopinavir and ritonavir, where 11% of drug penetrated breast milk when evaluating single concentrations versus 21% when including the entire dosing interval. It does not seem that single concentration time points accurately predict exposures over the dosing interval when using the latter as the gold standard. This could be concerning if in fact concentrations are lower than expected during the dosing interval leading to viral replication and conversely if concentrations are higher than expected leading to toxicity in the infant if exposed. Therefore, evaluation of drug exposures over the dosing interval (area under the concentration time curve) may be required for determining correlations with outcomes such as viral load and transmission rates.

Implications for Differential Exposures of Antiretrovirals into Breast Milk

In summary, the NRTIs zidovudine and lamivudine have higher concentrations in breast milk compared to mother's plasma with only lamivudine penetrating into infant plasma at very low concentrations. The NNRTIs efavirenz and nevirapine have moderate exposure in breast milk with low but potentially effective penetration into infant's plasma. The PIs atazanavir, indinavir, lopinavir, nelfinavir, and ritonavir have low-level exposures in breast milk with no detectable concentrations in infant plasma. These differential exposures are somewhat predicted based on the degrees of lipophilicity and protein binding of each of these classes.

Further information on protein binding and intracellular NRTI concentrations in breast milk are needed to evaluate the best antiretroviral prevention strategies during breastfeeding. In addition, the concentration needed for viral load suppression in breast milk should be investigated as a common approach for all agents. Reasons for transmission events despite antiretroviral interventions and adequate adherence need to be identified. Low exposures of antiretrovirals in both breast milk and infant plasma need to be considered in order to prevent potential antiretroviral resistance in the event an infant is infected. More data correlating the already available pharmacokinetics in breast milk with outcomes such as plasma and breast milk viral load, adherence to antiretrovirals, and transmission of HIV to infants are needed to provide guidance on the best antiretroviral approaches for prevention of mother-to-child transmission during breastfeeding.

References

1. UNICEF: Child Info: Monitoring the situation of children and women. lhttp://www.childinfo.org/breastfeeding_challenge.html. Accessed 17 May 2011
2. Mofenson L (2010) Antiretroviral drugs to prevent breastfeeding HIV transmission. Antivir Ther 15:537–553

3. Chasela CS, Hudgens MG, Jaimeson DJ, Kayira D, Hosseinipour M, Kourtis AP et al (2010) Maternal or infant antiretroviral drugs to reduce HIV-1 transmission. N Engl J Med 362:2271–2281
4. The Kesho Bora Study Group (2011) Triple antiretroviral compared with zidovudine and single-dose nevirapine prophylaxis during pregnancy and breastfeeding for prevention of mother-to-child transmission of HIV-1 (Kesho Bora study): a randomized controlled trial. Lancet Infect Dis 11:171–180
5. Kilewo C, Karlsson K, Massawe A, Lyamaya E, Swai A, Mhalu F et al (2008) Prevention of mother-to-child transmission of HIV-1 through breastfeeding by treating infants prophylactically with lamivudine in Dar es Salaam, Tanzania: the Mitra Study. J Acquir Immune Defic Syndr 48:315–323
6. Kilewo C, Karlsson K, Ngarina M, Massawe A, Lyamuya E, Swai A et al (2009) Prevention of mother-to-child transmission of HIV-1 through breastfeeding by treating mothers with triple antiretroviral therapy in Dar es Salaam, Tanzania: the MITRA PLUS Study. J Acquir Immune Defic Syndr 52:406–416
7. Kumwenda NI, Hoover DR, Mofenson LM, Thigpen MC, Kafulafula G, Li Q et al (2008) Extended antiretroviral prophylaxis to reduce breast-milk HIV-1 transmission. N Engl J Med 359:119–129
8. Marazzi CM, Germano P, Liotta G, Guidotti G, Loureiro S, Gomes AC et al (2007) Implementing antiretroviral triple therapy to prevent HIV mother-to-child transmission: a public health approach in resource-limited settings. Eur J Pediatr 166:1305–1307
9. Palombi L, Marazzi MC, Voetberg A, Magid NA (2007) Treatment acceleration program and the experience of the DREAM program in prevention of mother-to-child transmission of HIV. AIDS 21(suppl 4):S65–S71
10. Peltier CA, Ndayisaba GF, Lepage P, van Griensven J, Leroy V, Pharm CO et al (2009) Breastfeeding with maternal antiretroviral therapy or formula feeding to prevent HIV postnatal mother to child infection in Rwanda. AIDS 23:2415–2423
11. Shapiro R, Hughes M, Ogwe A, Kitch D, Lockman S, Moffat C et al (2010) Antiretroviral regimens in pregnancy and breastfeeding in Botswana. N Engl J Med 362:2282–2294
12. Six Week Extended-Dose Nevirapine (SWEN) Study Team (2008) Extended-dose nevirapine to 6 weeks of age for infants to prevent HIV transmission via breastfeeding in Ethiopia, India, and Uganda: an analysis of three randomized controlled trials. Lancet 372:300–313
13. Thior I, Lockman S, Smeaton LM, Shapiro RL, Wester C, Heymann SJ et al (2006) Breastfeeding plus infant zidovudine prophylaxis for 6 months vs formula feeding plus infant zidovudine for 1 month to reduce mother-to-child HIV transmission in Botswana: a randomized trial: the Mashi Study. JAMA 296:794–805
14. Thomas TK, Masaba R, Borkowf CB, Ndivo R, Zeh C, Misore A, et al. Triple-antiretroviral prophylaxis to prevent mother-to-child HIV transmission through breastfeeding – the Kisumu Breastfeeding Study, Kenya: a clinical trial. PLoS Med 2011 Mar;8(3):e1001015. Epub 2011 Mar 29
15. Vyankandondera J, Luchters S, Hassink E, Pakker N, Mmiro F, Okong P et al SIMBA-stopping infection from mother to child via breastfeeding in Africa. 3rd international AIDS society conference on HIV pathogenesis. 24–27 July 2003, Paris, France. [Abstract LB07]
16. Begg E, Duffull S, Hackett L, Ilett KF (2002) Studying drugs in human milk: time to unify the approach. J Hum Lact 18:323–343
17. Anderson G (2006) Using pharmacokinetics to predict the effects of pregnancy and maternal-infant transfer of drugs during lactation. Expert Opin Drug Metab Toxicol 2:947–960
18. Dean M, Stock B, Patterson RJ, Levy G (1980) Serum protein binding of drugs during and after pregnancy in humans. Clin Pharmacol Ther 28:253–261
19. Tomson T, Lindbom U, Ekqvist B, Sundqvist A (1994) Epilepsy and pregnancy: a prospective study of seizure control in relation to free and total plasma concentrations of carbamazepine and phenytoin. Epilepsia 35:122–130
20. Yerby MS, Friel PN, McCormick K, Koerner M, van Allen M, Leavitt AM et al (1990) Pharmacokinetics of anticonvulsants in pregnancy: alterations in plasma protein binding. Epilepsy Res 5:223–228
21. Best BM, Stek AM, Mirochnick M, Hu C, Li H, Burchett SK et al (2010) Lopinavir tablet pharmacokinetics with an increased dose during pregnancy. J Acquir Immune Syndr 54:381–388
22. Patterson KB, Dumond JB, Prince HA, Jenkins A, Scarsci K, Wang R et al Pharmacokinetics of the lopinavir/ritonavir tablet in HIV-infected pregnant women: a longitudinal investigation of protein bound and unbound drug exposure with empiric dosage adjustment. 18th conference on retroviruses and opportunistic infections, Feb 2011, Boston, MA
23. Larsen LA, Ito S, Koren G (2003) Prediction of milk/plasma concentration ratio of drugs. Ann Pharmacother 37:1299–1306
24. Atkinson HC, Begg EJ (1990) Prediction of drug distribution into human milk from physiochemical characteristics. Clin Pharmacokinet 18:151–167
25. Notarianni LJ, Belk D, Aird SA, Bennett PN (1995) An in vitro technique for the rapid determination of drug entry into breast milk. Br J Clin Pharmacol 40:333–337
26. Riant P, Urien S, Albengres E, Duche JC, Tillement JP (1986) High plasma protein binding as a parameter in the selection of betablockers for lactating women. Biochem Pharmacol 35:4579–4581
27. da Costa TH, Haisma H, Wells JC, Mander AP, Whitehead RG, Bluck LJ (2010) How much human milk do infants consume? Data from 12 countries using a standardized stable isotope methodology. J Nutr 140:2227–2232

28. Powe CE, Puopolo KM, Newburg DS, Lonnerdal B, Chen C, Allen M et al (2011) Effects of recombinant human prolactin on breast milk composition. Pediatrics 127:e359–e366
29. Rossi DT, Wright DS (1997) Analytical considerations for trace determinations of drugs in breast milk. J Pharm Biomed Anal 15:495–504
30. Rezk N, White N, Bridges A, Abdel-Megeed MF, Mohamed TM, Moselhy SS et al (2008) Studies on antiretroviral drug concentrations in breast milk: validation of a liquid chromatography-tandem mass spectrometric method for determination of 7 anti-human immunodeficiency virus medications. Ther Drug Monit 30:611–619
31. Bulterys M, Weidle PJ, Abrams EJ, Fowler MG (2005) Combination antiretroviral therapy in African nursing mothers and drug exposure in their infants: new pharmacokinetic and virologic findings. J Infect Dis 192:709–712
32. Colebunders R, Hodossy B, Burger D, Daems T, Roelens K, Coppens M et al (2005) The effect of highly active antiretroviral treatment on viral load and antiretroviral drug levels in breast milk. AIDS 19:1912–1914
33. Corbett A, Martinson F, Rezk N, Kashuba A, Jamieson D, Chasela C et al Antiretroviral drug concentrations in breast milk and breastfeeding infants. 15th conference on retroviruses and opportunistic infections, Feb 2008, Boston, MA [Poster 648]
34. Corbett A, Chasela C, Martinson F, Rezk N, Kashuba A, Jamieson D et al Lopinavir/ritonavir concentrations in breast milk and breastfeeding infants: the BAN study. 16th Conference on retroviruses and opportunistic infections, Feb 2009, Montreal, Canada [Poster T-139]
35. Corbett A, Kayira D, Kashuba A, White N, Kourtis A, Chasela C et al Pharmacokinetics (PK) of HAART in Mothers and Breastfeeding infants from 6–24 wks post partum. 51st interscience conference on antimicrobial agents and chemotherapy, Sep 2010, Boston, MA. [Poster A1-2017]
36. Mirochnick M, Thomas T, Capparelli E, Zeh C, Holland D, Masaba R et al (2009) Antiretroviral concentrations in breast-feeding infants of mothers receiving highly active antiretroviral therapy. Antimicrob Agents Chemother 53:1170–1176
37. Schneider S, Peltier A, Gras A, Arendt V, Karasi-Omes C, Mujawamariwa A et al (2008) Efavirenz in human breast milk, mothers', and newborns' plasma. J Acquir Immune Defic Syndr 48:450–454
38. Shapiro R, Holland D, Capparelli E, Lockman S, Thior I, Wester C et al (2005) Antiretroviral concentrations in breast-feeding infants of women in Botswana receiving antiretroviral treatment. J Infect Dis 192:720–727
39. Spencer L, Neely M, Mordwinkin N, Leon T, Fredericks T, Karim R et al Intensive pharmacokinetics of zidovudine, lamivudine and atazanavir and HIV-1 viral load in breast milk and plasma of HIV + women receiving HAART. 16th conference on retroviruses and opportunistic infections, Feb 2009, Montreal, Canada
40. Salazar-Gonzalez JF, Salazar MG, Learn GH, Fouda GG, Kang HH, Mahlokozera T et al (2011) Origin and evolution of HIV-1 in breast milk determined by single-genome amplification and sequencing. J Virol 85:2751–2763
41. Bennetto-Hood C, Aldrovandi G, King J, Woodman K, Ashouri N, Acosta EP (2007) Persistence of nevirapine in breast milk after discontinuation of treatment. Clin Infect Dis 45:391–394
42. Hudelson SE, McConnell MS, Bagenda D, Piwowar-Manning E, Parsons TL, Nolan ML et al (2010) Emergence and persistence of nevirapine resistance in breast milk after single-dose nevirapine administration. AIDS 24:557–561

Part III
Mechanisms of HIV-1 Transmission Through Breast Milk: Immunology

Chapter 9
The Immune System of Breast Milk: Antimicrobial and Anti-inflammatory Properties

Philippe Lepage and Philippe Van de Perre

Introduction

Human breast milk is recognized as the optimal infant feeding. Human milk contains the nutrients necessary to support the infant's development [1–3]. Breast milk also contains components that protect young children against various infectious diseases and, as more recently described, constituents necessary to support the development of the infant's immune system (ontogeny) [1–3]. This includes various antimicrobial substances, constituents that promote tolerance and priming of the infant immune system, as well as anti-inflammatory components. It has recently become clearer that protection provided through breast milk against some infections extends well beyond weaning [4].

The immune system of the neonate differs considerably from that of an adult [5]. At birth, cells of the innate immune system (dendritic cells, macrophages, and neutrophils) and IgM- and IgG-producing cells are present in the infant's intestinal mucosae, but IgA secreting cells are extremely rare [6]. Because at least 90% of microorganisms infecting human beings use the mucosae as portal of entry [7, 8], young infants are exposed to a large number of microorganisms and are at increased risk for infection [9]. However, the intestinal immune system develops rapidly in the early postnatal period. During this critical period of immunological vulnerability, resistance to infection relies both on the protective factors in milk and on the infant developing his own innate and adaptive immunity. Breast milk provides the missing components while also actively stimulating maturation of the infant's own intestinal defense [3].

In addition to eliminating infectious agents and minimizing the damage they cause, the neonatal immune system must develop the ability to acquire tolerance. This means to discriminate between antigens that are harmless and those that are potentially dangerous. Induction of tolerance is believed to occur primarily in the gut and is facilitated by the specialized B and T cells, the production of secretory IgA (SIgA), and the skewed Th2 response [10, 11]. Failure to regulate tolerance and

P. Lepage (✉)
Department of Pediatrics, Hôpital Universitaire des Enfants Reine Fabiola,
Université Libre de Bruxelles, 15, av JJ Crocq, 1020 Brussels, Belgium
e-mail: philippe.lepage@huderf.be

P. Van de Perre
Department of Bacteriology-Virology, INSERM U 1058 "Infection by HIV and by Agents with Mucocutaneous Tropism: From Pathogenesis to Prevention", University Montpellier 1, CHRU Montpellier, 371 Avenue du Doyen Gaston Giraud, 34295 Montpellier Cedex 5, France
e-mail: p-van_de_perre@chu-montpellier.fr

active immune responses is hypothesized to contribute to food-related allergy, autoimmunity, and inflammatory bowel disorders. Human milk promotes oral tolerance in the infant [12] and exclusive breast-feeding may even prevent or delay the onset of atopic illness such as allergies.

Immunological Components of Human Milk

Breast Milk Cells

The cellular content of breast milk has certainly not yet revealed all its surprises. Recent studies demonstrated that, particularly in mature milk, epithelial cell adhesion molecule (EPCAM+) cells from the mammary gland are the most represented cell type in breast milk [13]. The exact role of these cells ingested by neonates and breast-feeding infants remains unknown. Stem/progenitor cells have also been identified by their specific surface markers in breast milk [14]. Their physiological role, if any, is equally mysterious.

Excluding epithelial cells, neutrophils comprise roughly 80% of breast milk cells, macrophages 15%, and lymphocytes 4% [15].

Lymphocytes

Various lymphocyte types are coexisting in breast milk: CD3+ T cells (representing roughly 83% of lymphocytes, almost equally distributed in CD4+ and CD8+ lymphocytes), gdT cells (11%), CD16+ NK cells (3–4%), and B cells (2%). Breast milk T and B lymphocytes share three major characteristics that differentiate them from circulating lymphocytes. First, memory T and B lymphocytes are overrepresented in breast milk, the majority of cells failing to express CD45RA receptor characterizing naïve cells [16, 17]. More than 70% of breast milk B cells are IgD-CD27+ memory B cells [18]. Second, many T and B lymphocytes from breast milk are activated, as they frequently express markers such as HLA-DR and CD38 [16–20]. This is paradoxical since human milk per se is remarkably not prone to immune activation. As compared with other media including plasma, blood lymphocytes incubated in the presence of human milk display considerably less frequently activation markers at their surface (E. Tuaillon, personal communication). Most likely, breast milk lymphocytes become activated through the extravasation process and/or their transepithelial migration [18–21]. Breast milk B cells are frequently switched memory cells primed to secrete antibodies with overrepresentation of large-sized B cells, plasmablasts, and plasma cells [18] and are not expressing complement receptor [22]. Third, breast milk T [16] and B cells [18] predominantly harbor mucosal homing markers (CD49f, b7 integrin, CD103, CD44) confirming that most of these cells migrated from the highly compartmentalized mucosal-associated lymphoid tissue (MALT) to the mammary gland as an effector site. At least in milk-derived B cells, migration seems to have occurred preferentially from gut-associated lymphoid tissue (GALT) confirming the predominant entero-mammary axis of the mucosal immune system. Migrating B cells, mainly those primed to secrete IgA antibodies, are attracted to and anchored within the mammary acini by the mucosae-associated epithelial chemokine CCL28 [23], other cells being released in breast milk.

CD8+ cytotoxic T lymphocytes can be found in human milk. These cells result from antigenic exposure of maternal mucosal surfaces and are thought to be functional [24–27].

These characteristics of breast milk lymphocytes reinforce the idea that human milk provides neonates and infants with a supplemental, highly immunoactive system primed to recognize the maternal environment and to protect from potential pathogens the mother–infant dyad may encounter.

Table 9.1 Distribution of immunoglobulins and other soluble substances in the colostrum and milk delivered to the breast-fed infant during a 24-h period

Soluble product	Concentration in mg/day at postpartum			
	<1 week	1–2 weeks	3–4 weeks	>4 weeks
IgG	50	25	25	10
IgA	5,000	1,000	1,000	1,000
IgM	70	30	15	10
Lysozyme	50	60	60	100
Lactoferrin	1,500	2,000	2,000	1,200

Adapted from ref. 34

Macrophages

Macrophages have long been considered to represent the major cell type in colostrum and breast milk consisting of 40% of breast milk leukocytes in early lactation and up to 85% in mature milk [19, 28, 29]. However, in these studies, macrophages were defined by morphologic characteristics and by using forward and scatter flow cytometry analyses. More recent studies strongly suggest that the number of breast milk macrophages has been overestimated since only a small proportion of breast milk cells harbor surface markers of macrophages such as CD14 [13–15]. Indeed, macrophages may represent only 15% of breast milk leukocytes [15].

Also, breast milk macrophages are distinct in terms of phenotype and functions from blood monocytes/macrophages conferring them a higher phagocytic capacity and a more efficacious defense against pathogens [30]. Breast milk macrophages are frequently activated [31] and their motility is much enhanced as compared with blood monocytes/macrophages [32]. Breast milk macrophages are morphologically distinct from blood macrophages and spontaneously secrete granulocyte–macrophage colony-stimulating factor (GMCSF), an important cell growth factor that enhances effector functions and cell signaling pathways [30]. They express DC-SIGN gene and protein (a DC-specific lectin that mediates interaction with HIV-1), and differentiate into CD1+ dendritic cells after incubation with IL-4 [30]. In addition, breast milk macrophages contain secretory IgA that can be released during the phagocytosis process [33] and are profuse secretors of soluble factors such as lactoferrin and complement factors C3 and C4 [29].

Immunoglobulins

The mammary gland produces SIgA in high quantity with lesser amounts of SIgM and SIgG (Table 9.1). They are one of the predominant proteins in breast milk and found at highest level in colostrum (5 mg/mL in colostrum, about 1 mg/mL in mature milk, [35]). IgA is synthesized in human milk by resident B-cells anchored in the mammary gland tissue through CCL28. These cells have migrated from the mother's intestine via the enteromammary axis [36]. IgA are synthesized as a dimer and linked to a secretory component (SC). SIgA is relatively resistant to proteolytic enzymes from the infant's gastrointestinal tract, allowing to provide a supply of IgA antibodies to the infant's gut [2]. Human milk is not only a rich source of SIgA that could be involved in situ in immune exclusion, but the specificity of the SIgA is directed against microbes common to both the mother and her infant [1].

SIgA provides antimicrobial defense in three ways (Fig. 9.1): preventing bacteria and viruses from attaching to mucosal surfaces by immune exclusion and mucosal painting, as well as neutralizing microbial toxins. Through these mechanisms, SIgA would prevent the establishment of bacterial colonies in the intestine and/or translocation across the mucosal barrier, thereby preventing an inflammatory response that would be damaging to the infant. This provides a potential mechanistic

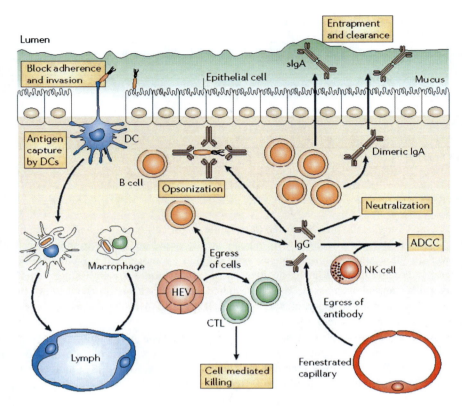

Fig. 9.1 Mechanisms of immune protection of human mucosae. From ref. 37, with permission. Abbreviations: *ADCC* antibody-dependent cellular cytotoxicity, *NK cell* natural killer cell, *CTL* cytotoxic lymphocyte responses, *DC* dendritic cells, *sIgA* secretory immunoglobulin A, *HEV* high endothelial venules

explanation for the reduced neonatal septicemia associated with breast-feeding [1]. SIgA have been identified in breast milk against many bacterial pathogens including *Escherichia coli*, *Vibrio cholerae*, *Campylobacter*, *Shigella*, *Giardia lamblia*, *Haemophilus influenzae*, *Clostridium difficile*, and *Streptococcus pneumoniae* [4], viruses such as Rotavirus, CMV, HIV, Influenza virus, respiratory syncytial virus, and yeast [36]. SIgA against maternal microflora and dietary proteins have also been identified in human milk [38]. Vaccination of mothers in late gestation will produce specific SIgA in her breast milk that may protect her infant against exposure.

Animal experiments strongly suggest that SigA is the crucial protective component of breast milk [39]. Knock-out mice lacking SIgA and SIgM show reduced protection against certain epithelial infections [6]. However, in striking contrast to humans, immunoglobulins from breast milk in many animal species (rodents, bovines, cats, ferrets, etc.) are transported across the intestinal epithelium into the neonatal circulation [40]. This is an evident limitation in translating observations made from animal models into humans.

Secretory Component

The secretory component (SC) is a glycoprotein present in many external secretions that is usually coupled with polymeric immunoglobulins. SC is also abundant in human milk [41]. It may interfere with the mucosal adhesion of enterotoxigenic *E. coli*, salmonellae, and *C. difficile* toxin A, as well as interact with a pneumococcal surface antigen [6]. SC may inhibit certain toxins, such as *C. difficile* toxin A [6].

Lactoferrin

Lactoferrin is a predominant whey protein in mature milk (≈1 g/L), its maximal concentration is found in colostrum (≈7 g/L). The concentration of lactoferrin in breast milk is controlled by the reproductive hormones prolactin and estrogen [1, 3].

Lactoferrin is an iron-binding protein. It is believed that its ability to withhold iron from pathogens explains a great part of its antimicrobial potency [42]. However, this protein has antibacterial activities that are independent of iron chelation. It may also act against some pro-inflammatory cytokines such as IL-1β, TNF-α, and IL-6 [1].

Lactoferrin may inhibit the attachment of bacteria to intestinal cells and has a direct activity against a wide range of Gram-positive and Gram-negative bacteria, fungi, viruses, including HIV-1 [43], and tumor cells [44, 45].

Lysozyme

Lysozyme is also a major whey protein of human milk. Together with SIgA, lactoferrin, and other antimicrobial compounds, this protein plays a major role in the immune defense of the infant. Lactoferrin removes lipopolysaccharide from the outer cell membrane so that lysozyme can enter and degrade the inner proteoglycan matrix of the membrane of invading cells [46].

Oligosaccharides

Nondigestible oligosaccharides are an abundant constituent of human milk. They are found in high concentration in colostrum (≈22 g/L) and are less abundant in mature milk (≈12 g/L). The types of oligosaccharides also change during lactation. Oligosaccharides are produced by an antigen-independent mechanism in epithelial cells in the mammary gland. Oligosaccharides resist hydrolysis by gastro-intestinal enzymes, indicating that they would remain intact in the small intestine.

These compounds function like bacterial receptor analogues by competing with pathogens for binding sites on the epithelial surfaces of the intestine. They can bind pathogens present on the mucosal epithelia, such as *E. coli*, *Campylobacter jejuni*, and *S. pneumoniae* [44].

Other Nonantibody Protein Defense Agents and Other Innate Immune Components

Many microorganisms express surface molecules that are believed to act like oligosaccharides [47]. These include lactadherin, mucins, and antisecretory lectins.

Lactadherin is a mucin-related glycoprotein produced by mammary epithelial cells during lactation. Its concentration in milk peaks immediately postpartum and declines thereafter. It has been reported to bind human rotavirus by preventing viral attachment to host cell receptors [48, 49]. Its concentration has been inversely associated with rotavirus symptoms in infected breast-fed infants [50].

In addition, several hydrolysis products of casein and of α-lactalbumin have antimicrobial activity against *E. coli*, *Klebsiella pneumoniae*, staphylococci, streptococci, and *C. albicans* [51].

The protein haptocorrin has been shown in vitro to resist digestion and to inhibit the growth of enterotoxigenic *E. coli* [51].

The secretory leukocyte protease inhibitor (SLPI), a serine protease inhibitor, is present at potentially active concentrations (1,100 ng/mL) in colostrum and transition milk, and it can inhibit HIV-1 entry into host cells in vitro [52].

Human milk contains active lactoperoxidase. The antibacterial properties of lactoperoxidase are closely connected to the oxidation of thiocyanate, both in vitro and in milk.

Mature milk is reported to contain a soluble form of CD14 (sCD14), a cell receptor expressed on macrophages. sCD14 is a glycoprotein that binds bacterial wall components; it is produced by mammary epithelial cells in concentration 20-fold higher than in serum [1].

Several components (C3 and C4), receptors (CF2, CD21), and activation fragments of complement are found in human milk. They might participate in immune bacteriolysis, neutralization of viruses, immune adherence, cytolysis, and enhanced phagocytosis in the infant's intestine.

Human β-defensin-1 has been demonstrated to have antimicrobial activity against *E. coli* and might also be involved in upregulating the adaptative immune system in the gut [53].

Toll-like receptors, lacking in the neonate's gut, have recently been reported in breast milk. They play a crucial "sensing" role in the early innate immune response to invading pathogens [1].

Fatty Acids

The fat content of breast milk is 3–4 g/L in colostrum, transitional and mature milk, with 93–97% of the lipids in the form of triglycerides. In addition to their nutritive and developmental benefits, milk fats have been demonstrated to provide antimicrobial activity in infant gut [54]. The mechanism for antimicrobial effects of fatty acids and monoglycerides has not been established, but it has been suggested that free fatty acids damage bacteria by disrupting their cell membranes or by changing intracellular pH.

Compounds that Promote "Healthy" Intestinal Microflora

It is well documented that the colon of breast-fed infants contains fewer potentially pathogenic bacteria, such as *E. coli*, *Bacteroides*, *Campylobacter*, and *Streptococcus*, and more beneficial bacteria, such as *Lactobacillus* and *Bifidobacterium*, as compared to formula-fed infants [55]. Activation of SIgA-producing plasma cells in the young infant's gut is dependent of the colonization of the gut by *Bifidobacteria* and *Lactobacilli* stimulated by fermentation of nondigestible oligosaccharides which are found in large amounts in human milk [10]. These are bifidus factors (compounds that enhance the growth of bifidobacteria in the intestine), prebiotics, and compounds with prebiotic activity that stimulate the growth of beneficial bacteria. Bifidogenic peptides from lactoferrin have been isolated from human milk. Lactoferricin, a cationic peptide derived from digested lactoferrin has been found to be structurally related to certain bifidogenic peptides and is hypothesized to participate in the colonization of newborn infants' flora [56]. The SC of IgA may also have some prebiotic properties [56]. It has also been hypothesized that the long-chain polyunsaturated fatty acids (LCPUFAs) in human milk facilitate the adhesion of probiotics to mucosal surface [57].

Effect of Human Milk on the Infant's Immune Development (Table 9.2)

The infant's intestinal immune system develops rapidly in the early postnatal period as it comes into contact with breast milk components as well as dietary and microbial antigens [6]. During the first 4–6 weeks of life, subepithelial plasma cells are unable to secrete a sufficient amount of SIgA to protect the infant mucosae [6]. This has to be compensated by breast milk intake.

Table 9.2 Compounds in human milk with possible influence on the infant's immune development

Maternal mammary epithelial cells
Maternal immune cells
Macrophages and dendritic cells
Neutrophils
Natural Killer cells
T-cells
B-cells and their immunoglobulins
Stem/progenitor cells
Cytokines
Nucleotides
Other immune components
Chemokines
Other soluble factors
Long-chain polyunsaturated fatty acids
Compounds that promote microbiological colonization of the infant's colon
Hormones and bioactive peptides

Adapted from ref. 1

Table 9.3 Cytokines found in human milk

TNF-α	IL-7
TGF-β	IL-8
IFN-γ	IL-10
IL-1β	IL-12
IL-2	IL-13
IL-4	IL-16
IL-6	IL-18

Adapted from ref. 1

Cells

The immune cells present in mother's milk play a pivotal role in bridging the gap between birth and the child's development of a fully functional immune system.

These cells may secrete immunoregulatory factors such as cytokines. The cytokines may stimulate IgA production by peripheral blood lymphocytes [2]. The same cytokines are found in breast milk, and the presence of transforming growth factor-β (TGF-β), IL-6, IL-7, and IL-10 is important for immunologic development and differentiation of IgA-producing cells in the infant [6, 58, 59].

Breast milk neutrophils are also present in activated form, but seem to have limited functional capacity once secreted into milk.

Cytokines

Human milk contains an array of cytokines, some in concentrations that could potentially influence immune function (Table 9.3). The exact source of the cytokines present in the aqueous fraction of breast milk remains to be determined, though it is suspected that the primary source is from cells present in the mammary gland [60]. The extent to which cytokines survive passage through the infant stomach is largely unknown, but recent work suggests that some cytokines may be sequestered and protected until they reach the intestine [61]. For example, IL-6 may be unable to survive the passage through the gastro-intestinal tract, unless it is released from milk macrophages in the neonatal gut and thereby enhance the development of the mucosal immunity.

In general, the concentration of cytokines varies widely in human milk, changes through lactation, and is influenced by the mother's health, making it difficult to assess their roles in the development of the infant's immune system.

The intake of cytokines through human milk has the potential to influence the maturation and development of immune cells in the infant. Neonates have a limited ability to produce many cytokines found in breast milk [1]. For example, TGF-β is present in breast milk and has been proposed to stimulate appropriate maturation of naïve infant intestinal immune system [4, 62, 63]. IL-6 is hypothesized to enhance the development of the infant's mucosal immunity [6]. Additionally, the cytokines present in milk assist both the transport of maternal leukocytes into milk and across the infant's gut epithelium [64]. There are many factors in breast milk that could either facilitate or inhibit cytokine activities that are not accounted for in studies conducted in vitro.

Hormones

Little is known about the activity of these compounds on the naïve immune system of an infant when delivered orally. However, it can be suspected that they impact immune development in some synergistic manner [1].

Nucleotides

Nucleotides are present in small amounts in human milk. It is likely that they play an important role for the rapid early expansion of the lymphoid system as well as the early rapid growth of the gut of the young infant. Dietary nucleotides are reported to benefit the immune system by promoting lymphocyte proliferation, NK activity, macrophage activation, and by producing a variety of other immunomodulatory factors [65]. It is suggested that nucleotides promote a Th1 response and modulate the differentiation of T and B cells [65].

Long-Chain Polyunsaturated Fatty Acids

Docosahexaenoic acid (DHA) and arachidonic acid (AA) constitute a small fraction of LCPUFAs in human breast milk, but have recently been suggested to participate in immune development [66]. Adding the LCPUFAs DHA and AA to preterm formula resulted in lymphocyte populations and cytokines more similar to human milk-fed infants than to infants who received unsupplemented formula [66].

Immune Tolerance

In addition to eliminating infectious agents and minimizing the damage they cause, the infant's immune system also has to process harmless antigens and has evolved intricate mechanisms to discriminate between antigens with the potential to cause damage and those without (tolerance). Induction of tolerance is believed to occur primarily in the gut by unique features of GALT, including specialized B- and

T-cells, the production of IgA, and the Th2 response [11, 67]. Failure to regulate tolerance and active immune responses can lead to diseases such as food-related allergy, autoimmunity, and inflammatory bowel disorders.

Although it is well established that dietary antigens/potential allergens are present in human milk (reviewed in ref. 68), the consequences on the infant's immune system remain unclear. It is hypothesized that breast milk promotes tolerance to dietary antigens and microflora via immunosuppressive cytokines (IL-10 and TGF-β) and antigens present in breast milk [6].

Compounds in Human Milk that May Be Involved in the Induction of Tolerance

1. TGF-β and IL-10
   Acquisition of tolerance involves the downregulation of the immune system through the secretion of cytokines such as TGF-β and IL-10. Low levels of TGF-β in breast milk have been related to an increased risk of atopic illness in infants, which supports the role of this cytokine in the development of tolerance [67]. Other dietary nutrients may influence the concentration of TGF-β in human milk, as levels of TGF-β were reported to be positively correlated to polyunsaturated fatty acids (PUFA) content [69]. IL-10 present in breast milk is bioactive. Human milk inhibits the proliferation of human blood lymphocytes in vitro and this inhibition is lessened by adding antihuman IL-10 antibodies in the milk [70]. IL-10 plays also a role in the synthesis of IgA [70].
2. $N-6$ and $n-3$ fatty acids
   In human milk, there are roughly ten different long chain (LC)-PUFAs consistently detected, representing both the $n-3$ and the $n-6$ series and including AA and DHA [71]. Maternal intake of PUFA prior to and during lactation is reflected in breast milk PUFA content [72]. Membrane phospholipids fatty acid composition in infants is strongly influenced by maternal diet and can alter the function of the immune cells [73]. Low $n-6/n-3$ ratios (due to higher content of $n-3$) in breast milk have been related to lower risk of atopic illness in the breast-fed population [4]. Animal studies suggest that the essential fatty acid content of the maternal diet significantly affects the development of immunological tolerance to food antigens in the suckling offspring [74]. Immunoregulatory benefits have been attributed to $n-3$ LCPUFA [68]. Dietary intervention during critical early stages of immune development before the establishment of allergic responses is thought to be most preferable [68]. Cohort studies have shown lower levels of $n-3$ PUFA in the milk of mothers whose infants showed symptoms of atopic illness before the age of 18 months [68].

Priming of the Immune System

A balance between tolerance and sensitization (priming) is necessary for the gut immune system to discriminate between harmless antigens and those associated with pathogenic and nonpathogenic microbes [67]. Antigen exposure via mother's milk has been shown to prime the immune response of the suckling pup against that antigen in rats [11]. Clinical trials have shown that breast-fed babies have enhanced specific antibody titer to some, but not all vaccines [4, 5]. The ability to transfer vaccinations from mother to infant via her milk is of great interest because it could eliminate potential problems associated with directly vaccinating the infant [76]. One possible explanation for the ability to immunize infants with their mother's milk has been attributed to the presence of anti-idiotypic antibodies in their breast milk [4]. Anti-idiotypic antibodies are antibodies with specificity against other autologous antibodies [10]. Therefore, anti-idiotypic antibodies, if present in breast milk, could have the capability of priming the infant's antibody response against the antigen the idiotype is directed to [40].

Anti-inflammatory Properties of Breast Milk

Inflammation is a necessary part of the immune response that helps protect the infant from infection. The inflammatory response traps pathogens and signals the arrival of immune cells to destroy the antigen. However, this process results in a great deal of adverse events on healthy tissue if it is not controlled [1].

The immature human intestine overreacts to both indigenous (IL-1β) and exogenous (Lipopolysaccharide-LPS) inflammatory stimuli to produce an excessive inflammatory (i.e., IL-8) response [3]. In preterm infants, this excessive inflammatory response may contribute to the high incidence of necrotizing enterocolitis. Furthermore, the immature human intestinal epithelium cannot distinguish between pathogens and commensal colonizing bacteria, and reacts to both with an inflammatory response [77]. This immature inflammatory response has to be controlled in order to prevent premature newborns from being in a chronic state of inflammation. Breast milk contains many protective anti-inflammatory components that prevent excessive inflammation until the infant can develop its own mature anti-inflammatory mechanisms.

In a human intestinal model, colostral whey incubated with a human small intestinal xenograft not only reduces the epithelial secretion of IL-8 (inflammatory response), but also downregulates the luminal expression of TLR-4 expression, thereby reducing innate inflammation.

These studies indicate that breast milk protection from infection could be due in part to anti-inflammatory components in the human milk that prevent inappropriate immune response of the infant against nondeleterious antigens.

There have been very few experimental studies on the anti-inflammatory properties of human milk. Neutrophils are the main immune cells involved in the inflammatory process and in vitro studies have shown that human milk can limit the oxidative injury produced by them (reduced cytochrome c and consumed H_2O_2) [78].

The explanation for the prophylactic nature of human milk is not currently known. However, some components of human milk have potential anti-inflammatory effects; these include cytokines (as well as their receptors and antagonists), antioxidants, antiproteases, and fatty acids [79].

Cytokines

IL-10

IL-10 inhibits the production of pro-inflammatory cytokines, providing the necessary balance to ensure that the inflammatory response is limited to destroying the pathogen and not healthy tissue. In vivo evidence of the necessity of IL-10 as an anti-inflammatory cytokine is provided by genetically altered mice that are not able to produce IL-10. These mice mount an immune response to the normal microflora in their gut, but without the IL-10 to suppress the inflammation, they develop enterocolitis (similar to ulcerative colitis and celiac disease in humans) [80]. This suggests that IL-10 in human milk might help regulate aberrant immune responses in the infant.

TGF-β

TGF-β inhibits the production of inflammatory cytokines and promotes the healing of intestinal cells damaged by injury or infection. A feeding trial examining the effectiveness of a diet with proteins and supplemented with TGF-β in the management of pediatric Crohn's disease provides the most convincing in vivo evidence of the anti-inflammatory properties of TGF-β [81]. The enteral diet containing high levels of TGF-β resulted in decreased mucosal IL-1 mRNA (pro-inflammatory cytokine) and clinical remission in 79% of the children [81].

IL-1 Receptor Antagonist

The IL-1 receptor antagonist (IL-1ra) is present in human milk. IL-1ra limits inflammation by competing with IL-1 (pro-inflammatory cytokine) for receptor binding. The reduced inflammatory response in rats with colitis fed human milk compared to formula was similar to the inflammatory response in rats fed infant formula supplemented with IL-1ra [82]. These results suggest that the IL-1ra content of human milk contributes to its anti-inflammatory properties.

Antioxidants

Free radicals, or reactive oxygen species, are produced during the normal metabolic activity of cells. These free radicals can damage cells by lipid peroxidation and alteration of protein and/or nucleic acid structures leading to oxidative stress [83]. Antioxidants in both milk and formula prevent significant lipid oxidation, breast milk suppresses oxidative DNA damage better than formula does. We have yet to identify all of the compounds in human milk that have antioxidant properties; however, there are several antioxidants in human milk that can scavenge free radicals and thereby limit the damage caused by oxidative stress. These compounds include α-tocopherol, β-carotene, cysteine, ascorbic acid, catalase, and glutathione peroxidase. The ability of human milk to resist oxidative stress is greater than formula [84]. In vitro studies have shown that human milk degrades the naturally occurring hydrogen peroxide as well as that produced by neutrophils. This is possibly due to the catalase content of human milk [79].

Lactoferrin has been shown to inhibit the production of proinflammatory cytokines (IL-6 and TNF-α) as well as inflammatory mediators (nitric oxide, GMCSF [4]). The anti-inflammatory activity of lactoferrin is generally attributed to its ability to search out free iron, which is a potent oxidizer.

Antiproteases

Inflammatory cells produce proteases, which allow the cells to enter the affected area. Some pathogens also produce proteases in order to enter the body. Human milk contains active protease inhibitors (e.g., α-1-antitrypsin, α-1-antichymotrypsin, and elastase inhibitor) that can limit the ability of pathogens to gain entry into the body and limit the inflammation caused by the inflammatory response.

Long-Chain Polyunsaturated Fatty Acids

It is hypothesized that the effects of LCPUFA $n-3$ fatty acids on immune function are mediated by their ability to compete with the metabolism of the $n-6$ fatty AA [85]. AA can be metabolized into the pro-inflammatory prostaglandin-E2 (PGE2) or leukotriene-B4 (LTB4) [86].

Prostaglandin E2 is one of the most important prostaglandins formed as it initiates the typical symptoms associated with inflammation: pain, fever, and swelling [87]. The metabolism of AA to yield PGE2 and LTB4 can be inhibited by DHA, thereby decreasing the capacity of immune cells to synthesize eicosanoids from AA. DHA will then give rise to PGE3 and LTB5, which are considered less biologically potent than the eicosanoids derived from AA. However, their activities have not yet been fully investigated.

Others

Other substances in breast milk can also produce anti-inflammatory effects in the newborn intestine. For example, lactoferrin can reduce the production of inflammatory cytokines in monocytes by inhibiting nuclear factor kappa light-chain enhancer of activated B cell (NFκB) activation.

Protection Against Infection

The adoption of exclusive breast-feeding for the first 6 months of life has been estimated to be the single most effective preventive strategy for saving lives of young children in low-income settings, with a potential estimated reduction of 13% in infant mortality rate [82]. A large part of this protective effect can contribute to a dramatic reduction in infectious diseases incidence, as it has been observed for decades [88–92].

Protection by breast-feeding against death, hospitalization, diarrheal diseases, and acute lower respiratory tract infections (ALRI) is undisputed in children living in conditions of poor hygiene, where mucosal pathogens are major killers in young children.

In developing nations, it has been demonstrated for many years that the risk of dying from diarrhea is dramatically reduced in breast-fed children [7, 93–95]. For example, Scrimshaw and Taylor [96], in the years 1955–1959, demonstrated that the infant mortality rate in villages from Punjab, India, was 950 per 1,000 live births in formula-fed babies compared with 120 in the breast-fed ones. In Rwanda, case-fatality rates for diarrhea, pneumonia, and measles were significantly lower in breastfed, than in completely weaned, hospitalized children, for all three diseases [92].

Additional data from developing countries have confirmed and extended these early observations [97–106]. In communities with a high prevalence of malnutrition, such as rural Bangladesh, breast-feeding may substantially enhance child survival up to 3 years of age [98]. Lack of breast-feeding is an important risk factor for pneumonia and ALRI mortality in developing countries [101, 103–105]. In a population-based study of infant mortality in two urban areas of southern Brazil, the type of milk in an infant's diet was found to be an important risk factor for deaths from diarrheal and respiratory infections [102]. Compared with infants who were breast-fed with no milk supplements, and after adjusting for confounding variables, those completely weaned had 3.6 times the risk of death from respiratory infections [102]. In a prospective observational study conducted in slum areas of Dhaka, Bangladesh [103], partial or no breast-feeding was associated with a 2.23-fold higher risk of infant deaths resulting from all causes and 2.40- and 3.94-fold higher risk of deaths attributable to ALRI and diarrhea, respectively, when compared with exclusive breast-feeding. Betrán et al. [104] have strongly suggested that 5% of infant deaths from diarrheal disease and acute respiratory infections in Latin America are preventable by exclusive breast-feeding among infants aged 0–3 months and partial breast-feeding throughout the remainder of infancy. A strong association between delayed initiation of breast-feeding and increased neonatal mortality has been shown in a large observational study in rural Ghana [99, 100].

A beneficial effect of breast-feeding has also been demonstrated in the industrialized world. Numerous careful studies in children from wealthy nations have now confirmed that breast-feeding protects against several common diseases such as diarrhea, acute otitis media (AOM), respiratory tract infections, invasive *Haemophilus influenzae* type B infections, neonatal septicaemia, and necrotising enterocolitis [6, 8, 107–109]. Chen and Rogan [110] have assessed the effect of breast-feeding on postneonatal mortality in the USA using recent (1988) National Maternal and Infant Health Survey data. Breast-feeding was associated with a statistically significant reduction in the risk for postneonatal death [104]. In a study of the epidemiology of middle ear disease in Massachusetts children, absence of breast-feeding was a risk factor for AOM and/or recurrent AOM (together with male gender, a sibling history of recurrent AOM, and early onset of otitis) [111].

In industrialized countries, breast-feeding has been shown to reduce the severity of lower respiratory infection in the first 6 months of life and an inverse relation between the duration of ALRI symptoms and the length of exclusive breast-feeding has been demonstrated [112]. In these countries, an effect is seen principally among infants living in crowded conditions in lower socioeconomic strata [112].

In the Republic of Belarus, an experimental intervention (modeled on the Baby-Friendly Hospital Initiative of WHO and UNICEF and emphasizing health care worker assistance with initiating and maintaining breast-feeding and lactation and postnatal breast-feeding support) was shown to decrease the risk of gastrointestinal tract infection in the first year of life [113]. In the UK, breastfeeding, particularly when exclusive and prolonged, protected against severe morbidity from diarrhea and ALRI in a large cohort of healthy, singleton, term infants born from 2000 to 2002 [114]. In Alicante, Spain, full breast-feeding was shown to lower the risk for hospital admission as a result of infections among infants younger than 1 year [115].

Conclusion

Human milk is extraordinary complex in composition. Notably, its content evolves over time. In early lactation phases, SIgA, SIgM, antiinflammatory factors and, most probably immunologically active cells provided by breast milk are substituting for the relatively immature neonatal immune system. Thereafter, breast milk continues to adapt remarkably to the infant ontogeny, to its immune protection needs and to its nutritional requirements. In particular, breast-milk antibodies are targeted against potentially deleterious infectious agents and antigens that are present in the maternal environment, which are those likely to be encountered by the infant. In that sense, it is legitimate to consider the maternal and the infant's immune systems as a continuum, with the placenta and the mammary glands/breast milk as interfaces, now known as the mother–offspring immune dyad [116].

References

1. Blewett HJH, Cicalo MC, Holland CD, Field CJ (2008) The immunological components of human milk. Adv Food Nutr 54:45–80
2. Brandtzaeg P (2010) The mucosal immune system and its integration in the mammary glands. J Pediatr 156:S8–15
3. Walker W (2010) Breast milk as the gold standard for protective nutrients. J Pediatr 156:S3–S7
4. Hanson LA, Korotkova M, Lundin S et al (2003) The transfer of immunity from mother to child. Ann NY Acad Sci 987:199–206
5. Kelly D, Coutts AG (2000) Early nutrition and the development of immune function in the neonate. Proc Nutr Soc 59:177–185
6. Brandtzaeg P (2003) Mucosal immunity: integration between mother and the breast-fed infant. Vaccine 21: 3382–3388
7. Anonymous (1994) A warm chain for breastfeeding. Lancet 344:1239–1241
8. Evidence Report/Technology Assessment Report No. 153, U.S. Department of Health: breastfeeding and maternal and infant health outcomes in developed countries. Agency for Healthcare Research and Quality (AHRQ), Publication No. 07-E007, April 2007

9. Brandtzaeg P, Nilssen DE, Rognum TO, Thrane PS (1991) Ontogeny of the mucosal immune system and IgA deficiency. Gastroenterol Clin North Am 20:397–439
10. Field CJ (2005) The immunological components of human milk and their effect on the immune development in infants. J Nutr 135:1–4
11. Strobel S (2001) Immunity induced after a feed of antigen during early life: oral tolerance v. sensitisation. Proc Nutr Soc 60:437–442
12. van Odijk J, Kull I, Borres MP et al (2003) Breastfeeding and allergic disease: a multidisciplinary review of the literature (1966–2001) on the mode of early feeding in infancy and its impact on later atopic manifestations. Allergy 58:833–843
13. Petitjean G, Becquart P, Tuaillon E et al (2007) Isolation and characterization of resting CD4+ T lymphocytes in breast milk. J Clin Virol 39:1–8
14. Fan Y, Chong YS, Choolani MA, Cregan MD, Chan JKY (2010) Unravelling the mystery of stem/progenitor cells in human breast milk. PLoS One 5:e14421
15. Goldman AS, Chheda S, Garofalo R (1998) Evolution of immunologic functions of the mammary gland and the postnatal development of immunity. Pediatr Res 43:155–162
16. Bertotto A, Gerli R, Fabietti G et al (1990) Human breast milk T cells display the phenotype and functional characteristics of memory T cells. Eur J Immunol 20:1877–1880
17. Valéa D, Tuaillon E, Al Tabaa Y et al (2011) CD4+ T cells spontaneously producing human immunodeficiency virus type I in breast milk from women with or without antiretroviral drugs. Retrovirology 8:34
18. Tuaillon E, Valea D, Becquart P et al (2009) Human milk-derived B cells: a highly activated switched memory cell population primed to secrete antibodies. J Immunol 182:7155–7162
19. Wirt DP, Adkins LT, Palkowetz KH et al (1992) Activated and memory T lymphocytes in human milk. Cytometry 13:282–290
20. Richie ER, Steinmetz KD, Meistrich ML, Ramirez I, Hilliard JK (1980) T lymphocytes in colostrum and peripheral blood differ in their capacity to form thermostable E-rosettes. J Immunol 125:2344–2346
21. Richie ER, Bass R, Meistrich ML, Dennison DK (1982) Distribution of T lymphocyte subsets in human colostrum. J Immunol 129:1116–1119
22. Bush JF, Beer AE (1979) Analysis of complement receptors on B-lymphocytes in human milk. Am J Obstet Gynecol 133:708–712
23. Wilson E, Butcher EC (2004) CCL28 controls immunoglobulin (Ig)A plasma cell accumulation in the lactating mammary gland and IgA antibody transfer to the neonate. J Exp Med 200:805–809
24. Losonsky GA, Fishaut JM, Strussenberg J, Ogra PL (1982) Effect of immunization against rubella on lactation products. I. Development and characterization of specific immunologic reactivity in breast milk. J Infect Dis 145:654–660
25. Ruben FL, Holzman IR, Fireman P (1982) Responses of lymphocytes from human colostrum or milk to influenza antigens. Am J Obstet Gynecol 143:518–522
26. Sabbaj S, Edwards BH, Ghosh MK et al (2002) Human immunodeficiency virus-specific CD8(+) T cells in human breast milk. J Virol 76:7365–7373
27. Lohman BL, Slyker J, Mbori-Ngacha D et al (2003) Prevalence and magnitude of human immunodeficiency virus (HIV) type 1-specific lymphocyte responses in breast milk from HIV-1-seropositive women. J Infect Dis 188:1666–1674
28. Crago SS, Prince SJ, Pretlow TG et al (1979) Human colostral cells. I. Separation and characterization. Clin Exp Immunol 38:585–597
29. Pitt J (1979) The milk mononuclear phagocyte. Pediatrics 64:745–749
30. Ichikawa M, Sugita M, Takahashi M, Satomi M, Takeshita T, Araki T, Takahashi H (2003) Breast milk macrophages spontaneously produce granulocyte-macrophage colony-stimulating factor and differentiate into dendritic cells in the presence of exogenous interleukin-4 alone. Immunology 108:189–195
31. Rivas RA, El Mohandes AA, Katona IM et al (1994) Mononuclear phagocytic cells in human milk: HLA-DR and FcdR ligand expression. Biol Neonate 66:195–204
32. Ozkaragöz F, Rudloff HB, Rajaraman S, Mushtaha AA, Schmalstieg FC, Goldman AS (1988) The motility of human milk macrophages in collagen gels. Pediatr Res 23:449–452
33. Weaver EA, Goldblum RM, Davis CP, Goldman AS (1981) Enhanced immunoglobulin A release from human colostral cells during phagocytosis. Infect Immun 34:498–502
34. Ogra SS, Ogra PL (1978) Immunologic aspects of human colostrum and milk. I. Distribution characteristics and concentrations of immunoglobulins at different times after the onset of lactation. J Pediatr 92:546–549
35. Goldman AS, Garza C, Nichols BL, Goldblum RM (1982) Immunologic factors in human milk during the first year of lactation. J Pediatr 100:563–567
36. Hanson LA, Korotkova M (2002) The role of breast-feeding in prevention of neonatal infection. Semin Neonatol 7:275–281
37. Neutra MR, Kozlowski PA (2006) Mucosal vaccines: the promise and the challenge. Nat Rev Immunol 6:148–158

38. Hanson LA, Korotkova M, Telemo E (2003) Breast-feeding, infant formulas and the immune system. Ann Allergy Asthma Immunol 90:59–63
39. Dickinson EC, Gorga JC, Garrett M et al (1998) Immunoglobulin A supplementation abrogates bacterial translocation and preserves the architecture of the intestinal epithelium. Surgery 124:284–290
40. Van de Perre P (2003) Transfer of antibody via mother's milk. Vaccine 21:3374–3376
41. Brandtzaeg P (1983) The secretory immune system of lactating human mammary glands compared with other exocrine organs. Ann NY Acad Sci 409:353–381
42. Lonnerdal B (2003) Nutritional and physiologic significance of human milk proteins. Am J Clin Nutr 77:1537S–1543S
43. Harmsen MC, Swart PJ, de Bethune MP et al (1995) Antiviral effect of plasma and milk proteins: lactoferrin shows potent activity against both human immunodeficiency virus and cytomegalovirus replication in vitro. J Infect Dis 172:380–388
44. Shah NP (2000) Effects of milk-derived bioactives: an overview. Br J Nutr 84(suppl 1):S3–S10
45. Bernt KM, Walker WA (1999) Human milk as carrier of biochemical message. Acta Paediatr Suppl 88:27–41
46. Ellison RT III, Giehl TJ (1991) Killing of gram-negative bacteria by lactoferrin and lysozyme. J Clin Invest 88:1080–1091
47. Gopal PK, Gill HS (2000) Oligosaccharides and glycoconjugates in bovine milk and colostrum. Br J Nutr 84(suppl 1):S69–S74
48. Kvistgaard AS, Pallesen LT, Arias CF, Lopez S, Petersen TE, Heegaard CW, Rasmussen JT (2004) Inhibitory effects of human and bovine milk constituents on rotavirus infections. J Dairy Sci 87:4088–4096
49. Yolken RH, Peterson JA, Vonderfecht SL, Fouts ET, Midthun K, Newburg DS (1992) Human milk mucin inhibits rotavirus replication and prevents experimental gastroenteritis. J Clin Invest 90:1984–1991
50. Newburg DS, Peterson JA, Ruiz-Palacios GM et al (1998) Role of human-milk lactadherin in protection against symptomatic rotavirus infection. Lancet 351:1160–1164
51. Pellegrini A, Thomas U, Bramaz N, Hunziker P, von Fellenberg R (1999) Isolation and identification of three bactericidal domains in the bovine alpha-lactalbumin molecule. Biochim Biophys Acta 1426:439–448
52. Wahl SM, McNeely TB, Janoff EN et al (1997) Secretory leukocyte protease inhibitor (SLPI) in mucosal fluids inhibits HIV-1. Oral Dis 3(suppl 1):S64–69
53. Jia HP, Starner T, Ackermann M, Kirby P, Tack BF, McCray PB Jr (2001) Abundant human beta-defensin-1 expression in milk and mammary gland epithelium. J Pediatr 138:109–112
54. German JB, Dillard CJ (2006) Composition, structure and absorption of milk lipids: a source of energy, fat-soluble nutrients and bioactive molecules. Crit Rev Food Sci Nutr 46:57–92
55. Kleessen B, Bunke H, Tovar K, Noack J, Sawatzki G (1995) Influence of two infant formulas and human milk on the development of the faecal flora in newborn infants. Acta Paediatr 84:1347–1356
56. Liepke C, Adermann K, Raida M, Magert HJ, Forssmann WG, Zucht HD (2002) Human milk provides peptides highly stimulating the growth of bifidobacteria. Eur J Biochem 269:712–718
57. Das UN (2002) Essential fatty acids as possible enhancers of the beneficial actions of probiotics. Nutrition 18:786
58. Brandtzaeg P, Johansen F-E (2005) Mucosal B cells: phenotypic characteristics, transcriptional regulation, and homing properties. Immunol Rev 206:32–63
59. Hanson LA (2007) Breast-feeding and immune function. Proc Nutr Soc 66:384–396
60. Bryan DL, Forsyth KD, Gibson RA, Hawkes JS (2006) Interleukin-2 in human milk: a potential modulator of lymphocyte development in breastfed infants. Cytokine 33:289–293
61. Calhoun DA, Lunoe M, Du Y, Staba SL, Christensen RD (1999) Concentrations of granulocyte colony-stimulating factor in human milk after in vitro stimulations of digestion. Pediatr Res 46:767–771
62. Bottcher MF, Jenmalm MC, Garofalo RP, Bjoksten B (2000) Cytokines in breast milk from allergic and non allergic mothers. Pediatr Res 47:157–162
63. Donnet-Hughes A, Duc N, Serrant P, Vidal K, Schiffrin EJ (2000) Bioactive molecules in milk and their role in health and disease: the role of transforming growth factor-beta. Immunol Cell Biol 78:74–79
64. Ustundag B, Yilmaz E, Dogan Y et al (2005) Levels of cytokines (IL-1beta, IL-2, IL-6, IL-8, TNF-alpha) and trace elements (Zn, Cu) in breast milk from mothers of preterm and term infants. Mediators Inflamm 2005:331–336
65. Aggrett P, Leach JL, Rueda R, MacLean J (2003) Innovation in infant formula development: a reassessment of ribunucleotides in 2002. Nutrition 19:375–384
66. Field CJ, Clandinin MT, Van Aerde JE (2001) Polyunsaturated fatty acids and T-cell function: implications for the neonate. Lipids 36:1025–1032
67. Mowat AM (2003) Anatomical basis of tolerance and immunity to intestinal antigens. Nat Rev Immunol 3:331–341
68. Palmer DJ, Makrides M (2006) Diet of lactating women and allergic reactions in their infants. Curr Opin Clin Nutr Metab Care 9:284–288
69. Laiho K, Lampi AM, Hamalainen M, Moilanen E, Piironen V, Arvola T, Syrjanen S, Isolauri E (2003) Breast milk fatty acids, eicosanoids, and cytokines in mothers with and without allergic disease. Pediatr Res 53:642–647

70. Garofalo R, Chheda S, Mei F et al (1995) Interleukin-10 in human milk. Pediatr Res 37:444–449
71. Koletzko B, Rodriguez-Palmero M, Demmelmair H, Fidler N, Jensen R, Sauerwald T (2001) Physiological aspects of human milk lipids. Earl Hum Dev 65 Suppl S3-S18
72. Hawkes JS, Bryan DL, Gibson RA (2002) Cytokine production by hm cells and peripheral blood mononuclear cells from the same mothers. J Clin Immunol 22:338–344
73. Field CJ, Thomson CA, Van Aerde JE, Parrott A, Euler A, Lien E, Clandinin MT (2000) Lower proportion of CD45RO+ cells and deficient interleukin-10 production by formula-fed infants, compared with human-fed, is corrected with supplementation of long-chain polyunsaturated fatty acids. J Pediatr Gastroenterol Nutr 31:291–299
74. Korotkova M, Telemo E, Hanson LA, Strandvik B (2004) Modulation of neonatal immunological tolerance to ovalbumin by maternal essential fatty acid intake. Pediatr Allergy Immunol 15:112–122
75. Harbige LS, Fisher BA (2001) Dietary fatty acid modulation of mucosally-induced tolerogenic immune responses. Proc Nutr Soc 60:449–456
76. Gust DA, Strine TW, Maurice E et al (2004) Underimmunization among children: effects of vaccine safety concerns on immunization status. Pediatrics 114:e16–e22
77. Goldman AS (1993) The immune system of human milk: antimicrobial, anti-inflammatory and immunomodulating properties. Pediatr Infect Dis J 12:664–671
78. Grazioso CF, Buescher ES (1996) Inhibition of neutrophil function by human milk. Cell Immunol 168:125–132
79. Garofalo RP, Goldman AS (1999) Expression of functional immunomodulatory and anti-inflammatory factors in human milk. Clin Perinatol 26:361–377
80. Sydora BC, Tavernini MM, Wessler A, Jewell LD, Fedorak RN (2003) Lack of interleukin-10 leads to intestinal inflammation, independent of the time at which luminal microbial colonization occurs. Inflamm Bowel Dis 9:87–97
81. Fell JM, Paintin M, Arnaud-Battandier F et al (2000) Mucosal healing and a fall in mucosal pro-inflammatory cytokine mRNA induced by a specific oral polymeric diet in paediatric Crohn's disease. Aliment Pharmacol Ther 14:281–289
82. Grazioso CF, Werner AL, Alling DW, Bishop PR, Buescher ES (1997) Antiinflammatory effects of human milk on chemically induced colitis in rats. Pediatr Res 42:639–643
83. Sharda B (2006) Free radicals: emerging challenge in environmental health research in childhood and neonatal disorders. Int J Environ Res Public Health 3:286–291
84. Friel JK, Martin SM, Langdon M, Herzberg GR, Buettner GR (2002) Milk from mothers of both premature and full-term infants provides better antioxidant protection than does infant formula. Pediatr Res 51:612–618
85. Gottrand F (2008) Long-chain polyunsaturated fatty acids influence the immune system of infants. J Nutr 38:1807S–1812S
86. Calder PC, Grimble RF (2002) Polyunsaturated fatty acids, inflammation and immunity. Eur J Clin Nutr 56(suppl 3):S14–S19
87. Wahle KW, Heys SD, Rotondo D (2004) Conjugated linoleic acids: are they beneficial or detrimental to health? Prog Lipid Res 43:553–587
88. Jones G, Steketee RW, Black RE, Bhutta ZA, Morris SS (2003) How many child deaths can we prevent this year? Lancet 362:65–71
89. Chien PF, Howie PW (2001) Breast milk and the risk of opportunistic infection in infancy in industrialized and non-industrialized settings. Adv Nutr Res 10:69–104
90. Cunningham AS, Jelliffe DB, Jelliffe EF (1991) Breastfeeding and health in the 1980s: a global epidemiologic review. J Pediatr 118:659–666
91. Grulee C, Sandord H, Schwartz H (1935) Breast and artificially fed infants; study of the age incidence in the morbidity and mortality in 20,000 cases. JAMA 104:1986–1988
92. Lepage P, Munyakazi C, Hennart P (1981) Breastfeeding and hospital mortality in children in Rwanda. Lancet 2:409–11
93. Feachem R, Koblinski MA (1984) Interventions for the control of diarrhoeal diseases among young children: promotion of breast-feeding. Bull WHO 62:271–291
94. Plank SJ, Milanesi MI (1973) Infant feeding and infant mortality in rural Chile. Bull WHO 48:203–210
95. Puffer RR, Serrano CV (1973) Patterns of mortality in childhood. Scientific publication no. 262. Pan American Health Organization 1973, Washington DC
96. Scrimshaw NS, Taylor CE (1968) Interaction of nutrition and infection. WHO monograph 1968; No. 29, Geneva
97. Bennish ML, Harris JR, Wojtyniak BJ, Struelens M (1990) Death in shigellosis: incidence and risk factors in hospitalized patients. J Infect Dis 16:500–506
98. Briend A, Wojtyniak B, Rowland MGM (1988) Breast feeding, nutritional state, and child survival in rural Bangladesh. BMJ 296:879–882
99. Edmond KM, Zandoh C, Quigley MA, Amenga-Etego S, Owusu-Agyei S, Kirkwood BR (2006) Delayed breast-feeding initiation increases risk of neonatal mortality. Pediatrics 117:380–386

100. Edmond KE, Kirkwood BR, Amenga-Etego S, Owusu-Agyei S, Hurt LS (2007) Effect of early infant feeding practices on infection-specific neonatal mortality: an investigation of the causal links with observational data from rural Ghana. Am J Clin Nutr 86:1126–1131
101. Victora CG, Kirkwood BR, Ashworth A et al (1999) Potential interventions for the prevention of childhood pneumonia in developing countries: improving nutrition. Am J Clin Nutr 70:309–320
102. Victora CG, Smith PG, Vaughan JP et al (1987) Evidence for protection by breast-feeding against infant deaths from infectious diseases in Brazil. Lancet 2(8554):319–322
103. Arifeen S, Black RE, Antelman G, Baqui A, Caulfield L, Becker S (2001) Exclusive breastfeeding reduces acute respiratory infection and diarrhea deaths among infants in Dhaka slums. Pediatrics 108:e67
104. Betrán AP, de Onís M, Lauer JA, Villar J (2001) Ecological study of effect of breast feeding on infant mortality in Latin America. BMJ 323:1–5
105. Yoon PW, Black RE, Moulton LH, Becker S (1996) Effect of not breastfeeding on the risk of diarrheal and respiratory mortality in children under 2 years of age in Metro Cebu, The Philippines. Am J Epidemiol 143:1142–1148
106. Ashraf RN, Jalil F, Zaman S et al (1991) Breastfeeding and protection against neonatal sepsis in a high risk population. Arch Dis Child 66:488–490
107. Hanson LA, Silfverdal S-A, Stromback L, Erling V, Zaman S, Olcén P, Telemo E (2001) The immunological role of breast feeding. Pediatr Allergy Immunol 12(suppl 14):15–19
108. Hanson L, Telemo E (2008) Immunobiology and epidemiology of breastfeeding in relation to prevention of infections from a global perspective. In: Ogra P, Harris WS (eds) n-3 fatty acids in health: DaVinci's code. Am J Clin Nutr vol 88, pp. 595–596
109. Howie PW, Forsyth JS, Ogston SA, Clark A, Florey CD (1990) Protective effect of breast feeding against infection. BMJ 300:11–16
110. Chen A, Rogan WJ (2004) Breastfeeding and the risk of postneonatal death in the United States. Pediatrics 113: e435–e439
111. Pelton SI, Leibovitz E (2009) Recent advances in otitis media. Pediatr Infect Dis J 28:S133–S137
112. Simoes EAF (2003) Environmental and demographic risk factors for respiratory syncytial virus lower respiratory tract disease. J Pediatr 143:S118–S126
113. Kramer MS, Chalmers B, Hodnett ED et al (2001) Promotion of breastfeeding intervention trial (PROBIT). A randomized trial in the Republic of Belarus. JAMA 285:413–420
114. Quigley MA, Kelly YQ, Sacker A (2007) Breastfeeding and hospitalization for diarrheal and respiratory infection in the United Kingdom Millennium Cohort Study. Pediatrics 119:e837–e842
115. Talayero JMP, Lizán-García M, Puime ÁO et al (2006) Full breastfeeding and hospitalization as a result of infections in the first year of life. Pediatrics 118:e92–e99
116. Hanson LA (2000) The mother-offspring dyad and the immune system. Acta Paediatr 89:252–258

Chapter 10
B Lymphocyte-Derived Humoral Immune Defenses in Breast Milk Transmission of the HIV-1*

Laurent Bélec and Athena P. Kourtis

Human Immunodeficiency Virus Postnatal Transmission Via Breast Milk

The UNAIDS estimated that more than 370,000 (230,000–510,000) children were infected by human immunodeficiency virus (HIV) type 1 through mother-to-child transmission (MTCT) worldwide in 2009, with the majority (>90%) occurring in sub-Saharan Africa (a drop of 24% from 5 years earlier) [1]. The majority of MTCT occurs during pregnancy and birth. In addition, postnatal transmission of HIV from HIV-infected mother to her child through prolonged breastfeeding is well recognized, and may account for one-third to half of new infant HIV infections worldwide [2–10]. While studies of maternal or infant antiretroviral prophylaxis during the period of breastfeeding have shown substantial potential for reduction of infant HIV infections [11–14], postnatal virus transmissions may continue to occur even in the setting of optimal antiretroviral prophylaxis. Therefore, development of immunologic strategies to reduce HIV transmission via breast milk remains important to improving survival of infants born to HIV-infected mothers in the developing world.

The seemingly low efficiency of breast milk transmission (less than 10% of infants born to HIV-infected women and breastfed during the first 6 months of life become infected postnatally [15]) contrasts with the daily exposure to high amount of infectious viral particles, suggesting that anti-infective factors in breastfeeding HIV-infected mothers as well as in HIV-exposed breastfed children are involved [16]. Identifying these factors would provide important insights into the type of immune responses required to protect against infant HIV acquisition. Human breast milk from HIV-infected mothers contains high levels of HIV-specific IgG antibodies capable of HIV neutralization and antibody-dependent cell cytotoxicity (ADCC) [17]. In addition, recent advances from nonhuman primate models of transmission of the simian immunodeficiency virus (SIV) point to the unique protective role of autologous SIV-specific adaptive humoral immune response in breast milk as a major factor for limiting breast milk SIV transmission [18]. Thus, as breast milk is a very rich source of antibodies, the potential anti-HIV activity of breast milk could be mediated by adaptive humoral immune

---

*The findings and conclusions in this report are those of the authors and do not necessarily represent the official position of the Centers for Disease Control and Prevention.

L. Bélec, M.D., Ph.D., M.P.H. (✉)
Université Paris Descartes, Sorbonne Paris Cité (Paris V), and Laboratoire de Virologie,
Hôpital Européen Georges Pompidou, 15-20 rue Leblanc, 75 908 Paris Cedex 15, France
e-mail: laurent.belec@egp.aphp.fr

A.P. Kourtis, M.D., Ph.D., M.P.H.
Division of Reproductive Health, NCCDPHP, Centers for Disease Control and Prevention,
4770 Buford Highway, NE, MSK34, Atlanta, GA 30341, USA

responses [19]. We herein focus on the humoral preimmune and immune parameters in breast milk and in exposed children that modulate the transmission of HIV by breastfeeding, in a perspective of protective vaccine development.

Low Efficiency of HIV Transmission Via Breastfeeding Contrasts with High Quantity of Daily Ingested HIV Particles

Epidemiological studies have demonstrated a relatively low efficiency of HIV transmission via breastfeeding. For example, in Malawi, breastfeeding carried an estimated risk of transmission of 0.7%/month during age 1–5 months, 0.6%/month during age 6–11 months, and 0.3%/month during age 12–17 months [8]. Breastfeeding duration is clearly a major determinant of postnatal HIV transmission [20, 21]. In a large pooled analysis of individual data from West and South African cohorts, the 18-month HIV postnatal transmission rate was 3.9% among infants breastfed for less than 6 months, and 8.7% among children breastfed for more than 6 months [21]. It has been estimated that breastfeeding-related risk ranges from 30 to 45% of all pediatric HIV infections at 24 months postdelivery [22]. In addition, MTCT of HIV through breastfeeding was estimated to be 4% during the first 6 months of exclusive breastfeeding and close to 1% per additional month of breastfeeding thereafter [15, 23–25].

The overall probability of transmission via breastfeeding was estimated to be 0.00064/L of breast milk ingested [26]. However, this estimate did not account for breast milk viral load, the intermittent nature of virus RNA shedding in milk, and the intensity of breastfeeding [27]. By considering cumulative exposure to breast milk cell-free HIV and the pattern of feeding (exclusive versus mixed or partial), Neveu et al. [28] showed that postnatal acquisition of HIV is more strongly associated with cumulative exposure to cell-free particles in breast milk than with feeding mode, and estimated the probability of breast milk transmission at 0.0005/L of breast milk ingested. Thus, although relatively low levels of HIV RNA are present in breast milk, consumption of 0.5–1.0 L of breast milk daily provides continuous exposure to potentially infectious virus through the oral cavity and the gastrointestinal mucosa.

Finally, despite the infants' daily exposure via their oral and gastrointestinal mucosae to high amounts of cell-associated and cell-free HIV, estimated to be more than 700,000 viral particles per day [29], HIV acquisition in exposed breastfed children occurs infrequently. The observation that the majority of breastfed infants of HIV-infected mothers remain uninfected, even after many months of breastfeeding, constitutes one of the major paradoxes of HIV transmission via breast-milk. The infrequent breast milk HIV transmission despite prolonged exposure suggests that anti-infective properties of breast milk and natural and/or adaptive immunity to HIV in breastfed children could be involved in protection.

HIV Postnatal Transmission Via Breast Milk Is Multifactorial

The mechanisms of HIV transmission through breastfeeding remain speculative [2, 16, 30–36]. Because of the dynamic nature of the relationship between the source of HIV reservoirs (breast milk) and the potential target host (the maturing gastrointestinal tract of the young infant), multiple mechanisms are likely to be operating. The current model for breast milk postnatal transmission of HIV postulates the existence of several factors; in the mother (including the stage of maternal HIV disease, the frequency and duration of breastfeeding, the maternal nutritional status, co-infection with *Herpetoviridae* or other organisms, and the intake of antiretroviral treatment); in the breastfed child; and possibly in the virus, resulting in a complex and multifactorial process [10, 37] (Table 10.1).

Table 10.1 Factors associated with transmission of HIV through breastfeeding

Maternal
- Younger maternal age, lower parity
- Maternal seroconversion during lactation
- Clinical and/or immunological (CD4 cell count) disease progression
- RNA viral load in plasma
- RNA viral load in breast milk
- Local immune factors in breast milk
- Breast health (subclinical or clinical mastitis, abscess, cracked nipples)
- Maternal nutritional status
- Duration of breastfeeding

Infant
- Factors associated with the immune system
- Genetic factors (HLA concordance with mother, various single nucleotide polymorphisms)
- Pattern of infant feeding (exclusive breastfeeding versus mixed)
- Morbidity leading to less vigorous suckling, milk stasis and increased leakage of virus across milk ducts (oral thrush)

In the HIV-infected mother, cell-free and cell-associated HIV is found in breast milk [29, 38–41]. The virus could be originating in part from the systemic compartment and then be released into breast milk. Alternatively, HIV could be produced by local replication in breast milk CD4+ T cells, monocytes, and macrophages. Three major molecular form of the virus coexist in breast milk: cell-free viral RNA; proviral DNA as cell-associated virus integrated in latent T cells; and intracellular RNA representing cell-associated virus in activated virus-producing T cells [41–47]. Latently infected CD4+ T cells [48] and spontaneously activated CD4+ T cells [47] in breast milk may likely constitute the major reservoirs of HIV production [49]. A remaining question relates to the nature of HIV reservoirs in milk involved in transmission. Cell-associated and cell-free virus loads in milk both correlate with the risk of infant HIV acquisition [39, 43, 50, 51]. While breast milk cellular reservoirs may play a role in transmission [52], breast milk HIV RNA load appears to be a strong predictor of postnatal HIV transmission [46]. Higher cumulative exposure to cell-free HIV RNA in breast milk is associated with higher rates of postnatal infection in the infant, independent of maternal CD4+ T cell count and plasma viral load [28]. The relative contribution of exposure to cell-free and cell-associated HIV, e.g., which pool of virus initiates infection in infant, remains, however, to be determined. Resting, latently infected CD4+ T cells in breast milk might constitute a reservoir of virus responsible for the residual breast milk transmission despite maternal highly active antiretroviral therapy [45, 47]. Breast milk production of HIV is influenced by numerous mucosal and systemic modulatory factors. Mastitis, linked to elevated sodium (Na+) levels in breast milk, breast tissue inflammation, and nipple lesions have been associated with an increase in HIV RNA viral loads in breast milk, and have been linked to elevated postnatal HIV transmission risk [39, 53–55].

As counterpart of the HIV infectiousness in breast milk, anti-infective properties of breast milk against HIV reduce or control HIV production within the mammary reservoir itself, and/or reduce the risk of transmucosal crossing of the virus in the breastfed infant. Firstly, in the HIV-infected mother, anti-infective properties involve innate, preimmune and immune defenses, including high levels of HIV-specific humoral and cellular immunity, and may protect the breastfed child from further HIV infection, acting as negative modulatory cofactors of breast milk transmission, but yet not well understood. Secondly, in breastfed HIV-exposed children, modulary cofactors may include nonspecific salivary soluble factors acting against HIV, the integrity of the epithelial cell layer of the oral and gastrointestinal mucosae, the presence of concomitant infectious agents, as well as preimmune and acquired HIV-specific defenses present at the level of exposed mucosal surfaces and/or in the systemic compartment of the infant [16, 33, 56].

The balance between infective and anti-infective breast milk factors generally "profits" the exposed child, since successful acquisition of the virus, e.g., transmucosal crossing of HIV from mother's milk to child, is an infrequent event. The origin of the virus ingested (epithelial cells of mammary acini, macrophages, or other sources) could determine the tropism and transmissibility. In case of a large viral inoculum, the potentially protective mammary immunity directed to HIV could be insufficient to protect against prolonged exposure. It is to be noted that the composition of breast milk varies considerably over time, and the breast milk transmission cofactors may also change over time. It is also likely that early transmission mechanisms are very different from those of later transmission.

Breast Milk B Cell-Derived Immunity to HIV in the Mother

Human Breast Milk Immune Cells

Breast milk contains a relatively large number of immune cells, including viable T and B lymphocytes, macrophages and other mononuclear cells, ranging from 10^5 to 10^7/mL in colostrum and declining up to tenfold during the following 3 months of lactation [57–60]. The majority of lymphocytes in breast milk are CD3+ T cells (83%) and express mucosal homing markers such as CD103 [16, 61], indicating compartmentalization of T cells to the mammary gland. Breast milk CD4+ T and CD8+ T cells present an overall profile of immune activation [16, 62]. T lymphocytes expressing the CD40 ligand, a molecule involved in B cell isotype switching, are higher in the colostrum compared to blood of breastfeeding mothers [63]. B cells represent only 2% or less of the milk lymphocyte population [16, 64].

Most breast milk B lymphocytes display the phenotypic hallmarks of activated cells with high levels of CD38 expression and large size [65]. This is consistent with the lower expression of complement receptors observed on breast milk B cells, which might indicate that these cells are plasmablasts and/or antibody-producing plasma cells [66]. IgG B lymphocytes attracted to mammary tissue cross the mammary epithelium and enter milk secretions, producing IgG-secreting cells. In addition, IgA-secreting cells come from the mucosa-associated lymphoid tissue (MALT). Indeed, the mammary immune tissue is an effector site of the MALT, in which inductive sites comprise mucosa-associated B cell follicles and larger lymphoid aggregates as the origin of cells trafficking to mucosal effector sites [67]. Memory B cells and plasmablasts that colonize the mammary gland late in pregnancy originate from other mucosal areas where they have been exposed to antigens [67–69]. The integrated immune response in the MALT implies that following antigen exposure at one mucosal inductive site in the gastrointestinal or respiratory tracts, B cells migrate from MALT to colonize distant unexposed mucosal effector sites such as the lactating mammary gland, for subsequent extravasation and terminal plasma cell differentiation [69]. Specific homing receptors control this selective B cell migration through interactions with tissue-specific vascular addressins [70]. Breast milk is enriched in activated memory B cells that are distinct from those circulating in the blood and bear a particular profile of mucosal adhesion molecules ($\alpha 4\beta 7-/+$, $\alpha 4\beta 1+$, CD44+, CD62L−) [65], indicating that the homing pattern of breast milk B cells is similar to that of gut-associated lymphoid tissue (GALT) B cells [71]. Whereas within the breast milk lymphocyte population, the proportion of IgA-secreting cells does not exceed that of IgG-secreting cells [65], >90% of immunoglobulins secreted in breast milk are of the IgA class [72].

The mammary gland produces high levels of immunoglobulins. Breast milk contains maternal antibodies, with all basic forms of immunoglobulins IgG, IgM, IgA, IgD, and IgE present. The most abundant is usually secretory IgA [73]. Breast milk IgG and IgM are tenfold less concentrated than IgA [74].

Natural Polyreactive Humoral Immunity

Polyreactive, natural antibodies, thought to be important in selection of the preimmune B cell repertoire and in development of immune tolerance, are detected in colostrum and in breast milk [75, 76]. Polyreactivity is defined as the ability of an antibody molecule to bind several structurally unrelated antigens [77]. Natural antibodies are produced by B-1 cells, irrespective of any immunization procedure, and thus belong to the innate immune system [78]. In healthy individuals, at least 20% of circulating immunoglobulins are polyreactive. In contrast to antigen-primed antibodies, these low-affinity antibodies are polyreactive and may recognize different unrelated epitopes and autoantigens [78–80]. Polyreactive antibodies have been proposed as a first line of defense against pathogens [81]. Indeed, natural polyreactive antibodies have been demonstrated to synergize with the complement system in the opsonization of viruses and bacteria, thus directing the pathogens to secondary lymphoid organs and facilitating initiation of adaptive immune responses [82, 83].

The CCR5 chemokine receptor is the major coreceptor that is associated with mucosal transmission of CCR5-tropic HIV during postnatal transmission through breastfeeding [42]. Breast milk contains high amounts of immunoglobulins, predominantly of the secretory IgA and to a lesser degree of the IgG and secretory IgM isotypes that are produced in the absence of deliberate immunization and independently of exposure to antigens. Most natural antibodies in the breast milk of healthy women are self-reactive antibodies [75, 84]. Bouhlal et al. [76] demonstrated that breast milk of 66% and 83% of HIV-seronegative and seropositive women, respectively, contains natural antibodies of the secretory IgA and IgG isotypes directed against the CCR5 coreceptor for R5-tropic strains of HIV. Although the avidity differed, the amount of anti-CCR5 antibodies did not significantly differ between breast milk of HIV-seropositive and seronegative women. Purified anti-CCR5 antibodies inhibited up to 75% of infection of macrophages and dendritic cells with HIVBaL and HIVJR-CSF strains. These observations provide evidence for a role of natural antibodies to CCR5 in breast milk in possibly controlling transmissibility of HIV through breastfeeding.

In addition, Requena et al. [85] recently demonstrated that human breast milk and normal human polyclonal immunoglobulins purified from plasma [intravenous immunoglobulin (IVIg)] contain functional natural IgA and IgG antibodies directed against the carbohydrate recognition domain (CRD) of the dendritic cell-specific intercellular adhesion molecule-3-grabbing nonintegrin (DC-SIGN) molecule, which is involved in the binding of HIV to dendritic cells. Anti-DC-SIGN CRD peptide antibodies inhibited the attachment of virus to HeLa DC-SIGN+ cells by up to 78% and the attachment to immature monocyte-derived dendritic cells (iMMDCs) by 20%. Both breast milk- and IVIg-derived natural antibodies to the CRD peptide inhibited 60% of the transmission in *trans* of HIVJRCSF, a CCR5-tropic strain, from iMDDCs to CD4+ T lymphocytes. The inhibitory effect of natural breast milk antibodies on HIV may act in addition to the blocking of DC-SIGN-mediated transfection of CD4+ T lymphocytes with HIV by bile salt-stimulated lipase from human milk [67, 86]. Taken together, these observations suggest that the attachment of HIV to dendritic cells and transmission in *trans* to autologous CD4+ T lymphocytes occur through two independent mechanisms. These observations support a role for natural antibodies against DC-SIGN in the modulation of postnatal HIV transmission through breastfeeding and in the natural host defense against HIV in infected individuals [87].

Natural mucosal antibodies to CCR5 and to *tat* have been described in mucosal fluids [88]. Eslahpazir et al. [89] demonstrated that human cervicovaginal secretions also contain natural antibodies (e.g., polyreactive antibodies) directed against CCR5, capable to inhibit the infection of human macrophages and dendritic cells with primary CCR5-tropic HIV, and to limit the transfer of HIV from dendritic cells to T cells in vitro. In addition, Mouquet et al. [90] demonstrated that most of the anti-gp120 antibodies isolated from patients with high HIV-neutralizing titers are polyreactive. The authors propose that polyreactivity is utilized as a mechanism to increase the functional affinity (avidity) of the antibodies for the viral spikes. Thus, the simultaneous engagement (heteroligation) of gp140 (by one arm of an IgG) and of another yet unidentified structure on the viral membrane (by the other arm of the IgG) results in great improvement in binding avidity [90]. Finally, polyreactive HIV-neutralizing antibodies utilize the advantage of polyreactivity mostly as a way to gain in antigen-binding avidity, and may better tolerate the high mutability of HIV [91]. While all current HIV vaccination strategies, based on a structure-assisted rational vaccine design, aim at eliciting highly specific antibodies against sites that are vulnerable to neutralization, approaches to combat promiscuous mutable viruses such as HIV should also exploit the potential of promiscuous adaptive antibodies [91].

HIV-Specific Humoral Immune Response

Breast milk contains high levels of HIV-specific antibodies, mainly of the IgG isotype, but also of the IgA, and to a lesser extent of the IgM classes. Although HIV-specific breast milk antibodies demonstrate functional inhibitory properties against HIV in vitro, their role in vivo for limiting the transmission of the virus to the breastfed infant has never been clearly demonstrated.

Qualitative Detection of Antibodies Against HIV in Breast Milk

Antibodies against HIV have been detected in the breast milk of HIV-infected women [17, 35, 37, 92–97]. Belec et al. [93] in 1990 demonstrated the presence of HIV-specific antibody in breast milk of HIV-infected lactating mothers, including antibodies of the IgA isotype in 11/15 samples tested. There is now ample evidence that colostrum, as well as mature breast milk of HIV-infected women contains secretory IgA, secretory IgM and IgG against HIV antigens, including the *env*-encoded surface glycoproteins, and HIV *gag* and *pol* antigens [95]. Antibodies of the IgG class constitute the predominant isotype of HIV-specific antibodies in breast milk detected in nearly all samples [93–95]. HIV-specific secretory IgA was detected in differing proportions of breast milk samples studied, from 23% of 15 days postpartum breast milk and 41% of 18 months postpartum milk in the study by Van De Perre [94], to 59% of women in the study by Duprat et al. [98], and to all samples studied by Becquart et al. [95]. The majority of secretory IgA reacted with only one HIV antigen, most frequently p24 and gp160 [37]. Secretory IgM against gp160 was detected using specific ELISA assay only in colostral samples by Becquart et al. [95], but in 41–78% of samples using Western blot by Van De Perre et al. [94]. Recently, Fouda et al. [17] evaluated the HIV-specific humoral immune response in milk and plasma of 41 HIV-infected lactating women. HIV *env*-specific IgG antibodies were detected in the breast milk of all the women, whereas the seroprevalence of HIV *env*-specific IgA antibodies was variable between HIV *env* antigens, ranging from 10 to 73%. Furthermore, milk IgG responses directed against gp120 and gp140 were significantly higher than the milk HIV *env*-specific IgA responses, despite the tenfold-higher total IgA than IgG content of breast milk. The concentrations of anti-HIV gp120 IgG in milk and plasma were directly correlated ($r=0.75$; $P<0.0001$), yet the response in milk was two logs lower than in plasma.

To summarize, the breast milk of women with established HIV infection has been found to contain HIV-specific IgG with a wide spectrum of activity against HIV proteins, comparable to HIV-specific IgG in serum. The spectrum of activity of HIV-specific secretory IgA in breast milk is directed against only a limited number of viral proteins (env protein, gp 160, core proteins). The existence and spectrum of secretory IgM to HIV in breast milk remains controversial.

Origin of Humoral Mucosal Immunity to HIV in Breast Milk

Antibodies in milk are either transferred from plasma by transudation or locally produced by plasma cells that migrate to the mammary gland from other mucosal sites, in particular, the GALTs [99]. HIV-specific antibodies in breast milk are thus likely locally produced and also originating from plasma by transudation.

Local production of HIV-specific antibodies has been demonstrated by in vitro enumeration of anti-HIV-antibody-secreting cells [65], and by immunochemical studies [95]. Thus, the detection of plasma cells secreting HIV-specific IgG and more rarely IgA (following polyclonal activation using the CD40L ligation) can be detected in breast milk from HIV-infected mothers demonstrating the existence of immunoglobulin-secreting B cells specific to HIV antigens in breast milk [65]. Local HIV replication in the mammary gland leading to tissue-mediated selection of HIV variants might contribute to local differentiation of B lymphocytes and orientate the antibody production.

Antibodies in breast milk are mainly in the form of secretory IgA, a dimeric antibody containing the J chain peptide and the secretory component [69]. HIV-specific IgA antibodies in breast milk are likely produced by IgA-secreting cells belonging to the MALT.

Local production of HIV-specific IgG has also been demonstrated by immunochemical studies comparing the specific activities of HIV-specific IgG in paired sera and breast milk samples [95]. The specificity of HIV-specific IgG antibodies in the breast milk may differ from that of HIV antibodies in the serum of the same person [95], indicating compartmentalization of the IgG immune response against certain HIV antigens [95, 98]. Significant differences were observed between the patterns of IgG epitopic recognition against *env*-encoded surface HIV glycoproteins in paired colostrum and serum samples, supporting the notion of a compartmentalized IgG-producing immune system within the mammary gland [95].

By considering the close correlation in the magnitude of both neutralizing and antibody-dependent cellular cytotoxicity (ADCC) IgG immune responses in breast milk and plasma, researchers [17] hypothesized that the majority of the functional IgG antibodies in milk are derived from plasma. Moreover, IgG purified from breast milk and plasma had similar neutralization potencies, further supporting this hypothesis. In another study [95], anti-HIV IgG in the colostrum differed from serum IgG in its ability to inhibit transcytosis of cell-associated HIV through a tight epithelial barrier in vitro. These observations are consistent with the hypothesis of a functional compartmentalization of locally generated IgG immune response to HIV within the mammary gland [95].

Impairment of Mucosal Humoral Immunity in HIV Infection: The Mystery of the Predominance of HIV-Specific IgG Over IgA

HIV *env*-specific humoral response in milk is primarily of the IgG isotype. This is surprising considering that locally produced secretory IgA is the predominant immunoglobulin isotype in human colostrum and mature breast milk, as in most mucosal secretions [72]. This IgG predominance of anti-HIV antibodies in breast milk mirrors the antibody response to HIV [95, 96]. Indeed, in both plasma and breast milk, HIV-specific IgG predominate over HIV-specific antibodies of the IgA isotype [37]. These observations confirm and extend several previous studies, in which low levels of HIV-specific antibodies of the IgA isotype were reported in other mucosal sites such as the genital tract, saliva, tears, and duodenal fluid [100–109]. Plasma and mucosal HIV *env*-specific IgA antibody responses directed against gp120 and gp41 may be detected in individuals during acute HIV infection, although infrequently by comparison to IgM and IgG [108]. In contrast, in chronic HIV infection, the magnitude and frequency of HIV-specific IgA antibodies were low [108, 110].

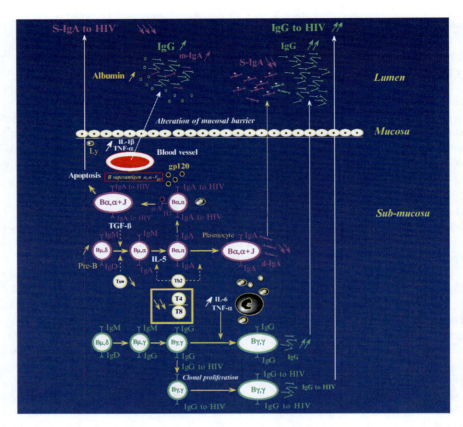

Fig. 10.1 Impaired mucosal humoral immunity during HIV infection. Increased levels of total IgG and monomeric IgA (m-IgA) occur in mucosal secretions, whereas the total level of secretory IgA (S-IgA) decreases progressively. In addition, the level of HIV-specific IgG is increased, whereas the level of HIV-specific S-IgA is decreased. Increased level of total IgG and m-IgA in mucosal secretions may be secondary to increased transudation of serum-borne immunoglobulins due to alteration of the mucosal barrier. Decreased level of total S-IgA in mucosal secretions is due to progressive decrease of IgA-secreting cells, likely secondary to the mucosal depletion in CD4 T lymphocytes, and "switch" T helper lymphocytes (Tws). Clonal proliferation of plasmacytes producing HIV-specific IgG may be secondary to increased production of mucosal TNF-α and IL-6. Progressive decrease of mucosal plasmacytes producing HIV-specific IgA may be due to the B superantigen activity of gp120, activating IgA-secreting B cells via its link with Ig V_{H3} followed by subsequent apoptosis [125]

In animal models, low or absent IgA responses have also been described in HIV-infected chimpanzees [111] and SIV-infected macaques [112, 113]. Thus, HIV appears markedly different from other viral and bacterial mucosal infections that stimulate vigorous albeit transient IgA [114, 115]. For example, antibodies against rotavirus [116], and respiratory syncytial virus [117] in breast milk are predominantly of the IgA isotype. Furthermore, mucosal immunization with a live influenza virus vaccine and infections with hepatitis B, cytomegalovirus, Epstein–Barr, or varicella viruses stimulate IgA responses shortly after immunization or acute infection, which, however, may not persist in the chronic stage [118–121].

A mechanism involved in this selective suppression of HIV-specific responses of the IgA isotype remains yet unclear. It may be fundamental to understanding how HIV is capable to be shed in body fluids, e.g., to be produced by mucosal reservoirs despite the existence of mucosal immune responses. Several hypotheses have been proposed.

The first hypothesis concerns a mucosal B cell dysregulation present in HIV 1 infection (Fig. 10.1). The numbers of mucosal IgA (and IgG and IgM) immunoglobulin-secreting cells increase in the asymptomatic stage of infection, consistent with early polyclonal activation of mucosal B cells [122]. The polyclonal B cell activation progressively decreases with disease, leading to a decrease of IgA immunoglobulin-secreting cells in the intestinal mucosa of patients with AIDS [123]. Additional strong evidence supporting mucosal B cell dysregulation in HIV-infected persons is the decreased proportion of IgA2 (relative to IgA1) immunoglobulin-secreting cells in intestinal tissues and fluids, and restriction in these IgA antibodies to gp120 and gp160 and less often pl7 and p24 (compared with IgG antibodies, which react to most HIV 1 antigens) [124]. Since IgA2 is produced preferentially in response to polysaccharide antigens, which are important determinants of intestinal bacterial pathogens, the dysregulation in IgA responses may play a role in predisposing HIV-infected persons to certain bacterial pathogens, including *Salmonella* species and *Shigella flexneri*. Finally, HIV infection may deplete B cells expressing Ig V_{H3} gene products, such as the majority of IgA plasmacytes, via the strong link between gp120 and Ig V_{H3} [125]. The glycoprotein gp120 could thus act as a B superantigen, activating IgA-secreting B cells via the link with Ig V_{H3}, followed by apoptosis [125].

More recently, Xu et al. [126] have shown that HIV evades virus-specific IgA and IgG2 responses by shuttling negative factor (*nef*) from infected macrophages to mucosal and systemic B cells through long-range intercellular conduits. As a result, B cell switching to IgA and IgG2 production is suppressed by *nef*.

Alternatively, the extensive mucosal replication of HIV and a profound depletion of CD4+ T cells involved in immunoglobulin-isotype switching and the generation of antibody diversity through somatic hypermutation [127] during acute infection may alter the capacity to mount an IgA response to new antigens.

The potential that IgA antibodies against HIV are not detectable due to the formation of HIV-IgA immune complexes in blood and external secretions [128], or that tolerance is induced, see less likely [129].

Secretory IgA as the dominant immunoglobulin isotype in most external secretions displays several structurally and functionally advantageous features when compared to IgG and IgM [130]. The lack of a vigorous or sustained IgA response to HIV may have particular implications for the development of protective immunity, although the antigen presented in a vaccine may differ considerably from the events during natural infection.

Functionality of HIV-Specific Humoral Immunity

Maternal antibodies transferred to the infant during breastfeeding are important for protection against pathogens encountered during the first months of life. Although HIV *env*-specific antibodies are detected in breast milk of infected mothers, their role in the prevention of infant HIV acquisition is still unclear. The role of breast-milk HIV-specific antibodies in HIV transmission through breastfeeding has been investigated by in vitro functional assays and by in vivo cohort studies of children exposed to HIV via breastfeeding.

In Vitro Evaluation of Breast Milk HIV-Specific Antibodies

The function, rather than the magnitude or class, of the HIV-specific breast milk antibody response may be the critical feature in protection against infant mucosal transmission.

The inhibitory properties towards viral transcytosis of secretory IgA and IgG antibodies purified from breast milk have been assessed in an in vitro model of transcytosis of cell-associated HIV through a tight monolayer of endometrial epithelial cells [37, 131, 132]. Such inhibitory activity was shown in all breast milk samples of transmitting and nontransmitting mothers [37]. Interestingly, despite a lower specific activity against gp160, purified secretory IgA from colostrum and breast milk exhibited similar inhibitory activity against HIV transcytosis as purified IgG [132]. Secretory IgA is considered to be more efficient than IgG in immune exclusion at mucosal surfaces, because it exhibits a dimeric structure with four antibody valencies.

Studies have reported an association between neutralization activity in maternal sera and protection against infant virus acquisition [133–135]. Furthermore, viruses that are resistant to neutralization by maternal plasma have been associated with HIV transmission via breastfeeding [136]. In addition, non-neutralizing antibodies, such as antibodies that mediate ADCC responses, may contribute to better clinical outcome and lower HIV/SIV load [137–139], and to protection of vaccinated rhesus macaques against SHIV challenge [140]. Fouda et al. [17] recently investigated autologous and heterologous virus neutralization and ADCC responses in the milk and plasma of 41 HIV-infected lactating women. Heterologous virus neutralization and ADCC activity in milk were directly correlated with that in the systemic compartment, but were two log units lower in magnitude. Autologous neutralization was rarely detected in milk. Heterologous virus neutralization titers in milk correlated with HIV gp120 *env*-binding IgG responses but not with IgA responses. These results suggest that plasma-derived IgG antibodies mediate the majority of the low-level HIV neutralization and ADCC activity in breast milk [17].

In Vivo Significance of Breast Milk HIV-Specific Antibodies

The in vivo significance of the presence of colostrum and breast milk anti-HIV antibodies is still unclear. Locally produced IgG may differ from its serum counterpart by a stronger adaptation to "mucosal" virus [35, 95], which could, in turn, confer protection to the infant from HIV transmission through breastfeeding. Indeed, locally produced IgG in the colostrum was able to inhibit transcytosis of HIV through a tight monolayer of epithelial cells in vitro [37]. However, no quantitative differences in milk HIV-specific antibody responses of transmitting and nontransmitting mothers have been identified [37, 97, 98].

In a study in Kigali, Rwanda, the presence of HIV-specific secretory IgA and secretory IgM, but not of IgG, in noncolostral breast milk was found to be associated with a decreased likelihood of HIV transmission to the infant [92]. Low levels of HIV-specific IgM in breast milk collected at 18 months were associated with a high risk of transmission of HIV [94]. More recently, Duprat et al. [98] found no association of anti-HIV sIgA in breast milk and protection from transmission to the infant, in a study of 63 mother–infant pairs. Bequart et al. [37] also found no association of mucosal humoral immunity to HIV and protection from transmission to the infant. Finally, in Zambia, HIV-specific secretory IgA was detected more often in breast milk of transmitting mothers (76.9%) than in breast milk of nontransmitting mothers (46.9%). The authors concluded that HIV-specific secretory IgA in breast milk did not appear to be a protective factor against HIV transmission among breastfed infants [97].

Whether HIV-specific neutralizing antibodies and ADCC activities of breast milk may be involved in protection against postnatal infant HIV acquisition remains unknown.

Nonhuman Primate Models of Breast Milk Transmission of SIV

Nonhuman primate models of breast milk transmission of the SIV or simian-human immunodeficiency viruses (SHIV) provide new insights on understanding the role of virus-specific humoral immunity in breast milk.

Passive Immunization Against SIV or SHIV Oral Challenge by Neutralizing Antibodies to HIV

The model of SHIV infection allows the evaluation of antiviral strategies that target the envelope glycoproteins of the HIV in macaques. Systemic administration of broadly neutralizing HIV-specific antibody to neonatal monkeys protected the infants against oral challenge with SHIV, indicating that passively transferred humoral immunity can protect infants from virus transmission through breastfeeding [141–146]. Van Rompay et al. [141] described protection in six of six neonates orally challenged with SIVmac251 after the administration of SIV hyperimmune serum just after birth. Ruprecht et al. [144, 145] showed protection of neonatal macaques from oral SHIV challenge after administration of monoclonal antibodies targeting neutralizing epitopes of the HIV envelope. This targeted specificity was likely responsible for the protection observed. Taken together, these observations indicate a potential role for neutralizing antibodies in preventing breast milk transmission of HIV. Passive administration of antibodies to the infant may only be protective if the antibodies are broadly neutralizing and capable of targeting transmissible viral variants.

Breast Milk Antibodies to SIV

SIV infection of lactating rhesus monkeys provides an excellent model to characterize virus-specific immune responses and virus evolution in milk, as the sequence of the virus inoculum, the timing of the infection, and the virus-specific immunodominant responses are well defined in this model. Furthermore, SIV-infected, lactating rhesus monkeys transmit the virus to their suckling infants via breastfeeding [147]. The milk produced by hormone-induced, lactating monkeys has immunoglobulin content and a lymphocyte phenotype similar to that produced by naturally lactating monkeys [148].

The SIV-macaque model provides a unique resource for deciphering the functional role of antibodies in breast milk transmission of HIV. Thus, humoral immune responses were analyzed in the milk and plasma of 14 female lactating rhesus macaques with suckling infants [149]. Total immunoglobulin levels in plasma and milk were similar in all females and could not be correlated with transmission to the infant. These females, however, had elevated milk IgG levels and decreased milk IgA levels, as compared with levels in seronegative controls. Furthermore, SIV envelope-specific antibody levels in milk and plasma were comparable in nontransmitting macaque females and those that transmitted during the chronic stage of infection.

Further studies have shown that virus-specific antibody responses are of low magnitude in milk of rhesus macaques, while robust virus-specific cellular immune responses are detected in the milk of SIV-infected rhesus macaques [113, 148]. While the principal antibody isotype in milk is locally produced secretory IgA, and hormone-induced lactating monkeys have a slightly higher secretory IgA content than naturally lactating monkeys [148], the SIV-specific humoral response in milk obtained from rhesus monkeys is predominantly of the IgG isotype [113, 149]. These observations indicate a specific impairment of the SIV envelope-specific IgA responses in breast milk [113]. In addition, breast milk antibody may not be effective at neutralizing autologous virus at its in vivo concentration, since inoculation virus-specific neutralizing antibody response could not be detected in breast milk samples obtained from SIV-infected lactating rhesus monkeys by 1 year after infection [113]. The low neutralizing potency of the locally produced or actively transported IgA response may be explained by impairment of mucosal lymphocyte responses [113]. In addition, the evaluation of the SIV envelope evolution of plasma and breast milk virus populations during chronic infection showed no difference in the number or location of acquired amino acid mutations in plasma and breast milk virus variants, demonstrating no compartment-specific SIV envelope evolution in milk [113]. This finding further supports the lack of a potent, functional SIV-specific antibody response in breast milk that would be expected to drive compartment-specific neutralization escape. The SIV envelope of milk virus is therefore likely to be under humoral immune pressure similar to that in plasma.

Finally, in the rhesus macaque model, as was observed in humans, no associations could be made between SIV-specific antibody levels and isotypes and transmission via breastfeeding [149].

Further insights in immune factors modulating SIV breast milk transmission have recently been reported from SIV-infected African Green monkeys [18]. These monkeys are naturally infected with SIV in the wild and do not or only extremely rarely transmit the virus via breastfeeding [150], in contrast to the high rate of postnatal SIV transmission in rhesus macaques. These contrasting transmission patterns provide a unique opportunity to study mechanisms that evolved to protect suckling infants from SIV infection [18].

The first point to note is that the apparent lack of SIV transmission via breastfeeding in African Green monkeys is not attributable to low-level virus exposure of the infants, as maternal milk virus RNA load in nonpathogenic SIV infection of African Green monkeys is comparable to that of pathogenic SIV infection in rhesus macaque [18]. Therefore, the breastfeeding infants of SIV-infected African Green monkeys are a highly exposed, uninfected population, warranting evaluation of the factors contributing to the impediment of virus transmission in this setting.

The second point is that a robust virus-specific CD8+ T lymphocyte response exists in the milk of chronically SIV-infected African Green monkeys [18], similar to that described for milk of chronically SIV-infected rhesus macaques [148]. Since both the transmitting and nontransmitting species display robust cellular immune responses in milk, these responses are not likely to be the ones to confer the observed protection against infant SIV transmission in the natural hosts.

The third and most important point is the possible role of SIV-specific neutralizing antibodies in breast milk to protect against SIV transmission in natural hosts. Thus, an autologous challenge virus-specific neutralization response is detected in milk of SIV-infected African Green monkeys that is comparable in magnitude to that in plasma [18]. In contrast, autologous challenge virus neutralization is not detectable in milk of SIV-infected rhesus macaques. Interestingly, the neutralization activity is only supported by the IgG fraction of milk of African Green monkeys [18]. The neutralization potency of breast milk of SIV-infected African Green monkeys may be provided from plasma IgG transudate (that predominantly contributes to the pool of milk IgG in primates), as well as from breast milk IgG-producing B cells that contribute to a portion of the IgG found in milk [18]. The autologous neutralization response detected in milk of chronically SIV-infected African Green monkeys may thus constitute a mechanism of the impedance of postnatal virus transmission in these natural hosts of SIV.

Taken together, the nonhuman primate models of SIV breast milk transmission: (1) deemphasize the importance of the role of milk virus-specific cellular immune response against SIV breast milk transmission, although virus-specific CD8+ T lymphocytes act on locally replicating virus in the breast milk compartment, and may decrease the breast milk viral inoculum [151] and (2) emphasize the protecting role of breast milk neutralizing humoral immunity against SIV, mainly of the IgG isotype. The possible protective role of milk non-neutralizing humoral immune response, such as ADCC, recently described during acute and chronic SIVmac251 infection of rhesus monkeys [152], should also be investigated. These findings would have major implications for the design of vaccines to prevent postnatal transmission of HIV.

Defenses in Children Exposed to HIV by Breastfeeding

The portal of entry of HIV and the factors that protect the infant from postnatal HIV transmission have so far received relatively little attention, and few studies on this subject have been published. During breastfeeding, HIV could theoretically be introduced into the gastrointestinal tract submucosa of the infant by a breach in the integrity of the epithelial cell layer, or by concomitant infectious agents, transport across M (microfold) cells in Peyer's patches, or infection of epithelial cells lining the oral cavity [153] or intestinal mucosa [33]. Finally, a virion that successfully traverses the infant gastrointestinal mucosal barrier must evade the infant innate and adaptive mucosal immune responses and contact a CD4+ T target cell in order to establish a systemic infection.

The defenses that the child opposes to the penetration of the virus include a priori nonimmune (nonspecific), preimmune, and immune defenses.

Nonimmune defenses involve the integrity of the mucosal barrier, the nonimmune salivary defenses, and the acidic gastric environment. In the presence of damaged oral mucosa—e.g., erosions, and after abrasions and damage by chemical irritants—the risk of HIV transmission could be enhanced. The saliva provides a hostile milieu to infection through a range of mechanisms [154], that include physical factors, such as salivary hypotonicity, which is thought to disrupt HIV-infected cells and to inactivate HIV in leucocytes [155, 156], as well as various substances with anti-HIV activity, including mucins [157], soluble proteins such as agglutinin, mucins, lysozyme, lactoferrin [158], and secretory leucocyte protease inhibitor (SLPI), and cytokines and chemokines [159–161]. SLPI could prevent the transmucosal penetration of HIV by inhibiting infection of mononuclear cells that are target cells for HIV in the mucosa [162]. Since gastric acid production in the neonate is reduced, more virus, either cell-free or cell-associated, may reach the monostratified intestinal mucosa [163]. However, whatever the site of virus entry, the virus must remain viable and retain the ability to traverse a mucosal barrier upon exposure to the infant saliva, the gastric environment, and/or the mucus-covered intestinal tract.

Natural preimmune defenses to infection via breastfeeding in exposed children involve natural killer cells and/or natural polyreactive antibodies to HIV at the level of exposed mucosal surfaces and/or in the systemic compartment. The role of such preadaptive immunity in resistance against HIV is, however, unknown.

Acquired mucosal resistance can involve principally HIV-specific CD8+ T cytotoxic immune responses, and specific antibody production against HIV proteins.

Breastfeeding HIV-exposed uninfected infants frequently display HIV-specific IFN-γ responses. Thus, of more than 200 breastfeeding HIV-exposed uninfected infants, who were serially and prospectively assessed for HLA-selected HIV peptide-specific cytotoxic T lymphocyte interferon (IFN)-γ responses by means of enzyme-linked immunospot (ELISpot) assays, almost half had HIV-specific CTL IFN-γ responses despite the absence of HIV infection [164]. At any single time point, only 12–22% of infants had HIV-specific IFN-γ responses. These observations suggest that, rather than completely escaping viral exposure, many HIV-uninfected infants of HIV-infected mothers are exposed to cell-associated HIV and elicit immune recognition of HIV-infected cells. HIV-infected individuals harbor large populations of replication-incompetent virus [165]. Such viruses may elicit cellular responses but be limited in their ability to cause productive infection. Sacha et al. [166] have demonstrated rapid induction of CTL responses (within 2 h) that are capable of eliminating HIV-infected cells before protein synthesis or productive infection ensue. An alternative explanation for our observation is that infants had restricted viral replication in oropharyngeal or esophageal lymphoid tissue without systemic viral detection, similar to what has been observed in primate models after low-dose lentiviral exposure [167]. In addition, greater early HIV-specific IFN-γ responses were associated with decreased HIV acquisition. The association between levels of HIV-specific IFN-γ responses and protection from breast milk HIV transmission lends some support for vaccine strategies that include induction of cellular responses against HIV to prevent breast milk HIV transmission and possibly to prevent transmission in other settings. The HLA B*18 haplotype may favor the cellular immune response in children exposed to HIV by breastfeeding, and may protect breastfeeding infants against both early and late HIV acquisition [168].

The majority of exposures to HIV in breastfed children is across oral mucosa, tonsillar tissue, and gastrointestinal mucosa, which are immunocompetent tissues, belonging to the afferent branch of the MALT [69]. Induction of mucosal immunity against HIV following prolonged exposure to infected breast milk is an attractive hypothesis. In particular, infant immune responses to HIV in saliva may provide protection against HIV acquisition [169]. Because secretory IgA is not transported actively across the placenta, levels are generally low to absent at birth and increase with age, achieving adult levels near 6–8 years [170]. In support of infant IgA responses are non-HIV studies demonstrating pathogen-specific salivary IgA antibodies against toxoplasmosis and influenza virus in saliva from young infants [170]. More compelling data come from a recent pediatric HIV vaccine trial which found that a small proportion of infants who had been vaccinated with a recombinant canarypox virus (ALVAC) HIV vaccine developed salivary HIV-specific IgA in response to this immunization [171]. In a prospective cohort study, Farquhar et al. [172] explored whether HIV-exposed, uninfected infants make immune responses in saliva after natural challenge with maternal breast milk and cervicovaginal secretions containing HIV. Overall, only 8% of HIV-uninfected infants in the study tested positive for HIV-specific salivary IgA at one time-point, and all infants with IgA responses remained uninfected during 1 year of follow-up. Furthermore, infants were very young at the time of specimen collection; all were under 6 months of age and median age was 1 month. These observations support the hypothesis that HIV-specific humoral immunity can be generated in response to HIV exposure. While the proportion of infants with positive assays was relatively low, these results also support that salivary HIV-specific IgA can be elicited in infants by immunizing neonates. Taken together, these findings provide some evidence that natural HIV exposure via the oral route can stimulate a humoral immune response in infants younger than 6 months of age and this may enhance understanding of how children are protected against mucosal exposure to HIV [172].

Unpublished observations from Sandrine Moussa (Pasteur Institute of Bangui, Central African Republic) reported the presence of HIV-specific antibodies from stool of HIV-uninfected children breastfed by their HIV-infected mothers. Purified stool antibodies of exposed children inhibited the attachment of HIV to epithelial cells by up to 65%, and to macrophages by up to 35%, and the infection of mono-macrophages by 90% (S. Moussa, personal communication). These findings suggest that the intestinal mucosa of children exposed to HIV by breastfeeding produce HIV-specific antibodies harboring in vitro functional properties against HIV, and thus possibly protective in vivo.

Overall, the observations reporting that the HIV-specific cellular and humoral immune defenses are activated in breastfed uninfected children are still few, and need to be complemented by further studies. The confirmation that a child exposed to HIV through breastfeeding actually develops protective specific immunity against the acquisition of the virus could have major importance for the demonstration of correlates of protection, and for the design of a prophylactic vaccine.

Humoral Immune Defenses as Vector of Protection of the Exposed Child: A Synthesis

Nonspecific and Specific Modulatory Cofactors of HIV Transmission Via Breastfeeding Are Numerous, Complex, and Intermingled

Breast milk from HIV-infected mothers contains both infective and anti-infective properties with regards to HIV. The breast milk contains a high number of nonspecific as well as specific factors with possible anti-HIV activity [16]. The general concept is that breast milk is a fabric to produce soluble [37, 173–185], cellular [186–188], and humoral [65, 76, 92–95, 97, 98] immune components potentially beneficial for the breastfed child. In addition, soluble factors present in breast milk, such as hormones, cytokines, and erythropoietin, may maintain intestinal epithelial integrity in the breastfed neonate, and thus prevent ingested milk-borne virus from penetrating into the infant [189, 190]. Similarly, natural or acquired, nonspecific as well as specific factors with possible anti-HIV activity, are likely present in children exposed to HIV by breastfeeding.

All these nonspecific and specific modulatory factors are closely associated, vary over time, depending on the lactation stage, and on the growth and maturity of the newborn. They probably act in an integrated way. So there are significant methodological difficulties in demonstrating the contribution of each factor in protection of the exposed child. In this context, the contribution of animal models is important. The recent demonstration in the African Green monkey model of the potent inhibitory activity of SIV-specific neutralizing humoral immunity against virus transmission through breastfeeding [18] reopens the debate on the potential role of virus-specific humoral immunity in breast milk as a protective factor for exposed children.

Breastfeeding by infants of HIV-infected mothers provides an opportunity to assess the role played by repeated HIV exposure in eliciting HIV-specific immunity and in defining whether immune responses correlate with protection from infection. However, consideration of the many factors involved in transmission of HIV through breast milk makes demonstration of correlates of protection very difficult. The preliminary evidence showing systemic cytotoxic and mucosal (salivary and stool) antibody responses against HIV in uninfected children exposed to HIV through breast milk [163, 172] is encouraging.

Conclusions

Prevention of MTCT of HIV via breastfeeding remains a challenge in developing countries, despite scaling up of antiretroviral treatment access. A number of postnatal virus transmissions will continue to occur even in the setting of optimal antiretroviral prophylaxis. Furthermore, maternal and infant antiretroviral drug toxicities, barriers to antiretroviral drug implementation, and the impact of the development of antiretroviral-resistant virus strains during maternal or infant antiretroviral prophylaxis have not been fully evaluated. Therefore, development of immunologic strategies to reduce HIV transmission via breast milk remains critical to improving HIV-free survival of infants born to HIV-infected mothers in the developing world.

The understanding of the numerous modulatory cofactors of HIV transmission via breastfeeding will certainly aid in the design of a protective vaccine, either to reduce or annulate the breast milk infectiousness of the HIV-infected mother [191], or to bolster the preimmune and/or immune defenses against HIV mucosal crossing in the breastfed HIV-exposed infant. Conceptually, the discovery in human breast milk of switched memory B cells, plasmablasts, and plasma cells, likely originating from the GALT and migrating to breast milk through the mammary tissue, provides the basis to conceive mucosal protection for the newborn against transmission of milk-borne HIV [65]. Furthermore, neutralizing SIV-specific antibodies in breast milk may likely protect against SIV transmission in suckling infants of African Green monkeys [18], providing for the first time strong evidence for the protective capabilities of neutralizing antibodies against transmucosal crossing of retroviruses. Finally, the situation of breastfed children exposed to HIV via oral and intestinal exposure offers the unique opportunity to evaluate correlates of protection involving innate, preimmune, and immune factors against mucosal acquisition and control of HIV infection.

References

1. UNAIDS (2010) Report on the global AIDS epidemic. http://www.unaids.org/globalreport/Global_report.htm
2. Van De Perre P, Simonon A, Hitimara DG et al (1993) Infective and anti-infective properties of breastmilk from HIV-infected women. Lancet 341:914–918
3. Datta P, Embree JE, Kreiss JK et al (1994) Mother-to-child transmission of human immunodeficiency virus type 1: report from the Nairobi Study. J Infect Dis 170:1134–1140
4. Bulterys M, Chao A, Dushimimana A, Saah A (1995) HIV-1 seroconversion after 20 months of age in a cohort of breastfed children born to HIV-1-infected women in Rwanda. AIDS 9:93–94
5. Bertolli J, St Louis ME, Simonds RJ et al (1996) Estimating the timing of mother-to-child transmission of human immunodeficiency virus in a breast-feeding population in Kinshasa, Zaire. J Infect Dis 174:722–726

6. Bobat R, Moodley D, Coutsoudis A, Coovadia H (1997) Breastfeeding by HIV-1-infected women and outcome in their infants: a cohort study from Durban, South Africa. AIDS 11:1627–1633
7. Ekpini ER, Wiktor SZ, Satten GA et al (1997) Late postnatal mother-to-child transmission of HIV-1 in Abidjan, Côte d'Ivoire. Lancet 349:1054–1059
8. Miotti PG, Taha TE, Kumwenda NI et al (1999) HIV transmission through breastfeeding: a study in Malawi. J Am Med Assoc 282:744–749
9. Fowler MG, Newell ML (2002) Breast-feeding and HIV-1 transmission in resource-limited settings. J Acquir Immune Defic Syndr 30:230–239
10. John-Stewart G, Mbori-Ngacha D, Ekpini R et al (2004) Breast-feeding and transmission of HIV-1. J Acquir Immune Defic Syndr 35:196–202
11. Thior I, Lockman S, Smeaton LM et al (2006) Breastfeeding plus infant zidovudine prophylaxis for 6 months vs formula feeding plus infant zidovudine for 1 month to reduce mother-to-child HIV transmission in Botswana: a randomized trial: the Mashi Study. JAMA 296:794–805
12. Kumwenda NI, Hoover DR, Mofenson LM et al (2008) Extended antiretroviral prophylaxis to reduce breast-milk HIV-1 transmission. N Engl J Med 359:119–129
13. Chasela CS, Hudgens MG, Jamieson DJ et al (2010) Maternal or infant antiretroviral drugs to reduce HIV-1 transmission. N Engl J Med 362:2271–2281
14. Shapiro RL, Hughes MD, Ogwu A et al (2010) Antiretroviral regimens in pregnancy and breast-feeding in Botswana. N Engl J Med 362:2282–2294
15. Coovadia HM, Rollins NC, Bland RM et al (2007) Mother-to-child transmission of HIV-1 infection during exclusive breastfeeding in the first 6 months of life: an intervention cohort study. Lancet 369:1107–1116
16. Kourtis AP, Butera S, Ibegbu C, Belec L, Duerr A (2003) Breast milk and HIV-1: vector of transmission or vehicle of protection? Lancet Infect Dis 3:786–793
17. Fouda GG, Yates NL, Pollara J et al (2011) HIV-specific functional antibody responses in breast milk mirror those in plasma and are primarily mediated by IgG antibodies. J Virol 85:9555–9567
18. Wilks AB, Perry JR, Ehlinger EP et al (2011) High cell-free virus load and robust autologous humoral immune responses in breast milk of simian immunodeficiency virus-infected African Green monkeys. J Virol 85: 9517–9526
19. Hanson LA, Korotkova M, Lundin S et al (2003) The transfer of immunity from mother to child. Ann N Y Acad Sci 987:199–206
20. Leroy V, Newell ML, Dabis F et al (1998) International multicentre pooled analysis of late postnatal mother to child transmission of HIV-1 infection. Lancet 352:597–600
21. Becquet R, Bland R, Leroy V et al (2009) Duration, pattern of breastfeeding and postnatal transmission of HIV: pooled analysis of individual data from West and South African cohorts. PLoS One 4:e7397
22. Gaillard P, Fowler MG, Dabis F et al (2004) Use of antiretroviral drugs to prevent HIV-1 transmission through breast-feeding: from animal studies to randomized clinical trials. J Acquir Immune Defic Syndr 35:178–187
23. The Breastfeeding and HIV International Transmission Study Group (2004) Late postnatal transmission of HIV-1 in breast-fed children: an individual patient data meta-analysis. J Infect Dis 189:2154–2166
24. Iliff PJ, Piwoz EG, Tavengwa NV et al (2005) Early exclusive breastfeeding reduces the risk of postnatal HIV-1 transmission and increases HIVfree survival. AIDS 19:699–708
25. Rollins NC, Becquet R, Bland RM, Coutsoudis A, Coovadia HM, Newell ML (2008) Infant feeding, HIV transmission and mortality at 18 months: the need for appropriate choices by mothers and prioritization within programmes. AIDS 22:2349–2357
26. Richardson BA, John-Stewart GC, Hughes JP et al (2003) Breast-milk infectivity in human immunodeficiency virus type 1-infected mothers. J Infect Dis 187:736–740
27. Willumsen JF, Newell ML, Filteau SM et al (2001) Variation in breastmilk HIV-1 viral load in left and right breasts during the first 3 months of lactation. AIDS 15:1896–1898
28. Neveu D, Viljoen J, Bland RM et al (2011) Cumulative exposure to cell-free HIV in breast milk, rather than feeding pattern per se, identifies postnatally infected infants. Clin Infect Dis 52:819–825
29. Lewis P, Nduati R, Kreiss JK et al (1998) Cell free HIV-1 in breast milk. J Infect Dis 177:34–39
30. Van De Perre P (1995) Postnatal transmission of the human immunodeficiency virus type 1 and the breast feeding dilemma. Am J Obstet Gynecol 173:483–487
31. Van De Perre P, Cartoux M (1996) Retroviral infection and breast-feeding. J Clin Microbiol Infect 1:6–12
32. Tranchat C, Van de Perre P, Simonon-Sorel A et al (1999) Maternal humoral factors associated with perinatal HIV-1 transmission in a cohort from Kigali, Rwanda, 1988–94. J Infect 39:213–220
33. Van De Perre P (1999) Mother to child transmission of HIV-1: the all mucosal hypothesis as a predominant mechanism of transmission. AIDS 13:1133–1138
34. Van De Perre P (1999) Transmission of human immunodeficiency virus type 1 through breast-feeding: how can it be prevented? J Infect Dis 179:S405–S407

35. Van De Perre P (2000) Breast milk transmission of HIV-1. Laboratory and clinical studies. Ann N Y Acad Sci 918:122–127
36. Page-Shafer K, Sweet S, Kassaye S, Ssali C (2006) Saliva, breast milk, and mucosal fluids in HIV transmission. Adv Dent Res 19:152–157
37. Becquart P, Hocini H, Levy M, Sepou A, Kazatchkine MD, Belec L (2000) Secretory anti-human immunodeficiency virus (HIV) antibodies in colostrum and breast milk are not a major determinant of the protection of early postnatal transmission of HIV. J Infect Dis 181:532–539
38. Nduati RW, John GC, Richardson BA et al (1995) HIV-1 infected cells in breast milk: association with immunosuppression and vitamin A deficiency. J Infect Dis 172:1461–1468
39. Semba RD, Kumwenda N, Taha TE et al (1999) Mastitis and immunological factors in breast milk of human immunodeficiency virus-infected women. J Hum Lact 15:301–306
40. Willumsen J, Filteau SM, Coutsoudis A et al (2003) Breastmilk RNA viral load in HIV-1 infected South African women: effects of subclinical mastitis and infant feeding. AIDS 17:407–414
41. Becquart P, Foulongne V, Willumsen J, Rouzioux C, Segondy M, Van de Perre P (2006) Quantitation of HIV-1 RNA in breast milk by real time PCR. J Virol Methods 133:109–111
42. Becquart P, Chomont N, Roques P et al (2002) Compartmentalization of HIV-1 between breast milk and blood of HIV-infected mothers. Virology 300:109–117
43. Rousseau CM, Nduati RW, Richardson BA et al (2004) Association of levels of HIV-1-infected breast milk cells and risk of mother-to-child transmission. J Infect Dis 190:1880–1888
44. Koulinska IN, Villamor E, Chaplin B et al (2006) Transmission of cell-free and cell-associated HIV-1 through breast-feeding. J Acquir Immune Defic Syndr 41:93–99
45. Petitjean G, Becquart P, Tuaillon E et al (2007) Isolation and characterization of HIV-1-infected resting CD4+ T lymphocytes in breast milk. J Clin Virol 39:1–8
46. Lunney KM, Iliff P, Mutasa K et al (2010) Associations between breast milk viral load, mastitis, exclusive breast-feeding, and postnatal transmission of HIV. Clin Infect Dis 50:762–769
47. Valea D, Tuaillon E, Al Tabaa Y et al (2011) CD4+ T cells spontaneously producing human immunodeficiency virus type I in breast milk from women with or without antiretroviral drugs. Retrovirology 8:34
48. Becquart P, Petitjean G, Tabaa YA et al (2006) Detection of a large T-cell reservoir able to replicate HIV-1 actively in breast milk. AIDS 20:1453–145
49. Pillay K, Coutsoudis A, York D, Kuhn L, Coovadia HM (2000) Cell-free virus in breast milk of HIV-1-seropositive women. J Acquir Immune Defic Syndr 24:330–336
50. Richardson BA, John-Stewart GC, Hughes J et al (2003) Breast milk infectivity in HIV-1-infected mothers. J Infect Dis 187:736–740
51. Rousseau CM, Nduati RW, Richardson BA et al (2003) Longitudinal analysis of human immunodeficiency virus type 1 RNA in breast milk and of its relationship to infant infection and maternal disease. J Infect Dis 187:741–747
52. Dimitrov DS, Willey RL, Sato H, Chang LJ, Blumenthal R, Martin MA (1993) Quantitation of human immunodeficiency virus type 1 infection kinetics. J Virol 67:2182–90
53. Semba RD, Neville MC (1999) Breast-feeding, mastitis, and HIV transmission: nutritional implications. Nutr Rev 57:146–153
54. Embree JE, Njenga S, Datta P et al (2000) Risk factors for postnatal mother-child transmission of HIV-1. AIDS 14:2535–2541
55. Fawzi W, Msamanga G, Spiegelman D et al (2002) Transmission of HIV-1 through breastfeeding among women in Dar es Salaam, Tanzania. J Acquir Immune Defic Syndr 31:331–338
56. Farquhar C, VanCott T, Bosire R et al (2008) Salivary human immunodeficiency virus (HIV)-1-specific immunoglobulin A in HIV-1-exposed infants in Kenya. Clin Exp Immunol 153:37–43
57. Ogra SS, Weintraub D, Ogra PL (1977) Immunologic aspects of human colostrums and milk. III. Fate and absorption of cellular and soluble components in the gastrointestinal tract of the newborn. J Immunol 119:245–248
58. Ogra SS, Ogra PL (1978) Immunologic aspects of human colostrums and milk. II. Characteristics of lymphocyte reactivity and distribution of E-rosette forming cells at different times after the onset of lactation. J Pediatr 92:550–555
59. Crago SS, Prince SJ, Pretlow TG et al (1979) Human colostral cells. I. Separation and characterization. Clin Exp Immunol 38:585–597
60. Xanthou M (1997) Human milk cells. Acta Paediatr 86:890–891
61. Bertotto A, Gerli R, Fabietti G et al (1990) Human breast milk T lymphocytes display the phenotype and functional characteristics of memory T cells. Eur J Immunol 20:1877–1880
62. Kourtis AP, Ibegbu CC, Theiler R et al (2007) Breast milk CD4+ T cells express high levels of C chemokine receptor 5 and CXC chemokine receptor 4 and are preserved in HIV-infected mothers receiving highly active antiretroviral therapy. J Infect Dis 195:965–972

63. Bertotto A, Castellucci G, Radicioni M, Bartolucci M, Vaccaro R (1996) CD40 ligand expression on the surface of colostral T cells. Arch Dis Child Fetal Neonatal 74:F135–F136
64. Davis MK (1991) Human milk and HIV infection: Epidemiologic and laboratory data. In: Mestecky J (ed) Immunology of milk and neonates. Plenum Press, NewYork, pp 271–280
65. Tuaillon E, Valea D, Becquart P et al (2009) Human Milk-derived B cells: a highly activated switched memory cell population primed to secrete antibodies. J Immunol 182:7155–7162
66. Bush JF, Beer AE (1979) Analysis of complement receptors on B-lymphocytes in human milk. Am J Obstet Gynecol 133:708–712
67. Macpherson AJ, McCoy KD, Johansen FE, Brandtzaeg P (2008) The immune geography of IgA induction and function. Mucosal Immunol 1:11–22
68. Roux ME, McWilliams M, Phillips-Quagliata JM, Weisz-Carrington P, Lamm ME (1977) Origin of IgA secreting plasma cells in the mammary gland. J Exp Med 146:1311–1322
69. Brandtzaeg P (2003) Mucosal immunity: integration between mother and the breast-fed infant. Vaccine 24:3382–3388
70. Kunkel EJ, Butcher EC (2003) Plasma-cell homing. Nat Rev Immunol 3:822–829
71. Williams MB, Rosé JR, Rott LS, Franco MA, Greenberg HB, Butcher EC (1998) The memory B cell subset responsible for the secretory IgA response and protective humoral immunity to rotavirus expresses the intestinal homing receptor, α4β7. J Immunol 161:4227–4235
72. Goldman SS (1993) The immune system of the human milk. Antimicrobial, anti-inflammatory and immuno-modulating properties. Pediatr J Infect Dis 12:664–671
73. Lawrence RM, Lawrence RA (2004) Breast milk and infection. Clin Perinatol 31:501–528
74. Goldman AS, Garza C, Nichols BL, Goldblum RM (1982) Immunologic factors in human milk during the first year of lactation. J Pediatr 100:563–567
75. Vassilev TL, Veleva KV (1996) Natural polyreactive IgA and IgM autoantibodies in human colostrum. Scand J Immunol 44:535–539
76. Bouhlal H, Latry V, Requena M et al (2005) Natural antibodies to CCR5 from breast milk block infection of macrophages and dendritic cells with primary R5-tropic HIV-1. J Immunol 174:7202–7209
77. Notkins AL (2004) Polyreactivity of antibody molecules. Trends Immunol 25:174–179
78. Duan B, Morel L (2006) Role of B-1a cells in autoimmunity. Autoimmun Rev 5:403–408
79. Avrameas S, Dighiero G, Lymberi P, Guilbert B (1983) Studies on natural antibodies and autoantibodies. Ann Immunol (Paris) 134D:103–113
80. Baumgarth N, Tung JW, Herzenberg LA (2005) Inherent specificities in natural antibodies: a key to immune defense against pathogen invasion. Springer Semin Immunopathol 26:347–362
81. Ehrenstein MR, Notley CA (2010) The importance of natural IgM: scavenger, protector and regulator. Nat Rev Immunol 10:778–786
82. Ochsenbein AF, Fehr T, Lutz C et al (1999) Control of early viral and bacterial distribution and disease by natural antibodies. Science 286:2156–2159
83. Zhou ZH, Zhang Y, Hu YF et al (2007) The broad antibacterial activity of the natural antibody repertoire is due to polyreactive antibodies. Cell Host Microbe 1:51–61
84. Quan CP, Berneman A, Pires R, Avrameas S, Bouvet JP (1997) Natural polyreactive secretory immunoglobulin A autoantibodies as a possible barrier to infection in humans. Infect Immun 65:3997–4004
85. Requena M, Bouhlal H, Nasreddine N et al (2008) Inhibition of HIV-1 transmission in trans from dendritic cells to CD4+ T lymphocytes by natural antibodies to the CRD domain of DC-SIGN purified from breast milk and intravenous immunoglobulins. Immunology 123:508–518
86. Naarding MA, Dirac AM, Ludwig IS et al (2006) Bile salt-stimulated lipase from human milk binds DC-SIGN and inhibits human immunodeficiency virus type 1 transfer to CD4+ T cells. Antimicrob Agents Chemother 50:3367–3374
87. Stax MJ, Naarding MA, Tanck MW et al (2011) Binding of human milk to pathogen receptor DC-SIGN varies with bile salt-stimulated lipase (BSSL) gene polymorphism. PLoS One 6:e17316
88. Lehner T, Bergmeier L, Wang Y, Tao L, Mitchell EA (1999) A rational basis for mucosal vaccination against HIV infection. Immunol Rev 170:183–196
89. Eslahpazir J, Jenabian MA, Bouhlal B et al (2008) Infection of macrophages and dendritic cells with primary R5-tropic human immunodeficiency virus type 1 inhibited by natural polyreactive anti-CCR5 antibodies purified from cervicovaginal secretions. Clin Vaccine Immunol 15:872–884
90. Mouquet H, Scheid JF, Zoller MJ et al (2010) Polyreactivity increases the apparent affinity of anti-HIV antibodies by heteroligation. Nature 467:591–595
91. Dimitrov JD, Kazatchkine MD, Kaveri SV, Lacroix-Desmazes S (2011) "Rational vaccine design" for HIV should take into account the adaptive potential of polyreactive antibodies. PLoS Pathog 7:e1002095
92. Van de Perre P, Hitimana DG, Lepage P (1988) Human immunodeficiency virus antibodies of IgG, IgA, and IgM subclasses in milk of seropositive mothers. J Pediatr 113:1039–1041

93. Belec L, Bouquety JC, Georges A et al (1990) Antibodies to HIV-1 in the breast milk of healthy, seropositive women. Pediatrics 85:1022–1026
94. Van De Perre P, Simonon A, Hitimana DG et al (1993) Infective and anti-infective properties of breastmilk from HIV-1-infected women. Lancet 341:914–918
95. Becquart P, Hocini H, Garin B et al (1999) Compartmentalization of the IgG immune response to HIV-1 in breast milk. AIDS 13:1323–1331
96. Lü FX (2000) Predominate HIV1-specific IgG activity in various mucosal compartments of HIV1-infected individuals. Clin Immunol 97:59–68
97. Kuhn L, Trabattoni D, Kankasa C et al (2006) HIV-specific secretory IgA in breast milk of HIV-positive mothers is not associated with protection against HIV transmission among breast-fed infants. J Pediatr 149:611–616
98. Duprat C, Mohammed Z, Datta P et al (1994) HIV-1 IgA antibody in breast milk and serum. Pediatr Infect Dis J 13:603–608
99. Hanson LA, Ahlstedt S, Andersson B et al (1984) The immune response of the mammary gland and its significance for the neonate. Ann Allergy 53:576–582
100. Belec L, Georges AJ, Steenman G, Martin PM (1989) Antibodies to HIV in vaginal secretions of heterosexual women. J Infect Dis 160:385–391
101. Wolff H, Mayer K, Seage G, Politch J, Horsburgh CR, Anderson D (1992) A comparison of HIV-1 antibody classes, titers, and specificities in paired semen and blood samples from HIV-1 seropositive men. J Acquir Immune Defic Syndr 5:65–69
102. Belec L, Dupre T, Prazuck T et al (1995) Cervicovaginal overproduction of specific IgG to human immunodeficiency virus (HIV) contrasts with normal or impaired IgA local response in HIV infection. J Infect Dis 172:691–697
103. Artenstein AW, VanCott TC, Sitz KV et al (1997) Mucosal immune responses in four distinct compartments of women infected with human immunodeficiency virus type 1: a comparison by site and correlation with clinical information. J Infect Dis 175:265–271
104. Haimovici F, Mayer KH, Anderson DJ (1997) Quantitation of HIV-1-specific IgG, IgA, and IgM antibodies in human genital tract secretions. J Acquir Immune Defic Syndr Hum Retrovirol 15:185–191
105. Raux M, Finkielsztejn L, Salmon-Ceron D et al (1999) Development and standardization of methods to evaluate the antibody response to an HIV-1 candidate vaccine in secretions and sera of seronegative vaccine recipients. J Immunol Methods 222:111–124
106. Lu FX (2000) Predominate HIV-1-specific IgG activity in various mucosal compartments of HIV-1-infected individuals. Clin Immunol 97:59–68
107. Wright PF, Kozlowski PA, Rybczyk GK et al (2002) Detection of mucosal antibodies in HIV type 1-infected individuals. AIDS Res Hum Retroviruses 18:1291–1300
108. Mestecky J, Jackson S, Moldoveanu Z et al (2004) Paucity of antigen-specific IgA responses in sera and external secretions of HIV-type 1-infected individuals. AIDS Res Hum Retroviruses 20:972–988
109. Alexander R, Mestecky J (2007) Neutralizing antibodies in mucosal secretions: IgG or IgA? Curr HIV Res 5:588–593
110. Kozlowski PA, Chen D, Eldridge JH, Jackson S (1994) Contrasting IgA and IgG neutralization capacities and responses to HIV type 1 gp120 V3 loop in HIV-infected individuals. AIDS Res Hum Retroviruses 10:813–822
111. Israel ZR, Marx PA (1995) Nonclassical mucosal antibodies predominate in genital secretions of HIV-1 infected chimpanzees. J Med Primatol 24:53–60
112. Schafer F, Kewenig S, Stolte N et al (2002) Lack of simian immunodeficiency virus (SIV) specific IgA response in the intestine of SIV infected rhesus macaques. Gut 50:608–614
113. Permar SR, Wilks AB, Ehlinger EP et al (2010) Limited contribution of mucosal IgA to simian immunodeficiency virus (SIV)-specific neutralizing antibody response and virus envelope evolution in breast milk of SIV-infected, lactating rhesus monkeys. J Virol 84:8209–8218
114. Holmgren J, Svennerholm A-M (2005) Mucosal immunity to bacteria. In: Mestecky J, Bienenstock J, Lamm M, Mayer L, McGhee J, Strober W (eds) Mucosal immunology, 3rd edn. Elsevier/Academic Press, Amsterdam, pp 783–797
115. Murphy BR (2005) Mucosal immunity to viruses. In: Mestecky J, Bienenstock J, Lamm ME, Mayer L, McGhee JR, Strober W (eds) Mucosal immunology, 3rd edn. Elsevier/Academic Press, Amsterdam, pp 799–813
116. Rahman MM, Yamauchi M, Hanada N, Nishikawa K, Morishima T (1987) Local production of rotavirus specific IgA in breast tissue and transfer to neonates. Arch Dis Child 62:401–405
117. Fishaut M, Murphy D, Neifert M, McIntosh K, Ogra PL (1981) Bronchomammary axis in the immune response to respiratory syncytial virus. J Pediatr 99:186–191
118. Nomura M, Imai M, Tsuda F et al (1985) Immunoglobulin A antibody against hepatitis B core antigen in the acute and persistent infection with hepatitis B virus. Gastroenterology 89:1109–1113
119. Engelhard D, Weinberg M, Or R et al (1991) Immunoglobulins A, G, and M to cytomegalovirus during recurrent infection in recipients of allogeneic bone marrow transplantation. J Infect Dis 163:628–630

120. Fachiroh J, Schouten T, Hariwiyanto B et al (2004) Molecular diversity of Epstein-Barr virus IgG and IgA antibody responses in nasopharyngeal carcinoma: a comparison of Indonesian, Chinese, and European subjects. J Infect Dis 190:53–62
121. Moldoveanu Z, Clements ML, Prince SJ, Murphy BR, Mestecky J (1995) Human immune responses to influenza virus vaccine administered by systemic or mucosal routes. Vaccine 13:1006–1012
122. Eriksson K, Kilander A, Hagberg L, Norkrans G, Homgren J, Czerkinsky C (1995) Virus-specific antibody production and polyclonal B-cell activation in the intestinal mucosa of HIV-infected individuals. AIDS 9:695–700
123. Kotler DP, Scholes JV, Tierney AR (1987) Intestinal plasma cell alterations in acquired immunodeficiency syndrome. Dig Dis Sci 38:1119–1127
124. Janoff EN, Jackson S, Wahl SM, Thomas K, Peterman JH, Smith PD (1994) Intestinal immunoglobulins during human immunodeficiency virus type 1 infection. J Infect Dis 170:299–307
125. Berberian L, Goodglick L, Kipps TJ, Braun J (1993) Immunoglobulin VH3 gene products: natural ligands for HIV gp120. Science 261:1588–1591
126. Xu W, Santini PA, Sullivan JS et al (2009) HIV-1 evades virus-specific IgG2 and IgA responses by targeting systemic and intestinal B cells via long-range intercellular conduits. Nat Immunol 10:1008–1017
127. Honjo T, Kinoshita K, Muramatsu M (2002) Molecular mechanisms of class switch recombination: linkage with somatic hypermutation. Annu Rev Immunol 20:165–196
128. Tomaras GD, Yates NL, Liu P et al (2008) Initial B-cell responses to transmitted human immunodeficiency virus type 1: virion-binding immunoglobulin M (IgM) and IgG antibodies followed by plasma anti-gp41 antibodies with ineffective control of initial viremia. J Virol 82:12449–12463
129. Mestecky J, Wright PF, Lopalco L et al (2011) Scarcity or absence of humoral immune responses in the plasma and cervicovaginal lavage fluids of heavily HIV-1-exposed but persistently seronegative women. AIDS Res Hum Retroviruses 27:469–486
130. Russell MW, Kilian M (2005) Biological activities of IgA. In: Mestecky J, Bienenstock J, Lamm ME, Mayer L, McGhee JR, Strober W (eds) Mucosal immunology, 3rd edn. Elsevier/Academic Press, Amsterdam, pp 267–289
131. Bomsel M (1997) Transcytosis of infectious human immunodeficiency virus across a tight human epithelial cell line barrier. Nat Med 3:42–47
132. Hocini H, Belec L, Iscaki S et al (1997) High-level ability of secretory IgA to block HIV type 1 transcytosis: contrasting secretory IgA and IgG responses to glycoprotein 160. AIDS Res Hum Retroviruses 13:1179–1185
133. Scarlatti G, Albert J, Rossi V et al (1993) Mother to child transmission of HIV-1: correlation with neutralizing antibodies against primary isolates. J Infect Dis 168:207–210
134. Barin F, Jourdain G, Brunet S et al (2006) Revisiting the role of neutralizing antibodies in mother-to-child transmission of HIV-1. J Infect Dis 193:1504–1511
135. Dickover R, Garratty E, Yusim K, Miller C, Korber B, Bryson Y (2006) Role of maternal autologous neutralizing antibody in selective perinatal transmission of human immunodeficiency virus type 1 escape variants. J Virol 80:6525–6536
136. Rainwater SM, Wu X, Nduati R et al (2007) Cloning and characterization of functional subtype A HIV-1 envelope variants transmitted through breastfeeding. Curr HIV Res 5:189–197
137. Banks ND, Kinsey N, Clements J, Hildreth JE (2002) Sustained antibody-dependent cell-mediated cytotoxicity (ADCC) in SIV-infected macaques correlates with delayed progression to AIDS. AIDS Res Hum Retroviruses 18:1197–1205
138. Forthal DN, Landucci G, Keenan B (2001) Relationship between antibody-dependent cellular cytotoxicity, plasma HIV type 1 RNA, and CD4+ lymphocyte count. AIDS Res Hum Retroviruses 17:553–561
139. Nag P, Kim J, Sapiega V et al (2004) Women with cervicovaginal antibody-dependent cell-mediated cytotoxicity have lower genital HIV-1 RNA loads. J Infect Dis 190:1970–1978
140. Xiao P, Zhao J, Patterson LJ et al (2010) Multiple vaccine-elicited nonneutralizing antienvelope antibody activities contribute to protective efficacy by reducing both acute and chronic viremia following simian/human immunodeficiency virus SHIV89.6P challenge in rhesus macaques. J Virol 84:7161–7173
141. Van Rompay KK, Berardi CJ, Dillard-Telm S et al (1998) Passive immunization of newborn rhesus macaques prevents oral simian immunodeficiency virus infection. J Infect Dis 177:1247–1259
142. Baba TW, Liska V, Hofmann-Lehmann R et al (2000) Human neutralizing monoclonal antibodies of the IgG1 subtype protect against mucosal simian-human immunodeficiency virus infection. Nat Med 6:200–2006
143. Hofmann-Lehmann R, Rasmussen RA, Vlasak J et al (2001) Passive immunization against oral AIDS virus transmission: an approach to prevent mother-to-infant HIV-1 transmission? J Med Primatol 30:190–196
144. Ruprecht RM, Hofmann-Lehmann R, Smith-Franklin BA et al (2001) Protection of neonatal macaques against experimental SHIV infection by human neutralizing monoclonal antibodies. Transfus Clin Biol 8:350–358
145. Ruprecht RM, Ferrantelli F, Kitabwalla M et al (2003) Antibody protection: passive immunization of neonates against oral AIDS virus challenge. Vaccine 21:3370–3373
146. Ferrantelli F, Buckley KA, Rasmussen RA et al (2007) Time dependence of protective post-exposure prophylaxis with human monoclonal antibodies against pathogenic SHIV challenge in newborn macaques. Virology 358:69–78

147. Amedee AM, Lacour N, Ratterree M (2003) Mother-to-infant transmission of SIV via breast-feeding in rhesus macaques. J Med Primatol 32:187–193
148. Permar SR, Kang HH, Carville A et al (2008) Potent simian immunodeficiency virus-specific cellular immune responses in the breast milk of simian immunodeficiency virus-infected, lactating rhesus monkeys. J Immunol 181:3643–3650
149. Rychert J, Amedee AM (2005) The antibody response to SIV in lactating rhesus macaques. J Acquir Immune Defic Syndr 38:135–141
150. Otsyula MG, Gettie A, Suleman M, Tarara R, Mohamed I, Marx P (1995) Apparent lack of vertical transmission of simian immunodeficiency virus (SIV) in naturally infected African green monkeys, *Cercopithecus aethiops*. Ann Trop Med Parasitol 89:573–576
151. Wilks AB, Christian EC, Seaman MS et al (2010) Robust vaccine-elicited cellular immune responses in breast milk following systemic simian immunodeficiency virus DNA prime and live virus vector boost vaccination of lactating rhesus monkeys. J Immunol 185:7097–7106
152. Sun Y, Asmal M, Lane S et al (2011) Antibody-dependent cell-mediated cytotoxicity in simian immunodeficiency virus-infected rhesus monkeys. J Virol 85:6906–6912
153. Moore JS, Rahemtulla F, Kent LW et al (2003) Oral epithelial cells are susceptible to cell-free and cell-associated HIV-1 infection in vitro. Virology 313:343–353
154. Malamud D, Wahl SM (2010) The mouth: a gateway or a trap for HIV? AIDS 24:5–16
155. Baron S, Poast J, Cloyd MW (1999) Why is HIV rarely transmitted by oral secretions? Saliva can disrupt orally shed, infected leukocytes. Arch Intern Med 159:303–310
156. Baron S, Poast J, Richardson CJ et al (2000) Oral transmission of human immunodeficiency virus by infected seminal fluid and milk: a novel mechanism. J Infect Dis 181:498–504
157. Rothenberg R, Scarlett M, del Rio C, Reznik D, O'Daniels C (1998) Oral transmission of HIV. AIDS 12:2095–2105
158. Carthagena L, Becquart P, Hocini H, Kazatchkine MD, Bouhlal H, Belec L (2011) Modulation of HIV binding to epithelial cells and HIV transfer from immature dendritic cells to CD4 T lymphocytes by human lactoferrin and its major exposed LF-33 peptide. Open Virol J 5:27–34
159. McNeely TB, Dealy M, Dripps DJ, Orenstein JM, Eisenberg SP, Wahl SM (1995) Secretory leukocyte protease inhibitor: a human saliva protein exhibiting anti-human immunodeficiency virus 1 activity in vitro. J Clin Invest 96:456–464
160. Shugars DC (1999) Endogenous mucosal antiviral factors of the oral cavity. J Infect Dis 179:S431–S435
161. Leigh JE, Steele C, Wormley FL Jr et al (1998) Th1/Th2 cytokine expression in saliva of HIV-positive and HIV-negative individuals: a pilot study in HIV-positive individuals with oropharyngeal candidiasis. J Acquir Immune Defic Syndr Hum Retrovirol 19:373–380
162. McNeely TB, Shugars DC, Rosendahl M, Tucker C, Eisenberg SP, Wahl SM (1997) Inhibition of human immunodeficiency virus type 1 infectivity by secretory leukocyte protease inhibitor occurs prior to viral reverse transcription. Blood 90:1141–1149
163. Meng G, Wei X, Wu X et al (2002) Primary intestinal epithelial cells selectively transfer R5 HIV-1 to CCR5+ cells. Nat Med 8:150–156
164. John-Stewart GC, Mbori-Ngacha D, Payne BL et al (2009) HIV-1-specific cytotoxic T lymphocytes and breast milk HIV-1 transmission. J Infect Dis 199:889–898
165. Chun TW, Carruth L, Finzi D et al (1997) Quantification of latent tissue reservoirs and total body viral load in HIV-1 infection. Nature 387:183–188
166. Sacha JB, Chung C, Rakasz EG et al (2007) Gag-specific CD8+ T lymphocytes recognize infected cells before AIDS-virus integration and viral protein expression. J Immunol 178:2746–2754
167. Miller CJ, Marthas M, Torten J et al (1994) Intravaginal inoculation of rhesus macaques with cell-free simian immunodeficiency virus results in persistent or transient viremia. J Virol 68:6391–6400
168. Farquhar C, Rowland-Jones S, Mbori-Ngacha D et al (2004) Human leukocyte antigen (HLA) B*18 and protection against mother-to-child HIV type 1 transmission. AIDS Res Hum Retroviruses 20:692–697
169. Farquhar C, John-Stewart G (2003) The role of infant immune responses and genetic factors in preventing HIV-1 acquisition and disease progression. Clin Exp Immunol 134:367–377
170. Wilson C (2005) Developmental immunology and role of host defenses in neonatal susceptibility. In: RemingtonJ KleinJ (ed) Infectious diseases of the fetus and newborn infant, 6th edn. WB Saunders Company, Philadelphia, PA, pp 37–49
171. Johnson DC, McFarland EJ, Muresan P et al (2005) Safety and immunogenicity of an HIV-1 recombinant canarypox vaccine in newborns and infants of HIV-1-infected women. J Infect Dis 192:2129–2133
172. Farquhar C, Van Cott T, Bosire R et al (2008) Salivary human immunodeficiency virus (HIV)-1-specific immunoglobulin A in HIV-1-exposed infants in Kenya. Clin Exp Immunol 153:37–43
173. Ebrahim GJ (1995) Breast milk immunology. J Trop Pediatr 4:2–4
174. Michie CA, Tantscher E, Schall T, Rot A (1998) Physiological secretion of chemokines in human breast milk. Eur Cytokine Netw 9:123–129

175. Jia HP, Starner T, Ackermann M et al (2001) Abundant human beta-defensin-1 expression in milk and mammary gland epithelium. J Pediatr 138:109–112
176. Tunzi CR, Harper PA, Bar-Oz B et al (2000) Beta-defensin expression in human mammary gland epithelia. Pediatr Res 48:30–35
177. Farquhar C, Van Cott TC, Mbori-Ngacha DA et al (2002) Salivary secretory leukocyte protease inhibitor is associated with reduced transmission of HIV-1 through breast milk. J Infect Dis 186:1173–1176
178. Farquhar C, Mbori-Ngacha DA, Redman MW et al (2005) CC and CXC chemokines in breastmilk are associated with mother-to-child HIV-1 transmission. Curr HIV Res 3:361–369
179. Saidi H, Eslahpazir J, Carbonneil C et al (2006) Differential modulation of human lactoferrin activity against both R5 and X4-HIV-1 adsorption on epithelial cells and dendritic cells by natural antibodies. J Immunol 177: 5540–5549
180. Bosire R, Guthrie BL, Lohman-Payne B et al (2007) Longitudinal comparison of chemokines in breastmilk early postpartum among HIV-1-infected and uninfected Kenyan women. Breastfeed Med 2:129–138
181. Bosire R, John-Stewart GC, Mabuka JM et al (2007) Breast milk alpha-defensins are associated with HIV type 1 RNA and CC chemokines in breast milk but not vertical HIV type 1 transmission. AIDS Res Hum Retroviruses 23:198–203
182. Walter J, Kuhn L, Ghosh MK et al (2007) Low and undetectable breast milk interleukin-7 concentrations are associated with reduced risk of postnatal HIV transmission. J Acquir Immune Defic Syndr 46:200–207
183. Lyimo MA, Howell AL, Balandya E, Eszterhas SK, Connor RI (2009) Innate factors in human breast milk inhibit cell-free HIV-1 but not cell-associated HIV-1 infection of CD4+ cells. J Acquir Immune Defic Syndr 51:117–124
184. Saeland E, de Jong MA, Nabatov AA, Kalay H, Geijtenbeek TB, van Kooyk Y (2009) MUC1 in human milk blocks transmission of human immunodeficiency virus from dendritic cells to T cells. Mol Immunol 46:2309–2316
185. Walter J, Ghosh MK, Kuhn L et al (2009) High concentrations of interleukin 15 in breast milk are associated with protection against postnatal HIV transmission. J Infect Dis 200:1498–1502
186. Lohman BL, Slyker J, Mbori-Ngacha D et al (2003) Prevalence and magnitude of human immunodeficiency virus (HIV) type 1-specific lymphocyte responses in breast milk from HIV-1-seropositive women. J Infect Dis 188: 1666–1674
187. Sabbaj S, Edwards BH, Ghosh MK et al (2002) Human immunodeficiency virus-specific CD8(+) T cells in human breast milk. J Virol 76:7365–7373
188. Sabbaj S, Ghosh MK, Edwards BH et al (2005) Breast milk-derived antigen-specific CD8+ T cells: an extralymphoid effector memory cell population in humans. J Immunol 174:2951–2956
189. Miller M, Iliff P, Stoltzfus RJ, Humphrey J (2002) Breast milk erythropoietin and mother to child HIV transmission through breastmilk. Lancet 360:1246–1248
190. Arsenault JE, Webb AL, Koulinska IN, Aboud S, Fawzi WW, Villamor E (2010) Association between breast milk erythropoietin and reduced risk of mother-to-child transmission of HIV. J Infect Dis 202:370–373
191. Van De Perre P (2003) Transfer of antibody via mother's milk. Vaccine 21:3374–3376

Chapter 11
Cellular Immunity in Breast Milk: Implications for Postnatal Transmission of HIV-1 to the Infant*

Steffanie Sabbaj, Chris C. Ibegbu, and Athena P. Kourtis

Breastfeeding accounts for up to 40% of all infant human immunodeficiency virus (HIV) type 1 infections in resource-limited settings, where prolonged breastfeeding is the only available and safe infant feeding option [1, 2]. However, most breastfed infants remain uninfected even after prolonged exposure to breast milk [3–5]. The factors in breast milk that protect the majority of breastfed infants of HIV-infected mothers from infection remain largely undetermined. Breast milk contains a multitude of immune parameters, including immunoglobulins, antimicrobial substances, pro-and anti-inflammatory cytokines, and leukocytes [6]. Moreover, breast milk not only provides passive protection, but also can directly modulate the immunological development of the infant [7, 8]. Cell-mediated immunity in breast milk has not been as extensively studied as humoral immunity, described in Chapter 10. There is, however, increasing interest in the role that lymphocytes, macrophages, and other immune cell types play, both in innate and in adaptive breast milk immunity [3].

Transmission of HIV-1 to the breastfeeding infant is thought to occur via both cell-free and cell-associated HIV [9, 10]. Increased levels of HIV-1 RNA in milk directly correlates with virus transmission [10]; however, recent studies have suggested that cell-associated virus in milk may be a more significant predictor of transmission to the infant during breastfeeding [6, 9, 10]. HIV-infected cells in breast milk include both CD4+ lymphocytes and macrophages, and possibly also breast milk epithelial cells [11]. Entry of HIV into the milk requires passage across mucosal surfaces and interaction of HIV-infected cells with epithelial cells lining the epithelial barrier in the mammary gland. In the infant, infected milk carries virus across the infant's oral, nasopharyngeal, and/or gastrointestinal tract. Contact between HIV-1-infected cells and epithelial cells are postulated to occur with binding of HIV envelope glycoproteins to the epithelial receptor galactosyl ceramide and stabilization of the interaction with integrin-dependent engagement [12]. There is increasing appreciation of the fact that

---

*The findings and conclusions in this chapter are those of the authors and do not necessarily represent the official views of the Centers for Disease Control and Prevention.

S. Sabbaj, Ph.D. (✉)
Department of Medicine, University of Alabama at Birmingham,
908 20th Street South, CCB 334, Birmingham, AL 35294, USA
e-mail:sabbaj@uab.edu

C.C. Ibegbu, Ph.D.
Emory Vaccine Center, Emory University School of Medicine, Atlanta, GA, USA

A.P. Kourtis, M.D., Ph.D., M.P.H.
Division of Reproductive Health, NCCDPHP, Centers for Disease Control and Prevention,
4770 Buford Highway, NE, MSK34, Atlanta, GA 30341, USA

epithelial cells are not simply passive barriers but have reciprocal regulating interactions with innate and adaptive immune factors against foreign antigens [13, 14]. Whether mammary epithelial cells themselves support HIV replication remains controversial [14, 15].

In this chapter, we discuss what is currently known about cells of the immune system present in breast milk, their phenotypes, activation status, chemokine expression, and their functional responses. We review maternal HIV infection-induced changes in the phenotype and function of breast milk cells, and discuss the role of these cells in modulating the risk of infant HIV infection.

Cellular Composition of Breast Milk

Breast milk, unlike most other secretions, contains a large number of lymphocytes, macrophages, and other mononuclear cells (ranging from 10^5 to 10^7/mL in colostrum and declining tenfold or more during the subsequent 2–3 months of lactation) [14, 16–28] and reviewed in ref. [2]. Estimates of the cellular composition of breast milk have varied according to the methodology used. Earlier studies using sheep red-blood-cell rosetting and immunoglobulin-coated beads yielded different proportions than more recent studies using flow cytometry [29]. Further complicating this issue are the dynamic compositional changes that occur with different stages of lactation, as well as technical variability introduced by factors such as the decreased adherence of breast milk macrophages [30] and the high lipid content of breast milk. Storage conditions, time since expression, and temperature may also affect measurements of the biological components of human milk [31].

Early studies suggested that the cell population of breast milk contains around 40–70% macrophages, 20–50% polymorphonuclear cells, and 7–30% lymphocytes [29, 32, 33]. More recent studies indicate that neutrophils and macrophages comprise roughly 95% of breast milk cells and 4% of lymphocytes [22, 34]. Various epithelial cells and their fragments are also seen later in lactation [14, 28]. The majority of CD4+ cells in the colostrum are CD14+ macrophages expressing chemokine receptors, CXCR4 and CCR5, and DC-SIGN, a dendritic cell-specific receptor for HIV [35, 36]. Within the lymphocyte population, the distribution of cell types is as follows: CD3+ T cells, 83%, γδ T cells, 11% [24, 34, 37], CD16+ NK cells, 3–4% [25, 37] and B cells, 6% or less [20, 34, 37, 38]. The presence of regulatory T cells in breast milk has not, to the best of our knowledge, been evaluated.

Breast Milk Macrophages

The majority of cells in colostrum and breast milk are breast milk macrophages expressing dendritic cell-specific intercellular adhesion molecule 3 (ICAM3) grabbing nonintegrin (DC-SIGN) [36]. Breast milk macrophages also express Toll-like receptor 3 (TLR3), in contrast with peripheral blood monocytes [36]. The functional characteristics of breast milk macrophages seem to be distinct from those of peripheral blood: breast milk macrophages spontaneously produce granulocyte–macrophage colony stimulating factor (GM-CSF) and can differentiate into CD1+ dendritic cells in the presence of exogenous interleukin 4 (IL-4) alone, in contrast with peripheral blood monocytes, which require GM-CSF in addition to IL-4 [39]. Interleukin-4 stimulated breast milk macrophages are efficient in stimulating T cells, but they also display enhanced expression of DC-SIGN, a dendritic cell-specific receptor for HIV [35, 39]. These findings indicate that macrophages can have a dual role, both in mediating immune responses, through stimulation of T cells, and in facilitating entry of HIV. This role is particularly relevant in the case of mastitis, when local production of IL-4 may upregulate the expression of DC-SIGN in breast milk macrophages, offering one mechanism to explain why mastitis is linked with higher HIV load in breast milk and with greater risk of transmission of infection to the infant [39]. Other recent findings indicate that HIV virions captured by DC-SIGN may be more efficiently transmitted to the gastrointestinal tract [35], suggesting a role in transmission to the infant. IL-4-treated breast milk macrophages also displayed enhanced expression of TLR3 [36]. When TLR3 was stimulated with its specific ligand, DC-SIGN expression on breast milk macrophages was reduced even in the IL-4-mediated enhanced state [36], suggesting that TLR3 signaling may offer a way to reduce transmission of mother-to-child transmission (MTCT) of HIV via breast milk.

Breast Milk T Lymphocytes

The proportions of CD4+ and CD8+ T lymphocytes in breast milk are similar, unlike blood where CD4 usually out number CD8 two to one [21, 23, 34, 37, 40]. For example, in one study, it was observed that the mean percentage of CD4+ T lymphocytes in the breast milk was 47% and that of CD8+ T lymphocytes was 46%; this compared with mean percentages of 64% and 37%, respectively, in the peripheral blood of the same women [21]. Expression of the lymph-node homing marker CCR7 is lower in both CD4+ and CD8+ breast milk T cells, compared with peripheral blood [37, 40]. By contrast, expression of the intestinal homing marker CD103 is significantly higher in breast milk CD4+ and CD8+ T cells, compared with peripheral blood [37, 40]. A large percentage of breast milk T cells express other mucosal homing markers such as $\alpha 4$ integrin (CD49f), $\beta 7$ integrin, and CCR9 [17, 40]. There is a marked preponderance of activated CD4+ and CD8+ T cells, as evidenced by the high expression of the markers HLA-DR, CD25, CD45RO, and CCR5 [17, 37, 40, 41]. Indeed, expression of the activation marker HLA-DR on breast milk CD4+ T cells was almost tenfold as high, and that on CD8+ T cells almost fivefold higher compared with peripheral blood in one study [37]. Co-expression of CCR5 and CXCR4 was observed in 26–73% of breast milk CD4+ T cells, in contrast with peripheral blood, where such co-expression was seen in 1–20% of CD4+ T cells [37]. Breast milk CD4+ T cells also express the proliferation marker Ki67 at a higher percentage than in peripheral blood (median of 2.6% vs. 1.2%, respectively) [37]. The phenotypic characteristics of CD8+ T lymphocytes in breast milk indicate that these cells are predominantly an antigen-experienced but not terminally differentiated effector cell population [37, 40]. T lymphocytes expressing CD40 ligand, a molecule involved in B cell isotype switching, and CD26 and CD31, two adhesion/activation molecules, were also seen in colostrum [16, 18]. Thus, given the increased proportion of CD8+ T cells in breast milk, and their activated and "mucosal" phenotype, breast milk T cells are more similar to T cells in other mucosal sites such as the gastrointestinal tract and vagina [42–48], supporting the notion that breast milk lymphocytes migrate to the breast from distant mucosal sites such as the gastrointestinal or genital tract.

Breast Milk B Lymphocytes

Breast milk B lymphocytes also display a phenotype that is different from that of peripheral blood. They are predominantly class-switched memory B cells, with few IgD+ memory and naïve B cells [38]. They bear a unique profile of adhesion molecules (CD44+ CD62L−, $a_4\beta_7^{+/-}$, $a_4\beta_1^+$), and a mucosal homing profile similar to B cells located in gut-associated lymphoid tissue, suggesting that these cells originate from the gut. In addition, breast milk contains higher percentages of activated CD38+ B cells, large-size B cells, plasmablasts, and plasma cells compared with blood [38]. Milk B lymphocytes do not express complement receptors [49], which is another indication that these cells are plasmablasts and/or antibody-producing plasma cells. These findings suggest that a significant proportion of breast milk B cells have undergone terminal plasma cell differentiation. Cells secreting immunoglobulin spontaneously were observed in breast milk [38], predominantly IgG rather than IgA.

As the mammary gland represents an effector site of the mucosal immune system, memory B cells and plasmablasts that colonize the breast late in pregnancy likely originate from other mucosal areas where they have been exposed to various antigens.

T and B Lymphocytes in the Breast Milk of HIV-Infected Women

HIV infection depletes CD4 T cells in the gastrointestinal tract [47, 50, 51]; however, a study of breast milk lymphocytes in HIV-infected women revealed a relative preservation of CD4+ T cells, despite the fact that the peripheral blood CD4+ T cells were markedly decreased [37]. The proportion of CCR5+CD4+ T cells was also preserved, compared with peripheral blood [37]. This finding concurs with the preservation of CD4+ memory T lymphocytes in breast milk during acute simian immunodeficiency virus (SIV) infection in rhesus monkeys [52]. In addition, a study demonstrated that HIV-infected women had a higher proportion of HLA-DR-expressing CD4+ T lymphocytes in their breast milk, compared with uninfected women, consistent with the preservation of activated CD4+ T cells [37]. By contrast, a different study reported a decrease in the CD4+ T cells in the breast milk of HIV-infected women when compared to the breast milk of uninfected women [40]. The differences in the results from these two studies could be due to differences in the cohorts. All HIV-infected women enrolled in the study showing preservation of CD4+ T cells were on highly active antiretroviral therapy (HAART), while the women enrolled in the study demonstrating a decrease in CD4+ T cells were not all on HAART.

As discussed previously, co-expression of CCR5 and CXCR4 was observed in breast milk CD4+ T cells [37]. Since CCR5 and CXCR4 are the major coreceptors required for HIV attachment and entry [53, 54], it appears that breast milk T lymphocytes offer the optimal cell target for HIV infection and transmission. In contrast with other mucosal sites, where a profound depletion of these cells is observed very early in HIV (or SIV) infection [50, 55], there is no apparent loss of CD4+ CCR5+ T cells in breast milk. Whether these CCR5+CD4+ T cells are resting cells that might be resistant to HIV killing is not known. Indeed, a study indicated that resting CD4+ T cells represent more than 90% of purified viable breast milk lymphocytes [56].

With regards to CD8+ T cell changes in breast milk during HIV infection, reports indicate that these cells are also predominantly effector memory cells [37, 40] similar to CD8+ T cells from breast milk from uninfected women. This same study showed a higher frequency of CD8+CD57−CD45RO+ T cells (cells usually retaining their proliferative capacity and relatively refractory to apoptosis [57]) from breast milk, whether infected or uninfected with HIV, when compared to matched samples from PBMC. Together these data may imply enhanced functional competence of breast milk CD8+ T lymphocytes, which could be a factor responsible for the lower HIV load in breast milk, compared with blood [58] and the relatively low transmission via this route.

Whether the presence of these activated CD4+ and CD8+ lymphocytes in the breast milk might increase the risk of HIV transmission to the infant or help protect from mucosal infection is currently uncertain and likely multifaceted. With regards to B lymphocytes in the breast milk of HIV-infected women, anti-HIV-1 antibody secreting cells were found, most of them producing IgG [38].

T-Cell Function in Breast Milk

The presence of antigen-specific T cells in human breast milk was first described in 1968 [59], years before the identification of the T-cell receptor [60–65]. In this first study, tetanus toxoid, diphtheria, and purified protein derivative (PPD)-specific T cells were detected using ^{3}H-thymidine incorporation by radioautograph and visually counting silver grains as a measure of DNA synthesis. Since then, T-cell responses to various pathogens including rubella, *Escherichia coli*, influenza, rotavirus, and respiratory syncytial virus (RSV) have been detected in breast milk by lymphocyte proliferation assays [26, 29, 33, 66–68]. In these studies, total T-cell responses were measured using mononuclear cells isolated from breast milk, but the T-cell subset responsible for the reactivity was not defined.

With the emergence of the HIV epidemic in sub-Saharan Africa, where breastfeeding is responsible for a large proportion of neonatal HIV infection, and reports implicating peripheral blood CD8+ T cells in the control of HIV [69, 70] and correlating their activity with a slower course of disease progression [71–75], interest arose in the role of CD8+ T cells and MTCT of HIV. Initial studies concentrated on demonstrating the presence of HIV-specific T-cell responses in breast milk. In these studies using the IFN-γ ELISPOT assay, it was demonstrated that the frequency of HIV-specific responses in breast milk was increased when compared to peripheral blood [40, 76, 77]. Conversely, the early studies using lymphocyte proliferation demonstrated that cells stimulated with either antigen or phytohemagglutinin (PHA) were decreased in breast milk compared to blood [26, 27, 29, 66, 67, 78]. These differences may have to do with the fact that breast milk-derived T cells are highly activated and express low levels of CD45RA and CCR7 [34, 37, 40]. In assays such as the IFN-γ ELISPOT or tetramer staining where short culture or no culture of the T cells are used, these highly activated cells are easily detected. However, in assays such as the lymphocyte proliferation assay, the cells require extensive culture and it is likely that due to the hyperactivation of these cells, mitogenic and/or antigenic stimulation led to apoptosis of the cells, resulting in what appeared as hyporesponsiveness. In studies of IFN-γ ELISPOT after CD8+ T cell depletion, it was demonstrated that CD8+ T cells mediated the HIV-specific responses detected [77]. These same investigators demonstrated that these cells have cytolytic activity. They showed, using the ^{51}Cr-release assay, that autologous B lymphoblastoid cell lines (BLCL) infected with vaccinia virus expressing HIV genes could be lysed with these T cells. Furthermore, these cells bound class I MHC tetramers specific for optimized HIV-peptides, conclusively demonstrating that this activity was CD8 T-cell-mediated [40, 77]. CD8+ T-cell responses in breast milk have also been mapped to HLA-restricted epitopes, and the repertoire of the specificity between breast milk and the peripheral blood is different suggesting different homing patterns [77] or variant HIV-1 species [79]. This data has implications for compartmentalization of HIV evolution and escape.

The immunologic benefits of breastfeeding for the infant conferred from antibodies in the milk are well established. However, the benefit of CD8+ T cells in breast milk for the infant in preventing MTCT of HIV remains to be elucidated. In humans, there is only one report demonstrating that immunization of mother results in antigen-specific cells in breast milk that may play a role in protection of the infant from infection. In this study, trafficking of rubella-specific T cells into breast milk after systemic and intranasal immunization was demonstrated [80]. In order to begin to address the question as to whether maternal immunization with an HIV vaccine that stimulates cell-mediated immunity can protect breastfed infants from HIV-1 infection, animal models need to be established. For this purpose, a group of investigators have established a hormone-induced lactation model in rhesus monkeys [81]. In these studies, they demonstrate that the composition of lymphocytes in hormonally induced milk was similar to normal lactation. They also demonstrated that the frequency of CD8+ SIV-specific T cells in breast milk was at least twice as high as in blood, replicating the human data [40, 76]. Finally, and most relevant to MTCT of HIV, they demonstrated that the appearance of SIV-specific CD8+ T cells in breast milk was associated with a reduction in breast milk viral load, lending direct evidence that breast milk-derived CD8+ T cells play a role in the local control of HIV found in the breast milk compartment [52]. In these same studies, an increase in SIV-specific effector CD8+ T cells in milk during acute infection was also demonstrated.

There are no direct data from vaccine studies in humans to demonstrate protection of a nursing infant from infection due to T cells in breast milk. However, a report studying SIV systemic vaccination (SIV DNA prime and live attenuated poxvirus vector expressing SIV mac239 *gag-pol* and *envelope* genes boost, followed by an Adenovirus serotype 5 vector containing matching immunogens) of lactating rhesus monkeys found that vaccine-elicited CD8 T cells were present in milk at similar or higher magnitude than in blood [82]. These studies have set the stage to study MTCT of SIV in the context of vaccination that may lead to strategies that can prevent breast milk transmission of HIV. HIV-specific CD4+ T cell responses have not been studied in the breast milk of HIV-infected women.

These studies have opened the door for further examination of other T-cell subsets in breast milk such as regulatory T (Tregs) and Th17 cells, as well as NK cells, and their role in transmission of HIV from mother-to-child via breast milk and protection thereof. In addition, whether these cells are adoptively transferred to the infant via the gastrointestinal tract for protection against HIV and other pathogens or play other roles within the infant is yet to be determined.

Conclusions and Future Directions

As noted above, many of the phenotypic and functional characteristics of breast milk lymphocytes are different than those in the peripheral blood. This compartmentalization may indicate selective homing of certain cells to the mammary gland. There is evidence to indicate that expression of particular adhesion molecules by epithelial cells in different tissues controls selective lymphocyte recruitment and localization [83]. According to the entero-mammary axis hypothesis, breast milk lymphocytes have homed to the breast from distant mucosal sites such as the genital or gastrointestinal tract, where they may have encountered mucosal-tropic strains of HIV and thus have already been activated against them. Direct labeling studies have shown that the mammary gland can draw from both circulating pools of T cells, intestinal and peripheral [84–86]. These studies further our understanding of the mechanisms underlying the dynamics of mucosal immune activation and lymphocyte recruitment and homeostasis in the mammary gland. Moreover, the kinetics and function of virus-specific cellular immunity in the breast milk, as well as the dynamic changes induced by HIV infection, are important and have implications for the study of immune correlates of protection and for the development of passive and active immunoprophylactic approaches to prevent transmission of HIV to the infant. The importance of regulatory lymphocytes in inducing immune tolerance is just beginning to be appreciated; their presence in breast milk and their role in modulating HIV infection in the infant needs to be studied, along with the complex interplay between epithelial, innate, and adaptive immunity.

References

1. Kourtis AP, Bulterys M (2010) Mother-to-child transmission of HIV: pathogenesis, mechanisms and pathways. Clin Perinatol 37(4):721–737
2. Kourtis AP, Butera S, Ibegbu C, Beled L, Duerr A (2003) Breast milk and HIV-1: vector of transmission or vehicle of protection? Lancet Infect Dis 3(12):786–793
3. Aldrovandi GM, Kuhn L (2010) What infants and breasts can teach us about natural protection from HIV infection. J Infect Dis 202(Suppl 3):S366–S370
4. Leroy V, Newell ML, Dabis F, Peckham C, Van de Perre P, Bulterys M et al (1998) International multicentre pooled analysis of late postnatal mother-to-child transmission of HIV-1 infection. Ghent International Working Group on Mother-to-Child Transmission of HIV. Lancet 352(9128):597–600
5. Nduati R, John G, Mbori-Ngacha D, Richardson B, Overbaugh J, Mwatha A et al (2000) Effect of breastfeeding and formula feeding on transmission of HIV-1: a randomized clinical trial. JAMA 283(9):1167–1174
6. Van de Perre P, Simonon A, Hitimana DG, Dabis F, Msellati P, Mukamabano B et al (1993) Infective and anti-infective properties of breastmilk from HIV-1-infected women. Lancet 341(8850):914–918
7. Garofalo R (2010) Cytokines in human milk. J Pediatr 156(2 Suppl):S36–S40

8. Goldman AS, Chheda S, Garofalo R, Schmalstieg FC (1996) Cytokines in human milk: properties and potential effects upon the mammary gland and the neonate. J Mammary Gland Biol Neoplasia 1(3):251–258
9. Koulinska IN, Villamor E, Chaplin B, Msamanga G, Fawzi W, Renjifo B et al (2006) Transmission of cell-free and cell-associated HIV-1 through breast-feeding. J Acquir Immune Defic Syndr 41(1):93–99
10. Rousseau CM, Nduati RW, Richardson BA, Steele MS, John-Stewart GC, Mbori-Ngacha DA et al (2003) Longitudinal analysis of human immunodeficiency virus type 1 RNA in breast milk and of its relationship to infant infection and maternal disease. J Infect Dis 187(5):741–747
11. Toniolo A, Serra C, Conaldi PG, Basolo F, Falcone V, Dolei A (1995) Productive HIV-1 infection of normal human mammary epithelial cells. AIDS 9(8):859–866
12. Alfsen A, Yu H, Magerus-Chatinet A, Schmitt A, Bomsel M (2005) HIV-1-infected blood mononuclear cells form an integrin- and agrin-dependent viral synapse to induce efficient HIV-1 transcytosis across epithelial cell monolayer. Mol Biol Cell 16(9):4267–4279
13. Bulek K, Swaidani S, Aronica M, Li X (2010) Epithelium: the interplay between innate and Th2 immunity. Immunol Cell Biol 88(3):257–268
14. Dorosko SM, Connor RI (2010) Primary human mammary epithelial cells endocytose HIV-1 and facilitate viral infection of CD4+ T lymphocytes. J Virol 84(20):10533–10542
15. Lyimo MA, Howell AL, Balandya E, Eszterhas SK, Connor RI (2009) Innate factors in human breast milk inhibit cell-free HIV-1 but not cell-associated HIV-1 infection of CD4+ cells. J Acquir Immune Defic Syndr 51(2):117–124
16. Bertotto A, Castellucci G, Radicioni M, Bartolucci M, Vaccaro R (1996) CD40 ligand expression on the surface of colostral T cells. Arch Dis Child Fetal Neonatal Ed 74(2):F135–F136
17. Bertotto A, Gerli R, Fabietti G, Crupi S, Arcangeli C, Scalise F et al (1990) Human breast milk T lymphocytes display the phenotype and functional characteristics of memory T cells. Eur J Immunol 20(8):1877–1880
18. Bertotto A, Spinozzi F, Gerli R, Castellucci G, Bassotti G, Crupi S et al (1995) CD26 and CD31 surface antigen expression on human colostral T cells. Biol Neonate 68(4):259–263
19. Crago SS, Prince SJ, Pretlow TG, McGhee JR, Mestecky J (1979) Human colostral cells. I. Separation and characterization. Clin Exp Immunol 38(3):585–597
20. Davis MK (1991) Human milk and HIV infection: epidemiologic and laboratory data. Adv Exp Med Biol 310:271–280
21. Eglinton BA, Roberton DM, Cummins AG (1994) Phenotype of T cells, their soluble receptor levels, and cytokine profile of human breast milk. Immunol Cell Biol 72(4):306–313
22. Goldman AS, Chheda S, Garofalo R (1998) Evolution of immunologic functions of the mammary gland and the postnatal development of immunity. Pediatr Res 43(2):155–162
23. Jarvinen KM, Suomalainen H (2002) Leucocytes in human milk and lymphocyte subsets in cow's milk-allergic infants. Pediatr Allergy Immunol 13(4):243–254
24. Lindstrand A, Smedman L, Gunnlaugsson G, Troye-Blomberg M (1997) Selective compartmentalization of gammadelta-T lymphocytes in human breastmilk. Acta Paediatr 86(8):890–891
25. Moro I, Abo T, Crago SS, Komiyama K, Mestecky J (1985) Natural killer cells in human colostrum. Cell Immunol 93(2):467–474
26. Ogra SS, Ogra PL (1978) Immunologic aspects of human colostrum and milk. II. Characteristics of lymphocyte reactivity and distribution of E-rosette forming cells at different times after the onset of lactation. J Pediatr 92(4):550–555
27. Ogra SS, Weintraub D, Ogra PL (1977) Immunologic aspects of human colostrum and milk. III. Fate and absorption of cellular and soluble components in the gastrointestinal tract of the newborn. J Immunol 119(1):245–248
28. Xanthou M (1997) Human milk cells. Acta Paediatr 86(12):1288–1290
29. Parmely MJ, Beer AE, Billingham RE (1976) In vitro studies on the T-lymphocyte population of human milk. J Exp Med 144(2):358–370
30. Thorpe LW, Rudloff HE, Powell LC, Goldman AS (1986) Decreased response of human milk leukocytes to chemoattractant peptides. Pediatr Res 20(4):373–377
31. Williamson MT, Murti PK (1996) Effects of storage, time, temperature, and composition of containers on biologic components of human milk. J Hum Lact 12(1):31–35
32. Ho FC, Wong RL, Lawton JW (1979) Human colostral and breast milk cells. A light and electron microscopic study. Acta Paediatr Scand 68(3):389–396
33. Keller MA, Turner JL, Stratton JA, Miller ME (1980) Breast milk lymphocyte response to K1 antigen of *Escherichia coli*. Infect Immun 27(3):903–909
34. Wirt DP, Adkins LT, Palkowetz KH, Schmalstieg FC, Goldman AS (1992) Activated and memory T lymphocytes in human milk. Cytometry 13(3):282–290
35. Satomi M, Shimizu M, Shinya E, Watari E, Owaki A, Hidaka C et al (2005) Transmission of macrophage-tropic HIV-1 by breast-milk macrophages via DC-SIGN. J Infect Dis 191(2):174–181

36. Yagi Y, Watanabe E, Watari E, Shinya E, Satomi M, Takeshita T et al (2010) Inhibition of DC-SIGN-mediated transmission of human immunodeficiency virus type 1 by Toll-like receptor 3 signalling in breast milk macrophages. Immunology 130(4):597–607
37. Kourtis AP, Ibegbu CC, Theiler R, Xu YX, Bansil P, Jamieson DJ et al (2007) Breast milk CD4+ T cells express high levels of C chemokine receptor 5 and CXC chemokine receptor 4 and are preserved in HIV-infected mothers receiving highly active antiretroviral therapy. J Infect Dis 195(7):965–972
38. Tuaillon E, Valea D, Becquart P, Al Tabaa Y, Meda N, Bollore K et al (2009) Human milk-derived B cells: a highly activated switched memory cell population primed to secrete antibodies. J Immunol 182(11):7155–7162
39. Ichikawa M, Sugita M, Takahashi M, Satomi M, Takeshita T, Araki T et al (2003) Breast milk macrophages spontaneously produce granulocyte–macrophage colony-stimulating factor and differentiate into dendritic cells in the presence of exogenous interleukin-4 alone. Immunology 108(2):189–195
40. Sabbaj S, Ghosh MK, Edwards BH, Leeth R, Decker WD, Goepfert PA et al (2005) Breast milk-derived antigen-specific CD8+ T cells: an extralymphoid effector memory cell population in humans. J Immunol 174(5):2951–2956
41. Rivas RA, el-Mohandes AA, Katona IM (1994) Mononuclear phagocytic cells in human milk: HLA-DR and Fc gamma R ligand expression. Biol Neonate 66(4):195–204
42. Farstad IN, Halstensen TS, Lien B, Kilshaw PJ, Lazarovits AI, Brandtzaeg P (1996) Distribution of beta 7 integrins in human intestinal mucosa and organized gut-associated lymphoid tissue. Immunology 89(2):227–237
43. Hladik F, Lentz G, Delpit E, McElroy A, McElrath MJ (1999) Coexpression of CCR5 and IL-2 in human genital but not blood T cells: implications for the ontogeny of the CCR5+ Th1 phenotype. J Immunol 163(4):2306–2313
44. Hussain LA, Kelly CG, Fellowes R, Hecht EM, Wilson J, Chapman M et al (1992) Expression and gene transcript of Fc receptors for IgG, HLA class II antigens and Langerhans cells in human cervico-vaginal epithelium. Clin Exp Immunol 90(3):530–538
45. Johansson EL, Rudin A, Wassen L, Holmgren J (1999) Distribution of lymphocytes and adhesion molecules in human cervix and vagina. Immunology 96(2):272–277
46. Quayle AJ, Kourtis AP, Cu-Uvin S, Politch JA, Yang H, Bowman FP et al (2007) T-lymphocyte profile and total and virus-specific immunoglobulin concentrations in the cervix of HIV-1-infected women. J Acquir Immune Defic Syndr 44(3):292–298
47. Veazey RS, DeMaria M, Chalifoux LV, Shvetz DE, Pauley DR, Knight HL et al (1998) Gastrointestinal tract as a major site of CD4+ T cell depletion and viral replication in SIV infection. Science 280(5362):427–431
48. Zabel BA, Agace WW, Campbell JJ, Heath HM, Parent D, Roberts AI et al (1999) Human G protein-coupled receptor GPR-9-6/CC chemokine receptor 9 is selectively expressed on intestinal homing T lymphocytes, mucosal lymphocytes, and thymocytes and is required for thymus-expressed chemokine-mediated chemotaxis. J Exp Med 190(9):1241–1256
49. Bush JF, Beer AE (1979) Analysis of complement receptors on B-lymphocytes in human milk. Am J Obstet Gynecol 133(6):708–712
50. Brenchley JM, Schacker TW, Ruff LE, Price DA, Taylor JH, Beilman GJ et al (2004) CD4+ T cell depletion during all stages of HIV disease occurs predominantly in the gastrointestinal tract. J Exp Med 200(6):749–759
51. Mattapallil JJ, Douek DC, Hill B, Nishimura Y, Martin M, Roederer M (2005) Massive infection and loss of memory CD4+ T cells in multiple tissues during acute SIV infection. Nature 434(7037):1093–1097
52. Permar SR, Kang HH, Carville A, Wilks AB, Mansfield KG, Rao SS et al (2010) Preservation of memory CD4(+) T lymphocytes in breast milk of lactating rhesus monkeys during acute simian immunodeficiency virus infection. J Infect Dis 201(2):302–310
53. Alkhatib G (2009) The biology of CCR5 and CXCR4. Curr Opin HIV AIDS 4(2):96–103
54. Wu Y, Yoder A (2009) Chemokine coreceptor signaling in HIV-1 infection and pathogenesis. PLoS Pathog 5(12):e1000520
55. Veazey RS, Marx PA, Lackner AA (2003) Vaginal CD4+ T cells express high levels of CCR5 and are rapidly depleted in simian immunodeficiency virus infection. J Infect Dis 187(5):769–776
56. Petitjean G, Becquart P, Tuaillon E, Al Tabaa Y, Valea D, Huguet MF et al (2007) Isolation and characterization of HIV-1-infected resting CD4+ T lymphocytes in breast milk. J Clin Virol 39(1):1–8
57. Brenchley JM, Karandikar NJ, Betts MR, Ambrozak DR, Hill BJ, Crotty LE et al (2003) Expression of CD57 defines replicative senescence and antigen-induced apoptotic death of CD8+ T cells. Blood 101(7):2711–2720
58. Lewis P, Nduati R, Kreiss JK, John GC, Richardson BA, Mbori-Ngacha D et al (1998) Cell-free human immunodeficiency virus type 1 in breast milk. J Infect Dis 177(1):34–39
59. Smith CW, Goldman AS (1968) The cells of human colostrum. I. In vitro studies of morphology and functions. Pediatr Res 2(2):103–109
60. Allison JP, McIntyre BW, Bloch D (1982) Tumor-specific antigen of murine T-lymphoma defined with monoclonal antibody. J Immunol 129(5):2293–2300
61. Haskins K, Kubo R, White J, Pigeon M, Kappler J, Marrack P (1983) The major histocompatibility complex-restricted antigen receptor on T cells. I. Isolation with a monoclonal antibody. J Exp Med 157(4):1149–1169

62. Hedrick SM, Cohen DI, Nielsen EA, Davis MM (1984) Isolation of cDNA clones encoding T cell-specific membrane-associated proteins. Nature 308(5955):149–153
63. Hedrick SM, Nielsen EA, Kavaler J, Cohen DI, Davis MM (1984) Sequence relationships between putative T-cell receptor polypeptides and immunoglobulins. Nature 308(5955):153–158
64. Meuer SC, Fitzgerald KA, Hussey RE, Hodgdon JC, Schlossman SF, Reinherz EL (1983) Clonotypic structures involved in antigen-specific human T cell function. Relationship to the T3 molecular complex. J Exp Med 157(2):705–719
65. Yanagi Y, Yoshikai Y, Leggett K, Clark SP, Aleksander I, Mak TW (1984) A human T cell-specific cDNA clone encodes a protein having extensive homology to immunoglobulin chains. Nature 308(5955):145–149
66. Ruben FL, Holzman IR, Fireman P (1982) Responses of lymphocytes from human colostrum or milk to influenza antigens. Am J Obstet Gynecol 143(5):518–522
67. Totterdell BM, Patel S, Banatvala JE, Chrystie IL (1988) Development of a lymphocyte transformation assay for rotavirus in whole blood and breast milk. J Med Virol 25(1):27–36
68. Scott R, Scott M, Toms GL (1985) Cellular reactivity to respiratory syncytial virus in human colostrum and breast milk. J Med Virol 17(1):83–93
69. Borrow P, Lewicki H, Hahn BH, Shaw GM, Oldstone MB (1994) Virus-specific CD8+ cytotoxic T-lymphocyte activity associated with control of viremia in primary human immunodeficiency virus type 1 infection. J Virol 68(9):6103–6110
70. Koup RA, Safrit JT, Cao Y, Andrews CA, McLeod G, Borkowsky W et al (1994) Temporal association of cellular immune responses with the initial control of viremia in primary human immunodeficiency virus type 1 syndrome. J Virol 68(7):4650–4655
71. Carrington M, Nelson GW, Martin MP, Kissner T, Vlahov D, Goedert JJ et al (1999) HLA and HIV-1: heterozygote advantage and B*35-Cw*04 disadvantage. Science 283(5408):1748–1752
72. Hendel H, Caillat-Zucman S, Lebuanec H, Carrington M, O'Brien S, Andrieu JM et al (1999) New class I and II HLA alleles strongly associated with opposite patterns of progression to AIDS. J Immunol 162(11):6942–6946
73. Kaslow RA, Carrington M, Apple R, Park L, Munoz A, Saah AJ et al (1996) Influence of combinations of human major histocompatibility complex genes on the course of HIV-1 infection. Nat Med 2(4):405–411
74. Keet IP, Tang J, Klein MR, LeBlanc S, Enger C, Rivers C et al (1999) Consistent associations of HLA class I and II and transporter gene products with progression of human immunodeficiency virus type 1 infection in homosexual men. J Infect Dis 180(2):299–309
75. Nelson GW, Kaslow R, Mann DL (1997) Frequency of HLA allele-specific peptide motifs in HIV-1 proteins correlates with the allele's association with relative rates of disease progression after HIV-1 infection. Proc Natl Acad Sci USA 94(18):9802–9807
76. Lohman BL, Slyker J, Mbori-Ngacha D, Bosire R, Farquhar C, Obimbo E et al (2003) Prevalence and magnitude of human immunodeficiency virus (HIV) type 1-specific lymphocyte responses in breast milk from HIV-1-seropositive women. J Infect Dis 188(11):1666–1674
77. Sabbaj S, Edwards BH, Ghosh MK, Semrau K, Cheelo S, Thea DM et al (2002) Human immunodeficiency virus-specific CD8(+) T cells in human breast milk. J Virol 76(15):7365–7373
78. Diaz-Jouanen E, Williams RC Jr (1974) T and B lymphocytes in human colostrum. Clin Immunol Immunopathol 3(2):248–255
79. Becquart P, Chomont N, Roques P, Ayouba A, Kazatchkine MD, Belec L et al (2002) Compartmentalization of HIV-1 between breast milk and blood of HIV-infected mothers. Virology 300(1):109–117
80. Losonsky GA, Fishaut JM, Strussenberg J, Ogra PL (1982) Effect of immunization against rubella on lactation products. I. Development and characterization of specific immunologic reactivity in breast milk. J Infect Dis 145(5):654–660
81. Permar SR, Kang HH, Carville A, Mansfield KG, Gelman RS, Rao SS et al (2008) Potent simian immunodeficiency virus-specific cellular immune responses in the breast milk of simian immunodeficiency virus-infected, lactating rhesus monkeys. J Immunol 181(5):3643–3650
82. Wilks AB, Christian EC, Seaman MS, Sircar P, Carville A, Gomez CE et al (2010) Robust vaccine-elicited cellular immune responses in breast milk following systemic simian immunodeficiency virus DNA prime and live virus vector boost vaccination of lactating rhesus monkeys. J Immunol 185(11):7097–7106
83. Kunkel EJ, Campbell JJ, Haraldsen G, Pan J, Boisvert J, Roberts AI et al (2000) Lymphocyte CC chemokine receptor 9 and epithelial thymus-expressed chemokine (TECK) expression distinguish the small intestinal immune compartment: epithelial expression of tissue-specific chemokines as an organizing principle in regional immunity. J Exp Med 192(5):761–768
84. Jain L, Vidyasagar D, Xanthou M, Ghai V, Shimada S, Blend M (1989) In vivo distribution of human milk leucocytes after ingestion by newborn baboons. Arch Dis Child 64(7 Spec No):930–933
85. Manning LS, Parmely MJ (1980) Cellular determinants of mammary cell-mediated immunity in the rat. I. The migration of radioisotopically labeled T lymphocytes. J Immunol 125(6):2508–2514
86. Schnorr KL, Pearson LD (1984) Intestinal absorption of maternal leucocytes by newborn lambs. J Reprod Immunol 6(5):329–337

Part IV
Prevention of Breast Milk Transmission of HIV-1

Chapter 12
Antiretroviral Drugs During Breastfeeding for the Prevention of Postnatal Transmission of HIV-1*

Athena P. Kourtis, Isabelle de Vincenzi, Denise J. Jamieson, and Marc Bulterys

The global pediatric human immunodeficiency virus (HIV) type 1 epidemic is fueled to a large extent by postnatal transmission from mother to infant through breastfeeding. As many as 90% of the estimated 430,000 new HIV infections in children less than 15 years of age in 2008 were due to mother-to-child transmission (MTCT) [1]. MTCT can occur in utero, intrapartum, or postpartum through breastfeeding; among children with known timing of infection, as much as 30–40% of MTCT of HIV-1 is attributable to breastfeeding; this proportion may be even higher in settings where effective interventions that decrease in utero and intrapartum transmission are being implemented [2–4].

The benefits of breastfeeding are well recognized and include providing the infant with optimal nutrition, reducing infant morbidity and mortality due to diarrheal and lower respiratory infections, protecting against common childhood infections, and promoting child spacing, which is associated with higher maternal and child survival [5]. These benefits are particularly important in areas where the water supply is unsafe and infant mortality high. In fact, promotion of breastfeeding has been one of the World Health Organization's (WHO) cornerstone approaches for enhancing infant survival [5, 6].

On the other hand, transmission of HIV through breastfeeding mitigates these benefits. Given the risk of HIV transmission, the US Centers for Disease Control and Prevention (CDC) has recommended since 1985 that HIV-1-infected women in the USA avoid breastfeeding [7]. As replacement feeding is safe, affordable and culturally acceptable for women in the USA and other resource-rich settings, postnatal MTCT of HIV-1 is nearly zero in such settings. For most HIV-1-infected mothers in resource-limited settings, however, breastfeeding remains the only feasible option for infant feeding, given the unsafe water and poor hygiene, cultural norms which stigmatize mothers who do not breastfeed, and the prohibitive costs and lack of availability of infant formula. The WHO recommends

---

*The findings and conclusions in this report are those of the authors and do not necessarily represent the official position of the Centers for Disease Control and Prevention.

A.P. Kourtis, M.D., Ph.D., M.P.H. (✉) • D.J. Jamieson, M.D.
Division of Reproductive Health, NCCDPHP, Centers for Disease Control and Prevention,
4770 Buford Highway, NE, MSK34, Atlanta, GA 30341, USA
e-mail: apk3@cdc.gov

I. de Vincenzi
World Health Organization, Geneva, Switzerland

M. Bulterys, M.D., Ph.D.
Division of Global HIV/AIDS (DGHA), Center for Global Health, Centers for Disease Control and Prevention (CDC), 1600 Clifton Road, NE, Atlanta, GA 30333, USA

CDC Global AIDS Program, Beijing, China

Adjunct Professor of Epidemiology, UCLA School of Public Health, Los Angeles, CA, USA

that in resource-limited settings, HIV-1-infected women breastfeed their infants if safe formula feeding is not possible [8]. This issue has presented HIV-infected mothers in resource-limited settings with a difficult dilemma: how to balance the benefits of breastfeeding in reducing malnutrition as well as infectious morbidity and mortality with the risk of transmitting HIV to the infant. This dilemma is now substantially relieved given the realization (gleaned from observational data and proven by recent clinical trials) that antiretroviral (ARV) drugs given to the mother or the infant during breastfeeding can significantly reduce postnatal HIV transmission to the infant. In fact, this realization has already resulted in a major paradigm shift in the WHO recommendations that advise, for the first time in 2010, that ARV prophylaxis be used during the entire period of breastfeeding in settings where replacement infant feeding is unsafe. This chapter summarizes the current state of the art in ARV use to either mother or infant during breastfeeding as a measure to prevent postnatal transmission of HIV.

Antiretroviral Drugs for the HIV-Infected Lactating Mother Who Needs Treatment for Her Own Infection

The WHO now recommends that all pregnant, HIV-1-infected women with CD4 count ≤350 cells/mm^3 receive antiretroviral therapy (ART) for their own health and that includes women who are breastfeeding (WHO 2010). Data from the Zambia Exclusive Breastfeeding Study (ZEBS) showed that mothers whose CD4 T-cell count was <350 cells/mm^3—approximately half of all mothers—accounted for approximately 90% of maternal deaths and infant HIV infections [9].

Since 2007, results from cohorts of women with CD4+ T-cell counts <200–250/mm^3 have suggested that maternal ARVs reduced the risk of transmitting HIV-1 to the infant through breast milk [10–13]. In the postexposure prophylaxis of infants (PEPI) trial follow-up, it was shown that women who were eligible for and received ARV treatment postpartum had a postnatal transmission rate significantly lower than those eligible for treatment but untreated, and even those ineligible for treatment (CD4 >350 cells/mm^3) (postnatal transmission rates of 1.8% vs. 10.6% vs. 3.7%, respectively) [14], underscoring the need for ARV treatment-eligible women to start such treatment as early as possible. In the observational arm of the Mma Bana trial, women with CD4 <200 cells/mm^3 were started on triple NVP-based ARV at 26–34 weeks of gestation and continued through 6 months of breastfeeding; the MTCT rate at 6 months was 0.6% with no postnatal infections [15]. In the observational part of the Kesho Bora trial, where women with CD4 <200 mm^3 were started on triple-ARV at 34–36 weeks of gestation, the MTCT rate at 6 months was 5.6%. The difference between the two studies is probably largely due to the earlier initiation of ART in Mma Bana than in Kesho Bora. The earlier the ART initiation, the lower is the MTCT rate [16], probably due to a higher likelihood of achieving undetectable viral load by the time of delivery.

The only controlled data in a breastfeeding population comparing the approach of triple-ARV vs. a short prophylaxis regimen of zidovudine (ZDV) and single-dose nevirapine (NVP) among women eligible for ART (the stratum of women with CD4+ T-cell counts of 200–350/mm^3) come from the Kesho Bora trial, where the MTCT risk reduction of triple-ARV in this subgroup of women was 48% at 6 months (5.5% vs. 10.5%) [13].

Provision of ART treatment to women who need it for their own health is critical, not only for preventing postnatal transmission of HIV, but also to decrease maternal mortality. Maternal mortality is a strong predictor of infant mortality, whether the infant is HIV-infected or not [17].

Antiretroviral Approaches for Prevention of MTCT of HIV-1 Through Breastfeeding in Mothers Who Do Not Need Treatment for Their Own Health (According to Current Guidelines)

For about 15 years, ARV regimens administered to the mother during pregnancy and delivery and to the infant at birth have proven very effective in reducing in utero and intrapartum transmission of HIV-1; this has included simplified short-course regimens in resource-limited settings [18–23]. Development of effective strategies to reduce postnatal transmission in breastfeeding populations has comparatively lagged behind. Recently, however, several studies have shown that ARV prophylaxis given either to the infant or to the mother during breastfeeding is highly effective in reducing breast-feeding transmission of HIV-1 to infants [10–12, 15, 24–27]. We discuss these two approaches below. ARV drugs given to the breastfed infant provide preexposure and postexposure prophylaxis through pharmacologically active ARV drug concentrations in the infant's plasma before and after exposure to the virus, whereas ARVs given to the lactating mother act though lowering the breast milk HIV load and thus rendering the milk less infectious. The effect of maternal ART on the HIV-1 load in breast milk has been reported in several recent studies from Africa. In Mozambique, ART in breastfeeding women decreased cell-free HIV-1 RNA load in breast milk [28]. A study in Botswana among women with CD4+ T-cell counts <200 cells/mm^3 showed that ART decreased HIV-1 RNA but had no apparent effect on HIV-1 DNA load in breast milk [29]. A third study from Kenya similarly showed the suppression of cell-free HIV-1 RNA in breast milk without suppression of HIV-1 DNA [30].

Infant ARV Prophylaxis

Results from an observational and several open-label randomized clinical trials that evaluated the use of extended ARV prophylaxis among infants of breastfeeding, HIV-1-infected mothers, are now available [11, 24, 27, 31]. The Mitra study was an observational study in which daily lamivudine given to the breastfeeding infant for up to 6 months resulted in a low cumulative HIV-1-infection rate of 4.9% at 6 months, with an HIV-1 infection risk of 1.2% between 6 weeks and 6 months [31]. The SIMBA trial (this study has not been published to-date) was a randomized open-label study in Uganda and Rwanda, where mothers were enrolled from 2001 to 2002, given short-course zidovudine and didanosine from 36 weeks of gestation until 1 week postpartum, and were counseled to exclusively breastfeed for 3–6 months. Infants were randomized to either NVP or 3TC during breastfeeding. The postnatal HIV transmission rate was only 2% (http://www.mujhu.org/simba.html, accessed 3 November 2010). The Six Week Extended-Dose Nevirapine (SWEN) trials, which evaluated the effectiveness of 6 weeks of extended nevirapine given to breastfeeding infants, found a significant reduction in HIV-1 transmission at 6 weeks, compared to the control arm, from 5.3 to 2.5% [24]. However, there was no statistically significant difference of HIV-1 transmission at 6 months between the two arms, suggesting that a longer duration of prophylaxis would be required to protect the infant from ongoing breastfeeding transmission [24]. The PEPI trial in Malawi found that extended prophylaxis with either nevirapine or nevirapine plus zidovudine through 14 weeks of infant age significantly reduced postnatal HIV-1 transmission compared to the control arm of single-dose nevirapine and 1 week of zidovudine [11]. There were no significant differences in the effectiveness of the extended prophylaxis arms in PEPI (transmission rate at 9 months of 5.2% in the extended NVP vs. 6.4% in the dual prophylaxis group); however, the nevirapine plus zidovudine regimen was associated with significantly more adverse events, primarily neutropenia [11]. Finally, the recently completed Breastfeeding, Antiretrovirals, and Nutrition (BAN) Study, also in Malawi, evaluated 6 months of extended nevirapine infant prophylaxis; the estimated risk of HIV-1 transmission through breastfeeding at 28 weeks was 1.7%, compared with 5.7% for infants in the control arm of single-dose nevirapine and 7 days of zidovudine and lamivudine [27].

The results from the SWEN, PEPI, and BAN trials showed infant nevirapine prophylaxis to be a safe and effective strategy in reducing breastfeeding transmission of HIV-1. However, infants who continued to breastfeed after prophylaxis ended were at continued risk of becoming HIV-1-infected in both the SWEN and PEPI studies. Combined, these results suggest greater protection with longer duration of nevirapine prophylaxis; infants may indeed benefit from daily nevirapine throughout the entire duration of breastfeeding. Nevirapine was found to be safe when used as prophylaxis in breastfed infants, with the BAN study reporting a rash (possible sign of hypersensitivity) within a few weeks of initiation in 1.9% of infants [11, 24, 27]. Extended infant nevirapine prophylaxis is a low-cost, single daily regimen associated with little toxicity and feasible for many resource-limited settings [32]. There is, however, concern that failure of such prophylaxis will result in an increased rate of nevirapine resistance in the HIV-1-infected infant, thereby reducing future treatment options for the infant [33]. Indeed, 92% of infants who became HIV-1-infected during the first 6 weeks of life (period of NVP prophylaxis) in the SWEN study had NVP resistance; the risk, however, was much lower (15%) for infants who became infected after prophylaxis had stopped [24, 34].

An additional infant prophylaxis study, PROMISE-PEP (peri-exposure prophylaxis), began enrollment in December 2009. The PROMISE-PEP study is a multisite, double-blinded, randomized clinical trial comparing twice daily lopinavir–ritonavir to twice daily lamivudine. Both regimens will be administered to the infant from day 7 through 4 weeks after cessation of breastfeeding with a maximum duration of 50 weeks of prophylaxis [35].

Maternal ARV Prophylaxis

Mothers with CD4+ T-cell counts higher than 350 cells/mm^3 and without clinical signs of advanced HIV disease or active tuberculosis do not need ARV therapy for their own health according to the current WHO guidelines [36]. For these mothers, ARV prophylaxis is an approach to prevent postnatal transmission of HIV-1 to the infant. Results are now available from several randomized clinical trials, proving that maternal ARVs are an effective prophylactic strategy to reduce postnatal HIV-1 transmission [26, 27]. Furthermore, both observational and randomized studies have shown that when ARV prophylaxis is started during the antenatal period, very low rates of postnatal HIV-1 transmission are achievable at 6 months [10, 12, 15, 25, 26, 37]. For example, in the Mitra Plus study, an observational study in which most women with CD4+ T-cell counts >200/mm^3 started triple-ARV prophylaxis at 34 weeks of gestation and continued through breastfeeding, the risk of infant HIV-1 infection between 6 weeks and 6 months was just 1.0%; and between 6 weeks and 18 months it was 2.1% [10]. The AMATA study was a nonrandomized study of women who received NNRTI-based combination ARV after the second trimester of gestation, and then offered the choice of breastfeeding for 6 months with continued ARV or formula feeding. A very low rate of HIV transmission was noted at 9 months of age (1.3%), with a breastfeeding transmission rate of 0.5% [37]. The Kisumu Breastfeeding Study (KiBS) was a prospective noncontrolled study in Kenya in which women received HAART from 34 weeks of pregnancy until 6 months of breastfeeding. A low postnatal HIV transmission rate of 2.6% was achieved in this unpublished study [25].

Results from randomized clinical trials are now also available. The BAN study showed a 50% reduction in postnatal transmission in women who received HAART starting after delivery and continued through 6 months of breastfeeding (2.9% postnatal transmission by 6 months of age vs. 5.7% in the control arm that received sdNVP and 1 week of zidovudine/lamivudine) [27]. The Kesho Bora study conducted in several African countries showed a significant benefit of maternal ARV during breastfeeding in preventing postnatal transmission of HIV among women with CD4+ T-cell counts of 200–500/mm^3 (cumulative rates of 4.9 vs. 8.4% transmission rate at 6 months, a 43% reduction compared with AZT/sdNVP; the postnatal transmission rate was reduced by 47%, from 5.9 to 3.1%) [26].

In children HIV-1-uninfected at age 2 weeks, the risk of (postnatal) transmission was reduced by about half both in BAN and Kesho Bora. However, the overall risk reduction at age 28 weeks was greater in Kesho Bora than in BAN (42 vs. 25%) due to the earlier start of maternal triple-ARV prophylaxis in Kesho Bora (28–36 weeks of gestation) than in BAN (after delivery). Of all PMTCT studies conducted in breastfeeding populations, the Mma Bana trial which initiated ARV between 26 and 34 weeks of gestation in women with CD4+ T-cell counts >200 cells/mm^3 had the lowest rate of infant HIV-1 infection at 6 months, with a cumulative infant transmission rate of just 1% [15]. Taken together, these results suggest that maternal prophylaxis may need to start antenatally—early enough to reach undetectable viral load during the period most at risk (last weeks of pregnancy and delivery)—in order to achieve maximal virologic suppression in the mother and maximal reduction of prepartum, intrapartum, and postnatal HIV-1 transmission to the infant.

While maternal ARV prophylaxis is effective in reducing postnatal HIV-1 transmission, this benefit needs to be considered along with any harm or discomfort the regimen may pose to the mother. Such considerations include maternal toxicities, ARV interruption at the end of breastfeeding, and poor adherence that may increase the mother's risk of developing resistance and limit her future treatment options [38]. (This issue is further addressed in Chapter 6) Evidence from the trials so far indicates that maternal ARV prophylaxis during pregnancy and breastfeeding is well tolerated [15, 27]. Also, failure of such prophylaxis has the potential for transmission of a resistant virus, limiting the infant's future treatment options as well—indeed 67% of infants infected postnatally in the KiBS trial had drug-resistant virus [25, 38]. Some ARV agents enter breast milk—zidovudine appears to be present at levels similar or slightly lower than those in maternal plasma, nevirapine levels are about 70% those in maternal plasma, 3TC appears to concentrate in breast milk at levels 3–5 times those of maternal plasma, while protease inhibitors seem to have very limited penetration [39]. Breastfed infants may thus be ingesting subtherapeutic levels of ARV present in breast milk, leading to the potential for the development of resistance. Other prophylactic ARV regimens in the nursing HIV-1-infected mothers, including tenofovir, tenofovir/emtricitabine, efavirenz, and lopinavir/ritonavir combinations, will be evaluated for their safety and efficacy in a number of ongoing and planned trials [34, 40, 41] (Table 12.1).

Lessons Learned from the Recent Clinical Trials: Input into WHO Guidelines

The WHO recommendations have undergone many revisions over the years as scientific knowledge has evolved and clinical trial results have increasingly become available in this rapidly evolving field. In November 2009, WHO revised the breastfeeding and MTCT prevention guidelines for HIV-1-infected mothers in resource-limited settings once again, based on the most recent evidence mentioned above [42]. The new guidelines recommend that women who received a 3-drug ARV regimen during pregnancy should continue this regimen through breastfeeding and for 1 week after all exposure of the infant to breast milk has ended. If a woman received only zidovudine or did not receive any ARV, daily nevirapine is recommended for her child from birth until 1 week after the end of breastfeeding.

WHO further recommended that national health authorities should decide whether they will counsel HIV-1-infected mothers either to avoid all breastfeeding or to breastfeed and receive either maternal or infant prophylaxis taking into account the probability of HIV-free survival associated with BF and replacement feeding for the average infant in a given setting [42]. If breastfeeding is recommended, then the HIV-1-infected mothers should exclusively breastfeed their infants for the first 6 months of life, introducing appropriate complementary foods thereafter, and continue breastfeeding for the first 12 months. Breastfeeding should only be stopped once a nutritionally adequate and safe diet without

Table 12.1 Ongoing and planned clinical trials for prevention of mother-to-child transmission of HIV-1 via breastfeeding through use of antiretroviral agents [45]

Trial (location, status)	Study arm	Antepartum	Intrapartum	Postpartum (mother)	Postnatal (infant) regimen
HPTN 046 (ongoing, $N=1,670$)	Arm 1	Local ARV standard of care for PMTCT	Local ARV standard of care for PMTCT	No drug	Open-label NVP×6 weeks then daily NVP through 6 months or for the duration of BF (if <6 months) EBF×6 months and encouraged to wean
Phase III, multisite, double-blind clinical trial in South Africa, Tanzania, Uganda, and Zimbabwe, closed to accrual, study completion expected March 2011	Arm 2 (control)	Local ARV standard of care for PMTCT	Local ARV standard of care for PMTCT	No drug	Open-Label NVP×6 weeks then placebo through 6 months or for the duration of BF (if <6 months) EBF×6 months and encouraged to wean
PROMISE-PEP (planned $N=1,500$)	Arm 1 (3TC)	Any perinatal ARV prophylaxis, per local standard	Any perinatal ARV prophylaxis, per local standard	No drug	sdNVP plus ZDV×1 week then 3TC from day 7 until 4 weeks after BF cessation for a maximum of 50 weeks EBF recommended for 6 months-weaning over 8 weeks
Phase III, multisite, randomized, double-blind clinical trial in Burkina Faso, Uganda, Zambia, and South Africa, ongoing	Arm 2 (LPV/rv)	Any perinatal ARV prophylaxis, per local standard	Any perinatal ARV prophylaxis, per local standard	No drug	sdNVP plus ZDV×1 week then LPV/rv from day 7 until 4 weeks after BF cessation for a maximum of 50 weeks EBF recommended for 6 months-weaning over 8 weeks

Study	Arm	Antepartum	Intrapartum	Postpartum (mother)	Postpartum (infant)
IMPAACT PROMISE 1077 (postpartum component – planned N = 12,536 for all components)	Arm 1 (maternal prophylaxis)	Triple-ARV regimen starting at 28 weeks (dependent on randomization in antepartum portion)	Triple-ARV regimen (dependent on randomization in antepartum portion)	Truvada plus LPV/rv for the duration of BF or 18 months (whichever occurs first)	NVP × 6 weeks
	Arm 2 (infant prophylaxis)	Triple-ARV regimen starting at 28 weeks (dependent on randomization in antepartum portion)	Triple-ARV regimen (dependent on randomization in antepartum portion)	No drug	Daily NVP for the duration of BF or 18 months (whichever occurs first)
UMA (planned N = 960)	Arm 1 (maternal prophylaxis)	EFV/TDF/FTC (formulated as Atripla) from 20 weeks	EFV/TDF/FTC	EFV/TDF/FTC for the duration of BF with advice to cease at 6 months	Daily ZDV syrup from birth through 1 week or updated with a prophylaxis regimen recommended by WHO. Advise BF cessation at 6 months
Phase III, multisite, randomized clinical trial in Côte D'Ivoire and Zambia	Arm 2 (maternal prophylaxis)	ZDV/3TC/LPV/rv from 20 weeks	ZDV/3TC/LPV/rv	ZDV/3TC/LPV/rv for the duration of BF with advice to cease at 6 months	Daily ZDV syrup from birth through 1 week or updated with a prophylaxis regimen recommended by WHO. Advise BF cessation at 6 months

For further information on some of these clinical trials, please see http://www.clinicaltrials.gov

EFV efavirenz, TDF tenofovir, FTC emtricitabine, NVP nevirapine, ZDV zidovudine, 3TC lamivudine, LPV/rv lopinavir/ritonavir, Trizivir zidovudine/lamivudine/abacavir, sd single dose, ARV antiretroviral drugs, Truvada tenofovir/emtricitabine, BF breastfeeding, EBF exclusive breastfeeding, MTCT mother-to-child transmission, HPTN HIV Prevention Trials Network

breast milk can be provided. Enabling breastfeeding in the presence of ARV interventions to continue to 12 months avoids many of the challenges associated with stopping breastfeeding and with providing a safe and adequate diet without breast milk to the infant between 6 and 12 months of age. HIV-1-infected mothers who decide to stop breastfeeding at any time should stop gradually within 1 month.

In infants and young children known to be HIV-1-infected, mothers are strongly encouraged to exclusively breastfeed for the first 6 months of life and continue breastfeeding as per the recommendations for the general population, that is up to 2 years.

In a randomized controlled trial in Zambia in which infants of HIV-1-infected breastfeeding mothers either stopped all breastfeeding at 4 months of age or continued to breastfeed, mortality at 24 months was 55% among those infants who were already HIV-1-infected and were randomized to continued breastfeeding, compared with 74% among those HIV-1-infected who stopped breastfeeding early [43]. In a study in Botswana that randomized HIV-1-exposed infants to either breast milk or infant formula, among infants who were already HIV-1-infected, mortality at 6 months of age was 7.5% in those who breastfed, compared with 33% in those randomized to receive infant formula [44].

Lessons Learned from the BAN and Kesho Bora Trials: The Investigators' Perspective

The Mitra and Mitra Plus observational studies of infant prophylaxis and maternal prophylaxis, respectively, had similar rates of postnatal transmission at 6 months [10, 31]. However, the studies were observational and were conducted sequentially at the same site. The BAN study is the only randomized controlled trial to date to have both a maternal and an infant prophylaxis arm, each of which was compared with the control arm. Although the study was not powered to directly compare the two interventions, there was a suggestion that HIV-1-free survival at 28 weeks may be greater with infant, compared with maternal prophylaxis ($p=0.07$) but efficacy of maternal prophylaxis should be maximized by initiating it antenatally as now recommended [27]. A more direct comparison of the two approaches is planned [40] (Table 12.1). While effective maternal and infant ARV prophylactic strategies are now available to reduce postnatal HIV-1 transmission through breastfeeding, more research is needed to determine the optimal method that balances benefits and risks for mothers and infants and is feasible for a particular resource-limited setting. In addition to the ARV interventions, BAN also evaluated the effect of a high-energy, micronutrient-fortified supplement given to breastfeeding mothers for prevention of maternal depletion. As breastfeeding guidance evolves, the effect of breastfeeding on maternal health also needs to be considered, and innovative strategies to ensure and promote maternal health need to be developed. In the Kesho Bora study, the preventive efficacy was much greater for women with CD4 <350 cells/mm^3 than in those with CD4 ≥350 cells/mm^3. Among women with CD4 >500/mm^3 (who received the ZDV/sdNVP regimen), the postnatal transmission rate was 0 between 6 weeks and 6 months. These results emphasize the importance of providing early ART to all women with CD4 <350/mm^3 with the potential to prevent up to 90% of MTCT while preserving the health of the mothers.

The 4.9% MTCT rate obtained at 6 months in Kesho Bora is higher than the Mma Bana rate (below 2%). Mma Bana achieved maternal viral load at delivery <400 copies/mL in >90% of women compared to 70% in Kesho Bora. Of note, the viral load at enrollment was lower and the duration of ARV prophylaxis before delivery longer in Mma Bana than in Kesho Bora. Obtaining an undetectable viral load by the time of delivery and during breastfeeding is clearly the goal of maternal ARV prophylaxis. At least 8 weeks of prophylaxis were needed before undetectable viral load was achieved in >75% of women in Kesho Bora.

A summary of ongoing and planned studies on maternal or infant ARV prophylaxis during breastfeeding is shown in Fig. 12.1.

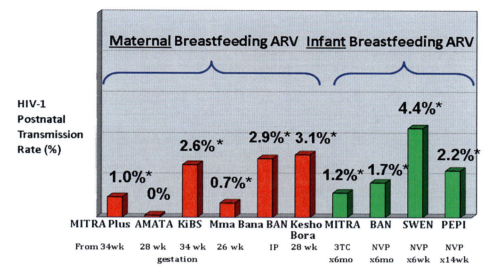

Fig. 12.1 Summary of the results of studies that used antiretroviral prophylaxis to the mother or to the infant during breastfeeding to prevent HIV-1 transmission via breastfeeding. Results are shown as postnatal transmission rate of HIV-1 from 1–6 weeks until 6–7 months of infant age. *ARV* antiretroviral, *IP* intrapartum, *BF* breastfeeding, *mo* month, *wk* week, *asterisk* a published study

Questions for the Future

As the most recent guidance from WHO highlights, ARV drugs in the mother or infant should be used throughout breastfeeding to reduce postnatal transmission of HIV-1 to the infant in resource-limited settings; breastfeeding should be completely avoided in resource-rich settings where safe infant feeding alternatives exist. However, several questions remain for the future. The distinction between resource-poor and resource-rich settings is not always clear-cut. For example, in many poor countries, there is a section of the population which may be able to provide safe replacement feeding; therefore, national recommendations will have to be nuanced. The safety for mothers and infants and the efficacy of other, newer ARV drugs or drugs of newer ARV classes during breastfeeding needs to be tested. Resistance development as a result of the new, prolonged prophylactic regimens during breastfeeding needs to be assessed for mothers and for infants who become infected despite prophylaxis. The safety of stopping ARV prophylaxis for the mothers after delivery and cessation of breastfeeding in light of their HIV disease progression and future treatment needs has to be evaluated. Whether maternal, vs. infant, prophylaxis is preferable for different settings needs careful cost/benefit assessment. Ways to improve adherence and monitor toxicities of the ARV regimens during pregnancy and breastfeeding will be important in translating research results into real-world effectiveness. Another important issue is the implementation of these findings in the particular settings. As countries scale up national programs, how the new components of recommended perinatal prevention fit into the context will be critical. For example, when ART regimens for adults are chosen, are considerations for safety during pregnancy taken into account? Also, as more adults are treated at earlier stages of disease, will infant prophylaxis still be an option, or will countries focus exclusively on maternal prophylaxis? Finally, whether breastfeeding with ARV prophylaxis is an acceptable strategy in resource-rich settings for the HIV-1-infected women with undetectable viral load in the plasma who strongly wish to breastfeed their infants has not been evaluated; a formal assessment of its risk/benefit ratio may deserve more study. Linking HIV-1 prevention, care, and treatment services with family planning, prenatal and child health services, and building the health infrastructure required to implement MTCT prevention programs with prolonged ARV prophylactic regimens remains a critical need for resource-limited settings.

References

1. UNAIDS. AIDS Epidemic Update: November 2009. http://data.unaids.org/pub/Report/2009/jc1700_epi_update_2009_en.pdf. Nov 24 2009, Geneva, Switzerland. Accessed 14 Dec 2010
2. Horvath T, Madi BC, Iuppa IM, Kennedy GE, Rutherford G, Read JS (2009) Interventions for preventing late postnatal mother-to-child transmission of HIV. *Cochrane Database Syst Rev* CD006734
3. Nduati R, John G, Mbori-Ngacha D et al (2000) Effect of breastfeeding and formula feeding on transmission of HIV-1: a randomized clinical trial. JAMA 283:1167–1174
4. World Health Organization (2010) HIV transmission through breastfeeding: a review of available evidence. http://www.unfpa.org/upload/lib_pub_file/276_filename_HIV_PREV_BF_GUIDE_ENG.pdf. Accessed 20 Nov 2010
5. World Health Organization. Report of a WHO technical consultation on birth spacing. http://www.who.int/making_pregnancy_safer/documents/birth_spacing.pdf. 15 Jun 2005. Accessed 20 Nov 2010
6. Jelliffe DB, Jelliffe EF (1978) Human milk in the modern world. Oxford University Press, New York
7. Achievements in public health (2006) Reduction in perinatal transmission of HIV infection – United States, 1985–2005. MMWR Morb Mortal Wkly Rep 55:592–597
8. World Health Organization (2010) Antiretroviral drugs for treating pregnant women and preventing HIV infection in infants. Recommendations for a public health approach. 2010 Version. http://www.who.int, Accessed: 15 Dec 2010.
9. Kuhn L, Aldrovandi GM, Sinkala M et al (2008) Effects of early, abrupt weaning on HIV-free survival of children in Zambia. N Engl J Med 359:130–141
10. Kilewo C, Karlsson K, Ngarina M et al (2009) Prevention of mother-to-child transmission of HIV-1 through breastfeeding by treating mothers with triple antiretroviral therapy in Dar es Salaam, Tanzania: the Mitra Plus study. J Acquir Immune Defic Syndr 52:406–416.10
11. Kumwenda NI, Hoover DR, Mofenson LM et al (2008) Extended antiretroviral prophylaxis to reduce breast-milk HIV-1 transmission. N Engl J Med 359:119–12911
12. Palombi L, Marazzi MC, Voetberg A, Magid NA (2007) Treatment acceleration program and the experience of the DREAM program in prevention of mother-to-child transmission of HIV. AIDS 21(Suppl 4):S65–S711
13. Kesho Bora Study Group (2010) Eighteen month follow-up of HIV-1-infected mothers and their children enrolled in the Kesho Bora study observational cohorts. J Acquir Immune Defic Syndr 54:533–541
14. Taha T, Kumwenda J, Cole S et al (2009) Postnatal HIV-1 transmission after cessation of infant extended antiretroviral prophylaxis and effect of maternal highly active antiretroviral therapy. J Infect Dis 200:1490–1497
15. Shapiro RL, Hughes M, Ogwu A et al (2010) Antiretroviral regimens in pregnancy and breast-feeding in Botswana. N Engl J Med 362:2282–2294
16. Townsend CL, Cortina-Borja M, Peckham CS et al (2008) Low rates of mother to child transmission of HIV following effective pregnancy interventions in the United Kingdom and Ireland, 2000–2006. AIDS 22:973–981
17. Nduati R, Richardson BA, John G et al (2001) Effect of breastfeeding on mortality among HIV-1 infected women: a randomised trial. Lancet 26(357):1651–1655
18. Dabis F, Msellati P, Meda N et al (1999) 6-month efficacy, tolerance, and acceptability of a short regimen of oral zidovudine to reduce vertical transmission of HIV in breastfed children in Cote d'Ivoire and Burkina Faso: a double-blind placebo-controlled multicentre trial. DITRAME Study Group. DIminution de la Transmission Mere-Enfant. Lancet 353:786–792
19. Guay LA, Musoke P, Fleming T et al (1999) Intrapartum and neonatal single-dose nevirapine compared with zidovudine for prevention of mother-to-child transmission of HIV-1 in Kampala, Uganda: HIVNET 012 randomised trial. Lancet 354:795–802
20. Connor EM, Sperling RS, Gelber R et al (1994) Reduction of maternal-infant transmission of human immunodeficiency virus type 1 with zidovudine treatment. Pediatric AIDS Clinical Trials Group Protocol 076 Study Group. N Engl J Med 331:1173–1180
21. Cooper ER, Charurat M, Mofenson L et al (2002) Combination antiretroviral strategies for the treatment of pregnant HIV-1-infected women and prevention of perinatal HIV-1 transmission. J Acquir Immune Defic Syndr 29:484–494
22. Shaffer N, Chuachoowong R, Mock PA et al (1999) Short-course zidovudine for perinatal HIV-1 transmission in Bangkok, Thailand: a randomised controlled trial. Bangkok Collaborative Perinatal HIV Transmission Study Group. Lancet 353:773–780
23. Wiktor SZ, Ekpini E, Karon JM et al (1999) Short-course oral zidovudine for prevention of mother-to-child transmission of HIV-1 in Abidjan, Cote d'Ivoire: a randomised trial. Lancet 353:781–785
24. Bedri A, Gudetta B, Isehak A et al (2008) Extended-dose nevirapine to 6 weeks of age for infants to prevent HIV transmission via breastfeeding in Ethiopia, India, and Uganda: an analysis of three randomised controlled trials. Lancet 372:300–313

25. Thomas T, Masaba R, Ndivo R, Kisumu Breastfeeding Study Team (2008) Prevention of mother-to-child transmission of HIV-1 among breastfeeding mothers using HAART: The Kisumu Breastfeeding Study, Kisumu, Kenya, 2003–2007. Abstract 45aLB. In: 15th Conference on retroviruses and opportunistic infections, Boston, MA. http://www.retroconference.org/2008/Abstracts/33397.htm. Accessed 20 Apr 2010
26. The Kesho Bora study Group (2010) Triple-antiretroviral compared with zidovudine and single-dose nevirapine prophylaxis during pregnancy and breastfeeding for prevention of mother-to-child transmission of HIV-1 (Kesho Bora study): a randomised controlled trial. Lan Infect Dis 2011;11(3):171–180
27. Chasela C, Hudgens M, Jamieson D et al (2010) Maternal or infant antiretroviral drugs to reduce HIV-1 transmission. N Engl J Med 362:2271–2281
28. Giuliano M, Guidotti G, Andreotti M et al (2007) Triple antiretroviral prophylaxis administered during pregnancy and after delivery significantly reduces breast milk viral load: a study within the Drug Resource Enhancement Against AIDS and Malnutrition Program. J Acquir Immune Defic Syndr 44:286–291
29. Shapiro RL (2005) Ndung'u T, Lockman S et al. Highly active antiretroviral therapy started during pregnancy or postpartum suppresses HIV-1 RNA, but not DNA, in breast milk. J Infect Dis 192:713–719
30. Lehman DA, Chung MH, John-Stewart GC et al (2008) HIV-1 persists in breast milk cells despite antiretroviral treatment to prevent mother-to-child transmission. AIDS 22:1475–1485
31. Kilewo C, Karlsson K, Massawe A et al (2008) Prevention of mother-to-child transmission of HIV-1 through breast-feeding by treating infants prophylactically with lamivudine in Dar es Salaam, Tanzania: the Mitra Study. J Acquir Immune Defic Syndr 48:315–323
32. Bulterys M, Fowler MG, Van Rompay KK, Kourtis AP (2004) Prevention of mother-to-child transmission of HIV-1 through breast-feeding: past, present, and future. J Infect Dis 189:2149–2153
33. Palumbo P, Lindsey JC, Hughes MD et al (2010) Antiretroviral treatment failure for children with peripartum nevirapine exposure. N Engl J Med 363:1510–1520
34. Moorthy A, Gupta A, Bhosale R et al (2009) Nevirapine resistance and breast-milk HIV transmission: effects of single and extended-dose nevirapine prophylaxis in subtype C HIV-infected infants. PLoS One 4:e4096
35. ClinicalTrials.gov (2010) Comparison of efficacy and safety of infant peri-exposure prophylaxis with lopinavir/ritonavir versus lamivudine to prevent HIV-1 transmission by breastfeeding. http://clinicaltrials.gov/ct2/show/NCT00640263. Updated 8 Apr 2010. Accessed 10 Jan 2011
36. World Health Organization (2010) Antiretroviral therapy for HIV infection in adults and adolescents. Recommendations for a public health approach. 2010 Revision. www.who.int. Accessed 10 Nov 2010.
37. Peltier CA, Ndayisaba GF, Lepage P et al (2009) Breastfeeding with maternal antiretroviral therapy or formula feeding to prevent HIV postnatal mother-to-child transmission in Rwanda. AIDS 23:2415–2423
38. Zeh C, Weidle P, Nafisa L, Musuluma H, Okonji J, Anyango E (2008) Emergence of HIV-1 drug resistance among breastfeeding infants born to HIV-infected mothers taking antiretrovirals for prevention of mother-to-child transmission of HIV: The Kisumu Breastfeeding Study, Kenya. In: 15th Conference on retroviruses and opportunistic infections. http://www.retroconference.org/AbstractSearch/Default.aspx?Conf=19. Accessed 20 Apr 2010
39. Mirochnick M, Thomas T, Capparelli E et al (2009) Antiretroviral concentrations in breast-feeding infants of mothers receiving highly active antiretroviral therapy. Antimicrob Agents Chemother 53:1170–1176
40. ClinicalTrials.gov (2010) Evaluating strategies to reduce mother-to-child transmission of HIV infection in resource-limited countries (PROMISE). http://clinicaltrials.gov/ct2/show/NCT01061151. Updated 14 Apr 2010. Accessed 30 Apr 2010
41. ClinicalTrials.gov (2009) Universal use of EFV-TDF-FTC and AZT-3TC-LPV/r combinations for HIV-1 PMTCT in pregnant and breastfeeding women: a phase 3 trial (UMA). http://clinicaltrials.gov/ct2/show/NCT00936195?term=UMA&rank=1. Updated 6 Aug 2009. Accessed 20 Jan 2011
42. World Health Organization (2010) Guidelines on HIV and infant feeding. Principles and recommendations for infant feeding in the context of HIV and a summary of evidence. www.who.int. Accessed 10 Nov 2010
43. Kuhn L, Aldrovandi MG, Sinkala M et al (2010) Potential impact of new World Health Organization criteria for antiretroviral treatment for prevention of mother to child HIV transmission. AIDS 24:1374–1377
44. Thior I, Lockman S, Smeaton LM et al (2006) Breastfeeding plus infant zidovudine prophylaxis for 6 months vs formula feeding plus infant zidovudine for 1 month to reduce mother-to-child HIV transmission in Botswana: a randomized trial: the Mashi Study. JAMA 296:794–805
45. Bulterys M, Ellington S, Kourtis AP (2010) HIV-1 and breastfeeding: biology of transmission and advances in prevention. Clin Perinatol 37:807–824

Chapter 13
Immune Approaches for the Prevention of Breast Milk Transmission of HIV-1

Barbara Lohman-Payne, Jennifer Slyker, and Sarah L. Rowland-Jones

Introduction

Mother-to-child transmission (MTCT) of HIV-1 infection remains a significant cause of new HIV-1 infections, despite the increasing implementation of prevention strategies using antiretroviral therapy (ART) and the resulting decline in infections across the developing world. In 2009, the UNAIDS global report estimated 370,000 children under the age of 15 years were newly infected with HIV-1 (refer UNAIDS Report on the global AIDS epidemic, 2010 http://www.unaids.org/globalreport/Global_report.htm), most of whom acquired the infection from their mothers in low- and middle-income countries. Even with substantial progress, challenges remain for poor countries in providing comprehensive screening programs for pregnant women and implementing the full range of prevention services for those identified as HIV-1-infected. Although antiretroviral regimens and risk reduction counseling have been successfully used for pregnant women and their infants in many parts of the developing world, full implementation of these programs remains a challenge in many countries, especially where antenatal clinical attendance and HIV-1 screening is not yet widespread. In addition, potential toxicities of and development of drug resistance to ART in both mother and child are concerns. Therefore, the development of a safe effective immunoprophylaxis regimen begun at birth and continuing during breastfeeding, perhaps alongside neonatal chemoprophylaxis, remains an area of active research interest. An ideal pediatric vaccine for prevention of MTCT (PMTCT) would combine the immediacy of passive immunization designed to protect the infant during the first vulnerable

---

B. Lohman-Payne (✉)
Department of Pediatrics and Child Health, University of Nairobi, Nairobi, Kenya 00202

Departments of Medicine, University of Washington, Seattle, WA 98104, USA

Departments of Global Health, University of Washington, Seattle, WA 98104, USA
e-mail: blpayne@iconnect.co.ke

J. Slyker
Departments of Global Health, University of Washington, Seattle, WA 98104, USA

S.L. Rowland-Jones (✉)
Nuffield Department of Medicine, John Radcliffe Hospital,
Oxford OX3 9DS, UK
e-mail: sarah.rowland-jones@nmd.ox.ac.uk

weeks of life with the durability of active immunization to protect against the repeated low-dose homologous virus exposure delivered multiple times a day via breastfeeding.

In the absence of maternal ART, up to one half of all MTCT occurs through breast milk [1, 2]. In the first month of life, it can be challenging to define the route of infection as early breast milk transmission as opposed to late in utero or intrapartum transmission. After 1 month of age, the route of transmission is clearly breast milk and risk of transmission remains relatively constant throughout the duration of breastfeeding [2]. Increased risks of breast milk transmission are associated with longer duration of breastfeeding [2] and high levels of HIV-1 in breast milk [3, 4].

Infants who acquire perinatal infection, particularly in developing countries, experience unusually rapid disease progression compared to that of adults, and mortality rates as high as 20–52% have been reported in the first 2 years of life [5, 6]. A key factor contributing to the rapid disease progression observed in infants may be the persistently high levels of HIV-1 viremia observed throughout the first year of life, with the "set-point" viral load (VL) rarely falling more than 1 log below the peak VL [7, 8]. However, there are some data to suggest that breast milk infection is associated with a better clinical outcome than in utero or intrapartum transmission. Both the peak VL and viral set-point are reported to be higher in babies infected at birth or in the first few weeks of life (early infection) than in those infected after 3 months of age (late infection) [7, 9]. These factors are associated with increased risk of disease progression and death in early infant HIV-1 infection [10]. Overall, after adjusting for length of infection, the risk of progression to AIDS and death is at least twofold higher for infants infected early compared with those infected through breast milk [11–13]. The most likely explanation for this is the relative immaturity of the infant immune system at birth, which is dominated by regulatory T-cell (Treg) [14, 15] and Th2 responses [16–18]. Following birth, there is a shift toward Th1 responses as the immune system switches from one predominantly adapted to tolerance of self and maternal antigens to a system better able to mount effective innate and acquired responses to external pathogens such as HIV-1 [19]. It is plausible that infants infected through breastfeeding may be more likely to become long-term survivors than those infected early, but there have been no natural history cohort studies assessing disease progression beyond 5 years to address this issue.

It is likely that co-infections acquired by babies in the developing world may play a role in rapid disease progression, such as human cytomegalovirus (CMV), which infects most West African children in the first year of life [20] and is associated with high VL in HIV-1 co-infected infants [21]. Co-infections may also contribute to the likelihood of transmission. Therefore, it is important to investigate and define the potential role of co-infections in breast milk transmission of HIV-1 so that these may be modified in preventive strategies.

Breast Milk Transmission of HIV-1: An Immunological Perspective

Mother-to-child transmission of HIV-1 is surprisingly inefficient: in the absence of treatment, 55–80% of HIV-1 exposed infants remain HIV-1 uninfected [22, 23]. This is striking when considering the large volumes (hundreds of liters) of HIV-1-contaminated milk ingested by infants breastfed by untreated HIV-1 uninfected mothers [24]. Why does such a large proportion of HIV-1-exposed uninfected children resist infection? Breast milk may confer protection from transmission through immunologic mechanisms. Oral exposure to HIV-1 has been reported to result in both systemic cellular and antibody immune responses detectable in ~13–28% of uninfected partners within HIV-1 discordant couple pairs [25, 26]. We have recently found that HIV-1 specific interferon-gamma (IFN-γ) secreting cells are present in the breast milk of the majority of HIV-1 infected mothers and confer a 70% reduction in the likelihood of early breast milk transmission, indicating that maternal HIV-1 specific immunity may protect against breast milk transmission (Lohman-Payne et al., manuscript submitted).

There are also some data that suggest that uninfected infants exposed to HIV in utero, at delivery or postnatally, can develop HIV-1-specific T-cell responses [27–30], which suggests that there has been sufficient exposure to replicating virus to prime such a response. Although the detection of an HIV-specific response in the absence of persistent infection does not necessarily imply that T-cell immunity contributes to protection, resistance to postnatal HIV-1 transmission through breast milk was shown to correlate with the magnitude of the HIV-1-specific T-cell response in a prospective study in Nairobi, Kenya [24]. This finding provides some encouragement that enhancing the immune responses to HIV-1 through immunotherapeutic strategies in uninfected infants could confer protection against breast milk infection.

What are the consequences of later transmission for the infected infant? As previously mentioned, epidemiological studies suggest that the risk of disease progression to AIDS or death by the age of 12 months is halved in children infected later in infancy, compared to those who acquire infection before 1 month of age [6, 11–13]. This is consistent with the clinical experience of congenital infection with pathogens such as CMV, *Toxoplasma gondii*, and rubella, where disease is much more severe when infants are infected in utero.

During the first year of life, the infant immune system undergoes a transition from a predominantly regulated [14, 15], Th2 biased system [16–18], designed to maintain fetal life in the intrauterine environment, toward a more adult-like immune system, capable of mounting successful immune responses against a range of pathogens encountered in early life [19]. The precise timings of these changes in human infancy have not been well defined and much of the data come from studies comparing the phenotype and function of immune cells in cord blood with those in adult samples. Cord blood T cells have lower basal expression of CD3 and adhesion molecules, defects in cytokine production and CD8+ T-cell activity [31], while monocytes and dendritic cells (DC) express lower levels of costimulatory molecules, have altered differentiation pathways and reduced cytokine/chemokine production [32, 33]. Nevertheless, neonates generally develop immune responses to vaccines given at birth, including cell-mediated responses to several acute viral infections [34]. Bacilli Calmette-Guerin (BCG) vaccination, which is widely given at birth throughout the developing world as part of the Extended Program of Immunization (EPI), elicits a Th1-type response in newborns of similar magnitude to that achieved later in life [35]. Moreover, BCG itself can act as an adjuvant for other vaccines [36]. Pre- and postnatal exposure to environmental microbial products that activate innate immunity might accelerate this maturation, diminishing Th2 and/or enhancing Th1 cell polarization, and this could be incorporated into adjuvant design. Both natural and inducible CD4+CD25+ Treg numbers are increased in infancy and are thought to be required for maintaining peripheral T-cell tolerance through inhibition of Th1 cell immunity [37–39]. It is intriguing to note that the cord blood of HIV-1 exposed uninfected infants contains large of populations of Treg cells and depletion of these cells ex vivo leads to an increased frequency of detection of IFN-γ secreting HIV-1 specific CD8+ T cells [38], consistent with Treg downregulation of antigen-specific cellular responses in cord blood. Exposure to foreign antigens, both allo-antigens [15] and antigens from *Plasmodium falciparum* present in placental malaria [40], promotes the development of regulatory T-cell populations at birth: this may mean that the infant exposed to HIV-1 and other pathogens in utero is less able to mount an efficient response to infections encountered in early life. Alternatively, perhaps depending on the timing of exposure, prenatal exposure might also promote Th1 responses in a similar manner to BCG given at birth [35].

The B-cell compartment is also affected in newborns, with responses to vaccines characterized by lower antibody levels, restricted diversity of antibody repertoire, and lower levels of IgG2 isotype as compared with responses induced in adults [41]. The implications of these deficiencies in pediatric vaccine development may include a requirement for enhanced stimulation of antigen-presenting cells with increased avidity of CD3/T-cell receptor (TCR) and co-stimulation molecule interactions to ensure immune activation.

There is a dearth of studies looking at the immune response and clinical course of infants infected through breast milk compared to those with early infection. We have shown that infants infected after the first month of life develop HIV-1-specific CD8+ T-cell IFN-γ responses more rapidly than infants infected early in life and the emergence of the response coincides with a decline in VL [42]; however, the overall magnitude of the response did not differ between early and late infection groups. Viral and host factors associated with the improved clinical outcome of breast milk infected infants have not been defined.

Co-infections and Breast Milk HIV-1 Transmission

HIV-1 infected infants in sub-Saharan Africa have poorer control over VLs and a higher risk of mortality compared to infants in Europe and the USA (~20–40 vs. ~10–20% mortality, respectively) [5, 7, 43–45]. Co-infections during infancy are thought to explain a large proportion of these cohort differences. In addition to causing significant morbidity as opportunistic infections, co-infections have the potential to increase both the risk of vertical HIV-1 transmission and infant HIV-1 progression. CMV [46–48], herpes simplex virus type-2 (HSV-2) [49, 50], and malaria [51, 52] may accelerate HIV-1 disease progression by causing transient or sustained elevations in HIV-1 VL. Since breast milk HIV-1 load is closely correlated with systemic VL, any infection that raises maternal HIV-1 VL could potentially affect the risk of breast milk transmission. Despite these associations, there have been few prospective studies examining the role co-infections play in breast milk transmission and downstream infant disease progression. CMV, which causes persistent asymptomatic infection in a large proportion of healthy adults, has been the most intensively studied co-infection in HIV-1. Longitudinal cohort studies conducted in West and East Africa suggest more than 80% of children may acquire CMV during their first year of life, regardless of HIV-1 status [20, 21]. Maternal CMV reactivation was associated with a fourfold increased risk of mortality in HIV-infected infants independent of maternal immunosuppression [53]. Among HIV-1 infected infants, co-infection with CMV is associated with higher peak CMV VLs, prolonged detection of CMV in the plasma [21], and a greater than twofold increased risk of mortality [47]. In co-infected women, there is a strong correlation between HIV-1 and CMV loads in both the blood and breast milk [21, 53, 54]. It is yet unclear whether a cause-and-effect relationship links these two pathogens; CMV could potentially act as a co-factor to increase HIV-1 replication directly (reviewed in ref. [55]) or indirectly by increasing cellular activation of breast milk lymphocytes. Further studies are needed to understand better the long-term effects of other highly prevalent persistent viral infections acquired during infancy, particularly those that are associated with the development of malignancies after years of HIV-1 infection, such as Epstein–Barr virus (EBV), human herpesvirus-8 (HHV-8), and hepatitis B virus.

Vaccine Strategies to Prevent Breast Milk Transmission of HIV-1

HIV-1 Vaccine Studies in Adults

Three large-scale human efficacy studies have been completed to date and, although not designed to evaluate mother-to-child transmission, they provide an immunologic foundation for future studies to include breast milk transmission. The first trial, from Vaxgen, employed an envelope subunit protein construct with the aim of inducing a protective antibody response, but a large phase III clinical trial showed no evidence of efficacy [56]. A move toward testing T-cell-inducing vaccines led to the phase 2b STEP trial (tested in high-risk volunteers) of a replication-defective adenovirus type 5 (Ad5)-based recombinant vaccine produced by Merck that appeared to be the most immunogenic available construct in human studies. This study was terminated prematurely, as study subjects were neither protected from HIV-1 infection nor experienced reduced VL when infection occurred [57, 58]. However, progressive improvements in the constructs used to elicit T-cell responses have provided some encouragement in the macaque model. For example, a heterologous rAd26 prime/rAd5 boost vaccine regimen expressing SIV Gag elicited broader and stronger cellular immune responses than had been seen with the homologous rAd5 regimen, which led to significantly lower set-point VLs as well as decreased AIDS-related mortality compared with control animals following challenge with a pathogenic simian immunodeficiency virus strain (SIVmac251) [59]. More recently, a novel approach using rhesus CMV constructs induced effector-memory T cells at mucosal sites that correlated with protection from infection in 4 of 12 macaques repeatedly challenged with the highly pathogenic SIVmac239 strain [60]. Further encouragement to the HIV vaccine field came from the results of the rv144 phase III efficacy trial of a combination of canary pox priming and HIV envelope protein boost tested in a low-risk population in Thailand [61]. Although the incidence of new infections was low in this study, meaning that significant efficacy was only seen in one of the three analyses (the modified intention-to-treat analysis), the vaccine appeared to confer 31% protection against infection, the first indication of efficacy in any human study. The mechanisms of protection are not yet known, but are unlikely to include either neutralizing antibodies or CTL, which are rarely induced by this vaccine approach.

Mucosal Vaccines

Mucosal immunization has been demonstrated to induce mucosal responses, and as the first sites of SIV and recombinant simian-human immunodeficiency virus (SHIV) infection in neonatal rhesus macaques are mucosal surfaces [62], similar tissues are likely to be exposed in MTCT of HIV-1 [63]. Both intrapartum and breast milk HIV-1 transmission are likely to be acquired through the oral route, and therefore HIV-1 specific immunity present at this site is most relevant to preventing breast milk transmission. Many cellular and cytokine properties of the neonatal oral and intestinal tissues are known to differ from the adult: these could either represent protective or susceptibility factors and include the levels of $\gamma\delta$ T cells, NK cells, macrophages, DC, IgG, IgA and secretory IgA levels, and levels of cytokines such as TNF-α and IFN-γ [64].

Mucosal vaccines developed for PMTCT of HIV-1 could exploit the common mucosa-associated lymphoid tissue to explore oral or nasal delivery. Oral delivery, although attractive for neonates, has challenges such as induction of tolerance, limitations in the choice of safe effective adjuvants, requirement for large doses of antigen, and the need for antigen stability in the gut. This last concern has been addressed through the development of lipid vesicles or polymeric nanoparticles that act as immunostimulants while preserving immunogens from intestinal enzymes. So far only a limited number of orally administered vaccines against HIV have been tested in humans. An attenuated canarypox vector (vCP 205) and Salmonella vaccine vector (CKS257) vaccine platforms were both reportedly well tolerated in humans but with less than expected mucosal immunogenicity [65, 66]. As opposed to oral vaccine delivery, the main advantage of nasally administered vaccine is the requirement for smaller doses of antigen. Several vaccine formulations have been tested in adults including peptides, DNA, and live bacterial and viral vectors [64]. Mucosal vaccine development against HIV-1 appears to require the deployment of stronger adjuvants, the use of which may be associated with safety concerns in young children.

In summary, a pediatric vaccine to prevent HIV infection would rapidly induce both antibody and cellular responses detectable at mucosal surfaces that would remain at effective levels during the duration of breastfeeding. A greater understanding of the most important inductive and effector sites in the newborn would guide research to address such questions as to whether systemic immunization can induce sufficient mucosal immunity, if oral or nasal vaccines can be effective, and if appropriate adjuvants can be developed for an effective vaccine development strategy for use in HIV-1-exposed infants.

SIV and SHIV Models for Vaccine Strategies for PMTCT of HIV Infection—Active and Passive Immunization

Animal models, especially the rhesus macaque/SIV–SHIV model, have been used to address areas of uncertainty in pediatric vaccine design, including the most appropriate vaccine, the timing of immunizations, and the duration of vaccine elicited responses. The SIV models for vertical transmission of HIV-1 and for neonatal vaccine development were validated in the 1990s and many important proof-of-concepts have been demonstrated including protection of newborn macaques from oral SIV/SHIV challenge through maternal vaccination and through administration of passive hyperimmune serum (SIVIG/HIVIG) or neutralizing monoclonal antibodies at birth [67–71].The mechanism of protection from infection could include antibody binding alone, classic viral neutralization, or NK cell-mediated antibody dependent cellular cytotoxicity. But will passive immunization to prevent MTCT of HIV-1 work? There are many challenges to this approach, particularly the composition of the antibody cocktail, which would need to be effective against more than one HIV clade and circulating recombinant forms, and the scale-up delivery logistics are daunting. Alternatively, boosting of maternal antibody levels by vaccination represents an important strategy to augment passive immunity during the infant's early months of life, until the infant can be actively immunized. However, the role of maternal antibody in reducing the effectiveness of immune responses in the newborn, especially T-cell responses, is a concern. This may be addressed in the results of a phase III randomized clinical trial that was completed in Uganda in 2006 (http://www.mujhu.org/hiviglob2.html). The trial was designed to compare the standard single-dose mother/infant nevirapine regimen for PMTCT with the addition of HIV immune globulin (HIVIGLOB) or a second arm of extended infant NVP dosing compared with the standard single-dose NVP regimen alone without HIVIGLOB and enrolled 722 mothers with 204 in the HIVIGLOB arm. The data from the HIVIGLOB arm were pooled with other data from Johns Hopkins trials in Ethiopia and India, and results are not yet available.

Live attenuated vaccines have the advantage of prolonged antigen delivery and stimulation of both innate and adaptive immunity. Modified vaccinia virus Ankara (MVA) and canarypox viral vectors (ALVAC) have been tested in the rhesus macaque/SIV–SHIV models in advance of human trials. MVA expressing SIV gag, pol, and env or expressing SIVmac1A11 was used to immunize infant macaques at birth and at 3 weeks of age [72]. The infants were challenged at 4 weeks of age with uncloned SIVmac251, using a multiple low-dose challenge model to deliver virus three times a day over 5 days to mimic breast milk exposure. Although the immunization regimen was unable to prevent infection, immunized infants mounted antibody responses and had improved clinical outcome compared to controls. An attenuated recombinant canary pox vector expressing SIV gag, pol, and env (ALVAC-SIV) was used to immunize infant macaques at birth, 2 and 3 weeks of age, followed by repeated oral low-dose challenge [73]. In this experiment, significantly fewer immunized infants were infected (6/16) compared to the unimmunized controls (14/16), demonstrating that neonatal immunization provided partial protection from infection. Finally, a topical DNA vaccine containing HIV-1 gag and env (DermaVir) represents an immunization strategy that targets lymph node DC. Rhesus macaques immunized with DermaVir generated HIV-1 specific Th1 and Th2 cytokines and antigen-specific memory T cells, while serum antibody levels were boosted after p27/gp140 protein boosting. Following mucosal challenge, none of the animals were protected from infection; however 4/5 immunized monkeys had reduced peak and set-point viremia [74].

Pediatric Clinical Trials of HIV Vaccines

The most recently reported clinical trials of live attenuated vaccine constructs tested as PACTG 326 part 1 and part 2 include the ALVAC constructs vCP205 and ALVAC-HIV vCP1452. Immunization of neonates was well tolerated and induced lymphoproliferative and/or cytotoxic T-cell responses in vaccines: ~40% of infants immunized with ALVAC vCP205 and 75% of infants immunized with ALVAC vCP1452 [75]. An MVA-vectored vaccine is also currently under evaluation in an open randomized phase I/II study evaluating safety and immunogenicity of a candidate HIV-1 vaccine, MVA. HIVA, administered to healthy infants born to HIV-1 infected mothers in Nairobi, Kenya. This active study aims to enroll 72 HIV-1 uninfected infants by the end of 2010, with infants in follow-up for 18 months. Infants will be randomized to receive MVA.HIVA or to remain unvaccinated. The design of the study will allow for multiple secondary aims for comparison of immunogenicity to other national immunization program vaccines in the MVA.HIVA vaccinated or unvaccinated infants (Tomas Hanke, personal communication).

Concerns Regarding Infant Vaccine Development

Many key questions regarding the development of a successful adult HIV-1 vaccine are equally valid for a neonatal immune-based intervention for PMTCT. However, pediatric vaccine development also faces a series of unique concerns. These include, but are not limited to, regulatory/ethical issues applicable to vulnerable populations, physiologic constraints of blood volumes that may limit the degree of safety and immunogenicity testing, existing immunization schedules and potential vaccine interference, and simultaneous exposure to both vaccine and pathogen in the presence of maternal antibodies and the developing neonatal immune system.

Currently, infants are immunized worldwide against an array of infections delivered over the first 2 years of life, including BCG, polio, hepatitis B, diphtheria, pertussis, tetanus, pneumococcus, and *Hemophilus influenza* b (Hib). Investigations into the sequence of exposure to murine viruses have demonstrated that the magnitude and specificity of the immune response elicited by the most recent infection are modified by the host's history of previous infections [76]. In newborn mice and humans, *Mycobacterium bovis* BCG immunization induces a potent immune response and this response has been shown to alter immunity to unrelated vaccines [77]. Also, individual components of multivalent vaccines may induce responses that differ when given individually or in combination (reviewed in ref. [78]). Therefore, the timing of introduction of new vaccines into the existing Expanded Program of Immunizations has the potential to modify responses to both previous and subsequent vaccines.

Vaccine constructs and adjuvants may also react differently in infants, although thus far, recombinant HIV-1 gp120 delivered either in alum or MF59, and recombinant canarypox vectored vaccines for HIV have proven safe in pediatric populations (PACTG 230, 326, and HPTN 027), while the testing of adjuvant CRM_{197} for use in multivalent pediatric vaccines has shown to improve immunogenicity of certain vaccines [79]. Live attenuated SIV vaccines tested in neonatal macaques have also proved safe, with the rare exception of a multiply deleted SIVmac239 that when administered to neonates, showed unexpected pathogenicity not initially observed in adults. Pathogenesis in adults was later documented in ~25% of vaccinated adults after a median of ~3 years of infection [80].

These concerns can be addressed readily through investment in the use of animal models, of continued testing in adults, and of basic research into underlying mechanisms of neonatal immune regulation, maternal antibody interference, and vaccine interference.

Conclusions and Future Directions

Although the HIV vaccine field still has some way to go before an effective vaccine to prevent infection becomes available, the special issues of mother-to-child transmission and infant immunization deserve further study. We have highlighted the approaches tested to date, as well as highlighting the potentially modifiable infectious co-factors that can facilitate transmission of HIV-1 from mother to child in the developing world and commenting on some of the issues that will need to be considered in the development of a pediatric HIV vaccine. Even if deployment of strategies to prevent mother-to-child transmission of HIV-1 becomes universal, a scenario that currently seems some distance away in resource-poor settings, it is very likely that a prophylactic HIV vaccine will ultimately need to be given as part of the EPI. Therefore, a better understanding of infant immune responses to candidate vaccine antigens and adjuvants is an important area for future investigation.

References

1. Datta P, Embree JE, Kreiss J, Ndinya-Achola JO, Braddick M, Temmerman M et al (1994) Mother-to-child transmission of HIV-1: report from the Nairobi study. J Infect Dis 170:1134–1140
2. Coutsoudis A, Dabis F, Fawzi W, Gaillard P, Haverkamp G, Harris DR et al (2004) Late postnatal transmission of HIV-1 in breast-fed children: an individual patient data meta-analysis. J Infect Dis 189(12):2154–2166
3. Rousseau CM, Nduati RW, Richardson BA, John-Stewart GC, Mbori-Ngacha DA, Kreiss JK et al (2004) Association of levels of HIV-1-infected breast milk cells and risk of mother-to-child transmission. J Infect Dis 190(10): 1880–1888
4. Rousseau CM, Nduati RW, Richardson BA, Steele MS, John-Stewart GC, Mbori-Ngacha DA et al (2003) Longitudinal analysis of human immunodeficiency virus type 1 RNA in breast milk and of its relationship to infant infection and maternal disease. J Infect Dis 187(5):741–747
5. Obimbo EM, Mbori-Ngacha DA, Ochieng JO, Richardson BA, Otieno PA, Bosire R et al (2004) Predictors of early mortality in a cohort of human immunodeficiency virus type 1-infected African children. Pediatr Infect Dis J 23(6):536–543
6. Newell ML, Coovadia H, Cortina-Borja M, Rollins N, Gaillard P, Dabis F (2004) Mortality of infected and uninfected infants born to HIV-infected mothers in Africa: a pooled analysis. Lancet 364(9441):1236–1243
7. Richardson BA, Mbori-Ngacha D, Lavreys L, John-Stewart GC, Nduati R, Panteleeff DD et al (2003) Comparison of human immunodeficiency virus type 1 viral loads in Kenyan women, men, and infants during primary and early infection. J Virol 77(12):7120–7123
8. Mphatswe W, Blanckenberg N, Tudor-Williams G, Prendergast A, Thobakgale C, Mkhwanazi N et al (2007) High frequency of rapid immunological progression in African infants infected in the era of perinatal HIV prophylaxis. AIDS 21(10):1253–1261
9. Rouet F, Sakarovitch C, Msellati P, Elenga N, Montcho C, Viho I et al (2003) Pediatric viral human immunodeficiency virus type 1 RNA levels, timing of infection, and disease progression in African HIV-1-infected children. Pediatrics 112(4):e289
10. Shearer WT, Quinn TC, LaRussa P, Lew JF, Mofenson L, Almy S et al (1997) Viral load and disease progression in infants infected with human immunodeficiency virus type 1. Women and Infants Transmission Study Group. N Engl J Med 336(19):1337–1342
11. Lepage P, Spira R, Kalibala S, Pillay K, Giaquinto C, Castetbon K et al (1998) Care of human immunodeficiency virus-infected children in developing countries. International Working Group on Mother-to-Child Transmission of HIV. Pediatr Infect Dis J 17(7):581–586
12. Zijenah LS, Moulton LH, Iliff P, Nathoo K, Munjoma MW, Mutasa K et al (2004) Timing of mother-to-child transmission of HIV-1 and infant mortality in the first 6 months of life in Harare, Zimbabwe. AIDS 18(2):273–280
13. Marinda E, Humphrey JH, Iliff PJ, Mutasa K, Nathoo KJ, Piwoz EG et al (2007) Child mortality according to maternal and infant HIV status in Zimbabwe. Pediatr Infect Dis J 26(6):519–526
14. Michaelsson J, Mold JE, McCune JM, Nixon DF (2006) Regulation of T cell responses in the developing human fetus. J Immunol 176(10):5741–5748
15. Mold JE, Michaelsson J, Burt TD, Muench MO, Beckerman KP, Busch MP et al (2008) Maternal alloantigens promote the development of tolerogenic fetal regulatory T cells in utero. Science 322(5907):1562–1565

16. Chheda S, Palkowetz KH, Garofalo R, Rassin DK, Goldman AS (1996) Decreased interleukin-10 production by neonatal monocytes and T cells: relationship to decreased production and expression of tumor necrosis factor – [alpha] and its receptors. Pediatr Res 40:475–483
17. Lilic D, Cant AJ, Abinun M, Calvert JE, Spickett GP (1997) Cytokine production differs in children and adults. Pediatr Res 42:237–240
18. Prescott SL, Macaubas C, Holt BJ, Smallacombe TB, Loh R, Sly PD et al (1998) Transplacental priming of the human immune system to environmental allergens: universal skewing of initial T cell responses toward the Th2 cytokine profile. J Immunol 160(10):4730–4737
19. Holt PG, Jones CA (2000) The development of the immune system during pregnancy and early life. Allergy 55(8):688–697
20. Miles DJ, van der Sande M, Jeffries D, Kaye S, Ismaili J, Ojuola O et al (2007) Cytomegalovirus infection in Gambian infants leads to profound CD8 T-cell differentiation. J Virol 81(11):5766–5776
21. Slyker JA, Lohman-Payne BL, John-Stewart GC, Maleche-Obimbo E, Emery S, Richardson B et al (2009) Acute cytomegalovirus infection in Kenyan HIV-infected infants. AIDS 23(16):2173–2181
22. John-Stewart G, Mbori-Ngacha D, Ekpini R, Janoff EN, Nkengasong J, Read JS et al (2004) Breast-feeding and Transmission of HIV-1. J Acquir Immune Defic Syndr 35(2):196–202
23. Aldrovandi GM, Kuhn L (2010) What infants and breasts can teach us about natural protection from HIV infection. J Infect Dis 202(Suppl 3):S366–S370
24. John-Stewart GC, Mbori-Ngacha D, Payne BL, Farquhar C, Richardson BA, Emery S et al (2009) HIV-1-specific cytotoxic T lymphocytes and breast milk HIV-1 transmission. J Infect Dis 199:889–898
25. Hasselrot K, Bratt G, Hirbod T, Saberg P, Ehnlund M, Lopalco L et al (2010) Orally exposed uninfected individuals have systemic anti-HIV responses associating with partners' viral load. AIDS 24(1):35–43
26. Perez CL, Hasselrot K, Bratt G, Broliden K, Karlsson AC (2010) Induction of systemic HIV-1-specific cellular immune responses by oral exposure in the uninfected partner of discordant couples. AIDS 24(7):969–974
27. Cheynier R, Langlade-Demoyen P, Marescot M-R, Blanche S, Blondin G, Wain-Hobson S et al (1992) Cytotoxic T lymphocyte responses in the peripheral blood of children born to HIV-1-infected mothers. Eur J Immunol 22:2211–2217
28. Rowland-Jones SL, Nixon DF, Aldhous MC, Gotch F, Ariyoshi K, Hallam N et al (1993) HIV-specific CTL activity in an HIV-exposed but uninfected infant. Lancet 341:860–861
29. Aldhous MC, Watret KC, Mok JY, Bird AG, Froebel KS (1994) Cytotoxic T lymphocyte activity and CD8 subpopulations in children at risk of HIV infection. Clin Exp Immunol 97(1):61–67
30. Kuhn L, Coutsoudis A, Moodley D, Trabattoni D, Mngqundaniso N, Shearer GM et al (2001) T-helper cell responses to HIV envelope peptides in cord blood: protection against intrapartum and breast-feeding transmission. AIDS 15(1):1–9
31. Adkins B, Leclerc C, Marshall-Clarke S (2004) Neonatal adaptive immunity comes of age. Nat Rev 4:553–564
32. Levy O (2007) Innate immunity of the newborn: basic mechanisms and clinical correlates. Nat Rev Immunol 7:379–389
33. Velilla PA, Rugeles MT, Chougnet CA (2006) Defective antigen-presenting cell function in human neonates. Clin Immunol 121:251–259
34. Marchant A, Appay V, van der Sande M, Dulphy N, Liesnard C, Kidd M et al (2003) Mature CD8+ T lymphocyte response to viral infection during foetal life. J Clin Invest 111:1747–1755
35. Marchant A, Goetghebuer T, Ota MO, Wolfe I, Ceesay SJ, De Groote D et al (1999) Newborns develop a Th1-type immune response to *Mycobacterium bovis* bacillus Calmette-Guerin vaccination. J Immunol 163(4):2249–2255
36. Ota MO, Vekemans J, Schlegel-Haueter SE, Fielding K, Sanneh M, Kidd M et al (2002) Influence of *Mycobacterium bovis* bacillus Calmette-Guerin on antibody and cytokine responses to human neonatal vaccination. J Immunol 168(2):919–925
37. Godfrey WR, Spoden DJ, Ge YG, Baker SR, Liu B, Levine BL et al (2005) Cord blood CD4+CD25+-derived T regulatory cell lines express FoxP3 protein and manifest potent suppressor function. Blood 105:750–758
38. Legrand FA, Nixon DF, Loo CP, Ono E, Chapman JM, Miyamoto M et al (2006) Strong HIV-1-specific T cell responses in HIV-1-exposed uninfected infants and neonates revealed after regulatory T cell removal. PLoS One 1:e102
39. Hartigan-O'Connor DJ, Abel K, McCune JM (2007) Suppression of SIV-specific CD4+ T cells by infant but not adult macaque regulatory T cells: implications for SIV disease progression. J Exp Med 204:2679–2692
40. Flanagan KL, Halliday A, Burl S, Landgraf K, Jagne YJ, Noho-Konteh F et al (2010) The effect of placental malaria infection on cord blood and maternal immunoregulatory responses at birth. Eur J Immunol 40(4):1062–1072
41. Gans H, Yasukawa L, Rinki M, DeHovitz R, Forghani B, Beeler J et al (2001) Immune responses to measles and mumps vaccination of infants at 6, 9, and 12 months. J Infect Dis 184:817–826
42. Lohman-Payne B, Slyker JA, Richardson BA, Farquhar C, Majiwa M, Maleche-Obimbo E et al (2009) Infants with late breast milk acquisition of HIV-1 generate interferon-gamma responses more rapidly than infants with early peripartum acquisition. Clin Exp Immunol 156(3):511–517

43. Abrams EJ, Weedon J, Steketee RW, Lambert G, Bamji M, Brown T et al (1998) Association of human immunodeficiency virus (HIV) load early in life with disease progression among HIV-infected infants. New York City Perinatal HIV Transmission Collaborative Study Group. J Infect Dis 178(1):101–108
44. Biggar RJ, Janes M, Pilon R, Miotti P, Taha TE, Broadhead R et al (1999) Virus levels in untreated African infants infected with human immunodeficiency virus type 1. J Infect Dis 180(6):1838–1843
45. Newell M-L, Peckham C, Dunn D, Ades T, Giaquinto C, The European Collaborative Study (1994) Natural history of vertically acquired human immunodeficiency virus-1 infection. Pediatrics 94(6 Pt 1):815–819
46. Doyle M, Atkins JT, Rivera-Matos IR (1996) Congenital cytomegalovirus infection in infants infected with human immunodeficiency virus type 1. Pediatr Infect Dis J 15(12):1102–1106
47. Kovacs A, Schluchter M, Easley K, Demmler G, Shearer W, La Russa P et al (1999) Cytomegalovirus infection and HIV-1 disease progression in infants born to HIV-1-infected women. Pediatric Pulmonary and Cardiovascular Complications of Vertically Transmitted HIV Infection Study Group. N Engl J Med 341(2):77–84
48. Nigro G, Krzysztofiak A, Gattinara GC, Mango T, Mazzocco M, Porcaro MA et al (1996) Rapid progression of HIV disease in children with cytomegalovirus DNAemia. AIDS 10(10):1127–1133
49. Mole L, Ripich S, Margolis D, Holodniy M (1997) The impact of active herpes simplex virus infection on human immunodeficiency virus load. J Infect Dis 176(3):766–770
50. Schacker T, Zeh J, Hu H, Shaughnessy M, Corey L (2002) Changes in plasma human immunodeficiency virus type 1 RNA associated with herpes simplex virus reactivation and suppression. J Infect Dis 186(12):1718–1725
51. Kublin JG, Patnaik P, Jere CS, Miller WC, Hoffman IF, Chimbiya N et al (2005) Effect of *Plasmodium falciparum* malaria on concentration of HIV-1-RNA in the blood of adults in rural Malawi: a prospective cohort study. Lancet 365(9455):233–240
52. Hoffman IF, Jere CS, Taylor TE, Munthali P, Dyer JR, Wirima JJ et al (1999) The effect of *Plasmodium falciparum* malaria on HIV-1 RNA blood plasma concentration. AIDS 13(4):487–494
53. Slyker JA, Lohman-Payne BL, Rowland-Jones SL, Otieno P, Maleche-Obimbo E, Richardson B et al (2009) The detection of cytomegalovirus DNA in maternal plasma is associated with mortality in HIV-1-infected women and their infants. AIDS 23(1):117–124
54. Gantt S, Carlsson J, Shetty AK, Seidel KD, Qin X, Mutsvangwa J et al (2008) Cytomegalovirus and Epstein–Barr virus in breast milk are associated with HIV-1 shedding but not with mastitis. AIDS 22(12):1453–1460
55. Griffiths PD (2006) CMV as a cofactor enhancing progression of AIDS. J Clin Virol 35(4):489–492
56. Pitisuttithum P, Gilbert P, Gurwith M, Heyward W, Martin M, van Griensven F et al (2006) Randomized, double-blind, placebo-controlled efficacy trial of a bivalent recombinant glycoprotein 120 HIV-1 vaccine among injection drug users in Bangkok, Thailand. J Infect Dis 194(12):1661–1671
57. Buchbinder SP, Mehrotra DV, Duerr A, Fitzgerald DW, Mogg R, Li D et al (2008) Efficacy assessment of a cell-mediated immunity HIV-1 vaccine (the Step Study): a double-blind, randomised, placebo-controlled, test-of-concept trial. Lancet 372(9653):1881–1893
58. McElrath MJ, De Rosa SC, Moodie Z, Dubey S, Kierstead L, Janes H et al (2008) HIV-1 vaccine-induced immunity in the test-of-concept Step Study: a case-cohort analysis. Lancet 372(9653):1894–1905
59. Liu J, O'Brien KL, Lynch DM, Simmons NL, La Porte A, Riggs AM et al (2009) Immune control of an SIV challenge by a T-cell-based vaccine in rhesus monkeys. Nature 457(7225):87–91
60. Hansen SG, Vieville C, Whizin N, Coyne-Johnson L, Siess DC, Drummond DD et al (2009) Effector memory T cell responses are associated with protection of rhesus monkeys from mucosal simian immunodeficiency virus challenge. Nat Med 15(3):293–299
61. Rerks-Ngarm S, Pitisuttithum P, Nitayaphan S, Kaewkungwal J, Chiu J, Paris R et al (2009) Vaccination with ALVAC and AIDSVAX to prevent HIV-1 infection in Thailand. N Engl J Med 361(23):2209–2220
62. Abel K, Pahar B, Van Rompay KKA, Fritts L, Sin C, Schmidt K et al (2006) Rapid virus dissemination in infant macaques after oral simian immunodeficiency virus exposure in the presence of local innate immune responses. J Virol 80:6357–6367
63. Kumar RB, Maher DM, Herzberg MC, Southern PJ (2006) Expression of HIV receptors, alternate receptors and co-receptors on tonsillar epithelium: implications for HIV binding and primary oral infection. Virol J 3:25–38
64. Azizi A, Ghunaim H, Diaz-Mitoma F, Mestecky J (2010) Mucosal HIV vaccines: a holy grail or a dud? Vaccine 28:4015–4026
65. Wright PF, Mestecky J, McElrath MJ, Keefer MC, Gorse GJ, Goepfert PA et al (2004) Comparison of systemic and mucosal delivery of 2 canarypox virus vaccines expressing either HIV-1 genes or the gene for rabies virus G protein. J Infect Dis 189:1221–1231
66. Kotton CN, Lankowski AJ, Scott N, Sisul D, Chen LM, Raschke K et al (2006) Safety and immunogenicity of attenuated *Salmonella enterica* serovar *Typhimurium* delivering an HIV-1 gag antigen via the *Salmonella* type III secretion system. Vaccine 24:6216–6224
67. Van Rompay KKA, Otsyula MG, Tarara RP, Canfield DR, Berardi CJ, McChesney MB et al (1996) Vaccination of pregnant macaques protects newborns against mucosal simain immunodeficiency virus infection. J Infect Dis 173:1327–1335

68. Van Rompay KKA, Berardi CJ, Dillard-Telm S, Tarara RP, Canfield DR, Valverde CR et al (1998) Passive immunization of newborn rhesus macaques prevents oral simian immunodeficiency virus infection. J Infect Dis 177:1247–1259
69. Hofmann-Lehmann R, Vlasak J, Rasmussen RA, Smith BA, Baba TW, Liska V et al (2001) Postnatal passive immunization of neonatal macaques with a triple combination of human monoclonal antibodies against oral simian-human immunodeficiency virus challenge. J Virol 75:7470–7480
70. Ferrantelli F, Rasmussen RA, Buckley KA, Li PL, Wang T, Montefiori DC et al (2004) Complete protection of neonatal rhesus macaques against oral exposure to pathogenic simian-human immunodeficiency virus by human anti-HIV monoclonal antibodies. J Infect Dis 189:2149–2153
71. Ferrantelli F, Buckley KA, Rasmussen RA, Chalmers A, Wang T, Li PL et al (2007) Time dependence of protective post-exposure prophylaxis with human monoclonal antibodies against pathogenic SHIV challenge in newborn macaques. Virology 358:69–78
72. Van Rompay KKA, Greenier JL, Cole KS, Earl P, Moss B, Steckbeck JD et al (2003) Immunization of newborn rhesus macaques with simian immunodeficiency virus (SIV) vaccines prolongs survival after oral challenge with virulent SIVmac251. J Virol 77:179–190
73. Van Rompay KKA, Abel K, Lawson JR, Singh RP, Schmidt KA, Evans T et al (2005) Attenuated poxvirus-based simian immunodeficiency virus (SIV) vaccines given in infancy partially protect infant and juvenile macaques against repeated oral challenge with virulent SIV. J Acquir Immune Defic Syndr 38:124–134
74. Cristillo AD, Lisziewicz J, He L, Lori F, Galmin L, Trocio JN et al (2007) HIV-1 prophylactic vaccine comprised of topical DermaVir prime and protein boost elicits cellular immune responses and controls pathogenic R5 SHIV162P3. Virology 366:197–211
75. McFarland EJ, Johnson DC, Muresan P, Fenton T, Tomaras GD, McNamara J et al (2006) HIV-1 vaccine induced immune responses in newborns of HIV-1 infected mothers. AIDS 20(11):1481–1489
76. Selin LK, Lin M-L, Kraemer KA, Pardoll DM, Schneck JP, Varga SM et al (1999) Attrition of T cell memory: selective loss of LCMV epitope-specific memory CD8 T cells following infections with heterologous viruses. Immunity 11:733–742
77. Ota MO, Vekemans J, Schlegel-Haueter SE, Fielding K, Sanneh M, Kidd M et al (2002) Influence of *Mycobacterium bovis* bacillus Calmette-Guerin on antibody and cytokine responses to human neonatal vaccine. J Immunol 168:919–925
78. Vidor E (2007) The nature and consequences of intra- and inter-vaccine interference. J Comp Pathol 137(Suppl 1):S62–S66
79. Shinefield HR (2010) Overview of the development and current use of CRM197 conjugate vaccines for pediatric use. Vaccine 28:4335–4339
80. Baba TW, Liska V, Khimani AH, Ray NB, Dailey PJ, Penninck D et al (1999) Live attenuated, multiply deleted simian immunodeficiency virus causes AIDS in infant and adult macaques. Nat Med 5:194–203

Chapter 14
Non-antiretroviral Approaches to Prevention of Breast Milk Transmission of HIV-1: Exclusive Breastfeeding, Early Weaning, Treatment of Expressed Breast Milk

Jennifer S. Read

Introduction

Identification of risk factors for breast milk transmission of HIV-1 has led to the development of interventions to prevent such transmission [1]. Mixed feeding of an infant, i.e., provision of other fluids and/or solids to the infant in addition to breast milk, has been associated with an increased risk of mother-to-child transmission of HIV-1 [2, 3]. Therefore, exclusive breastfeeding of the infant has been emphasized, not only for the benefits such feeding provides to infants in general [4–7], but also specifically to prevent mother-to-child transmission of HIV-1 [8]. The longer the duration of exposure to breast milk from an HIV-1-infected woman, the greater the transmission of HIV-1 to the infant [9]. Although complete avoidance of breastfeeding has been shown to be efficacious in preventing breast milk transmission of HIV-1 [10], such an intervention is not feasible in many settings. Therefore, the concept of early weaning from breastfeeding was developed, and this intervention has been evaluated [11]. The higher the maternal HIV-1 RNA concentration (viral load), including viral load in breast milk, the greater the HIV-1 transmission to the infant [12–15]. Treatment of expressed breast milk with microbicidal agents or with heat to decrease breast milk viral load, and thus decrease mother-to-child transmission of HIV-1, has been assessed [16, 17]. This chapter reviews these three approaches to the prevention of breast milk transmission of HIV-1: exclusive breastfeeding, early weaning, and treatment of expressed breast milk.

Exclusive Breastfeeding

In the general population, exclusive breastfeeding during the first few months of life is associated with less morbidity and mortality as compared to mixed feeding, i.e., feeding the infant other liquids and/or solids in addition to breast milk [4–7]. Initial studies of feeding modality among children of HIV-1-infected women suggested, but did not definitively demonstrate, a lower risk of mother-to-child transmission of HIV-1 with exclusive breastfeeding compared to mixed feeding [2, 18, 19]. In a study of 168 breastfed infants of HIV-1-infected women in São Paulo State, Brazil, breastfed infants who

---

J.S. Read, M.D., M.S., M.P.H., D.T.M.H. (✉)
Global Health Sciences, University of California, 50 Beale Street, Suite 1200, San Francisco, CA 94105, USA
e-mail: Jennifer.Read@ucsf.edu

ingested other fluids were at higher risk, albeit not statistically significantly, of acquisition of HIV-1 infection [other milk: adjusted odds ratio=1.7 (95% CI: 0.6–4.7); tea or fruit juice: adjusted odds ratio=1.7 (95% CI: 0.5–5.7)] [18]. In an analysis of 551 HIV-1-infected women and their infants enrolled in a randomized clinical trial of vitamin A supplementation in South Africa, the cumulative probability of HIV-1 infection at 3 months of age for three groups was assessed: 157 formula fed infants who were never breastfed, 276 infants with mixed feeding, and 118 infants exclusively breast-fed for 3 months or more [2]. Exclusive or mixed breastfeeding was defined as without or with water, other fluids, or food. At 3 months, the estimated proportion of infants with HIV-1 infection was lower for infants exclusively breastfed for 3 months (14.6%; 95% CI=7.1–21.4%) than for those who had mixed feeding before 3 months (24.1%; 95% CI=19.0–29.2%) ($p=0.03$). In analyses adjusted for maternal CD4/CD8 ratio, syphilis, and preterm birth, the hazard ratio for exclusive breastfeeding compared to that for mixed feeding was 0.52 (0.28–0.98).

Subsequently, additional studies were undertaken to assess if and how exclusive breastfeeding was protective against mother-to-child transmission of HIV-1. The ZVITAMBO Study Group [3] evaluated infant feeding modality and HIV-1 transmission among HIV-1-infected women and their infants enrolled in a vitamin A supplementation trial in Zimbabwe. Infant feeding modality was categorized according to maternal report. Exclusive breastfeeding was defined as infant consumption of only breast milk and no other liquids or solids except prescribed medications and vitamins at three time points (6 weeks, 3 months, and 6 months), or at two of the three time points if the nonbreast milk item consumed at the third time point was a nonmilk liquid. Predominant breastfeeding was defined as infant consumption of predominantly breast milk, but nonmilk liquids (e.g., water, juice, tea, cooking oil) also were consumed at all three time points (or, if consumed at two of the three time points, with exclusive breastfeeding reported at the third time point). Mixed breastfeeding was defined as infant consumption, at one or more time points, of breast milk and either non-human milk (e.g., infant formula or cow's milk), or of solid or semisolid food, or of both. Of 2,060 infants of HIV-1-infected women with negative HIV-1 DNA PCR results at 6 weeks of age and complete feeding information, early mixed breastfeeding was associated with an increased risk of mother-to-child transmission compared to exclusive breastfeeding: at 6 months [hazard ratio (HR)=4.03 (95% CI: 0.98, 16.61); $p=0.05$], at 12 months [HR=3.79 (95% CI: 1.40, 10.29); $p=0.009$], and at 18 months [HR=2.60; 1.21, 5.55); $p=0.02$]. The hazard ratios for predominant breastfeeding were consistently elevated compared to the reference category of exclusive breastfeeding at each time point, but statistical significance was not demonstrated. Later, in an intervention cohort study conducted in South Africa [8], 1,405 infants born to HIV-1-infected women were evaluated. In this study, breastfed infants who received solids at any time after birth were significantly more likely to become HIV-1-infected by 6 months of age than exclusively breastfed infants [HR=10.87 (95% CI: 1.51–78.00); $p=0.02$]. Finally, in a prospective cohort study in Zambia [20], part of a randomized trial of early weaning [11], associations between infant feeding modalities and HIV-1 transmission were assessed. A total of 958 HIV-1-infected women and their infants were enrolled, and all of the women were encouraged to exclusively breastfeed for 4 months. HIV-1 transmission before 4 months was significantly lower among infants who were exclusively breastfed than among infants who were not ($p=0.004$). The relative hazard of nonexclusive breastfeeding was 3.48 (95% CI: 1.71–7.08). Importantly, there were no significant differences in the severity of HIV-1 disease between those mothers who did or did not exclusively breastfeed, and the relative hazard of nonexclusive breastfeeding remained significant [2.68 (95% CI: 1.28–5.62)] after adjustment for maternal CD4 count, maternal plasma viral load, and other variables.

The association of exclusive breastfeeding and a decreased risk of mother-to-child transmission of HIV-1 seems counterintuitive, since exclusively breastfed infants ingest a greater amount of HIV-1-infected breast milk [21]. Mechanisms that have been proposed for the association of mixed feeding and an increased risk of infant HIV-1 infection include (1) mixed feeding damages the intestinal mucosa, thus facilitating HIV-1 infection of the infant through increased permeability or intestinal immune activation; and (2) infant feeding modalities other than exclusive breastfeeding lead to subclinical mastitis, which in turn is associated with increased breast milk viral load, leading to an increased risk of HIV-1 infection of the infant [22]. However, investigation of the first hypothesis has so far yielded no evidence that mixed feeding damages the intestinal mucosa; among 272 infants of HIV-1-infected South African women, results of a lactulose/mannitol dual sugar test did not demonstrate increased intestinal permeability among infants with mixed feeding relative to exclusively breastfed infants, and urinary neopterin (an indicator of immune system activation) excretion was not associated with infant feeding modality [23]. Similarly, evaluation of the second hypothesis indicated mixed breastfeeding was not associated with mastitis or elevations of breast milk viral load [24]. In the absence of proven biological mechanisms for a causal relationship between infant feeding modality and mother-to-child transmission of HIV-1, it has been proposed that HIV-1-infected women who are relatively healthier, and thus at decreased risk of transmitting HIV-1 infection to their children, are more likely to exclusively breastfeed their infants than HIV-1-infected women with more advanced disease [25, 26]. However, mothers who exclusively breastfed their children did not appear to be healthier than mothers who provided mixed breastfeeding to their children in different studies [20, 24]. Thus, differences in maternal HIV-1 disease severity between those mothers who did or did not exclusively breastfeed their infants do not seem to explain the association of exclusive breastfeeding with a lower risk of transmission.

Early Weaning

A longer duration of breastfeeding is associated with a greater risk of mother-to-child transmission of HIV-1 [9, 10]. In an individual patient data meta-analysis of data regarding over 4,000 children of HIV-1-infected women with negative HIV-1 diagnostic assays at 4 weeks of age, the risk of breast milk transmission of HIV-1 was generally constant throughout breastfeeding [9]. The cumulative probability of HIV-1 infection increased with a longer duration of breastfeeding, at a rate of approximately 0.5% transmissions/month of breastfeeding between 1 and 18 months of age [9]. Similarly, in the randomized clinical trial of breastfeeding and formula feeding with over 400 mother–child pairs included in the analysis, the cumulative probability of HIV-1 infection increased as the duration of breastfeeding increased, from birth until follow-up ended at 24 months of age [10]. Complete avoidance of breastfeeding, although efficacious in preventing mother-to-child transmission of HIV-1 [10], is not feasible in many parts of the world. Therefore, early weaning from breastfeeding was considered as an intervention that would allow a child to experience the benefits of breastfeeding while limiting exposure to HIV-1-infected breast milk.

The Zambia Exclusive Breastfeeding Study (ZEBS) was an unblinded, randomized clinical trial conducted in Lusaka, Zambia, among HIV-1-infected women to determine whether exclusive breastfeeding for 4 months, followed by abrupt weaning, would decrease mother-to-child transmission of HIV-1 and mortality during the first 24 months after birth as compared to the standard practice of continuing breastfeeding for a longer period of time [11]. HIV-1-infected women were recruited from antenatal clinics in Lusaka where voluntary HIV-1 counseling and testing along with single-dose nevirapine prophylaxis [27] were offered. Between May 2001 and September 2004, 1,435 HIV-1-infected pregnant women at less than 38 weeks' gestation were recruited. Enrolled women who were still breastfeeding their infants at 1 month after delivery ($n=958$) were randomized. Those randomized to the intervention arm were encouraged to exclusively breastfeed for 4 months, and then to wean as rapidly as possible. A 3-month supply of infant formula and fortified weaning cereal were provided. Women randomized to the control group were encouraged to exclusively breastfeed for 6 months, and subsequently introduce complementary foods gradually. Complementary foods were not provided. The duration of breastfeeding in the control group was according to each woman's own choice. The median durations of breastfeeding were 5 months (intervention arm) and 16 months (control arm). There was no significant difference in the rate of HIV-1-free survival of the children to 24 months according to randomization arm (68.4% intervention arm and 64% control arm; $p=0.13$). Among children who were HIV-1-uninfected and still being breastfed at 4 months of age, there was no significant difference in HIV-1-free survival at 24 months (83.9% intervention arm and 80.7% control arm; $p=0.27$). However, children who were HIV-1-infected by 4 months of age were more likely to die by 24 months if their mothers were randomized to the intervention arm (73.6%) compared to the control arm (54.8%) ($p=0.007$). Therefore, early, abrupt weaning by HIV-1-infected women did not improve the rate of HIV-1-free survival among their children, and was detrimental to HIV-1-infected children [11].

An important feature of the intervention arm of this trial was that early weaning, when utilized, was encouraged to be performed abruptly, or as rapidly as possible. Women were provided counseling and other support and instruction prior to, during, and after weaning to try to ensure the safest transition from exclusive breastfeeding to replacement feeding, and to avoid prolongation of mixed feeding. However, abrupt weaning is associated with mastitis and elevations in HIV-1 RNA concentration (viral load) in breast milk [28], and it is possible that more gradual weaning from breast milk could reduce the risk of transmission of HIV-1.

A later analysis of data from the trial was undertaken to examine the relationship between early weaning and HIV-1-free survival according to the severity of maternal HIV-1 disease [29]. Early weaning was harmful and continued breastfeeding resulted in better outcomes among children of women with less severe HIV-1 disease during pregnancy. However, for children of women with more severe HIV-1 disease during pregnancy (who were eligible for antiretroviral therapy, but did not receive it), early weaning was associated with better outcomes.

Associations between actual breastfeeding duration and mortality among HIV-1-exposed but uninfected children of women enrolled in the ZEBS trial were examined recently [30]. Among 749 uninfected children, 9.4% died by 12 months of age and 13.6% died by 24 months of age. Compared to weaning at ages over 18 months, weaning at earlier ages was associated with elevated risks of mortality among uninfected children: weaning at 0–3 months of age (HR=3.59), at 4–5 months (HR=2.03), at 6–11 months (HR=3.54), and at 12–18 months (HR=4.22). Maternal CD4+ count was a significant effect modifier, and the risk of mortality associated with weaning was greater among children of mothers with higher CD4+ cell counts.

Although limiting the duration of exposure to breast milk from an HIV-1-infected woman decreases the risk of transmission of HIV-1 to the infant, multiple studies in addition to ZEBS have shown an association between complete avoidance of breastfeeding or early weaning from breast milk and infant morbidity and mortality due to gastroenteritis [31–39]. In light of data regarding the risks associated with early weaning, this intervention has not been investigated further as an intervention to prevent mother-to-child transmission of HIV-1 and, in fact, is no longer recommended by the World Health Organization [40].

Microbicide/Heat Treatment of Expressed Breast Milk

A higher breast milk viral load is associated with a higher risk of mother-to-child transmission of HIV-1 [12–15]. Therefore, treatment of expressed breast milk with microbicidal agents or heat has been proposed as an intervention to decrease breast milk viral load and thus decrease the risk of transmission of HIV-1 through breast milk.

Alkyl sulfates, such as sodium dodecyl sulfate (SDS), are microbicides with activity against HIV-1 at low concentrations that are inexpensive, with little or no toxicity [41]. Inactivation of HIV-1 in breast milk with 0.1% SDS has been reported [42]. However, results of subsequent research (specifically, clinical studies of this intervention to prevent breast milk transmission of HIV-1) have not been reported.

Studies of heat treatment of breast milk to prevent HIV-1 transmission were initiated after an early study involving 17 breast milk samples from four HIV-1-infected women, in which simply allowing the milk to remain at room temperature for 6 hours in an attempt to inactivate HIV-1 was ineffective;

six of seven samples (86%) had positive HIV-1 DNA assays [43]. However, none of eight breast milk samples subjected to boiling had positive HIV-1 DNA assay results [43].

Heat treatment methods can be divided into two groups: direct heat treatment of breast milk and indirect heat treatment in hot water baths. Different methods of direct heat treatment of breast milk of HIV-1-infected women have been considered. Holder pasteurization (62°C for 30 min) is routinely practiced by milk banks worldwide to eradicate pathogens, including HIV-1, from breast milk samples. The advantages of this type of direct heat treatment are that both cell-free and cell-associated HIV-1 are undetectable after treatment, and lysozyme activity is conserved, but the disadvantages are that IgA concentrations are reduced by 20% and there is complete loss of IgM and most of lactoferrin [44]. Alternatively, with direct heat treatment of only 56°C for 30 min, immunoglobulins (IgA and IgM) and iron-binding proteins are conserved [45], but cell-associated HIV-1 remains detectable [44]. A third approach of direct heat treatment has been described which involves solar-powered pasteurization at 60°C for 30 min [46], but this approach is expensive, the extent of loss of nutritional components is unclear, and it involves a large machine which would be obvious to the community in which the HIV-1-infected woman and her infant live.

Indirect heat treatment of breast milk, by heating expressed breast milk in hot water baths, has been evaluated in several studies. In early studies, "Pretoria pasteurization," involving heat transfer from 450 mL of water heated to the boiling point in an aluminum pan to a smaller volume of milk in a glass jar placed into the water, maintained milk at a temperature of 56–62.5°C for 12–15 min [47]. In a study of 26 samples from known HIV-1-infected women and 25 samples from HIV-1-uninfected women or women of unknown HIV-1 status, there was no evidence of viral replication in breast milk samples that had undergone Pretoria pasteurization [48]. Finally, evaluation of 58 breast milk samples showed that Pretoria pasteurization killed pathogenic and commensal bacteria in hand-expressed breast milk, and expressed breast milk that had undergone this procedure was maintained without refrigeration for up to 12 hours with a minimal likelihood of bacterial contamination [49].

Subsequently, "flash heat" treatment of breast milk samples was evaluated. This procedure entails manually expressing 75–150 mL of breast milk into sterile jars. Then, 50 mL of expressed breast milk is placed in a glass jar heated in a 450-mL water jacket in an aluminum pan until water boils. Following this, the expressed breast milk is removed [50]. A pilot study evaluating the viral, nutritional, and bacterial safety of flash-heated and Pretoria-pasteurized breast milk ($n=5$ samples) suggested flash heating may be superior to Pretoria pasteurization in terms of viral neutralization, although both methods retain nutrients and destroy bacterial contamination [50]. In a subsequent study of flash heating of breast milk [17], 98 breast milk samples were collected from 84 HIV-1-infected women in South Africa and divided into two groups: unheated control and flash heating. At baseline, detectable HIV-1 was found in breast milk samples from 26/84 mothers (31%). Thirty breast milk samples with detectable viral load were subjected to flash heating, and subsequently none had a detectable viral load. More recently, multiple viral assays (including reverse transcriptase and peripheral blood mononuclear cell neutralization assays) were utilized to assess inactivation of HIV-1 in five flash-heated breast milk samples [51]. Flash heating inactivated a high titer of cell-free and cell-associated HIV-1, and 99.7% of cells were killed (suggesting that transmission of HIV-1 through latently infected lymphocytes would not occur).

Later studies evaluated the effect of flash heating on vitamin content, immunoglobulin concentrations, and antimicrobial activity in breast milk. First, expressed breast milk from 50 HIV-1-infected women in South Africa was assayed for vitamin content (vitamins A, C, B_2, B_6, B_{12}, and folate) [52]. Vitamin B_2 was decreased to 59%, and vitamin B_6 was decreased to 96%, of that in unheated milk. Three vitamins were increased significantly (vitamin B_{12}, vitamin C, and folate), and vitamin A was not significantly affected by flash heat. Next, breast milk samples from 50 HIV-1-infected women in South Africa underwent flash heating and then were assayed for immunoglobin concentrations [53]. Flash heating resulted in significantly decreased total IgA and IgG concentrations (decreased by 20 and 33%, respectively). However, IgA and IgG binding to influenza viruses increased by 13 and 15%, respectively. Finally, 50 breast milk samples from HIV-1-infected women in South Africa were divided into two aliquots, "spiked" with *Staphylococcus aureus* and *Escherichia coli*, and then subjected to flash heating or no heat [54]. The bacteriostatic activity of breast milk was unaffected (no difference in the rate of growth of either organism according to heat treatment). However, mean antibacterial activity of lactoferrin and lysozyme was decreased by 11.1 and 56.6%, respectively by flash heat treatment. Only 80% of heated lactoferrin survived digestion, compared to 100% of lactoferrin in unheated breast milk ($p<0.0001$). But, there was no significant difference between pre- and postheated digested breast milk samples ($p=0.12$), indicating heating did not affect the amount of lysozyme that survives digestion.

Although concerns have been raised regarding the practicality of interventions to prevent breast milk transmission of HIV-1 that involve treating expressed breast milk, qualitative data from Zimbabwe support the acceptability of heat treatment [55]. In this study [55], and in another, unpublished study (Caroline Chantry, personal communication, 31 March 2011), potential obstacles including time constraints as well as social and cultural stigma appeared to be overcome by its perceived affordability and the potential to prevent transmission. Further studies are required to more definitively address logistical issues related to treatment of expressed breast milk.

Summary

Based on these data regarding exclusive breastfeeding, early weaning, and treatment of expressed breast milk, the World Health Organization recently updated their guidelines on HIV-1 and infant feeding [40]. Thus, HIV-1-infected women whose infants are HIV-1-uninfected or of unknown HIV-1 infection status should exclusively breastfeed their infants for the first 6 months of life, and continue breastfeeding for the first 12 months of life with concomitant antiretroviral prophylaxis. Breastfeeding should only stop once nutritionally adequate and safe infant dietary intake is assured. Abrupt weaning from breast milk should be avoided; HIV-1-infected women who decide to stop breastfeeding should stop gradually within 1 month. Expressed, heat-treated breast milk could be used on an interim basis in certain circumstances: when a low birth weight or ill newborn is unable to breastfeed, when the mother has mastitis or is otherwise not well and is temporarily unable to breastfeed, or when the mother needs assistance in stopping breastfeeding. HIV-1-infected women with HIV-1-infected infants or young children should exclusively breastfeed for the first 6 months of life and continue breastfeeding up to 2 years or more.

Future research is required to address important issues regarding exclusive breastfeeding and treatment of expressed breast milk. Specifically, the biological mechanism underlying the observed association between exclusive breastfeeding and a lower risk of mother-to-child transmission of HIV-1 remains to be elucidated. In addition, operational research regarding the optimal strategies to increase the proportion of women who initiate and continue exclusive breastfeeding is still needed. Finally, a large-scale clinical trial to evaluate the efficacy of flash heat treatment of breast milk on infant health outcomes is planned.

References

1. Read JS (2003) Human milk, breastfeeding, and transmission of human immunodeficiency virus type 1 in the United States. Pediatrics 112(5):1196–1205
2. Coutsoudis A, Pillay K, Kuhn L, Spooner E, Tsai WY, Coovadia HM (2001) Method of feeding and transmission of HIV-1 from mothers to children by 15 months of age: prospective cohort study from Durham, South Africa. AIDS 15:379–387

3. Iliff P, Piwoz E, Tavengwa N et al (2005) Early exclusive breastfeeding reduces the risk of postnatal HIV-1 transmission and increases HIV-1-free survival. AIDS 19:699–708
4. WHO Collaborative Study Team on the Role of Breastfeeding on the Prevention of Infant Mortality (2000) Effect of breastfeeding on infant and child mortality due to infectious diseases in less developed countries: a pooled analysis. Lancet 355:451–455
5. Victora CG, Smith PG, Vaughan JP et al (1987) Evidence for protection by breast-feeding against infant deaths from infectious diseases in Brazil. Lancet 2:319–322
6. Brown KH, Black RE, Lopz de Romana G, Creed de Kanashiro H (1989) Infant-feeding practices and their relationship with diarrheal and other diseases in Huascar (Lima), Peru. Pediatrics 83:31–40
7. Koyanagi A, Humphrey JH, Moulton LH et al (2009) Effect of early exclusive breastfeeding on morbidity among infants born to HIV-1-negative mothers in Zimbabwe. Am J Clin Nutr 89(5):1375–1382
8. Coovadia HM, Rollins NC, Bland RM et al (2007) Mother-to-child transmission of HIV-1 infection during exclusive breastfeeding in the first 6 months of life: an intervention cohort study. Lancet 369:1107–1116
9. Breastfeeding and HIV-1 International Transmission Study Group (2004) Late postnatal transmission of HIV-1 in breast-fed children: an individual patient data meta-analysis. J Infect Dis 189:2154–2166
10. Nduati R, John G, Mbori-Ngacha D et al (2000) Effect of breastfeeding and formula feeding on transmission of HIV-1: a randomized clinical trial. JAMA 283:1167–1174
11. Kuhn L, Aldrovandi GM, Sinkala M et al (2008) Effects of early, abrupt weaning for HIV-1-free survival of children in Zambia. N Engl J Med 359:130–141
12. Semba RD, Kumwenda N, Hoover DR et al (1999) Human immunodeficiency virus load in breast milk, mastitis, and mother-to-child transmission of human immunodeficiency virus type 1. J Infect Dis 180:93–98
13. Richardson BA, John-Stewart GC, Hughes JP et al (2003) Breast-milk infectivity in human immunodeficiency virus type 1-infected mothers. J Infect Dis 187:739–740
14. Pillay K, Coutsoudis A, York D, Kuhn L, Coovadia HM (2000) Cell-free virus in breast milk of HIV-1-seropositive women. J Acquir Immune Defic Syndr 24:330–336
15. Rousseau CM, Nduati RW, Richardson BA et al (2003) Longitudinal analysis of human immunodeficiency virus type 1 RNA in breast milk and of its relationship to infant infection and maternal disease. J Infect Dis 187:741–747
16. Urdaneta S, Wigdahl B, Neely EB et al (2005) Inactivation of HIV-1 in breast milk by treatment with the alkyl sulfate microbicide sodium docecyl sulfate (SDS). Retrovirology 2:28
17. Israel-Ballard K, Donovan R, Chantry C et al (2007) Flash-heat inactivation of HIV-1 in human milk. J Acquir Immune Defic Syndr 45(3):318–323
18. Tess BH, Rodrigues LC, Newell M-L, Dunn DT, Lago TDG (1998) Infant feeding and risk of mother-to-child transmission of HIV-1 in Sao Paulo State, Brazil. Sao Paulo Collaborative Study for Vertical Transmission of HIV-1. J Acquir Immune Defic Syndr Hum Retrovirol 19(2):189–194
19. Coutsoudis A, Pillay K, Spooner E, Kuhn L, Coovadia HM (1999) Influence of infant feeding patterns on early mother-to-child transmission of HIV-1 in Durban, South Africa. Lancet 354:471–476
20. Kuhn L, Sinkala M, Kankasa C et al (2007) High uptake of exclusive breastfeeding and reduced early post-natal HIV-1 transmission. PLoS One 2(12):e1363
21. Kuhn L (2010) Milk mysteries: why are women who exclusively breast feed less likely to transmit HIV-1 during breast-feeding? [editorial]. Clin Infect Dis 50:770–772
22. Kasonka L, Makasa M, Marshall T et al (2006) Risk factors for subclinical mastitis among HIV-1-infected and uninfected women in Lusaka, Zambia. Paediatr Perinat Epidemiol 20(5):379–391
23. Rollins NC, Filteau SM, Coutsoudis A, Tomkins AM (2001) Feeding mode, intestinal permeability, and neopterin excretion: a longitudinal study in infants of HIV-1-infected South African women. J Acquir Immune Defic Syndr 28(2):132–139
24. Lunney KM, Iliff P, Mutasa K et al (2010) Associations between breast milk viral load, mastitis, exclusive breast-feeding, and postnatal transmission of HIV-1. Clin Infect Dis 50:762–769
25. Chisenga M, Kasonka L, Makasa M et al (2005) Factors affecting the duration of exclusive breastfeeding among HIV-1-infected and -uninfected women in Lusaka, Zambia. J Hum Lact 21(3):266–275
26. Phiri W, Kasonka L, Collin S et al (2006) Factors influencing breast milk HIV-1 RNA viral load among Zambian women. AIDS Res Hum Retroviruses 22(7):607–614
27. Guay LA, Musoke P, Fleming T et al (1999) Intrapartum and neonatal single-dose nevirapine compared with zidovudine for prevention of mother-to-child transmission of HIV-1 in Kampala, Uganda: HIV-1NET 012 randomised trial. Lancet 354:795–802
28. Thea DM, Aldrovandi G, Kankasa C et al (2006) Post-weaning breast milk HIV-1 viral load, blood prolactin levels and breast milk volume. AIDS 20(11):1539–1547
29. Kuhn L, Aldrovandi Gm, Sinkala M et al (2009) Differential effects of early weaning for HIV-1-free survival of children born to HIV-1-infected mothers by severity of maternal disease. PLoS One 4(6):e6059. doi:10.1371/journal.pone.0006059

30. Kuhn L, Sinkala M, Semrau K et al (2010) Elevations in mortality associated with weaning persist into the second year of life among uninfected children born to HIV-1-infected mothers. Clin Infect Dis 50:437–444
31. Kesho Bora Study Group (2011) Safety and effectiveness of antiretroviral drugs during pregnancy, delivery and breastfeeding for prevention of mother-to-child transmission of HIV-1: the Kesho Bora multicentre collaborative study rational, design, and implementation challenges. Contemp Clin Trials 32:74–85
32. Mbori-Ngacha D, Nduati R, John G et al (2001) Morbidity and mortality in bresatfed and formula-fed infants of HIV-1-infected women: a randomized clinical trial. JAMA 286(19):2413–2420
33. Becquet R, Bequet L, Ekouevi DK et al (2007) Two-year morbidity-mortality and alternatives to prolonged breast-feeding among children born to HIV-1-infected mothers in Cote d'Ivoire. PLoS Med 4(1):317. doi:10.1371/journal.pmed.0040017
34. Kagaayi J, Gray RH, Brahmbhatt H et al (2008) Survival of infants born to HIV-1-positive mothers, by feeding modality, in Rakai, Uganda. PLoS ONE 3(12):e3877. doi:10.1371/journal.pone.0003877
35. Harris JR, Greene SK, Thomas TK et al (2009) Effect of a point-of-use water treatment and safe water storage intervention on diarrhea in infants of HIV-1-infected mothers. J Infect Dis 200:1186–1193
36. Kafulafula G, Hoover DR, Taha TE et al (2010) Frequency of gastroenteritis and gastroenteritis-associated mortality with early weaning in HIV-1-uninfected children born to HIV-1-infected women in Malawi. J Acquir Immune Defic Syndr 53(1):6–13
37. Onyango-Makumbi C, Bagenda D et al (2010) Early weaning of HIV-1-exposed uninfected infants and risk of serious gastroenteritis: findings from two perinatal prevention trials in Kampala, Uganda. J Acquir Immune Defic Syndr 53(1):20–27
38. Creek TL, Kim A, Lu L et al (2010) Hospitalization and mortality among primarily nonbreastfed children during a large outbreak of diarrhea and malnutrition in Botswana, 2006. J Acquir Immune Defic Syndr 53(1):14–19
39. Homsy J, Moore D, Barasa A et al (2010) Breastfeeding, mother-to-child HIV-1 transmission, and mortality among infants born to HIV-1-infected women on highly active antiretroviral therapy in rural Uganda. J Acquir Immune Defic Syndr 53(1):28–35
40. World Health Organization. Guidelines on HIV-1 and infant feeding 2010. http://whqlibdoc.who.int/publications/2010/9789241599535_eng.pdf. Accessed 1 Apr 2011
41. Hartmann SU, Wigdahl B, Neely EB, Berlin CM, Schengrund C-L, Lin H-M, Howett MK (2006) Biochemical analysis of human milk treated with sodium dodecyl sulfate, an alkyl sulfate microbicide that inactivates human immunodeficiency virus type 1. J Hum Lact 22(1):61–74
42. Urdaneta S, Wigdahl B, Neely EB et al (2005) Inactivation of HIV-1 in breast milk by treatment with the alkyl sulfate microbicide sodium dodecyl sulfate (SDS). Retrovirology 2:28
43. Chantry CJ, Morrison P, Panchula J et al (2000) Effects of lipolysis or heat treatment on HIV-1 provirus in breast milk. J Acquir Immune Defic Syndr 24:325–329
44. Orloff SL, Wallingford JC, McDougal JS (1993) Inactivation of human immunodeficiency virus type 1 in human milk: effects of intrinsic factors in human milk and of pasteurization. J Hum Lact 9:13–17
45. Ford JE, Law BA, Marshall VM, Reiter B (1977) Influence of the heat treatment of human milk on some of its protective constituents. J Pediatr 90:29–35
46. Jorgensen AF, Boisen F (2000) Pasteurization of HIV-1 contaminated breast milk. In: Proceedings and abstracts of the XIII world AIDS conference, Durban, South Africa, July 2000; abstract LbPp122
47. Jeffery BS, Mercer KG (2000) Pretoria pasteurization: a potential method for the reduction of postnatal mother to child transmission of the human immunodeficiency virus. J Trop Pediatr 46:219–223
48. Jeffery BS, Webber L, Mokhondo KR, Erasmus D (2001) Determination of the effectiveness of inactivation of human immunodeficiency virus by pretoria pasteurization. J Trop Pediatr 47:345–349
49. Jeffery BS, Soma-Pillay P, Makin J, Moolman G (2003) The effect of Pretoria pasteurization on bacterial contamination of hand-expressed human breast milk. J Trop Pediatr 49(4):240–244
50. Israel-Ballard K, Chantry C, Dewey K et al (2005) Viral, nutritional, and bacterial safety of flash-heated and pretoria-pasteurized breast milk to prevent mother-to-child transmission of HIV-1 in resource-poor countries: a pilot study. J Acquir Immune Defic Syndr 40:175–181
51. Volk ML, Hanson CV, Israel-Ballard K, Chantry CJ (2010) Inactivation of cell-associated and cell-free HIV-1 by flash-heat treatment of breast milk [letter]. J Acquir Immune Defic Syndr 53(5):665–666
52. Israel-Ballard KA, Abrams BF, Coutsoudis A, Sibeko LN, Cheryk LA, Chantry CJ (2008) Vitamin content of breast milk from HIV-1-infected mothers before and after flash-heat treatment. J Acquir Immune Defic Syndr 48(4):444–449
53. Chantry CJ, Israel-Ballard K, Moldoveanu Z et al (2009) Effect of flash-heat treatment on immunoglobulins in breast milk. J Acquir Immune Defic Syndr 51(3):264–267
54. Chantry CJ, Wiedeman J, Buehring G et al (2011) Effect of flash-heat treatment on antimicrobial activity of breast milk. Breastfeed Med 6(3):111–116
55. Israel-Ballard KA, Maternowska MC, Abrams BF et al (2006) Acceptability of heat treating breast milk to prevent mother-to-child transmission of human immunodeficiency virus in Zimbabwe: a qualitative study. J Hum Lact 22(1):48–60

Chapter 15
Breast Milk Micronutrients and Mother-to-Child Transmission of HIV-1

Monal R. Shroff and Eduardo Villamor

Introduction

According to the World Health Organization/United Nations Programme on HIV/AIDS [1], there were 370,000 new cases of HIV infection among children in 2009. This is in addition to the estimated 2.5 million children already infected with HIV worldwide. Ninety percent of pediatric HIV infections are contracted from the mother. Mother-to-child transmission (MTCT) of HIV occurs during pregnancy, at delivery, or through breastfeeding in about 20–45% of babies born to HIV-infected mothers, in the absence of interventions. Because the availability of antiretroviral prophylaxis to prevent perinatal MTCT is expanding [2, 3], the number of infections through breastfeeding is increasing and their prevention remains a major challenge [2].

In resource-poor countries, where replacement feeding may not be affordable, feasible, acceptable, safe, or sustainable, HIV-positive women are recommended to exclusively breastfeed their infants during the first 6 months of life [4]. If national authorities support breastfeeding and antiretroviral therapy, breastfeeding can be extended until at least 12 months of age [4]. Not all children born to HIV-infected mothers become infected during breastfeeding, despite the fact that the HIV virus is excreted in milk. Thus, host characteristics and factors that are present in breast milk must play a protective effect against transmission. Breast milk contains important micronutrients with the potential to modulate immune responses of the mother and infant, and affect the risk of transmission through several pathways. In this review, we describe epidemiological and biological evidence available on the relation between breast milk micronutrients and MTCT of HIV through breastfeeding.

M.R. Shroff
Department of Epidemiology, University of Michigan, Ann Arbor, MI, USA

Center for Social Epidemiology and Population Health, University of Michigan School of Public Health, 4648 SPH Tower, 1415 Washington Heights, Ann Arbor, MI 48109-2029, USA
e-mail: mshroff@umich.edu

E. Villamor (✉)
Department of Epidemiology, University of Michigan, Ann Arbor, MI, USA

Breast Milk Micronutrients and MTCT

Vitamin A/β-Carotene

Vitamin A deficiency is still highly prevalent among women of reproductive age worldwide. In developing countries, the prevalence of vitamin A deficiency is estimated at 15% among pregnant women [5]. Among HIV-infected persons, low serum vitamin A levels are commonplace [6–10] and have been associated with HIV disease progression and mortality [11–13].

Vitamin A is essential for the proliferation and differentiation of immune cells, mainly lymphocytes [14], and for the maintenance of mucosal surfaces, especially in the gastrointestinal tract. Due to the immunological functions of vitamin A and the fact that supplemental vitamin A can be transferred from mother to infant through breastfeeding [15], it was suggested that improved vitamin A status in the mother could reduce MTCT through improved maternal immune competence [16].

An early observational study among 338 HIV-infected women from Malawi showed an inverse association between serum vitamin A levels during pregnancy and mother-to-child HIV transmission during the first year postpartum [17]. In this study, the highest rate of transmission (32.4%) occurred among women with vitamin A levels <0.70 µmol/L, whereas the lowest (7.2%) was observed in women with vitamin A levels >1.40 µmol/L. Additional studies conducted in the USA among non-breastfeeding HIV-infected women and their offspring offered conflicting results [18–20]. Greenberg et al. showed that lower maternal serum vitamin A levels (<0.70 µmol/L) were associated with a five-fold higher risk of perinatal MTCT. However, Burger et al. and Burns et al. did not find associations between low levels of serum vitamin A during late pregnancy and the risk of early MTCT of HIV-1.

The main limitation of observational studies is the role of potential confounding variables on the association between maternal vitamin A concentrations and risk of MTCT. Also the association could not be deemed causal because serum retinol levels are not necessarily representative of vitamin A status, especially in HIV-infected women. Because retinol binding protein is sensitive to the acute phase response, low serum retinol could be the result of advanced maternal HIV disease stage, rather than the cause. Despite their limitations, findings from these observational studies motivated the planning of randomized clinical trials to ascertain whether vitamin A supplementation could play a causal role in the prevention of MTCT. Two randomized trials conducted in South Africa [21] and Malawi [22] examined the effect of administering preformed vitamin A supplements to pregnant HIV-infected mothers. In the South Africa study, the vitamin supplement was given to 728 women during the third trimester of pregnancy and consisted of 5,000 IU retinyl palmitate and 30 mg β-carotene per day, plus 200,000 IU retinyl palmitate at delivery. Babies were followed for 15 months. In the Malawi study, 697 women received a daily dose of 10,000 IU vitamin A from 18 to 28 weeks' gestation until delivery and children were followed for HIV infection during a 24-month period. Vitamin A supplements did not have significant effects on MTCT of HIV in these studies: RR=0.98 (95% CI: 0.73, 1.31) in South Africa and RR=0.84 (95% CI: 0.64, 1.11) in Malawi. In a randomized placebo-controlled trial conducted in Tanzania [23], researchers assigned 1,078 HIV-infected women between 12 and 27 weeks' gestation, to one of four treatment groups: a daily oral dose of 5,000 IU preformed vitamin A and 30 mg of β-carotene; multivitamins (B-complex, C, E) excluding vitamin A/β-carotene; multivitamins in combination with vitamin A and β-carotene; or placebo. Supplements were administered during pregnancy and continued throughout the first 2 years postpartum. In this two-by-two factorial design study, vitamin A/β-carotene supplementation unexpectedly increased the risk of MTCT by age 24 months (RR=1.35; 95% CI: 1.10–21.65).

Several reasons could explain differences in the outcome of these studies. In the South Africa and Malawi trials, supplementation was given during the antenatal period only, whereas in Tanzania, supplementation was continued throughout the breastfeeding period. Also, the composition and dosage of the supplements differed among the three studies as noted before. Further, the occurrence of

HIV infection in the infant was assessed at 3 months in South Africa, whereas in the Malawi and Tanzania studies HIV infection was assessed throughout the first 24 months. In another trial conducted in Zimbabwe, 14,110 mother–infant pairs were enrolled within 96 h of birth. Mothers and/or infants received a single vitamin A dose in a two-by-two factorial manner: 400,000 IU for the mother and 50,000 IU for the infant. Most of the participants initiated breastfeeding. Vitamin A supplementation given to either the mother or the infant, but not both, increased the risk of HIV infection or death by 2 years of age among infants who were uninfected at 6 weeks postpartum [24].

The unexpected finding of increased MTCT due to maternal vitamin A supplementation prompted a search for potential explanations. Increased viral shedding in breast milk [15, 25] and subclinical mastitis [26, 27] are well-known risk factors for MTCT. Thus, these mechanisms might mediate the effects of vitamin A on MTCT. An observational study in Kenya showed an association between low vitamin A status in HIV-infected women and higher detectable HIV-1 DNA in breast milk, especially in those with CD4 cell counts of <400/mm^3 [28]. However, observational studies are subject to confounding, reverse causality, and other biases. Villamor et al. [29] examined the effect of vitamin A and β-carotene supplementation on HIV shedding in breast milk through the first 2 years postpartum in the context of the Tanzania trial. Women who received vitamin A and β-carotene supplementation had higher cell-free HIV load in breast milk at or after 6 months postpartum, compared with women who did not receive these supplements. Because vitamin A and β-carotene were administered together, it was not possible to determine the effects of each nutrient separately. Nevertheless, breast milk concentrations of β-carotene, but not those of retinol, were positively related to HIV shedding. Biological explanations for these findings are speculative. High dose β-carotene supplementation has been related to other adverse health outcomes including cancer [30, 31], possibly due to the prooxidant activity of β-carotene at high concentrations, and could be in part responsible for the adverse effects observed on MTCT. However, an effect of preformed vitamin A cannot be ruled out in light of the results from the Zimbabwe trial. Retinoids may increase expression of the chemokine (C-C) receptor 5 (CCR5), a co-receptor for HIV in target cells [32]. It is unclear whether this mechanism may be at play in breast milk macrophages or in mucosal cells of the infant gut.

The same group [33] examined the effect of vitamin A and β-carotene supplementation on subclinical mastitis which was defined as a ratio of the sodium to potassium (Na:K) concentration in breast milk >0.6 [34]. Among women who received vitamin A and β-carotene supplements, there was a 45% increase in the risk of severe subclinical mastitis, defined as a Na:K ratio >1. Increased incidence of mastitis could contribute to explain the unexpected rise in HIV MTCT due to vitamin A and β-carotene supplementation. In summary, evidence to date does not support a beneficial effect of vitamin A supplementation to prevent MTCT of HIV.

Vitamins B, C, and E

HIV-infected persons may experience profound deficiencies of water and fat-soluble vitamins. Low dietary intake or serum levels of vitamins B, C, and E have been associated with disease progression and mortality among HIV-infected people in observational studies [12, 35]. Some clinical trials indicate beneficial effects of supplementation with these vitamins on HIV disease progression [36]. These benefits might be mediated through enhancement of the immune function [37, 38]. For example, vitamin B6 is involved in proliferation of lymphocytes and T-cell-mediated natural killer cell function [38]. Low vitamin B12 levels are associated with impaired neutrophil function and CD4 cell count decline in HIV disease [39]. Vitamin C is a potent antioxidant that protects against oxidative damage caused by reactive oxygen species produced during phagocytosis. Vitamin E deficiency is associated with impaired cell-mediated immunity, neutrophil phagocytosis, and decreased lymphocyte proliferation in human and animal studies [40]. Vitamin E supplementation to AIDS-infected mice increases

IL-2 production and natural killer cell cytotoxicity, and reduces production of inflammatory cytokines such as tumor necrosis factor (TNF)-α and IL-6 [41].

The immune modulating properties of these vitamins suggest that they could offer benefits against MTCT. The efficacy of supplementation with vitamins B, C, and E on MTCT was tested in a randomized trial conducted in Tanzania [23]. In this study, 1,078 HIV-infected women were recruited when they were 12–27 weeks' gestation and randomly assigned to receive a daily oral dose of multivitamins with or without vitamin A/β-carotene in a two-by-two factorial manner. The multivitamins consisted of vitamins B-complex, C, and E, administered at several multiples of the recommended dietary allowance. Transmission through breastfeeding was defined as infection after 6 weeks of age among infants who were not known to be infected at 6 weeks. Multivitamins excluding vitamin A/β-carotene resulted in a nonstatistically significant reduction in transmission of HIV through breastfeeding, and reduced mortality among the infants who were not infected at 6 weeks of age. Among women with low lymphocyte counts or low hemoglobin at baseline, multivitamins significantly reduced breastfeeding transmission of HIV by 63% and 52%, respectively. Similarly, among low birth weight babies, multivitamins significantly decreased the risk of breastfeeding transmission by 73%. In women with low lymphocytes, serum retinol, or serum vitamin E concentrations, multivitamins significantly decreased the risk of infant death by 24 months. This study suggests that supplementation with vitamins B, C, and E could offer benefits against MTCT in immunologically or nutritionally compromised HIV-infected women.

Vitamin D

Vitamin D deficiency has been associated with a number of adverse health outcomes including bone pathology, some cancers, and infections [42, 43]. Low vitamin D levels are common in HIV-infected adults and adolescents [44–46], but the consequences of vitamin D deficiency on HIV-related outcomes remain poorly understood. In an observational study of 884 HIV-infected pregnant women who participated in the Tanzania trial of vitamins [46], serum levels of vitamin D at gestational weeks 20–27 were inversely associated with the babies' risk of dying or being HIV-infected at birth (RR=1.49; 95% CI: 1.07–2.09). Also, the risk of HIV infection by 6 weeks of age increased by 50% (RR=1.50; 95% CI: 1.02–2.20) in children born to mothers with low vitamin D levels at baseline, compared to mothers with higher levels. Among children who were uninfected at 6 weeks, maternal vitamin D deficiency was associated with a twofold higher risk of MTCT through breastfeeding.

Several mechanisms could explain the potential effect of vitamin D on MTCT of HIV. Vitamin D is a strong immune modulator and regulates activation of several immune cell lines by binding to vitamin D receptors that are abundantly present on these cells [47, 48]. Specifically, vitamin D increases the phagocytic capacity of macrophages and the number of natural killer cells and CD8 lymphocytes; it also contributes to activation of effector cells during inflammation [49–51]. A randomized trial of vitamin D supplementation to HIV-infected women and/or their babies would be warranted to determine the effectiveness of this intervention in preventing MTCT of HIV and other adverse outcomes.

Zinc

Zinc deficiency affects approximately one-third of the global population [52]. Among HIV-infected women the prevalence of zinc deficiency can be as high as 76% [53]. The importance of zinc in the prevention of diarrheal and respiratory morbidity in children has been widely investigated [54, 55]; however, less is known on the potential role of zinc in the prevention of breastfeeding MTCT of HIV.

Several mechanisms could support protective effects of zinc on MTCT. Zinc is involved in the maintenance of cell-mediated immunity and in the regeneration of CD4 T cells [56]. It may also inhibit the production of proinflammatory cytokines, including TNF-α, which have been implicated in the pathophysiology of HIV infection [57]. Zinc contributes to maintaining gut integrity and innate immunity on gastrointestinal epithelia, which can be compromised during HIV infection [58]. Zinc is part of the copper-zinc superoxide dismutase enzyme, which is important for protection against overproduction of free radicals seen during early stage of infection [59]. On the other hand, zinc is part of the HIV nucleocapsid protein, which is fundamental for viral assembly and infectivity [60]. Thus, HIV-1 could be characterized as a zinc-dependent virus.

Low serum levels or low dietary zinc intake were associated with accelerated HIV disease progression [61] and mortality [35, 62, 63] in some observational cohort studies. In others, however, positive associations were reported between zinc status and adverse HIV-related outcomes [12, 35]. Although zinc supplementation trials have been conducted among nonpregnant adults [64, 65] and in children [66], the potential effects of zinc supplementation on MTCT of HIV have only been examined in one trial [67]. Four hundred HIV-infected women were recruited between 12 and 27 weeks of gestation in Dar es Salaam, Tanzania. They were randomized to receive oral daily zinc supplements (25 mg) or placebo from the time of recruitment till 6 weeks postpartum. The authors found no effect of zinc supplementation on early mother-to-child HIV transmission. It is uncertain whether zinc could have an effect on breastfeeding MTCT since follow-up ended at 6 weeks postpartum.

Considering the protective effects of zinc supplements against diarrhea and immunological failure among HIV-infected persons suggested by previous trials [65, 66], a potential effect on MTCT cannot be discounted. Whether this effect exists would need to be examined in the context of antiretroviral prophylaxis against MTCT of HIV.

Selenium

Selenium deficiency is prevalent among HIV-infected persons and has been associated with progression of HIV disease and mortality [68, 69]. Low selenium levels have been reported among HIV-infected pregnant women [53, 70]. In children infected with HIV, low levels of selenium are related to increased risk of mortality [71].

In an observational study among 670 Tanzanian women, low selenium serum levels at weeks 20–27 of gestation were related to HIV transmission through the intrapartum route (RR=2.51; 95% CI: 1.19–5.30) [70]. In a study of 318 women from Kenya [72], serum selenium levels were inversely related to vaginal HIV-1 shedding. This finding could contribute to explain the association with intrapartum MTCT observed in Tanzania.

The effect of selenium supplementation on MTCT has not been examined in randomized trials. However, one trial reported the effect of selenium supplementation to HIV-infected women on infant mortality in Tanzania [73]. In this study, 915 HIV-infected pregnant women between 12 and 27 weeks' gestation were assigned to receive a daily oral dose of 200 μg selenium or placebo from the time of recruitment to 6 months postpartum. Selenium supplementation reduced the risk of child mortality after 6 weeks by 57% (RR=0.43; 95% CI: 0.19–0.99). In another randomized, placebo-controlled trial from Kenya [74], 400 HIV-1-infected nonpregnant women were supplemented for 6 weeks with micronutrients including selenium (B-complex vitamins, vitamin C, E, and 200 μg of selenium). There was a statistically significant, twofold increase in vaginal shedding of HIV-infected cells in the supplementation group compared to placebo. While these results suggested higher infectivity in women assigned to the micronutrients group, it is not possible to attribute this effect to selenium. In summary, selenium supplementation to HIV-infected pregnant women may decrease early infant mortality but it is unknown whether this effect could be mediated through decreased MTCT.

Conclusion

Human breast milk is a rich source of micronutrients that have strong immune modulating properties and anti-infective potential. Some of these nutrients could modify the risk of MTCT of HIV through breastfeeding. Evidence from randomized trials indicates that maternal supplementation with high dose vitamin A and β-carotene may increase the risk of MTCT of HIV and cannot be safely recommended. On the other hand, supplementation with vitamins B-complex, C, and E may be protective against MTCT in HIV-infected women who are nutritionally or immunologically compromised. The role of other micronutrients on the risk of MTCT through breastfeeding has not been evaluated in clinical trials; however, observational studies suggest that vitamin D could be protective. Possible benefits from zinc and selenium cannot be ruled out, but current available evidence is insufficient to recommend routine supplementation with these nutrients. Some other breast milk factors that are potentially amenable to modification through supplementation have been recently identified as modulators of the risk of MTCT of HIV. These include complex oligosaccharides that carry the Lewis antigen glycan [75], long chain $n-6$ polyunsaturated fatty acids [76], and erythropoietin [77]. The efficacy of these factors in preventing MTCT is still to be tested in randomized clinical trials. In the meanwhile, additional efforts are required to increase the availability of prophylactic antiretroviral regimens to prevent MTCT in the areas most hardly hit by the HIV epidemic. Whether micronutrient supplementation could offer additional benefits when administered together with antiretroviral prophylaxis or treatment is an important research question that is currently under evaluation in randomized clinical trials.

References

1. UNAIDS (2010) Global report: UNAIDS report on the global AIDS epidemic 2010
2. Becquet R, Ekouevi DK, Arrive E, Stringer JSA, Meda N, Chaix ML, Treluyer JM, Leroy V, Rouzioux C, Blanche S, Dabis F (2009) Universal antiretroviral therapy for pregnant and breast-feeding HIV-1-infected women: towards the elimination of mother-to-child transmission of HIV-1 in resource-limited settings. Clin Infect Dis 49(12):1936–1945
3. WHO PMCT Strategic Vision 2010–2015: preventing mother-to-child transmission of HIV to reach the UNGASS and millennium development goals. Moving towards the elimination of paediatric HIV. World Health Organizantion
4. WHO (2010) Guidelines on HIV and infant feeding 2010: principles and recommendations for infant feeding in the context of HIV and a summary of evidence. World Health Organization, Geneva
5. WHO (2009) Global prevalence of vitamin A deficiency in populations at risk 1995–2005. WHO global database on vitamin A deficiency. World Health Organization, Geneva
6. Skurnick JH, Bogden JD, Baker H, Kemp FW, Sheffet A, Quattrone G, Louria DB (1996) Micronutrient profiles in HIV-1-infected heterosexual adults. J Acquir Immune Defic Syndr Hum Retrovirol 12(1):75–83
7. Baum MK, Shor-Posner G, Zhang G, Lai H, Quesada JA, Campa A, Jose-Burbano M, Fletcher MA, Sauberlich H, Page JB (1997) HIV-1 infection in women is associated with severe nutritional deficiencies. J Acquir Immune Defic Syndr Hum Retrovirol 16(4):272–278
8. Kassu A, Andualem B, Van Nhien N, Nakamori M, Nishikawa T, Yamamoto S. Ota F (2007) Vitamin A deficiency in patients with diarrhea and HIV infection in Ethiopia. Asia Pac J Clin Nutr 16(suppl 1):323–328
9. Monteiro JP, Freimanis-Hance L, Faria LB, Mussi-Pinhata MM, Korelitz J, Vannucchi H, Queiroz W, Succi RC, Hazra R (2009) Both human immunodeficiency virus-infected and human immunodeficiency virus-exposed, uninfected children living in Brazil, Argentina, and Mexico have similar rates of low concentrations of retinol, beta-carotene, and vitamin E. Nutr Res 29(10):716–722
10. Beach RS, Mantero-Atienza E, Shor-Posner G, Javier JJ, Szapocznik J, Morgan R, Sauberlich HE, Cornwell PE, Eisdorfer C, Baum MK (1992) Specific nutrient abnormalities in asymptomatic HIV-1 infection. AIDS 6(7):701–708
11. Semba RD, Graham NM, Caiaffa WT, Margolick JB, Clement L, Vlahov D (1993) Increased mortality associated with vitamin A deficiency during human immunodeficiency virus type 1 infection. Arch Intern Med 153(18):2149–2154

12. Tang AM, Graham NM, Kirby AJ, McCall LD, Willett WC, Saah AJ (1993) Dietary micronutrient intake and risk of progression to acquired immunodeficiency syndrome (AIDS) in human immunodeficiency virus type 1 (HIV-1)-infected homosexual men. Am J Epidemiol 138(11):937–951
13. Coodley GO, Coodley MK, Nelson HD, Loveless MO (1993) Micronutrient concentrations in the HIV wasting syndrome. AIDS 7(12):1595–1600
14. Ruhl R (2007) Effects of dietary retinoids and carotenoids on immune development. Proc Nutr Soc 66(3):458–469
15. Webb AL, Aboud S, Furtado J, Murrin C, Campos H, Fawzi WW, Villamor E (2009) Effect of vitamin supplementation on breast milk concentrations of retinol, carotenoids and tocopherols in HIV-infected Tanzanian women. Eur J Clin Nutr 63(3):332–339
16. Semba RD (1997) Overview of the potential role of vitamin A in mother-to-child transmission of HIV-1. Acta Paediatr Suppl 421:107–112
17. Semba RD, Miotti PG, Chiphangwi JD, Saah AJ, Canner JK, Dallabetta GA, Hoover DR (1994) Maternal vitamin A deficiency and mother-to-child transmission of HIV-1. Lancet 343(8913):1593–1597
18. Greenberg BL, Semba RD, Vink PE, Farley JJ, Sivapalasingam M, Steketee RW, Thea DM, Schoenbaum EE (1997) Vitamin A deficiency and maternal-infant transmissions of HIV in two metropolitan areas in the United States. AIDS 11(3):325–332
19. Burger H, Kovacs A, Weiser B, Grimson R, Nachman S, Tropper P, vanBennekum AM, Elie MC, Blaner WS (1997) Maternal serum vitamin A levels are not associated with mother-to-child transmission of HIV-1 in the United States. J Acquir Immune Defic Syndr Hum Retrovirol 14(4):321–326
20. Burns DN, FitzGerald G, Semba R, Hershow R, Zorrilla C, Pitt J, Hammill H, Cooper ER, Fowler MG, Landesman S (1999) Vitamin A deficiency and other nutritional indices during pregnancy in human immunodeficiency virus infection: prevalence, clinical correlates, and outcome. Women and Infants Transmission Study Group. Clin Infect Dis 29(2):328–334
21. Coutsoudis A, Pillay K, Spooner E, Kuhn L, Coovadia HM (1999) Randomized trial testing the effect of vitamin A supplementation on pregnancy outcomes and early mother-to-child HIV-1 transmission in Durban, South Africa. South African Vitamin A Study Group. AIDS 13(12):1517–1524
22. Kumwenda N, Miotti PG, Taha TE, Broadhead R, Biggar RJ, Jackson JB, Melikian G, Semba RD (2002) Antenatal vitamin A supplementation increases birth weight and decreases anemia among infants born to human immunodeficiency virus-infected women in Malawi. Clin Infect Dis 35(5):618–624
23. Fawzi WW, Msamanga GI, Hunter D, Renjifo B, Antelman G, Bang H, Manji K, Kapiga S, Mwakagile D, Essex M, Spiegelman D (2002) Randomized trial of vitamin supplements in relation to transmission of HIV-1 through breastfeeding and early child mortality. AIDS 16(14):1935–1944
24. Humphrey JH, Iliff PJ, Marinda ET, Mutasa K, Moulton LH, Chidawanyika H, Ward BJ, Nathoo KJ, Malaba LC, Zijenah LS, Zvandasara P, Ntozini R, Mzengeza F, Mahomva AI, Ruff AJ, Mbizvo MT, Zunguza CD, Group ZS (2006) Effects of a single large dose of vitamin A, given during the postpartum period to HIV-positive women and their infants, on child HIV infection, HIV-free survival, and mortality. J Infect Dis 193(6):860–871
25. Koulinska IN, Villamor E, Chaplin B, Msamanga G, Fawzi W, Renjifo B, Essex M (2006) Transmission of cell-free and cell-associated HIV-1 through breast-feeding. J Acquir Immune Defic Syndr 41(1):93–99
26. Kantarci S, Koulinska IN, Aboud S, Fawzi WW, Villamor E (2007) Subclinical mastitis, cell-associated HIV-1 shedding in breast milk, and breast-feeding transmission of HIV-1. J Acquir Immune Defic Syndr 46:651–654
27. Semba RD, Kumwenda N, Hoover DR, Taha TE, Quinn TC, Mtimavalye L, Biggar RJ, Broadhead R, Miotti PG, Sokoll LJ, van der Hoeven L, Chiphangwi JD (1999) Human immunodeficiency virus load in breast milk, mastitis, and mother-to-child transmission of human immunodeficiency virus type 1. J Infect Dis 180(1):93–98
28. Nduati RW, John GC, Richardson BA, Overbaugh J, Welch M, Ndinyaachola J, Moses S, Holmes K, Onyango F, Kreiss JK (1995) Human immunodeficiency virus type 1 infected cells in breast milk association with immunosuppression and vitamin A deficiency. J Infect Dis 172(6):1461–1468
29. Villamor E, Koulinska IN, Aboud S, Murrin C, Bosch RJ, Manji KP, Fawzi WW (2010) Effect of vitamin supplements on HIV shedding in breast milk. Am J Clin Nutr 92(4):881–886
30. Omenn GS, Goodman GE, Thornquist MD, Balmes J, Cullen MR, Glass A, Keogh JP, Meyskens FL, Valanis B, Williams JH, Barnhart S, Hammar S (1996) Effects of a combination of beta carotene and vitamin A on lung cancer and cardiovascular disease. N Engl J Med 334(18):1150–1155
31. Russell RM (2004) The enigma of beta-carotene in carcinogenesis: what can be learned from animal studies. J Nutr 134(1):262S–268S
32. MacDonald KS, Malonza I, Chen DK, Nagelkerke NJ, Nasio JM, Ndinya-Achola J, Bwayo JJ, Sitar DS, Aoki FY, Plummer FA (2001) Vitamin A and risk of HIV-1 seroconversion among Kenyan men with genital ulcers. AIDS 15(5):635–639
33. Arsenault JE, Aboud S, Manji KP, Fawzi WW, Villamor E (2010) Vitamin supplementation increases risk of subclinical mastitis in HIV-infected women. J Nutr 140(10):1788–1792
34. Willumsen JF, Filteau SM, Coutsoudis A, Uebel KE, Newell ML, Tomkins AM (2000) Subclinical mastitis as a risk factor for mother-infant HIV transmission. In: Koletzko B, Michaelsen KF, Hernell O (eds) Short and long term

effects of breast feeding on child health. vol 478. Advances in experimental medicine and biology. Kluwer Academic/Plenum, New York, pp 211–223
35. Tang AM, Graham NM, Saah AJ (1996) Effects of micronutrient intake on survival in human immunodeficiency virus type 1 infection. Am J Epidemiol 143(12):1244–1256
36. Fawzi WW, Msamanga GI, Spiegelman D, Wei R, Kapiga S, Villamor E, Mwakagile D, Mugusi F, Hertzmark E, Essex M, Hunter DJ (2004) A randomized trial of multivitamin supplements and HIV disease progression and mortality. N Engl J Med 351(1):23–32
37. Webb AL, Villamor E (2007) Update: effects of antioxidant and non-antioxidant vitamin supplementation on immune function. Nutr Rev 65(5):181–217
38. Baum MK, Manteroatienza E, Shorposner G, Fletcher MA, Morgan R, Eisdorfer C, Sauberlich HE, Cornwell PE, Beach RS (1991) Association of Vitamin-B6 status with parameters of immune function in early HIV-1 infection. J Acquir Immune Defic Syndr Hum Retrovirol 4(11):1122–1132
39. Baum MK, Shor-Posner G, Lu Y, Rosner B, Sauberlich HE, Fletcher MA, Szapocznik J, Eisdorfer C, Buring JE, Hennekens CH (1995) Micronutrients and HIV-1 disease progression. AIDS 9(9):1051–1056
40. Wintergerst ES, Maggini S, Hornig DH (2007) Contribution of selected vitamins and trace elements to immune function. Ann Nutr Metab 51(4):301–323
41. Wang YJ, Huang DS, Liang BL, Watson RR (1994) Nutritional status and immune responses in mice with murine AIDS are normalized by vitamin E supplementation. J Nutr 124(10):2024–2032
42. Holick MF (2007) Vitamin D deficiency. N Engl J Med 357(3):266–281
43. Holick MF, Chen TC (2008) Vitamin D deficiency: a worldwide problem with health consequences. Am J Clin Nutr 87(4):1080S–1086S
44. Stephensen CB, Marquis GS, Kruzich LA, Douglas SD, Aldrovandi GM, Wilson CM (2006) Vitamin D status in adolescents and young adults with HIV infection. Am J Clin Nutr 83(5):1135–1141
45. Villamor E (2006) A potential role for vitamin D on HIV infection? Nutr Rev 64(5 Pt 1):226–233
46. Mehta S, Hunter DJ, Mugusi FM, Spiegelman D, Manji KP, Giovannucci EL, Hertzmark E, Msamanga GI, Fawzi WW (2009) Perinatal outcomes, including mother-to-child transmission of HIV, and child mortality and their association with maternal Vitamin D status in Tanzania. J Infect Dis 200(7):1022–1030
47. Provvedini DM, Tsoukas CD, Deftos LJ, Manolagas SC (1983) 1,25-dihydroxyvitamin D3 receptors in human-leukocytes. Science 221(4616):1181–1182
48. Bhalla AK, Amento EP, Clemens TL, Holick MF, Krane SM (1983) Specific high-affinity receptors for 1,25-dihydroxyvitamin-D3 in human peripheral blood mononuclear cells in presence in monocytes and induction in T lymphocytes following activation. J Clin Endocrinol Metab 57(6):1308–1310
49. Adams JS, Liu PT, Chun R, Modlin RL, Hewison M (2007) Vitamin D in defense of the human immune response. In: Zaidi M (ed) Skeletal biology and medicine, Pt B – disease mechanisms and therapeutic challenges, vol 1117. Annals of the New York academy of sciences. Blackwell, Oxford, pp 94–105
50. Yang SL, Smith C, Prahl JM, Luo XL, Deluca HF (1993) Vitamin-D deficiency suppresses cell-mediated-immunity in vivo. Arch Biochem Biophys 303(1):98–106
51. Veldman CM, Cantorna MT, DeLuca HF (2000) Expression of 1,25-dihydroxyvitamin D-3 receptor in the immune system. Arch Biochem Biophys 374(2):334–338
52. Prasad AS (2009) Impact of the discovery of human zinc deficiency on health. J Am Coll Nutr 28(3):257–265
53. Kassu A, Yabutani T, Mulu A, Tessema B, Ota F (2008) Serum zinc, copper, selenium, calcium, and magnesium levels in pregnant and non-pregnant women in Gondar, Northwest Ethiopia. Biol Trace Elem Res 122(2):97–106
54. Bhutta ZA, Bird SM, Black RE, Brown KH, Gardner JM, Hidayat A, Khatun F, Martorell R, Ninh NX, Penny ME, Rosado JL, Roy SK, Ruel M, Sazawal S, Shankar A (2000) Therapeutic effects of oral zinc in acute and persistent diarrhea in children in developing countries: pooled analysis of randomized controlled trials. Am J Clin Nutr 72(6):1516–1522
55. Bhutta ZA, Black RE, Brown KH, Gardner JM, Gore S, Hidayat A, Khatun F, Martorell R, Ninh NX, Penny ME, Rosado JL, Roy SK, Ruel M, Sazawal S, Shankar A (1999) Prevention of diarrhea and pneumonia by zinc supplementation in children in developing countries: pooled analysis of randomized controlled trials. Zinc Investigators' Collaborative Group. J Pediatr 135(6):689–697
56. Beck FW, Prasad AS, Kaplan J, Fitzgerald JT, Brewer GJ (1997) Changes in cytokine production and T cell subpopulations in experimentally induced zinc-deficient humans. Am J Physiol 272(6 Pt 1):E1002–E1007
57. Baum MK, Shor-Posner G, Campa A (2000) Zinc status in human immunodeficiency virus infection. J Nutr 130(5S suppl):1421S–1423S
58. Hoque KM, Sarker R, Guggino SE, Tse CM (2009) A new insight into pathophysiological mechanisms of zinc in diarrhea. Ann N Y Acad Sci 1165:279–284
59. Evans P, Halliwell B (2001) Micronutrients: oxidant/antioxidant status. Br J Nutr 85(suppl 2):S67–S74
60. Muriaux D, Darlix JL (2010) Properties and functions of the nucleocapsid protein in virus assembly. RNA Biol 7(6):744–753

61. Graham NM, Sorensen D, Odaka N, Brookmeyer R, Chan D, Willett WC, Morris JS, Saah AJ (1991) Relationship of serum copper and zinc levels to HIV-1 seropositivity and progression to AIDS. J Acquir Immune Defic Syndr 4(10):976–980
62. Lai H, Lai S, Shor-Posner G, Ma F, Trapido E, Baum MK (2001) Plasma zinc, copper, copper:zinc ratio, and survival in a cohort of HIV-1-infected homosexual men. J Acquir Immune Defic Syndr 27(1):56–62
63. Baum MK, Campa A, Lai S, Lai H, Page JB (2003) Zinc status in human immunodeficiency virus type 1 infection and illicit drug use. Clin Infect Dis 37(suppl 2):S117–S123
64. Mocchegiani E, Veccia S, Ancarani F, Scalise G, Fabris N (1995) Benefit of oral zinc supplementation as an adjunct to zidovudine (AZT) therapy against opportunistic infections in AIDS. Int J Immunopharmacol 17(9):719–727
65. Baum MK, Lai S, Sales S, Page JB, Campa A (2010) Randomized, controlled clinical trial of zinc supplementation to prevent immunological failure in HIV-infected adults. Clin Infect Dis 50(12):1653–1660
66. Bobat R, Coovadia H, Stephen C, Naidoo KL, McKerrow N, Black RE, Moss WJ (2005) Safety and efficacy of zinc supplementation for children with HIV-1 infection in South Africa: a randomised double-blind placebo-controlled trial. Lancet 366(9500):1862–1867
67. Villamor E, Aboud S, Koulinska IN, Kupka R, Urassa W, Chaplin B, Msamanga G, Fawzi WW (2006) Zinc supplementation to HIV-1-infected pregnant women: effects on maternal anthropometry, viral load, and early mother-to-child transmission. Eur J Clin Nutr 60(7):862–869
68. Baum MK, Shor-Posner G, Lai S, Zhang G, Lai H, Fletcher MA, Sauberlich H, Page JB (1997) High risk of HIV-related mortality is associated with selenium deficiency. J Acquir Immune Defic Syndr Hum Retrovirol 15(5): 370–374
69. Kupka R, Msamanga GI, Spiegelman D, Morris S, Mugusi F, Hunter DJ, Fawzi WW (2004) Selenium status is associated with accelerated HIV disease progression among HIV-1-infected pregnant women in Tanzania. J Nutr 134(10):2556–2560
70. Kupka R, Garland M, Msamanga G, Spiegelman D, Hunter D, Fawzi W (2005) Selenium status, pregnancy outcomes, and mother-to-child transmission of HIV-1. J Acquir Immune Defic Syndr 39(2):203–210
71. Campa A, Shor-Posner G, Indacochea F, Zhang G, Lai H, Asthana D, Scott GB, Baum MK (1999) Mortality risk in selenium-deficient HIV-positive children. J Acquir Immune Defic Syndr Hum Retrovirol 20(5):508–513
72. Baeten JM, Mostad SB, Hughes MP, Overbaugh J, Bankson DD, Mandaliya K, Ndinya-Achola JO, Bwayo JJ, Kreiss JK (2001) Selenium deficiency is associated with shedding of HIV-1–infected cells in the female genital tract. J Acquir Immune Defic Syndr 26(4):360–364
73. Kupka R, Mugusi F, Aboud S, Msamanga GI, Finkelstein JL, Spiegelman D, Fawzi WW (2008) Randomized, double-blind, placebo-controlled trial of selenium supplements among HIV-infected pregnant women in Tanzania: effects on maternal and child outcomes. Am J Clin Nutr 87(6):1802–1808
74. McClelland RS, Baeten JM, Overbaugh J, Richardson BA, Mandaliya K, Emery S, Lavreys L, Ndinya-Achola JO, Bankson DD, Bwayo JJ, Kreiss JK (2004) Micronutrient supplementation increases genital tract shedding of HIV-1 in women: results of a randomized trial. J Acquir Immune Defic Syndr 37(5):1657–1663
75. Hong P, Ninonuevo MR, Lee B, Lebrilla C, Bode L (2008) Human milk oligosaccharides reduce HIV-1-gp120 binding to dendritic cell-specific ICAM3-grabbing non-integrin (DC-SIGN). Br J Nutr 101:474
76. Villamor E, Koulinska IN, Furtado J, Baylin A, Aboud S, Manji K, Campos H, Fawzi WW (2007) Long-chain n-6 polyunsaturated fatty acids in breast milk decrease the risk of HIV transmission through breastfeeding. Am J Clin Nutr 86(3):682–689
77. Arsenault JE, Webb AL, Koulinska IN, Aboud S, Fawzi WW, Villamor E (2010) Association between breast milk erythropoietin and reduced risk of mother-to-child transmission of HIV. J Infect Dis 202:370–373

Part V
Research Implementation and Policy Related to Breastfeeding by HIV-1-Infected Mothers

Chapter 16
Historical Perspective of African-Based Research on HIV-1 Transmission Through Breastfeeding: The Malawi Experience

Taha E. Taha

Introduction

Transmission of HIV-1 from the mother to the infant postnatally through breastfeeding remains an important concern in sub-Saharan Africa where breastfeeding is widely practiced. African women continue to breastfeed despite the risk of transmitting HIV to their infants for several reasons [1]: (a) breastfeeding is encouraged by family members and has been culturally adopted by generations of postpartum women; (b) not breastfeeding raises suspicion in the community about the HIV status of the woman and could potentially lead to discrimination; (c) breastfeeding is the most important nutritional source for the growing child; (d) breastfeeding is readily available and convenient for the mother to provide the infant whenever needed; and (e) in several African settings, substitutes of breast milk are either expensive or not safe to use due to lack of safe water to prepare these substitutes and vehicles for feeding the infant can easily become contaminated. Additionally, strong global evidence exists showing that breastfeeding protects against diarrheal and upper respiratory diseases of the infant [2–4]. Biologically, breast milk is known to contain several well-documented protective factors [5–7]. In a pooled analysis of data from multiple countries, the protective effects of breastfeeding were greatest during early infancy and declined with increasing age: the risk of death associated with infectious diseases among infants not breastfed compared to breastfed was 5.8 times higher during the first month, 4.1 times higher during 2–3 months, 2.6 times higher during 4–5 months, 1.8 times higher during 6–8 months, and 1.4 times higher during 9–11 months of age [8].

The 2010 report of the World Health Organization [9] showed encouraging trends of declining numbers of children born with HIV. Nonetheless, approximately 370,000 (range: 230,000–510,000) children were newly infected with HIV in 2009. Most of these infections occurred in Africa where HIV prevalence among women of reproductive age remains high (the UNAIDS report estimated that 1.8 million adults and children were newly infected with HIV and 1.3 million AIDS-related deaths occurred among adults and children in 2009). In some countries such as South Africa, even the rates of HIV acquisition (incidence) remain high ($\cong$9 per 100 person-years [pys] in S. Africa and $\cong$4 per 100 pys in Malawi [10, 11]). The recent successes of declines in mother-to-child transmission (MTCT) of HIV are the result of combination of factors including spread of antiretroviral coverage primarily for prophylaxis and also for treatment. This achievement, however, conceals some major

T.E. Taha, M.B.B.S., M.C.M., M.P.H., Ph.D. (✉)
Department of Epidemiology, Bloomberg School of Public Health,
Johns Hopkins University, Room E7138, 615 N. Wolfe Street, Baltimore, MD 21205, USA
e-mail: ttaha@jhsph.edu

historical attempts started in the early days of the HIV epidemic and through the mid-nineties to understand the magnitude of the HIV epidemic among women and children and to initiate preventive activities.

In this chapter, we provide a historical account of the research experience of a single African research site that systematically studied MTCT of HIV. This research site, the Johns Hopkins Research Project in Blantyre, Malawi, has been conducting research on MTCT of HIV since 1989 to this date to document rates of transmission of HIV, investigating the various risk factors associated with MTCT of HIV and conducting interventions to reduce transmission. Evaluation of this approximately 20 years of continuous research on MTCT of HIV is a unique opportunity to link the past and the present and explore future directions. This is an illustrative example of several on-going research collaborations at multiple sites in Africa, including Malawi (Lilongwe), South Africa, Kenya, Uganda, Tanzania, Zimbabwe, Rwanda, Ivory Coast, Ethiopia as well as other sites. These research sites provided important data that contributed to better understanding of perinatal transmission of HIV, the role of breastfeeding, and the effect of innovative interventions to decrease MTCT of HIV in sub-Saharan Africa.

The Johns Hopkins University Research Project, in collaboration with local institutions and communities in Blantyre, Malawi, started research on MTCT of HIV in 1989. The goal of this research project was to investigate and understand the magnitude of MTCT of HIV-1 in a breastfeeding community. In this report, we review the rates of MTCT of HIV since the early days of HIV research in Africa, examine risk factors associated with postnatal HIV transmission, and comment on the interventions that have been developed and implemented to decrease transmission of HIV from mother to infant primarily through breastfeeding. These research endeavors progressed over the years from observational studies to complex interventions and are still continuing. Review of these historical events will assist in documenting the progresses that have been made and their policy implications.

Malawi and the Johns Hopkins Research Project in Blantyre

Malawi, a southern African country with a population of approximately 12 million, is bordered by Mozambique, Tanzania, and Zambia. This landlocked country has 20% of its surface area covered by Lake Malawi. The capital is Lilongwe in the Central Region. Blantyre in the southern region of Malawi is a large commercial center and is the site of the Johns Hopkins Research Project at the Queen Elizabeth Central Hospital (QECH; Fig. 16.1). About 80% of the population of Malawi lives in rural areas engaged in subsistence farming. According to the Malawi Demographic and Health Survey, the infant and under-five child mortality rates (per 1,000 live birth) have been substantially declining during the past 20 years. For example, infant mortality rates per 1,000 live births were 135 in 1992, 104 in 2000, and 66 in 2010; and under-five child mortality rates per 1,000 live birth were 234 in 1992, 189 in 2000, and 112 in 2010 [12].

AIDS was first detected in Malawi in 1985 and currently the national HIV prevalence is estimated at 14% in adults (15–49 years old), with 900,000 adults and children living with HIV/AIDS [13]. The predominant virus is HIV-1, more than 95% subtype C. Figures 16.2 and 16.3 show HIV-1 prevalence and incidence among women of reproductive age in Malawi. The most recent prevalence estimate obtained among antenatal women attending QECH in Blantyre during the period 2004–2006 is 22% [14]. These data suggest that HIV prevalence has leveled off, and probably showing a trend of decline. However, based on longitudinal hospital-based research studies conducted by the Johns Hopkins University Research Project, HIV incidence during the period 1990–2009 has remained stable among women of reproductive age (see Fig. 16.3; an additional recent estimate of HIV incidence obtained in 2008 among women of reproductive age is 4.0 [95% CI 2.3–9.4] per 100 woman-years [11]). This has important implications regarding the potential for MTCT of HIV in Malawi.

Fig. 16.1 The Johns Hopkins Research Project clinical research site in Blantyre, Malawi

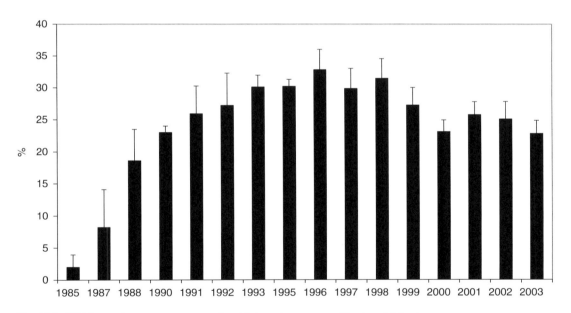

Fig. 16.2 HIV-1 prevalence among antenatal and intrapartum women, Blantyre, Malawi

An important strength of the Johns Hopkins Research Project is its collaboration with local partners. This has allowed access and enrollment of large sample sizes from a single research site for both observational and clinical trial studies. The findings from these studies are likely to be representative of the urban and suburban populations of Malawi. Through extensive, long-term research support, including clinical, laboratory, pharmacy, and community programs, the site has been able to build infrastructure, train investigators and other technical staff, and locally create an environment

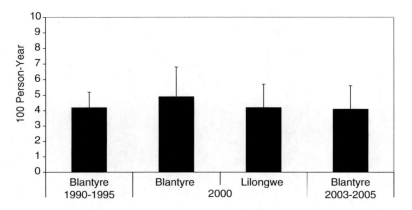

Fig. 16.3 HIV-1 incidence among antenatal and postnatal women, Blantyre and Lilongwe, Malawi

Table 16.1 Mother-to-child transmission of HIV studies conducted at the Johns Hopkins Research Project in Blantyre, Malawi, 1989–2009

Study name and objective	Year	# Screened [# enrolled]	References	Funding source
ICAR: Determine the HIV prevalence, incidence, rates of MTCT and risk factors	1989–1992	6,603 (1,366 mothers)	[16, 24]	NIH/NIAID
PAVE: Study rates and risk factors associated with MTCT	1993–1994	2,471 (1,100 mothers)	[15, 16, 24, 25, 27, 36]	NIH/NIAID
WASH: Effect of washing the birth canal to prevent MTCT of HIV and other infections	1994–1995	6,964 (6,964 women)	[31, 32, 60–63]	NIH/NCI
MCH: Determine the effect of prenatal vitamin A supplementation on MTCT of HIV and birth outcomes	1995–1998	3,949 (985 women)	[33, 36, 64–70]	NIH/NICHD
PAVE2 (HIVNET): Study of HIV transmission through breastfeeding	1995–1997	853 (853 babies)	[35, 71, 72]	NIH/NIAID/HIVNET
Study the epidemiology and microbiology of mastitis and its association with MTCT	1999–2002	1,000 (800 mothers)	[37, 38, 73, 74]	NIH/NICHD
HPTN 024: Determine the efficacy of chorioamnionitis antibiotics treatment in reducing MTCT of HIV and birth outcomes	1998–2002	3,100 (500 women)	[40, 75–89]	NIH/NIAID/HPTN
NVAZ: Determine the effect of short antiviral regimen prophylaxis of AZT and NVP on MTCT of HIV	2000–2003	2,650 (2,650 women)	[41, 42, 50–52, 66, 90–100]	NIH/Fogarty/Doris Duke Charitable Foundation
PEPI: Extended infant antiretroviral prophylaxis to reduce breast milk transmission of HIV	2004–2009	34,000 (3,300 mother–infant pairs)	[44, 48, 49, 101]	CDC

conducive to conduct of research. These research activities provided an opportunity to collaborate with investigators from other institutions in the USA and regionally within Africa.

The MTCT of HIV studies conducted in Blantyre, Malawi by the Johns Hopkins University Research Project and its collaborators are summarized in Table 16.1. These studies were approved and monitored by appropriate institutional review boards in the country and in the USA. In addition to

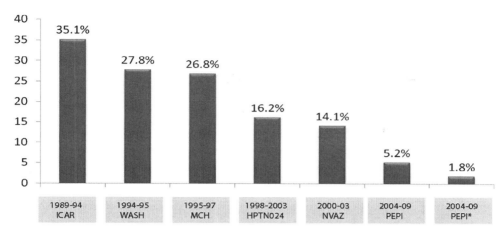

Fig. 16.4 Rates of mother-to-child transmission of HIV, Blantyre, Malawi, 1990–2009. ICAR: International Collaborative AIDS Research [15], WASH: Birth Canal Washing study [31], MCH: Vitamin A Antenatal Supplementation Study [36], HPTN024: Antibiotic treatment of Chorioamnionitis Study [40], NVAZ: Postexposure Prophylaxis with Nevirapine and Zidovudine Study [42], PEPI: Postexposure Prophylaxis of Infant Study [48], PEPI*: Postexposure Prophylaxis of Infant substudy—sub-study [49]

independent international monitors, the clinical trials conducted at the site were also monitored by appropriate Data and Safety Monitoring Committees. All investigators, clinicians, coordinators, managers, and nursing/laboratory staff have been trained and certified in Human Subjects Protection (HSP), and Good Clinical Practice/Good Laboratory Practice (GCP/GLP). The research site has an active community education program to keep the community well informed of on-going research and to solicit input in research activities. The research project provides routine clinical care including treatment for sexually transmitted infections, anemia, malaria and antiretroviral prophylaxis and treatment. Referral for specialized clinical care is also provided by project clinicians and nurses.

Figure 16.4 summarizes the trends of MTCT of HIV during the period 1989–2009 based on research studies conducted by the Johns Hopkins University Research Project in Blantyre, Malawi. As shown in this figure, the rates of MTCT of HIV substantially declined from approximately 35% in 1990 to low rates in recent years. These studies are summarized below to discuss the main objectives, study design, and main findings. These experiences are discussed under four time periods signifying some important progresses in MTCT of HIV research in Blantyre, Malawi: (1) *1989–1994*; (2) *1994–1999*; (3) *2000–2003*; and (4) *2004–2009*.

1989–1994: The Early Days of the HIV Epidemic in Malawi

Two major observational studies, the International Collaborative AIDS Research [ICAR] and Preparation for AIDS Vaccine Evaluation [PAVE] studies were initiated in Blantyre, Malawi in 1989 and 1993, respectively. These two studies were conventional cohort studies and in conjunction with other similar studies in selected countries in Africa provided most of the knowledge and data about transmission of HIV through breastfeeding in the African setting. Both ICAR and PAVE screened women for HIV infection during pregnancy and enrolled HIV positive and a sample of HIV-negative women at the time of delivery. Mothers and their children were followed postnatally for 2–3 years. These longitudinal studies provided the best estimates of rate of MTCT of HIV and allowed for investigation of risk factors associated with transmission [15–17]. Additionally, these studies assessed the impact of HIV on the health and survival of children and their mothers [18]. Despite their rigor in following mothers and infants, however, these studies suffered a major technological limitation: laboratory testing to confirm infant HIV infection using PCR (as commonly practiced today at the African sites) was not yet available and the investigators were entirely dependent on serology to document infant HIV infection. This approach requires following children to more than 12 months of age for maternal antibodies to disappear before performing HIV testing using serological tests (ELISA and Western blot). During this critical period of follow-up, infants may die or be lost to follow-up. In settings where infant/child mortality is high as in sub-Saharan Africa, this was a major challenge.

Measuring rates of transmission at or after 12 months also does not allow for estimation of perinatal HIV transmission. Due to this limitation, a team of investigators (the Ghent Working Group) developed a method to estimate the overall MTCT of HIV in these settings, taking into account the proportion of children HIV seropositive and survival probabilities at 12 months [19]. Based on this method, the overall rate of MTCT of HIV in the ICAR study in Blantyre, Malawi was estimated as 35.1% [15]; Fig. 16.4. At this time, it was not possible to determine how much of this transmission was in utero, intrapartum or postnatally through breastfeeding. In subsequent years, it became apparent that this estimate was consistent with other estimates from sub-Saharan Africa which showed that transmission could be as high as 45% in breastfeeding populations [20]. It was not until a randomized breastfeeding versus formula feeding clinical trial was conducted in Nairobi, Kenya (1992–1998), and the results became available in 2000, that it was confirmed that 44% of the MTCT of HIV is attributed to breastfeeding [21].

The ICAR and PAVE studies in Blantyre, Malawi, showed additional findings on the association of some important risk factors with MTCT of HIV. At this time, not even the ACTG 076 results were available from the USA and other developed countries showing that prophylaxis of the mother and infant with zidovudine (AZT) can reduce perinatal transmission by 76% [22]. Therefore, the idea of any antiretroviral-based intervention was some years away. Data from the ICAR study in Malawi showed that women with low levels of serum vitamin A were more likely to transmit HIV to their infants [23]. Data from the US also suggested that most of the MTCT of HIV occurs late during pregnancy and during the delivery process. Data from both ICAR and PAVE also showed strong associations with sexually transmitted infections (STIs) and HIV infection—both transmission and acquisition [16]. The importance of other genital tract infections such as bacterial vaginosis (BV) was also observed in these observational studies from Malawi [24–26].

In the early and mid-nineties when the adverse social, clinical, and economic impact of HIV among adults started to mount and it became clear that HIV infection is reversing some of the earlier major gains achieved in child survival in the African setting, investigators and policy makers started to seriously consider potential interventions to prevent HIV transmission. This debate benefited from the results of two important interventions (one from Africa and the other from the USA): the Mwanza study results from Tanzania showing that a community-based treatment of STIs can reduce heterosexual transmission [27] and an antiretroviral (ARV)-based intervention in the USA showing perinatal transmission can be reduced by the use of AZT prophylaxis to mother and infant [22]. The team of investigators working in Malawi, guided by available data locally and internationally, considered several interventions.

1994–1999: The Nonantiretroviral Interventions Era to Reduce MTCT of HIV in Malawi

The PACTG 076 regimen was determined to be too expensive and complex to implement in the African setting. Therefore, interventions (if successful in reducing MTCT of HIV) that can be easily implemented in resource-constrained settings and that can be sustained were a priority. The investigators in Blantyre, Malawi, initiated three innovative interventions. The main features of these interventions were their simplicity and low cost—their efficacy was yet to be determined. The interventions implemented in Blantyre to reduce MTCT were (1) cleaning the birth canal with an antiseptic solution (the WASH study); (2) supplementing antenatal women with vitamin A (MCH study); and (3) antibiotic treatment of chorioamionitis (HPTN 024 study). At this time, none of the interventions conducted in Africa, with the exception of the trial conducted in Kenya that later determined the rate of HIV transmission associated with breastfeeding, specifically targeted the breastfeeding mode of MTCT of HIV. Women continued to breastfeed and the WHO recommendations regarding exclusive breastfeeding were not issued until 2000 [28]. Table 16.2 and Fig. 16.6 show the frequency of breastfeeding among Malawian women who participated in studies conducted at the Johns Hopkins Research Project site in Blantyre, Malawi, 1989–2009. During the period 1989–1997, >90% of the woman still breastfed when the infant was 1-year old.

The WASH Study (1994–1995): Cleansing the Birth Canal with an Antiseptic Solution

In the early nineties, several reports confirmed that >50% of the perinatal transmission of HIV occurs around the time of delivery. Other studies also suggested that transmission of HIV among twins could vary by birth order—the first twin being more likely to be HIV infected than the second [29, 30]. Other studies targeting perinatal infections with Group B Streptococcus also suggested that cleansing the birth canal with chlorhexidine, a broad spectrum antiseptic, could reduce the frequency of several pathogens. These findings provided adequate biological and safety data to consider a clinical trial using a chlorhexidine vaginal wash to assess the efficacy of this simple intervention in reducing transmission of HIV and other perinatal infections. A secondary outcome of this trial was to assess the efficacy of chlorhexinde wash in reducing other perinatal infections associated with maternal and infant sepsis. A large clinical trial was conducted in a single hospital in Blantyre (the QECH) among approximately 7,000 mother/infant pairs. The HIV infection status of infants of 3,327 control women (conventional delivery procedures with no vaginal wash) was compared with that of 3,637 infants of intervention-delivered women (received vaginal wash). The infants' HIV status was determined by PCR conducted on samples collected at 6 and 12 weeks of age. The intervention consisted of manual cleansing of the birth canal with a cotton pad soaked in 0.25% chlorhexidine performed on admission in labor and every 4 h until delivery. The newborn was also cleansed (wiped) with chlorhexidine. No adverse reactions to the intervention procedure were seen. Overall, 30% of the women enrolled in this study were HIV infected. Among 982 vaginal vertex singleton deliveries to HIV-infected women, 269 (27%) infants were infected. The intervention had no significant impact on HIV transmission rates (27% in the intervention arm and 28% in the control arm; Fig. 16.4). However, there were statistically significant reductions in HIV transmission when membranes were ruptured more than 4 h before delivery (transmission of 25% in the intervention arm compared to that of 39% in the control arm, $p=0.02$) [31]. Despite its simplicity and low cost (chlorhexidine wash costs $\cong 5$ cents), this intervention was not effective in reducing perinatal transmission of HIV.

Although cleansing the birth canal with an antiseptic did not reduce MTCT of HIV, further analyses of the WASH study data showed substantial benefits to the infant and the mother. The WASH study trial was designed to allow for 2 months of no intervention, followed by 3 months of the intervention, and finally 1 month of no intervention. This design allowed comparison of several outcomes before and after the intervention during the 6 months of the study. Among infants born during the intervention phase of the trial, there were statistically significant reductions in the overall rate of neonatal admissions 16.9% versus 19.3% (634/3,743 vs. 661/3,417, $p<0.01$), admissions for neonatal sepsis (7.8 vs. 17.9 per 1,000 live births, $p<0.0002$), overall neonatal mortality (28.6 vs. 36.9 per 1,000 live births, $p<0.06$), and mortality due to infectious causes (2.4 vs. 7.3 per 1,000 live births, $p<0.005$). Among mothers receiving the intervention, there were significant reductions in rates of admissions related to postpartum problems (29.4 vs. 40.2 per 1,000 deliveries, $p<0.02$), rates of admissions due to postpartum sepsis

(1.7 vs. 5.1 per 1,000 deliveries, $p=0.02$), and duration of hospitalization ($p=0.008$) [32]. These data showed cleansing of the birth canal with chlorhexidine can reduce early neonatal and maternal postpartum infectious problems. Due to its safety, simplicity, and low cost, the procedure was suggested to be implemented as standard care in these settings to lower infant and maternal morbidity and mortality.

The MCH Study (1995–1997): Vitamin A Antenatal Supplementation Study

An important objective of the ICAR cohort study conducted in the late eighties and early nineties was to determine risk factors for transmission of HIV from mother to her child. As part of ICAR we assessed serum vitamin A in 338 HIV-infected women. MTCT of HIV (based on HIV serological tests) was 21.9% among mothers whose infants survived to 12 months of age. Mean vitamin A concentration (μmol/L) in 74 mothers who transmitted HIV to their infants was lower than that in 264 mothers who did not transmit HIV to their infants (0.86 vs. 1.07, $p<0.0001$). There was a clear gradient of increased MTCT when HIV positive mothers were divided into four groups: those with vitamin A concentrations of less than 0.70, between 0.70 and 1.05, between 1.05 and 1.40, and greater than or equal to 1.40. The MTCT rates for each group were 32.4%, 26.2%, 16.0%, and 7.2%, respectively ($p<0.0001$). Maternal CD4 cell count, CD4%, and CD4/CD8 ratio were also associated with increased MTCT of HIV. Based on the results of this study that suggested maternal vitamin A deficiency is strongly associated with MTCT of HIV, a phase three trial of vitamin A antenatal supplementation was started in Malawi in 1995. In the MCH trial, 697 women were enrolled and received iron and folate during pregnancy. Half of the enrolled women were randomly assigned antenatally to receive daily 10,000 IU of vitamin A. Despite improvements in pregnancy outcomes, especially higher infant birth weight, the intervention did not reduce MTCT of HIV; the rate of MTCT of HIV was $\cong$27% in both study arms [23, 33]; Fig. 16.4.

PAVE2-HIVNET Study (1995–1997): HIV Transmission Through Breastfeeding

In the mid-nineties, as better diagnostic technology (primarily PCR) became available in the African setting, studies to more precisely estimate transmission associated with breastfeeding became possible. In an observational study, a subgroup of HIV uninfected infants from the WASH study were followed longitudinally to determine rate of breastfeeding HIV transmission of HIV. A total of 672 infants, HIV-negative at birth, were enrolled (all women did not receive antiretroviral drugs during or after pregnancy). Forty-seven children became HIV infected while breastfeeding and none after breastfeeding had stopped. The cumulative infection rate while breastfeeding, from month 1 to the end of months 5, 11, 17, and 23, was 3.5%, 7.0%, 8.9%, and 10.3%, respectively. Incidence per month was 0.7% during age 1–5 months, 0.6% during age 6–11 months, and 0.3% during age 12–17 months ($p=0.01$ for trend). These data suggested that the risk of MTCT of HIV through breastfeeding is highest in the early months of breastfeeding. This trend was later observed in other studies conducted in Kenya and S. Africa [21, 34, 35].

The Epidemiology and Microbiology of Mastitis and Its Association with MTCT (1999–2002)

A risk factor that consistently showed importance in the MTCT of HIV studies in Malawi was inflammation of breast tissue (mastitis) [36]. In these studies, mastitis was assessed by elevated breast milk sodium concentration. Subclinical mastitis is associated with increased breast milk viral load. In an observational study of postpartum women, antibiotic treatment (oral amoxicillin/clavulanic acid) for subclinical mastitis was assessed. Seventy-five HIV-infected women were followed from 1 to 24 weeks. Breast milk RNA load and sodium concentration were measured and additionally, microbiological assessments were performed. Antibiotic treatment was associated with statistically significant decrease in subclinical mastitis. However, breast milk HIV load remained elevated suggesting that resumption of breastfeeding following treatment of inflammation of the breast may still increase transmission of HIV from the mother to the infant [37]. To assess microbiological factors associated with subclinical mastitis (assessed by breast milk leucocyte counts), 250 HIV-infected women were followed from 2 weeks to 12 months postpartum. Overall, 27% of the women had at least one episode of subclinical mastitis. *Staphylococcus aureus* was isolated in 30% of women with subclinical mastitis. This study showed that subclinical mastitis is prevalent among breastfeeding women [38].

2000–2003: The NVP Era—Short Postexposure Prophylaxis Trials in Malawi

In 1999 the results of the landmark study, HIVNET 012, were published. HIVNET 012 was a phase 2 clinical trial conducted in Uganda. In this trial, oral single-dose nevirapine (NVP), a NNRT inhibitor, was given to the mother at onset of labor (200 mg tablet) and to the infant within 72 h of birth (2 mg per kg syrup). This short course of prophylaxis reduced transmission of HIV from mother to infant by 47% [39]. The results of the HIVNET 012 were quickly adopted as the standard of care by researchers involved in PMTCT of HIV. This was also the decision taken by the Malawi investigators when considering the conduct of a trial that straddled this transitional period—the HPTN 024 study.

The HPTN 024 Study (1998–2003): Efficacy of Antibiotics to Reduce MTCT of HIV

Another nonantiretroviral intervention study conducted in Blantyre, Malawi, was the HPTN 024. This trial was based on the premise that several organisms associated with genital tract infections (especially the group of organisms associated with bacterial vaginosis) were associated with inflammation of the placenta and the fetal membranes. It was hypothesized that these inflammatory processes could be associated with transmission of HIV during pregnancy and intrapartum. Therefore, antibiotic treatment was expected to prevent these infections and reduce MTCT of HIV. HIV-infected women were randomized at 20–24 weeks gestation to receive either antibiotics (metronidazole plus erythromycin antenatally and metronidazole plus ampicillin intrapartum) or placebo. Maternal study procedures were performed at 20–24, 26–30, and 36 weeks antenatally, and at labor/delivery. All women and infants also received the HIVNET 012 prophylactic regimen. The primary efficacy endpoint was overall infant HIV infection. There were no differences in the proportion of infants HIV infected at birth (7.1% with intervention and 8.3% with placebo, $p=0.41$). Likewise, there were no statistically significant differences at 4–6 weeks in overall risk of MTCT of HIV (16.2% in antibiotic arm and 15.8% in placebo arm, $p=0.89$; Fig. 16.4) or HIV-1-free survival (79.4% in each study arm at 4–6 weeks). Postrandomization, the proportion of women with bacterial vaginosis was significantly lower in the antibiotic arm compared to placebo arm (23.8% vs. 39.7%; $p<0.001$), but the there were no differences in the frequency of histological chorioamnionitis (36.9% in antibiotic arm vs. 39.7% in placebo arm; $p=0.30$). This simple antibiotic regimen antenatally and intrapartum did not reduce the risk of MTCT of HIV [40]. The reduction in rates of MTCT from earlier studies was most likely due to the addition of NVP prophylaxis as in HIVNET 012.

The NVAZ Studies (2000–2003): Short Postexposure Prophylaxis of Newborns to Reduce MTCT of HIV

Two simultaneously randomized clinical trials were conducted in Blantyre, Malawi, during the period 2000–2003. These were known as the NVAZ studies, an acronym of the two drugs that have been used in these trials: *NVP* and *AZT* (zidovudine or ZDV). In NVAZ, one trial was conducted among women who presented late for delivery ["*late presenters*"] at the QECH with unknown HIV status and another trial was conducted simultaneously among women who presented early for delivery ["*early presenters*"] at QECH. For the late presenters, it was not possible to pretest counsel, perform HIV test, and post-test counsel these women before delivery; these were women in advance labor and anticipated to deliver within 2–4 h from time of admission. Therefore to establish their HIV status these women were counseled, consented, and tested for HIV after delivery. HIV testing was done on cord blood collected at time of delivery. For early presenters, there was adequate time to counsel and test for HIV prior to delivery (anticipated to deliver after 4 h or more from time of admission). A major difference between late and early presenters was that late presenting mothers did not receive the single-dose oral NVP at onset of labor (the intrapartum component of the HIVNET 012 regimen) because their HIV status was not known prior to delivery. The early presenting women did receive intrapartum NVP because their HIV status was established prior to delivery. We describe below the two NVAZ interventions.

NVAZ Late Presenters: NVP/ZDV Only to the Infant: Will Postexposure Prophylaxis Work?

In this trial, 1,119 infants born to late presenting women were randomly assigned to receive either NVP alone or NVP plus ZDV (see Fig. 16.5 for study design). Both drugs were given immediately after birth: NVP single-dose (2 mg/kg weight) and ZDV dosing twice daily for 1 week (4 mg/kg weight). Infant HIV infection was determined using PCR at birth and at 6–8 weeks. Primary outcome was HIV infection in infants at 6–8 weeks in those not infected at birth. The overall rate of MTCT of HIV at 6–8 weeks was 15.3% in 484 infants who received NVP/ZDV and 20.9% in 468 infants who received NVP only ($p=0.03$). At 6–8 weeks, in infants who were HIV negative at birth, 34 (7.7%) infants who had NVP/ZDV and 51 (12.1%) who received NVP only were infected ($p=0.03$) — a protective efficacy of 36%. These results remained significant after controlling for maternal viral load and other baseline factors [41]. This trial showed that postexposure prophylaxis of the infant alone using NVP single-dose and 1 week of ZDV can offer protection against HIV infection to infants where pregnant women missed opportunities to be counseled during pregnancy and present to the health facilities for delivery with unknown HIV infection status. The findings of this study assisted WHO and others to formulate guidelines for prevention of MTCT of HIV in Africa and elsewhere.

NVAZ Early Presenters — Adding ZDV to the HIVNET 012 Regimen: Will It Increase Efficacy?

The main objective of the NVAZ trial conducted in early presenting women was to determine the rate of MTCT of HIV when either the HIVNET 012 NVP to infant and mother regimen was given or

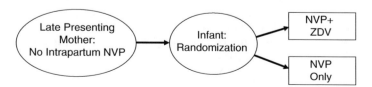

Fig. 16.5 NVAZ study late presenters study design randomized, open label phase III clinical

the HIVNET 012 regimen in combination with ZDV for 1 week. Similar to the NVAZ study among late presenters, a randomized, open-label, phase 3 trial was conducted between April 1, 2000 and March 15, 2003, at six clinics in Blantyre, Malawi. Overall, 894 infants born to HIV-positive women were enrolled. All women received NVP intrapartum, and were previously antiretroviral treatment–naive. Infants were randomly assigned to NVP or NVP plus ZDV. Infants HIV infection status was established at 6–8 weeks of age. The rate of MTCT of HIV at birth was 8.1% (36/445) in infants administered NVP only and 10.1% (45/444) in infants administered NVP plus ZDV ($p=0.30$). The rate of MTCT at 6–8 weeks was 14.1% (95% confidence interval [CI], 10.7–17.4%) in infants who received NVP (Fig. 16.4) and 16.3% (95% CI, 12.7–19.8%) in those who received NVP plus ZDV ($p=0.36$). All infants were breastfed at 1 week and 6–8 weeks visits [42]. This study showed that the rates of MTCT of HIV were similar with or without adding ZDV to the HIVNET 012 regimen when women had received single-dose NVP before delivery.

2004–2009: The Era of Extended Infant Antiretroviral Postexposure Prophylaxis to Prevent Breastfeeding-Associated HIV Transmission in Malawi

The NVAZ trials in Malawi provided evidence that postexposure prophylaxis of the infant is a practical and effective strategy to reduce MTCT of HIV. However, these regimens were provided for a short period postpartum. With better regimens to reduce HIV transmission around the time of birth, how to prevent postnatal HIV transmission through breastfeeding became the research focus in Malawi and elsewhere. Building on the successes of the NVAZ trials, a longer regimen of prophylaxis to prevent breastfeeding HIV transmission was justified. Therefore, the extended antiretroviral *PostExposure Prophylaxis of Infants* in Malawi (PEPI-Malawi) was conceived and initiated in Blantyre, Malawi.

An important feature of the NVAZ studies was that HIV-infected women continued to breastfeed and were not counseled to discontinue at a specific time point postpartum. The WHO recommendations from 2000 that HIV-infected women should exclusively breastfeed up to 6 months and then wean the infant [28] were not implemented until sometime later. In the PEPI-Malawi study, however, women were counseled to follow the WHO recommendations from 2000. There were other changes in breastfeeding recommendations during the conduct of the PEPI-Malawi trial, 2004–2009: the WHO recommendations were further revised [43] to allow for longer breastfeeding to avoid infant gastroenteritis and malnutrition as reported from Malawi and elsewhere in sub-Saharan Africa [44–47] during the conduct of the PEPI trial (Table 16.2 and Fig. 16.6 show the frequency of breastfeeding among HIV-infected women in the NVAZ and PEPI studies).

The PEPI-Malawi Trial

The PEPI-Malawi was a phase 3 clinical trial, conducted in a single site in Blantyre, Malawi, to assess the efficacy of an extended infant antiretroviral postexposure regimen to age 14 weeks in reducing risk of HIV infection through breastfeeding. Breastfeeding, HIV-infected women were enrolled. Infants were randomized at birth to receive either: (a) single-dose oral NVP syrup plus 1 week of oral ZDV syrup (*control regimen* similar to the previously tested NVAZ regimen), (b) the control regimen plus extended daily NVP to age 14 weeks, or (c) the control regimen plus extended daily NVP+ZDV to age 14 weeks. Overall, 3,126 infants HIV uninfected (based on DNA PCR) at birth were enrolled: 1,004 in control arm, 1,071 in extended NVP arm, and 1,051 in the extended NVP+ZDV arm. The rate of HIV infection was consistently lower in the two extended arms than in the control arm. At age 9 months (the primary end point of the trial), HIV infection rates (based on DNA PCR positivity on two separate samples) were 5.0% in the extended NVP arm (Fig. 16.4), 6.0% in the extended

Table 16.2 Frequency of breastfeeding in HIV-infected women in studies conducted in Blantyre, Malawi, 1989–2009

Infant's age	ICAR[a] (1989–1992)		PAVE2[a] (1994–1997)		HPTN024[a] (2000–2004)		NVAZ[a] (2000–2003)		PEPI[a] (2004–2007)	
	No.	%	No.	%	No.	%	No	%	No.	%
1 week	–	–	–	–	–	–	1,659	99.6	2,768	99.7
3 week	–	–	–	–	–	–	–	–	2,909	99.5
6 week	1,111	99.6	329	99.4	471	97.9	1,645	99.5	2,829	99.0
9 week	971	99.7	–	–	–	–	–	–	2,774	98.7
3 month	976	99.8	329	99.4	457	96.9	1,529	99.2	2,762	97.3
6 month	953	99.3	330	98.5	434	92.4	1,393	98.6	2,646	89.2
9 month	833	97.8	329	96.7	387	85.5	1,322	97.7	2,435	35.3
12 month	794	93.2	329	94.8	310	83.2	1,244	96.3	2,308	27.3
15 month	714	84.9	329	88.2	–	–	1,012	92.4	2,180	21.8
18 month	682	60.0	329	71.7	–	–	962	84.2	2,121	15.3
21 month	601	31.5	–	–	–	–	912	63.1	–	–
24 month	572	13.6	329	17.9	–	–	880	33.6	1,860	5.6

[a]Acronyms of studies conducted in Blantyre, Malawi (see text)

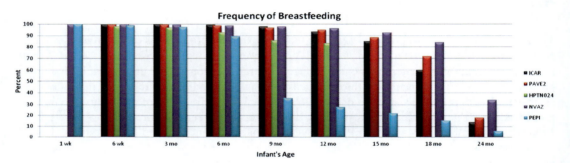

Fig. 16.6 Frequency of breastfeeding in HIV-infected women in studies conducted in Blantyre, Malawi, 1989–2009

NVP+ZDV, and 11.1% in the control arm ($p<0.001$ comparing extended regimens to control). At age 24 months HIV infection rates were ~11% in the extended arms compared to 15.6% in the control arms ($p<0.05$). The rates of HIV infection or death were also significantly lower in the extended arms. There were no differences in severe adverse events with the exception of higher possibly related events (mostly neutropenia) in the extended NVP+ZDV arm [14, 48].

PEPI-Malawi Substudy: Continuation of HIV Transmission Following
Cessation of Infant Postexposure Prophylaxis

In the PEPI-Malawi study, the extended infant antiretroviral prophylaxis was effective in reducing postnatal HIV infection by ~70% during the period of prophylaxis and ~50% at 9 months of age. However, after cessation of antiretroviral prophylaxis at 14 weeks, HIV transmission continued at approximately 1–2% every 3 months among breastfeeding infants not infected at 14 weeks [49]. Among infants at risk of HIV infection (685 in control, 757 in extended NVP, and 741 in extended NVP+ZDV) at 6 months, the rate of MTCT was 1.3% (95% 0.7–2.5) in control, 0.9% (95% CI 0.4–1.9) in extended NVP, and 1.8% (95% CI 1.1–3.1) in extended NVP+ZDV. At 9 months, the rate of MTCT was 2.1% (95% 1.2–3.4) in control, 2.0% (95% CI 1.2–3.3) in extended NVP, and 3.0% (95% CI 2.0–4.5) in extended NVP+ZDV. At 12 months, the rate of MTCT was 2.9% (95% 1.8–4.5) in control, 3.4% (95% CI 2.3–5.1) in extended NVP, and 4.3% (95% CI 3.1–6.1) in extended NVP+ZDV. At 15 months, the rate of MTCT was 4.3% (95% 2.9–6.3) in control, 4.3% (95% CI 3.0–6.2) in extended NVP and 5.1% (95% CI 3.7–7.0) in extended NVP+ZDV. At 18 months, the rate of MTCT was 6.1% (95% 4.4–8.5) in control, 6.6% (95% CI 4.9–9.0) in extended NVP, and 6.7% (95% CI 5.0–9.0) in extended NVP+ZDV. At 24 months, the rate of MTCT was 6.9% (95% 5.0–9.4) in control, 8.2% (95% CI 6.1–11.1) in extended NVP, and 7.9% (95% CI 5.9–10.4) in extended NVP+ZDV.

PEPI-Malawi Substudy: The Effect of Treating Eligible Breastfeeding HIV-Infected Women with Antiretrovirals

Further analysis of the PEPI study also showed some important results regarding antiretroviral treatment of HIV-infected women. At the time when the PEPI-Malawi started, antiretroviral treatment was not yet available at a wider scale. A government supported antiretroviral treatment program became available during the conduct of the study and eligible women were therefore referred to a specialized clinic within QECH. We assessed the association of postnatal HIV transmission with maternal highly active antiretroviral therapy (HAART) after infant extended antiretroviral prophylaxis was stopped at age 14 weeks. Maternal CD4 count (cells/mm^3) and HAART use were collected in the PEPI-Malawi study. Maternal HAART was categorized to *HAART-eligible-untreated* (CD4<250, no HAART), *HAART-eligible-treated* (CD4<250, received HAART), and *HAART-ineligible* (CD4≥250). Overall, 301 (~13%) women received HAART and 130 (5.6%) of 2,318 infants became HIV infected. The rate (95% CI) of HIV transmission per 100 person-years was HAART-eligible-untreated=10.56 (7.91–13.82); HAART-eligible-treated=1.79 (0.58–4.18); and HAART-ineligible=3.66 (2.86–4.61). Infant prophylaxis-adjusted rate ratios (95% CI) for HAART-eligible-treated and HAART-ineligible versus HAART-eligible-untreated were 0.18 (0.07–0.44) and 0.35 (0.25–0.50), respectively. This observational substudy of PEPI-Malawi showed that treating HAART-eligible HIV-infected women can reduce MTCT to very low levels, ~1.8% [49]; Fig. 16.4.

PEPI-Malawi Substudy: Emergence of NVP Resistance in Breastfed Infants in the PEPI-Malawi Study

Earlier NVP resistance studies from the NVAZ trial in Blantyre, Malawi, showed that NVP resistance mutations were detected in 87% of infants who received single-dose NVP and likewise the frequency of these mutations among HIV-infected women who received single-dose NVP intrapartum was 69% [50]. The rate of resistance mutations among children and mothers with subtype C HIV infection in Malawi was significantly higher than in those developing resistance and infected with subtypes A and D from Uganda [50, 51]. In the NVAZ studies, addition of ZDV to NVP to infant prophylaxis (and mother did not receive intrapartum NVP) compared to prophylaxis of infants with NVP alone (and mother received intrapartum NVP) significantly reduced development of resistance in infants who became infected at 6–8 weeks; 27% versus 87%, $p<0.001$ [52]. In the PEPI-Malawi study, the resistance studies were continued to examine if extended NVP prophylaxis could increase the risk of emergence of NVP resistance and whether the extended ZDV prophylaxis could decrease the emergence of NVP resistance among infants who become HIV infected despite prophylaxis. Consistent with what was observed in the NVAZ studies, at 14 weeks of age among infants infected at birth (in utero), the frequency of NVP resistance was lower in the extended NVP+ZDV arm compared with the extended NVP alone arm; 62% versus 86%, $p<0.02$ [53]. No ZDV resistance was detected. Addition of ZDV to NVP (extended NVP+ZDV arm) was associated with reduced risk of NVP resistance at 14 weeks if the extended prophylaxis was stopped by 6 weeks in infants already infected at birth (in settings such as Malawi testing to establish the infant HIV infection status on samples collected at birth would usually take 7–10 days); 55% versus 86%, $p=0.007$. There was no statistically significant difference in resistance by extended arm if infant prophylaxis was continued after age 6 weeks (83% vs. 88%). Another preliminary finding from the PEPI-Malawi study is the potential of inducing multi-class drug resistance (MCR) in breastfeeding and already HIV-infected infants whose mothers initiate antiretroviral treatment after delivery [54] In this substudy, 30 of 37 infants tested (81%) had NNRTI resistance and 11 (30%) of these infants had MCR. Earlier postpartum maternal HAART use was strongly associated with acquisition of MCR in the infants. MCR was detected in infants of mothers who were exclusively breastfeeding at the time of reported HAART use. These preliminary findings have practical implications related to infant antiretroviral treatment. It suggests that development of MCR in infants may limit their chances of treatment with drugs commonly available in these resource-constrained settings. Further evaluation of the timing of initiating HAART in breastfeeding women whose infants are already HIV infected is needed.

Although coverage with single-dose NVP for prevention of perinatal transmission is still below the desired levels (51% documented NVP intake in a study that tested for NVP in cord blood [55]), study of the implications of previous exposure to single-dose NVP and potential development of NVP resistance continues to be important with expansion of antiretroviral treatment in Africa. For example, in a recent multicenter trial conducted at six African research sites showed that among infants who became infected despite single-dose NVP prophylaxis, antiretroviral treatment outcomes were better if the infant received a regimen with no NVP (zidovudine plus lamivudine plus ritonavir-boosted lopinavir) compared to a regimen including NVP (zidovudine plus lamuvidine plus NVP) [56].

Conclusion

Substantial progress has been made in efforts to understand the dynamics of MTCT in sub-Saharan Africa. The contributions of the Johns Hopkins Research Project to these encouraging developments have assisted the research community interested in the fields of HIV prevention, treatment, and maternal child health to utilize these findings to develop appropriate interventions. In a systematic approach ranging from observational to clinical trials, the research site in Blantyre was able to quantitatively estimate the rates of MTCT of HIV, identify important risk factors associated with MTCT, and to subsequently conduct clinical trials to assess if modification of these factors can reduce transmission. Figure 16.4 shows a summary of the rates of MTCT during this period of ~20 years of research at a single African research site. These data show that it is possible to decrease transmission to low levels in a research setting despite continued breastfeeding by African women. A greater challenge for researchers and policy makers at this stage is how to translate this evidence into programs to achieve the ultimate goal of HIV-free survival among African children.

The Future

The PEPI-Malawi and subsequent studies of BAN in Lilongwe [57], Malawi and Mma Bana in Botswana [58] provided consistent evidence that extended prophylaxis with antiretrovirals to the infants and HIV-infected mothers can substantially reduce MTCT of HIV. Based on these data, the WHO recommended two alternative options: "*if a woman received AZT during pregnancy, daily nevirapine is recommended for her child from birth until the end of the breastfeeding period*" or "*if a woman received a three-drug regimen during pregnancy, a continued regimen of three-drug prophylaxis is recommended for the mother until the end of the breastfeeding period*" [59]. However, none of these options has been evaluated in a randomized clinical trial. A large multicountry randomized clinical trial (The PROMISE study: Clinical Trial # NCT01061151) is now in-progress and will provide this data in the years to come. The Johns Hopkins University Research Project in Blantyre, Malawi, is one of the participating sites in the PROMISE study.

References

1. Bulterys M, Fowler MG, Van Rompay KK, Kourtis AP (2004) Prevention of mother-to-child transmission of HIV-1 through breast-feeding: past, present, and future. J Infect Dis 189(12):2149–2153
2. Habicht JP, DaVanzo J, Butz WP (1986) Does breastfeeding really save lives, or are apparent benefits due to biases? Am J Epidemiol 123(2):279–290
3. Arifeen S, Black RE, Antelman G, Baqui A, Caulfield L, Becker S (2001) Exclusive breastfeeding reduces acute respiratory infection and diarrhea deaths among infants in Dhaka slums. Pediatrics 108(4):E67
4. Wright AL, Bauer M, Naylor A, Sutcliffe E, Clark L (1998) Increasing breastfeeding rates to reduce infant illness at the community level. Pediatrics 101(5):837–844
5. Pelto GH, Zhang Y, Habicht JP (2010) Premastication: the second arm of infant and young child feeding for health and survival? Matern Child Nutr 6(1):4–18
6. Goldman AS (1993) The immune system of human milk: antimicrobial, antiinflammatory and immunomodulating properties. Pediatr Infect Dis J 12(8):664–671
7. Goldman AS, Garza C, Nichols BL, Goldblum RM (1982) Immunologic factors in human milk during the first year of lactation. J Pediatr 100(4):563–567
8. (2000) Effect of breastfeeding on infant and child mortality due to infectious diseases in less developed countries: a pooled analysis. WHO collaborative study team on the role of breastfeeding on the prevention of infant mortality. Lancet 355(9202):451–455
9. UNAIDS (2010) AIDS epidemic update 2010. [Link to PDF]: UNAIDS World Health Organization. http://www.unaids.org/globalreport/Global_report.htm. Accessed 16 Dec 2010
10. Abdool Karim Q, Abdool Karim SS, Frohlich JA, Grobler AC, Baxter C, Mansoor LE et al (2010) Effectiveness and safety of tenofovir gel, an antiretroviral microbicide, for the prevention of HIV infection in women. Science 329(5996):1168–1174
11. Abdool Karim SS, Richardson BA, Ramjee G, Hoffman IF, Chirenje ZM, Taha T et al (2011) Safety and effectiveness of BufferGel and 0.5% PRO2000 gel for the prevention of HIV infection in women. AIDS 25(7):957–966
12. Malawi Demographic and Health Survey 2010, Preliminary Report. Zomba, Malawi: Demography and Social Statistics Division (DDS). National Statistical Office, Chimbiya Road. P.O. Box 333, Zomba, Malawi 2010 February 2011
13. Malawi HIV and AIDS monitoring and evaluation report. 2008. http://data.unaids.org/pub/Report/2008/malawi_2008_country_progress_report_en.pdf. Accessed 2 May 2011
14. Taha TE, Li Q, Hoover DR, Mipando L, Nkanaunena K, Thigpen MC et al (2011) Post-exposure prophylaxis of breastfeeding HIV-exposed Infants with antiretroviral drugs to age 14 weeks: updated efficacy results of the PEPI-Malawi trial. J Acquir Immune Defic Syndr 57(4):319–325
15. Taha TE, Dallabetta GA, Canner JK, Chiphangwi JD, Liomba G, Hoover DR et al (1995) The effect of human immunodeficiency virus infection on birthweight, and infant and child mortality in urban Malawi. Int J Epidemiol 24(5):1022–1029
16. Taha TE, Dallabetta GA, Hoover DR, Chiphangwi JD, Mtimavalye LA, Liomba GN et al (1998) Trends of HIV-1 and sexually transmitted diseases among pregnant and postpartum women in urban Malawi. AIDS 12(2): 197–203
17. Miotti PG, Dallabetta GA, Chiphangwi JD, Liomba G, Saah AJ (1992) A retrospective study of childhood mortality and spontaneous abortion in HIV-1 infected women in urban Malawi. Int J Epidemiol 21(4):792–799 [Research Support, U.S. Gov't, P.H.S.]
18. Taha TE, Canner JK, Chiphangwi JD, Dallabetta GA, Yang LP, Mtimavalye LA et al (1996) Reported condom use is not associated with incidence of sexually transmitted diseases in Malawi. AIDS 10(2):207–212
19. Dabis F, Msellati P, Dunn D, Lepage P, Newell ML, Peckham C et al (1993) Estimating the rate of mother-to-child transmission of HIV. Report of a workshop on methodological issues Ghent (Belgium), 17–20 Feb 1992. The Working Group on Mother-to-Child Transmission of HIV. AIDS 7(8):1139–1148
20. De Cock KM, Fowler MG, Mercier E, de Vincenzi I, Saba J, Hoff E et al (2000) Prevention of mother-to-child HIV transmission in resource-poor countries: translating research into policy and practice. JAMA 283(9):1175–1182
21. Nduati R, John G, Mbori-Ngacha D, Richardson B, Overbaugh J, Mwatha A et al (2000) Effect of breastfeeding and formula feeding on transmission of HIV-1: a randomized clinical trial. JAMA 283(9):1167–1174

22. Connor EM, Sperling RS, Gelber R, Kiselev P, Scott G, O'Sullivan MJ et al (1994) Reduction of maternal-infant transmission of human immunodeficiency virus type 1 with zidovudine treatment. Pediatric AIDS Clinical Trials Group Protocol 076 Study Group. N Engl J Med 331(18):1173–1180
23. Semba RD, Miotti PG, Chiphangwi JD, Saah AJ, Canner JK, Dallabetta GA et al (1994) Maternal vitamin A deficiency and mother-to-child transmission of HIV-1. Lancet 343(8913):1593–1597
24. Taha TE, Hoover DR, Dallabetta GA, Kumwenda NI, Mtimavalye LA, Yang LP et al (1998) Bacterial vaginosis and disturbances of vaginal flora: association with increased acquisition of HIV. AIDS 12(13):1699–1706
25. Taha TE, Gray RH, Kumwenda NI, Hoover DR, Mtimavalye LA, Liomba GN et al (1999) HIV infection and disturbances of vaginal flora during pregnancy. J Acquir Immune Defic Syndr Hum Retrovirol 20(1):52–59
26. Taha TE, Gray RH (2000) Genital tract infections and perinatal transmission of HIV. Ann N Y Acad Sci 918:84–98
27. Grosskurth H, Mosha F, Todd J, Mwijarubi E, Klokke A, Senkoro K et al (1995) Impact of improved treatment of sexually transmitted diseases on HIV infection in rural Tanzania: randomised controlled trial. Lancet 346(8974):530–536
28. WHO (2001) New data on the prevention of mother-to-child transmission of HIV and their policy implications: conclusions and recommendations: WHO Technical consultation on behalf of the UNFPA/UNICEF/WHO/UNAIDS Inter-Agency Task Team on Mother-to-Child Transmission of HIV, Geneva, 11–13 Oct 2000: Geneva: World Health Organization, 2001. http://whqlibdoc.who.int/hq/2001/WHO_RHR_01.28.pdf
29. Goedert JJ, Duliege AM, Amos CI, Felton S, Biggar RJ (1991) High risk of HIV-1 infection for first-born twins. The International Registry of HIV-Exposed Twins. Lancet 338(8781):1471–1475
30. Duliege AM, Amos CI, Felton S, Biggar RJ, Goedert JJ (1995) Birth order, delivery route, and concordance in the transmission of human immunodeficiency virus type 1 from mothers to twins. International Registry of HIV-Exposed Twins. J Pediatr 126(4):625–632
31. Biggar RJ, Miotti PG, Taha TE, Mtimavalye L, Broadhead R, Justesen A et al (1996) Perinatal intervention trial in Africa: effect of a birth canal cleansing intervention to prevent HIV transmission. Lancet 347(9016):1647–1650
32. Taha TE, Biggar RJ, Broadhead RL, Mtimavalye LA, Justesen AB, Liomba GN et al (1997) Effect of cleansing the birth canal with antiseptic solution on maternal and newborn morbidity and mortality in Malawi: clinical trial. BMJ 315(7102):216–219
33. Kumwenda N, Miotti PG, Taha TE, Broadhead R, Biggar RJ, Jackson JB et al (2002) Antenatal vitamin A supplementation increases birth weight and decreases anemia among infants born to human immunodeficiency virus-infected women in Malawi. Clin Infect Dis 35(5):618–624
34. Gray GE, Urban M, Chersich MF, Bolton C, van Niekerk R, Violari A et al (2005) A randomized trial of two postexposure prophylaxis regimens to reduce mother-to-child HIV-1 transmission in infants of untreated mothers. AIDS 19(12):1289–1297
35. Miotti PG, Taha TE, Kumwenda NI, Broadhead R, Mtimavalye LA, Van der Hoeven L et al (1999) HIV transmission through breastfeeding: a study in Malawi. JAMA 282(8):744–749 [Research Support, U.S. Gov't, P.H.S.]
36. Semba RD, Kumwenda N, Hoover DR, Taha TE, Quinn TC, Mtimavalye L et al (1999) Human immunodeficiency virus load in breast milk, mastitis, and mother-to-child transmission of human immunodeficiency virus type 1. J Infect Dis 180(1):93–98 [Research Support, Non-U.S. Gov't Research Support, U.S. Gov't, P.H.S.]
37. Nussenblatt V, Kumwenda N, Lema V, Quinn T, Neville MC, Broadhead R et al (2006) Effect of antibiotic treatment of subclinical mastitis on human immunodeficiency virus type 1 RNA in human milk. J Trop Pediatr 52(5):311–315
38. Nussenblatt V, Lema V, Kumwenda N, Broadhead R, Neville MC, Taha TE et al (2005) Epidemiology and microbiology of subclinical mastitis among HIV-infected women in Malawi. Int J STD AIDS 16(3):227–232
39. Guay LA, Musoke P, Fleming T, Bagenda D, Allen M, Nakabiito C et al (1999) Intrapartum and neonatal single-dose nevirapine compared with zidovudine for prevention of mother-to-child transmission of HIV-1 in Kampala, Uganda: HIVNET 012 randomised trial. Lancet 354(9181):795–802
40. Taha TE, Brown ER, Hoffman IF, Fawzi W, Read JS, Sinkala M et al (2006) A phase III clinical trial of antibiotics to reduce chorioamnionitis-related perinatal HIV-1 transmission. AIDS 20(9):1313–1321
41. Taha TE, Kumwenda NI, Gibbons A, Broadhead RL, Fiscus S, Lema V et al (2003) Short postexposure prophylaxis in newborn babies to reduce mother-to-child transmission of HIV-1: NVAZ randomised clinical trial. Lancet 362(9391):1171–1177
42. Taha TE, Kumwenda NI, Hoover DR, Fiscus SA, Kafulafula G, Nkhoma C et al (2004) Nevirapine and zidovudine at birth to reduce perinatal transmission of HIV in an African setting: a randomized controlled trial. JAMA 292(2):202–209
43. WHO (2009) HIV and infant feeding revised principles and recommendations. Rapid advice. World Health Organization. http://www.searo.who.int/linkfiles/HIV-AIDS_rapid_advice_infant_feeding(web).pdf. Accessed 21 Dec 2010
44. Kafulafula G, Hoover DR, Taha TE, Thigpen M, Li Q, Fowler MG et al (2010) Frequency of gastroenteritis and gastroenteritis-associated mortality with early weaning in HIV-1-uninfected children born to HIV-infected women in Malawi. J Acquir Immune Defic Syndr 53(1):6–13

45. Onyango-Makumbi C, Bagenda D, Mwatha A, Omer SB, Musoke P, Mmiro F et al (2010) Early weaning of HIV-exposed uninfected infants and risk of serious gastroenteritis: findings from two perinatal HIV prevention trials in Kampala, Uganda. J Acquir Immune Defic Syndr 53(1):20–27
46. Homsy J, Moore D, Barasa A, Were W, Likicho C, Waiswa B et al (2010) Breastfeeding, mother-to-child HIV transmission, and mortality among infants born to HIV-Infected women on highly active antiretroviral therapy in rural Uganda. J Acquir Immune Defic Syndr 53(1):28–35
47. Creek TL, Kim A, Lu L, Bowen A, Masunge J, Arvelo W et al (2010) Hospitalization and mortality among primarily nonbreastfed children during a large outbreak of diarrhea and malnutrition in Botswana, 2006. J Acquir Immune Defic Syndr 53(1):14–19
48. Kumwenda NI, Hoover DR, Mofenson LM, Thigpen MC, Kafulafula G, Li Q et al (2008) Extended antiretroviral prophylaxis to reduce breast-milk HIV-1 transmission. N Engl J Med 359(2):119–129
49. Taha TE, Kumwenda J, Cole SR, Hoover DR, Kafulafula G, Fowler MG et al (2009) Postnatal HIV-1 transmission after cessation of infant extended antiretroviral prophylaxis and effect of maternal highly active antiretroviral therapy. J Infect Dis 200(10):1490–1497
50. Eshleman SH, Hoover DR, Chen S, Hudelson SE, Guay LA, Mwatha A et al (2005) Resistance after single-dose nevirapine prophylaxis emerges in a high proportion of Malawian newborns. AIDS 19(18):2167–2169
51. Eshleman SH, Hoover DR, Chen S, Hudelson SE, Guay LA, Mwatha A et al (2005) Nevirapine (NVP) resistance in women with HIV-1 subtype C, compared with subtypes A and D, after the administration of single-dose NVP. J Infect Dis 192(1):30–36
52. Eshleman SH, Hoover DR, Hudelson SE, Chen S, Fiscus SA, Piwowar-Manning E et al (2006) Development of nevirapine resistance in infants is reduced by use of infant-only single-dose nevirapine plus zidovudine postexposure prophylaxis for the prevention of mother-to-child transmission of HIV-1. J Infect Dis 193(4):479–481
53. Lidstrom J, Li Q, Hoover DR, Kafulafula G, Mofenson LM, Fowler MG et al (2010) Addition of extended zidovudine to extended nevirapine prophylaxis reduces nevirapine resistance in infants who were HIV-infected in utero. AIDS 24(3):381–386
54. Fogel J, Li Q, Taha TE, Hoover DR, Kumwenda NI, Mofenson LM et al (2011) Initiation of antiretroviral treatment in women after delivery can induce multiclass drug resistance in breastfeeding HIV-infected infants. Clin Infect Dis 52(8):1069–1076
55. Stringer EM, Ekouevi DK, Coetzee D, Tih PM, Creek TL, Stinson K et al (2010) Coverage of nevirapine-based services to prevent mother-to-child HIV transmission in 4 African countries. JAMA 304(3):293–302 [Research Support, Non-U.S. Gov't Research Support, U.S. Gov't, P.H.S.]
56. Palumbo P, Lindsey JC, Hughes MD, Cotton MF, Bobat R, Meyers T et al (2010) Antiretroviral treatment for children with peripartum nevirapine exposure. N Engl J Med 363(16):1510–1520 [Multicenter Study Randomized Controlled Trial Research Support, N.I.H., Extramural]
57. Chasela CS, Hudgens MG, Jamieson DJ, Kayira D, Hosseinipour MC, Kourtis AP et al (2010) Maternal or infant antiretroviral drugs to reduce HIV-1 transmission. N Engl J Med 362(24):2271–2281
58. Shapiro RL, Hughes MD, Ogwu A, Kitch D, Lockman S, Moffat C et al (2010) Antiretroviral regimens in pregnancy and breast-feeding in Botswana. N Engl J Med 362(24):2282–2294
59. WHO (2009) World Health Organization, New recommendations: preventing mother-to-child transmission. World Health Organization. http://www.who.int/hiv/pub/mtct/mtct_key_mess.pdf
60. Mtimavalye L, Biggar RJ, Taha TE, Chiphangwi J (1995) Maternal-infant transmission of HIV-1. N Engl J Med 332(13):890–891 [Comment Letter]
61. Biggar RJ, Mtimavalye L, Justesen A, Broadhead R, Miley W, Waters D et al (1997) Does umbilical cord blood polymerase chain reaction positivity indicate in utero (pre-labor) HIV infection? AIDS 11(11):1375–1382 [Research Support, U.S. Gov't, P.H.S.]
62. Biggar RJ, Miley W, Miotti P, Taha TE, Butcher A, Spadoro J et al (1997) Blood collection on filter paper: a practical approach to sample collection for studies of perinatal HIV transmission. J Acquir Immune Defic Syndr 14(4):368–373 [Clinical Trial]
63. Ioannidis JP, Taha TE, Kumwenda N, Broadhead R, Mtimavalye L, Miotti P et al (1999) Predictors and impact of losses to follow-up in an HIV-1 perinatal transmission cohort in Malawi. Int J Epidemiol 28(4):769–775 [Clinical Trial]
64. Semba RD, Kumwenda N, Taha TE, Hoover DR, Quinn TC, Lan Y et al (1999) Mastitis and immunological factors in breast milk of human immunodeficiency virus-infected women. J Hum Lact 15(4):301–306 [Research Support, U.S. Gov't, Non-P.H.S. Research Support, U.S. Gov't, P.H.S.]
65. Semba RD, Kumwenda N, Taha TE, Hoover DR, Lan Y, Eisinger W et al (1999) Mastitis and immunological factors in breast milk of lactating women in Malawi. Clin Diagn Lab Immunol 6(5):671–674 [Research Support, U.S. Gov't, Non-P.H.S. Research Support, U.S. Gov't, P.H.S.]
66. Lan Y, Kumwenda N, Taha TE, Chiphangwi JD, Miotti PG, Mtimavalye L et al (1999) Carotenoid status of pregnant women with and without HIV infection in Malawi. East Afr Med J 76(3):133–137 [Comparative Study Research Support, U.S. Gov't, Non-P.H.S. Research Support, U.S. Gov't, P.H.S.]

67. Semba RD, Kumwenda N, Taha TE, Mtimavalye L, Broadhead R, Miotti PG et al (2000) Plasma and breast milk vitamin A as indicators of vitamin A status in pregnant women. Int J Vitam Nutr Res 70(6):271–277 [Clinical Trial Controlled Clinical Trial Research Support, U.S. Gov't, Non-P.H.S. Research Support, U.S. Gov't, P.H.S.]
68. Semba RD, Kumwenda N, Hoover DR, Taha TE, Mtimavalye L, Broadhead R et al (2000) Assessment of iron status using plasma transferrin receptor in pregnant women with and without human immunodeficiency virus infection in Malawi. Eur J Clin Nutr 54(12):872–877 [Comparative Study Research Support, Non-U.S. Gov't Research Support, U.S. Gov't, P.H.S.]
69. Semba RD, Taha TE, Kumwenda N, Mtimavalye L, Broadhead R, Miotti PG et al (2001) Iron status and indicators of human immunodeficiency virus disease severity among pregnant women in Malawi. Clin Infect Dis 32(10):1496–1499 [Research Support, U.S. Gov't, Non-P.H.S. Research Support, U.S. Gov't, P.H.S.]
70. Semba RD, Kumwenda N, Taha TE, Mtimavalye L, Broadhead R, Garrett E et al (2001) Impact of vitamin A supplementation on anaemia and plasma erythropoietin concentrations in pregnant women: a controlled clinical trial. Eur J Haematol 66(6):389–395 [Clinical Trial Randomized Controlled Trial Research Support, U.S. Gov't, Non-P.H.S. Research Support, U.S. Gov't, P.H.S.]
71. Taha TE, Graham SM, Kumwenda NI, Broadhead RL, Hoover DR, Markakis D et al (2000) Morbidity among human immunodeficiency virus-1-infected and -uninfected African children. Pediatrics 106(6):E77
72. Miotti PG, Taha TET, Kumwenda NI, Van der Hoeven L, Broadhead R, Mtimavalye LAR et al (2000) Risk of HIV transmission through breastfeeding – reply. JAMA 283(8):999–1000
73. Dancheck B, Nussenblatt V, Ricks MO, Kumwenda N, Neville MC, Moncrief DT et al (2005) Breast milk retinol concentrations are not associated with systemic inflammation among breast-feeding women in Malawi. J Nutr 135(2):223–226
74. Dancheck B, Nussenblatt V, Kumwenda N, Lema V, Neville MC, Broadhead R et al (2005) Status of carotenoids, vitamin A, and vitamin E in the mother-infant dyad and anthropometric status of infants in Malawi. J Health Popul Nutr 23(4):343–350 [Research Support, N.I.H., Extramural]
75. Aboud S, Msamanga G, Read JS, Mwatha A, Chen YQ, Potter D et al (2008) Genital tract infections among HIV-infected pregnant women in Malawi, Tanzania and Zambia. Int J STD AIDS 19(12):824–832 [Research Support, N.I.H., Extramural Research Support, Non-U.S. Gov't]
76. Aboud S, Msamanga G, Read JS, Wang L, Mfalila C, Sharma U et al (2009) Effect of prenatal and perinatal antibiotics on maternal health in Malawi, Tanzania, and Zambia. Int J Gynaecol Obstet 107(3):202–207 [Clinical Trial, Phase III Multicenter Study Randomized Controlled Trial Research Support, N.I.H., Extramural]
77. Brown E, Chi BH, Read JS, Taha TE, Sharma U, Hoffman IF et al (2008) Determining an optimal testing strategy for infants at risk for mother-to-child transmission of HIV-1 during the late postnatal period. AIDS 22(17):2341–2346 [Multicenter Study Randomized Controlled Trial Research Support, N.I.H., Extramural Research Support, Non-U.S. Gov't]
78. Chasela C, Chen YQ, Fiscus S, Hoffman I, Young A, Valentine M et al (2008) Risk factors for late postnatal transmission of human immunodeficiency virus type 1 in sub-Saharan Africa. Pediatr Infect Dis J 27(3):251–256 [Clinical Trial, Phase III Randomized Controlled Trial Research Support, N.I.H., Extramural Research Support, Non-U.S. Gov't]
79. Chi BH, Wang L, Read JS, Sheriff M, Fiscus S, Brown ER et al (2005) Timing of maternal and neonatal dosing of nevirapine and the risk of mother-to-child transmission of HIV-1: HIVNET 024. AIDS 19(16):1857–1864 [Comparative Study Multicenter Study Randomized Controlled Trial Research Support, N.I.H., Extramural Research Support, Non-U.S. Gov't]
80. Chi BH, Wang L, Read JS, Taha TE, Sinkala M, Brown ER et al (2007) Predictors of stillbirth in sub-saharan Africa. Obstet Gynecol 110(5):989–997 [Comment Randomized Controlled Trial Research Support, N.I.H., Extramural Research Support, Non-U.S. Gov't]
81. Chilongozi D, Wang L, Brown L, Taha T, Valentine M, Emel L et al (2008) Morbidity and mortality among a cohort of human immunodeficiency virus type 1-infected and uninfected pregnant women and their infants from Malawi, Zambia, and Tanzania. Pediatr Infect Dis J 27(9):808–814 [Randomized Controlled Trial Research Support, N.I.H., Extramural Research Support, Non-U.S. Gov't]
82. Goldenberg RL, Andrews WW, Hoffman I, Fawzi W, Valentine M, Young A et al (2007) Fetal fibronectin and adverse infant outcomes in a predominantly human immunodeficiency virus-infected African population: a randomized controlled trial. Obstet Gynecol 109(2 Pt 1):392–401 [Clinical Trial, Phase III Multicenter Study Randomized Controlled Trial Research Support, N.I.H., Extramural Research Support, Non-U.S. Gov't]
83. Goldenberg RL, Mudenda V, Read JS, Brown ER, Sinkala M, Kamiza S et al (2006) HPTN 024 study: histologic chorioamnionitis, antibiotics and adverse infant outcomes in a predominantly HIV-1-infected African population. Am J Obstet Gynecol 195(4):1065–1074 [Randomized Controlled Trial Research Support, N.I.H., Extramural Research Support, Non-U.S. Gov't]
84. Goldenberg RL, Mwatha A, Read JS, Adeniyi-Jones S, Sinkala M, Msmanga G et al (2006) The HPTN 024 study: the efficacy of antibiotics to prevent chorioamnionitis and preterm birth. Am J Obstet Gynecol 194(3):650–661

[Clinical Trial, Phase III Randomized Controlled Trial Research Support, N.I.H., Extramural Research Support, Non-U.S. Gov't]

85. Kafulafula G, Mwatha A, Chen YQ, Aboud S, Martinson F, Hoffman I et al (2009) Intrapartum antibiotic exposure and early neonatal, morbidity, and mortality in Africa. Pediatrics 124(1):e137–e144 [Randomized Controlled Trial Research Support, N.I.H., Extramural Research Support, Non-U.S. Gov't]
86. Mehta S, Manji KP, Young AM, Brown ER, Chasela C, Taha TE et al (2008) Nutritional indicators of adverse pregnancy outcomes and mother-to-child transmission of HIV among HIV-infected women. Am J Clin Nutr 87(6):1639–1649 [Multicenter Study Research Support, N.I.H., Extramural Research Support, Non-U.S. Gov't]
87. Msamanga GI, Taha TE, Young AM, Brown ER, Hoffman IF, Read JS et al (2009) Placental malaria and mother-to-child transmission of human immunodeficiency virus-1. Am J Trop Med Hyg 80(4):508–515 [Clinical Trial, Phase III Randomized Controlled Trial Research Support, N.I.H., Extramural Research Support, Non-U.S. Gov't]
88. Nelson JA, Loftis AM, Kamwendo D, Fawzi WW, Taha TE, Goldenberg RL et al (2009) Nevirapine resistance in human immunodeficiency virus type 1-positive infants determined using dried blood spots stored for up to six years at room temperature. J Clin Microbiol 47(4):1209–1211 [Research Support, N.I.H., Extramural Research Support, Non-U.S. Gov't]
89. Read JS, Mwatha A, Richardson B, Valentine M, Emel L, Manji K et al (2009) Primary HIV-1 infection among infants in sub-Saharan Africa: HPTN 024. J Acquir Immune Defic Syndr 51(3):317–322 [Research Support, N.I.H., Extramural]
90. Taha TE, Kumwenda N, Gibbons A, Hoover D, Lema V, Fiscus S et al (2002) Effect of HIV-1 antiretroviral prophylaxis on hepatic and hematological parameters of African infants. AIDS 16(6):851–858 [Clinical Trial Randomized Controlled Trial Research Support, Non-U.S. Gov't Research Support, U.S. Gov't, P.H.S.]
91. Taha TE, Kumwenda N, Kafulafula G, Kumwenda J, Chitale R, Nkhoma C et al (2004) Haematological changes in African children who received short-term prophylaxis with nevirapine and zidovudine at birth. Ann Trop Paediatr 24(4):301–309 [Clinical Trial Multicenter Study Randomized Controlled Trial Research Support, Non-U.S. Gov't Research Support, U.S. Gov't, P.H.S.]
92. Taha TE, Nour S, Kumwenda NI, Broadhead RL, Fiscus SA, Kafulafula G et al (2005) Gender differences in perinatal HIV acquisition among African infants. Pediatrics 115(2):e167–e172 [Research Support, Non-U.S. Gov't Research Support, U.S. Gov't, P.H.S.]
93. Fiscus SA, Chen S, Hoover D, Kerkau MG, Alabanza P, Siharath S et al (2005) Affordable, abbreviated roche monitor assay for quantification of human immunodeficiency virus type 1 RNA in plasma. J Clin Microbiol 43(8):4200–4202 [Research Support, N.I.H., Extramural Research Support, Non-U.S. Gov't Research Support, U.S. Gov't, Non-P.H.S. Research Support, U.S. Gov't, P.H.S.]
94. Taha TE, Kumwenda NI, Hoover DR, Kafulafula G, Fiscus SA, Nkhoma C et al (2006) The impact of breastfeeding on the health of HIV-positive mothers and their children in sub-Saharan Africa. Bull World Health Organ 84(7):546–554 [Research Support, N.I.H., Extramural Research Support, Non-U.S. Gov't]
95. Eshleman SH, Lie Y, Hoover DR, Chen S, Hudelson SE, Fiscus SA et al (2006) Association between the replication capacity and mother-to-child transmission of HIV-1, in antiretroviral drug-naive Malawian women. J Infect Dis 193(11):1512–1515
96. Church JD, Jones D, Flys T, Hoover D, Marlowe N, Chen S et al (2006) Sensitivity of the ViroSeq HIV-1 genotyping system for detection of the K103N resistance mutation in HIV-1 subtypes A, C, and D. JMD [Comparative Study Evaluation Studies Research Support, N.I.H., Extramural Research Support, Non-U.S. Gov't Research Support, U.S. Gov't, Non-P.H.S.] 8(4):430–432, quiz 527
97. Flys TS, Chen S, Jones DC, Hoover DR, Church JD, Fiscus SA et al (2006) Quantitative analysis of HIV-1 variants with the K103N resistance mutation after single-dose nevirapine in women with HIV-1 subtypes A, C, and D. J Acquir Immune Defic Syndr 42(5):610–613 [Research Support, N.I.H., Extramural Research Support, Non-U.S. Gov't Research Support, U.S. Gov't, Non-P.H.S.]
98. Taha TE, Hoover DR, Kumwenda NI, Fiscus SA, Kafulafula G, Nkhoma C et al (2007) Late postnatal transmission of HIV-1 and associated factors. J Infect Dis 196(1):10–14 [Randomized Controlled Trial Research Support, N.I.H., Extramural Research Support, Non-U.S. Gov't]
99. Church JD, Hudelson SE, Guay LA, Chen S, Hoover DR, Parkin N et al (2007) HIV type 1 variants with nevirapine resistance mutations are rarely detected in antiretroviral drug-naive African women with subtypes A, C, and D. AIDS Res Hum Retroviruses 23(6):764–768 [Clinical Trial Research Support, N.I.H., Extramural Research Support, Non-U.S. Gov't Research Support, U.S. Gov't, Non-P.H.S.]
100. Church JD, Towler WI, Hoover DR, Hudelson SE, Kumwenda N, Taha TE et al (2008) Comparison of LigAmp and an ASPCR assay for detection and quantification of K103N-containing HIV variants. AIDS Res Hum Retroviruses 24(4):595–605 [Comparative Study Research Support, N.I.H., Extramural Research Support, Non-U.S. Gov't]
101. Taha TE, Kumwenda N, Kafulafula G (2008) Antiretroviral prophylaxis to reduce breast-milk HIV-1 transmission – reply. N Engl J Med 359(17):1846–1847

Chapter 17
Breastfeeding and HIV Infection in China

Christine Korhonen, Liming Wang, Linhong Wang, Serena Fuller, Fang Wang, and Marc Bulterys

Introduction

Breast milk is the ideal food source for human infants, and breastfeeding is known to have many beneficial effects for both infants and mothers including providing proper nutrition, supporting the infant immune system, enhancing mother–infant bonding, and providing a decreased risk for maternal breast and ovarian cancer [1–5]. Longer term benefits of breastfeeding have also been observed including decreased risk of asthma and diabetes later in life [4]. However, breastfeeding carries a significant risk of transmission of HIV-1 (further referred to as HIV), especially in late stages of maternal disease [6–8]. In order to avoid transmitting HIV postnatally, women with HIV infection have been advised to avoid breastfeeding under certain conditions [9]. China has adopted a national policy of recommending replacement feeding for HIV-infected mothers where replacement feeding is acceptable, feasible, affordable, sustainable, and safe (AFASS) [10] through the national prevention of mother-to-child transmission of (PMTCT) HIV program. In 2008, 89% of the population were reported to have access to improved water sources [11], and AFASS conditions are met in most localities in China except certain remote, mountainous, and/or ethnic minority areas.

This chapter describes issues impacting breastfeeding and HIV infection in China. Current breastfeeding trends and issues surrounding postpartum incident HIV infection among breastfeeding women are outlined. The national PMTCT program recommendations on breastfeeding practices are provided, along with opportunities for PMTCT, including advising HIV-infected pregnant women to avoid breastfeeding. Two populations with particularly high risk for HIV MTCT, women with hepatitis B co-infection and female injecting drug users (IDU), require special consideration which are also addressed.

C. Korhonen • L. Wang • S. Fuller • M. Bulterys (✉)
Division of Global HIV/AIDS (DGHA), Center for Global Health, Centers for Disease Control and Prevention (CDC), 1600 Clifton Road, NE, Atlanta, GA 30333, USA

CDC Global AIDS Program, Beijing, China

Adjunct Professor of Epidemiology, UCLA School of Public Health, Los Angeles, CA, USA
e-mail: zbe2@cdc.gov

L. Wang • F. Wang
National Center for Women and Children's Health, China Center for Disease Control and Prevention, Beijing, China

Breastfeeding Prevalence and Infant Feeding Practices

In China there is historical precedence for early complementary feeding [11a]. Since then breastfeeding rates have followed the path of development. Breastfeeding declined in urban areas in the 1970s with the widespread adoption of formula feeding [12]. In the 1990s breastfeeding was strongly promoted by the government following WHO guidance with a 2010 goal of 85% of infants breastfed for at least 4 months [13]. Hospitals were certified Baby Friendly following the UNICEF model, and by 2002 over 6,300 hospitals were so designated [14]. Although breastfeeding rates vary substantially throughout the country and are generally lower in urban areas, multiple surveys conducted recently by Save the Children and UNICEF showed, on average, 29% of infants were exclusively breastfed for at least 6 months [15, 16]. A study in Eastern China showed that 97% of women initiated breastfeeding in the hospital, and 50% were still exclusively breastfeeding by hospital discharge (average stay 6 days). More mothers from rural (61%) than urban areas (38%) continued to exclusively breastfeed at hospital discharge [17].

A 2009 review of breastfeeding in predominantly urban areas of China found exclusive breastfeeding rates at 4 months ranging from 2% in urban and rural areas of Xinjiang Autonomous Region in Western China to 84% in urban Shenzhen City near Hong Kong. Any breastfeeding at 4 months ranged from 53% in urban Qiqihar, Heilongjiang to 92% in urban Guangzhou, Guangdong [12]. High rates of exclusive and any breastfeeding were found in Guangzhou and Shenzhen, both large urban areas in Guangdong Province, where studies were carried out in Baby Friendly Hospitals. The low rate of 2% exclusive breastfeeding in Xinjiang was balanced by higher rates of any breastfeeding at 4 months, 85%. Low rates of exclusive breastfeeding with higher rates and longer average duration of any breastfeeding were also found in urban and rural areas of Tibet Autonomous Region (3% exclusive at 4 months, 89% any, 14 month duration) [12] and in rural minority areas of Yunnan Province in Southwestern China (23% exclusive at 4 months, 98% any, 10 month duration) [18]. Water is commonly provided to breastfed infants and recommended by some medical staff in both Xinjiang and Tibet Autonomous Regions, following the traditional belief that it is needed to avoid dehydration and jaundice in the infant [19].

Formula use has been adopted into the Chinese culture and is now a common gift for new mothers. One study reported 32% of new mothers received presents of formula from friends or family [17]. This is similar to the exchange of cigarette cartons which are often given as presents in China, including to celebrate the birth of a child [20]. Nationally, close to 60% of Chinese men smoke and a rapidly increasing proportion of young women are initiating cigarette smoking, particularly in large cities [21]. Studies in both Xinjiang and Hong Kong showed infants were less likely to be breastfed in homes where the father smoked [22, 23].

The melamine scandal in China, where many parents unwittingly fed their babies formula milk contaminated by a toxic chemical, demonstrates another extreme way that infant health can be affected by feeding practices [24, 25]. Although the effects of this scandal on breastfeeding practices has not been studied in China, after the scandal breastfeeding rates increased temporarily in France among Chinese-speaking women [26].

National Prevention of Mother-to-Child Transmission of HIV Program

Approximately 75% of people living with HIV in China are living in rural areas (Fig. 17.1). In the past decade, the proportion of women infected has doubled to 30%. An estimated 7,000–9,000 HIV-infected pregnant women deliver each year [27–29]. In 2009, there were an estimated 230,000 women over 15 years of age living with HIV in China, 31% of the 740,000 total estimated people living with HIV/AIDS (PLHA) [30]. About two-thirds of the estimated PLHA in China do not know their

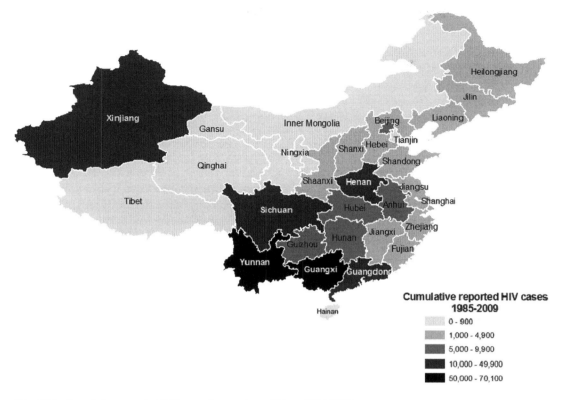

Fig. 17.1 Cumulative reported HIV cases by province, China, 1985–2009

HIV-positive status, putting these women at increased risk for MTCT. There are a number of localized areas in certain prefectures within Sichuan, Xinjiang, and Yunnan provinces where HIV prevalence among pregnant women is at or above 1% [31–34]. Documented instances of MTCT account for 1–1.5% of newly reported HIV cases, approximately 500 per year [35]. This is a substantial underestimate since many exposed infants will die before they are diagnosed with HIV and many HIV-infected mothers do not know their own HIV status [36]. Starting in 2003, the governmental China CARES program provided selected high-HIV prevalence counties with a package of services titled "Four Free and One Care," including free HIV testing and antiretroviral (ARV) prophylaxis for mothers and infants. The Four Free and One Care services include: (1) Free voluntary HIV counseling and testing, (2) Free antiretroviral treatment (ART), (3) Free PMTCT testing and prophylaxis, (4) Free schooling for children orphaned by AIDS, and (5) Care for people living with HIV and AIDS [37, 38]. At the same time, PMTCT services were initiated in other high-HIV prevalence counties by the National Center for Women and Children's Health (NCWCH). Over time, all PMTCT services were transitioned to NCWCH and broadened to include free HIV testing for mother and infant, ARV prophylaxis, "safe delivery" promoting the avoidance of unnecessary invasive procedures, and provision of free infant formula for 12 months in AFASS areas. Follow-up services are provided to mothers in their homes 1 and 4 weeks after birth which include a nutritional assessment of the infant [9]. In 2010, national policy makers decided to expand the PMTCT program to cover at least 44% of all pregnant women nationwide for routine screening of HIV, syphilis, and hepatitis B virus infection [10, 38]. Given the UNICEF estimate of 18 million births a year in China [16] and since many rural areas are still difficult to reach, this is a daunting task.

In the 271 counties where NCWCH first carried out the PMTCT program, 79% of 4.4 million pregnant women who attended antenatal care were tested for HIV [39]. By 2006, MTCT was reduced to 9% of HIV-exposed infants in participating areas from an estimated baseline of 33% [37]. Eighty-five percent of mothers enrolled in the program reported using exclusive replacement feeding [35].

Postpartum Incident Maternal HIV Infection and Transmission via Breastfeeding

With improved screening of blood supplies, HIV transmission through blood transfusion has decreased drastically in many countries [40]. In China, a large number of paid plasma donors were infected with HIV in the early to mid-1990s [41, 42]. More people were infected through transfusion of contaminated blood products. In a study of 104 women in Hubei Province who received a contaminated blood transfusion from 1994 to 1999 due to excessive bleeding after delivery and subsequently breastfed, 38 exposed infants (36%) became HIV-infected [43]. Duration of breastfeeding was not significantly associated with transmission. The expected high viral load of the women during the acute phase of infection likely led to the high rate of transmission to the infant [44]. The infected children followed in this study had a mortality rate of 11% after 5 years without treatment [43], a lower mortality rate than the 33% after 2 years seen in infants infected postnatally in Zimbabwe [45, 46]. Although this was a retrospective study and some mothers and children may have been excluded, the authors suggest that the short time spent HIV-free after birth was sufficient for the immune system of the infants to extend their life expectancy.

The national PMTCT program addresses potential late-term HIV infection in mothers by providing a second voluntary HIV test for pregnant women who initially tested negative, but it is unknown how many women take advantage of this opportunity. Women who test negative in antenatal care may receive a second HIV test at delivery at the discretion of the attending medical providers. Few are tested after delivery or during breastfeeding since China is considered a low-HIV prevalence country overall.

Breastfeeding Among HIV-Infected Women

Provision of infant formula to women who choose it is part of the national PMTCT program, and the 2011 PMTCT guidance follows 2010 WHO recommendations to provide maternal triple ARV prophylaxis. Infant feeding advice and provision of antiretroviral prophylaxis during breastfeeding also follow WHO recommendations [9, 10, 47]. As part of the national PMTCT program, most HIV-infected women in China are encouraged to avoid breastfeeding and exclusively formula feed since most areas reach AFASS standards. Special studies in Yunnan and Henan provinces have reported 96–100% avoidance of breastfeeding among known HIV-infected mothers [48–50].

Following WHO option B, ARVs are provided from 14 weeks of gestation or as soon as possible thereafter. Mothers who choose to breastfeed continue on ARVs until 1 week after breastfeeding is discontinued. Women with CD4+ T cell count <350 cells/μl who need antiretroviral therapy (ART) for their own health receive an efavirenz (EFV)-based regimen, zidovudine+lamivudine+efavirenz (AZT+3TC+EFV). For women with CD4+ T cell count <250 cells/μl, a nevirapine (NVP)-based regimen (AZT+3TC+NVP) can also be used. Women with CD4+ T cell counts over 350 cells/μl receive a lopinavir/ritonavir (LPV/r)-based regimen (AZT+3TC+LPV/r) or EFV-based regimen (AZT+3TC+EFV). After delivery, women receive advice regarding ARV therapy cessation if the infant is exclusively formula fed or continue with ARVs until 1 week after complete weaning.

Regardless of the maternal CD4+ T cell count or feeding method, infants receive NVP or AZT starting 6–12 hours after birth for 4–6 weeks.

Women determined HIV-infected at delivery receive WHO prophylaxis option A, single-dose NVP (sd-NVP) at labor in addition to AZT+3TC for 7 days postpartum if they choose not to breastfeed. Infants receive sd-NVP plus AZT or daily NVP from birth until 4–6 weeks of age. If mothers choose to breastfeed, there are two choices. Either (1) the mothers receive an LPV/r-based regimen (AZT+3TC+LPV/r) or an EFV-based regimen (AZT+3TC+EFV) until 1 week after weaning and the infant receives NVP from birth to 6 weeks or (2) the mother discontinues prophylaxis after delivery and the infant receives NVP from birth until 1 week after complete weaning from breastfeeding. If a woman is determined HIV-infected postdelivery, infants receive NVP through 1 week after weaning.

Breastfeeding Among Women with Hepatitis B Infection

Co-infection with HIV and hepatitis B virus (HBV) affects both diseases. HIV increases HBV replication, with the potential for increased risk of MTCT of HBV. There is increased liver-related mortality from chronic hepatitis B infection, and infants who are infected with HIV do not respond as well to the hepatitis B vaccine [51].

An estimated 7% of the population of China ages 1–59 (approximately 94 million) is positive for hepatitis B surface antigen (HBsAg) [52]. In one study carried out in four cities, 5% of HIV-positive women had HBsAg co-infection [53]. HBsAg rates among pregnant women are similar, 7–8% [54, 55], One study observed slightly higher HBsAg rates in rural areas (7.1%) compared to that in urban areas (5.8%) [55].

Breastfeeding is not thought to be a major MTCT route for HBV, especially in properly immunized neonates [56–58], even though HBV markers have been found in the breast milk of highly infectious women [59]. A 2010 meta-analysis of studies from China found no difference in HBV transmission between breastfed and nonbreastfed infants after proper immunization [60]. While the recommendation for women with HIV is to avoid breastfeeding where formula feeding is safe, women with chronic HBV infection are advised to continue breastfeeding if the infant has received full immunization and immunoglobin against hepatitis B [10]. For HBV and HIV co-infected women, breastfeeding is not recommended.

HIV Among Women of Childbearing Age Who Inject Drugs

Female IDU are a population with an inordinately high risk of HIV infection and with special needs related to PMTCT. Females comprise approximately 20% of the estimated 2.4 million IDU in China [61, 62]. Many female IDU, 40–60%, also engage in sex work and are at especially high risk for HIV infection [63–65]. Female IDU, more so than male IDU or female sex workers (FSW) who do not inject drugs, serve as a bridge population transmitting HIV through sexual contact to non-IDU partners and clients [65]. Female IDU who engage in sex work often participate in other high-risk behaviors such as sharing needles or syringes and not using condoms. They also tend to have less family support [63, 66–68].

HIV-infected drug users in China are less likely to receive antiretroviral treatment [27, 35] and more likely to have co-infections such as hepatitis B and C virus [53]. FSW are more likely to be co-infected with chlamydia, gonorrhea, or syphilis [69]. Female IDU, especially those with HIV or other infections, are in need of comprehensive health services, including opioid replacement therapy, and have a greater need for prenatal care and support services when pregnant. Proper prenatal care and methadone maintenance treatment (MMT) can reduce the risk of prematurity, low birth weight, and MTCT of HIV, HBV, and syphilis [70]. With the rapid expansion since 2004 of MMT clinics across China [71], priority enrollment to those who are HIV-positive, and linkages between MMT and PMTCT programs, an increasing number of these women are being reached. Methadone therapy is not a contraindication for breastfeeding, and breastfeeding is particularly beneficial to methadone-exposed neonates who are at high risk for morbidity and neurobehavioral difficulties [72]. However, many female IDU are not eligible or do not elect to enroll in MMT [73] and do not seek antenatal care. An urgent need exists for earlier HIV diagnosis and better access to HIV care and treatment for female (as well as male) IDU in China, especially before they become severely immunosuppressed [74].

Conclusion and Recommendations

With the widespread expansion of PMTCT services across China including HIV testing during antenatal care and delivery, ARV prophylaxis, and the use of replacement feeding, reductions in MTCT of HIV have already been observed and further reductions are expected. However, the national goal of virtual elimination of MTCT (defined provisionally as <5%) by 2015 will require an expansion of efforts.

Progress towards reaching the goals for pediatric HIV elimination will be determined in particular by population-wide coverage with intensive PMTCT-related interventions in the ten provinces with the highest HIV case load. PMTCT is particularly difficult among women who do not know they are HIV-infected, including those who seroconvert during pregnancy or the breastfeeding period. Where HIV prevalence among pregnant women is above or near 1%, consideration should be given to provision of regular HIV retesting during late pregnancy, labor and delivery, and the breastfeeding period, in order to identify new infections among women who tested negative earlier in pregnancy and to provide appropriate interventions. The addition of HBV and syphilis testing to PMTCT programs will locate more women with these co-infections and provide further opportunity to avoid complications of co-infection. Outreach is needed to women who inject drugs to encourage enrollment in MMT, timely prenatal care and routine HIV testing and PMTCT services, if needed. With these efforts, maternal and child health outcomes can be improved and transmission of HIV from mothers to their babies decreased.

References

1. Walker A (2010) Breast milk as the gold standard for protective nutrients. J Pediatr 156:S3–S7
2. Duijts L, Jaddoe VW, Hofman A, Moll HA (2010) Prolonged and exclusive breastfeeding reduces the risk of infectious diseases in infancy. Pediatrics 126:e18–e25
3. Ip S, Chung M, Raman G, Chew P, Magula N, DeVine D et al (2007) Breastfeeding and maternal and infant health outcomes in developed countries. Evid Rep Technol Assess 153:1–186
4. WHO (2007) Evidence on the long-term effects of breastfeeding: systematic review and meta-analyses. http://www.who.int/child_adolescent_health/documents/9241595230/en/index.html. Accessed 22 May 2011
5. Gartner LM, Morton J, Lawrence RA, Naylor AJ, O'Hare D, Schanler RJ et al (2005) Breastfeeding and the use of human milk. Pediatrics 115:496–506
6. de Vincenzi I (2011) Triple antiretroviral compared with zidovudine and single-dose nevirapine prophylaxis during pregnancy and breastfeeding for prevention of mother-to-child transmission of HIV-1 (Kesho Bora study): a randomised controlled trial. Lancet Infect Dis 11:171–180
7. Bulterys M, Wilfert CM (2009) HAART during pregnancy and during breastfeeding among HIV-infected women in the developing world: has the time come? AIDS 23:2473–2477
8. Kuhn L, Aldrovandi GM, Sinkala M, Kankasa C, Mwiya M, Thea DM (2010) Potential impact of new WHO criteria for antiretroviral treatment for prevention of mother-to-child HIV transmission. AIDS 24:1374–1377

9. WHO (2010) Guidelines on HIV and infant feeding. http://www.who.int/child_adolescent_health/documents/9789241599535/en/. Accessed 20 May 2011
10. Chinese National Center for Women's and Children's Health (2011) Implementation of the national program for prevention of AIDS, syphilis and hepatitis B mother to child transmission. Ministry of Health, Beijing (Chinese)
11. WHO/UNICEF (2010) Progress on sanitation and drinking water: 2010 update. http://whqlibdoc.who.int/publications/2010/9789241563956_eng.pdf. Accessed 1 June 2011
11a. Gartner LM, Stone C (1994) Two thousand years of medical advice on breastfeeding: comparison of Chinese and western texts. Semin Perinatol 18:532–536
12. Xu F, Qiu L, Binns CW, Liu X (2009) Breastfeeding in China: a review. Int Breastfeed J 4:6. http://www.internationalbreastfeedingjournal.com/content/4/1/6. Accessed 31 Jan 2012
13. All-China Women's Federation (2010). National program of action for child development in China (2001–2010). http://www.womenofchina.cn/html/report/514-1.htm. Accessed 27 May 2011
14. UNICEF. The baby-friendly hospital initiative. http://www.unicef.org/programme/breastfeeding/baby.htm. Accessed 20 May 2011
15. Save the children. Only one in three Chinese mothers exclusively breastfeed for six months. http://www.savethechildren.org.cn/index.php/en/news-room/525-36. Accessed 20 May 2011
16. UNICEF. China statistics. http://www.unicef.org/infobycountry/china_statistics.html. Accessed 20 May 2011
17. Qiu L, Zhao Y, Binns CW, Lee AH, Xie X (2009) Initiation of breastfeeding and prevalence of exclusive breastfeeding at hospital discharge in urban, suburban and rural areas of Zhejiang China. Int Breastfeed J 4:1. http://www.internationalbreastfeedingjournal.com/content/4/1/1. Accessed 31 Jan 2012
18. Yang C, Sangthong R, Chongsuvivatwong V, McNeil E, Lu L (2009) Effect of village income and household income on sanitation facilities, hygiene behaviours and child undernutrition during rapid economic growth in a rural cross-border area, Yunnan, China. J Epidemiol Community Health 63:403–407
19. Xu F, Binns C, Wu J, Yihan R, Zhao Y, Lee A (2007) Infant feeding practices in Xinjiang Uygur Autonomous Region, People's Republic of China. Public Health Nutr 10:198–202
20. Chu A, Jiang N, Glantz SA (2011) Transnational tobacco industry promotion of the cigarette gifting custom in China. Tob Control 30 January. http://tobaccocontrol.bmj.com/content/early/2011/01/30/tc.2010.038349.abstract. Accessed 27 Jan 2012
21. Chinese Center for Disease Control and Prevention. Global Adult Tobacco Survey (GATS) Fact Sheet China. http://www.cdc.gov/tobacco/global/gats/countries/wpr/fact_sheets/china/2010/index.htm. Accessed 2 June 2011
22. Xu F, Binns C, Zhang H, Yang G, Zhao Y (2010) Paternal smoking and breastfeeding in Xinjiang, PR China. J Hum Lact 26:242–247
23. Leung G, Ho L, Lam T (2002) Maternal, paternal and environmental tobacco smoking and breast feeding. Paediatr Perinat Epidemiol 16:236–245
24. Ingelfinger JR (2008) Melamine and the global implications of food contamination. N Engl J Med 359:2745–2748
25. Guan N, Fan Q, Ding J, Zhao Y, Lu J, Ai Y et al (2009) Melamine-contaminated powdered formula and urolithiasis in young children. N Engl J Med 360:1067–1074
26. Seror J, Amar A, Braz L, Rouzier R (2010) The Google news effect: did the tainted milk scandal in China temporarily impact newborn feeding patterns in a maternity hospital? Acta Obstet Gynecol Scand 89:823–827
27. Zhang F, Dou Z, Ma Y, Zhao Y, Liu Z, Bulterys M et al (2009) Five-year outcomes of the China National Free Antiretroviral Treatment Program. Ann Intern Med 151:241–251, W-52
28. Bulterys M, Vermund SH, Chen RY, Ou CY (2009) A public health approach to rapid scale-up of free antiretroviral treatment in China: an ounce of prevention is worth a pound of cure. Chin Med J (Engl) 122:1352–1355
29. UNAIDS (2010) China 2010 UNGASS country progress report (2008–2009). http://www.unaids.org/en/dataanalysis/monitoringcountryprogress/2010progressreportssubmittedbycountries/china_2010_country_progress_report_en.pdf. Accessed 9 June 2011
30. Wang N, Wang L, Wu Z, Guo W, Sun X, Poundstone K et al (2010) Estimating the number of people living with HIV/AIDS in China: 2003–09. Int J Epidemiol 39(Suppl 2):ii21–ii28
31. Duan S, Shen S, Bulterys M, Jia Y, Yang Y, Xiang L et al (2010) Estimation of HIV-1 incidence among five focal populations in Dehong, Yunnan: a hard hit area along a major drug trafficking route. BMC Public Health 10:180. http://www.biomedcentral.com/1471-2458/10/180. Accessed 31 Jan 2012
32. Wang F, Nie Y, Liu J, Hou Z, Jiao X, Li Z (2009) [Analysis of prevention of mother-to-child transmission of HIV (PMTCT) work in Zhumadian city, 2001–2009]. Zhonghua Yu Fang Yi Xue Za Zhi 43:988–990 (Chinese with English abstract)
33. Liu L, Luan R, Yang W, Zhang L, Zhang J, Nan L et al (2009) Projecting dynamic trends for HIV/AIDS in a highly endemic area of China: estimation models for Liangshan Prefecture, Sichuan Province. Curr HIV Res 7:390–397
34. Ni M, Wheeler KM, Cheng J, Dong Y, Chen W, Fitzwarryne C et al (2006) HIV/AIDS prevalence and behaviour in drug users and pregnant women in Kashgar Prefecture: case report. Harm Reduct J 3:7. http://www.harmreductionjournal.com/content/3/1/7. Accessed 31 Jan 2012

35. State Council AIDS Working Group, UN Theme Group on AIDS in China (2007) A joint assessment of HIV/AIDS prevention, treatment and care in China (2007). Beijing: China Ministry of Health. http://www.undp.org.cn/modules.php?op=modload&name=News&file=article&catid=18&topic=7&sid=499&mode=thread&order=0&thold=0. Accessed 20 May 2011
36. Fang L, Xing Z, Wang L, Wang Q, Zhang W, Sun D et al (2009) Influencing factors on the death of infants born to HIV infected mothers. Zhonghua Yu Fang Yi Xue Za Zhi 43:991–995 (Chinese with English abstract)
37. Han M, Chen Q, Hao Y, Hu Y, Wang D, Gao Y et al (2010) Design and implementation of a China comprehensive AIDS response programme (China CARES), 2003–08. Int J Epidemiol 39(Suppl 2):ii47–ii55
38. Wu Z, Sullivan SG, Wang Y, Rotheram-Borus MJ, Detels R (2007) Evolution of China's response to HIV/AIDS. Lancet 369:679–690
39. Wang F, Fang L, Wang L (2008) Action for prevention of mother-to-child transmission of HIV in China towards universal access for women and children. XVII international AIDS conference, 3–8 August 2008. Mexico City, Mexico; abstract THPE0269
40. (2010) HIV transmission through transfusion – Missouri and Colorado, 2008. MMWR Morb Mortal Wkly Rep 59: 1335–1339
41. Wu Z, Liu Z, Detels R (1995) HIV-1 infection in commercial plasma donors in China. Lancet 346:61–62
42. Dou Z, Chen RY, Wang Z, Ji G, Peng G, Qiao X et al (2010) HIV-infected former plasma donors in rural Central China: from infection to survival outcomes, 1985–2008. PLoS One 5:e13737. http://www.plosone.org/article/info%3Adoi%2F10.1371%2Fjournal.pone.0013737. Accessed 31 Jan 2012
43. Liang K, Gui X, Zhang Y, Zhuang K, Meyers K, Ho D (2009) A case series of 104 women infected with HIV-1 via blood transfusion postnatally: high rate of HIV-1 transmission to infants through breast-feeding. J Infect Dis 200: 682–686
44. John-Stewart GC (2009) Strategic approaches to decrease breast milk transmission of HIV-1: the importance of small things. J Infect Dis 200:1487–1489
45. Zhao Y, Sun X, He Y, Tang Z, Peng G, Liu A et al (2010) Progress of the National Pediatric Free Antiretroviral Therapy program in China. AIDS Care 22:1182–1188
46. Marinda E, Humphrey JH, Iliff PJ, Mutasa K, Nathoo KJ, Piwoz EG et al (2007) Child mortality according to maternal and infant HIV status in Zimbabwe. Pediatr Infect Dis J 26:519–526
47. WHO (2010) Antiretroviral drugs for treating pregnant women and preventing HIV infection in infants: recommendations for a public health approach. http://www.who.int/hiv/pub/mtct/antiretroviral2010/en/index.html. Accessed 7 June 2011
48. Zhou Z, Meyers K, Li X, Chen Q, Qian H, Lao Y et al (2010) Prevention of mother-to-child transmission of HIV-1 using highly active antiretroviral therapy in rural Yunnan, China. J Acquir Immune Defic Syndr 53(Suppl 1):S15–S22
49. Cheng W, Zi X, Zhang L (2009) Evaluation of effectiveness on prevention of HIV mother-to-child transmission. Mod Prev Med 36(1252):1252–1257 (Chinese with English abstract)
50. Chen Z, Wang Y, Sun D, Wang Q, Wang W, Peng Y (2007) Analysis on efficacy of intervention on mother to child transmission of HIV in Henan province. Chin J Public Health 23(1417):1417–1418 (Chinese with English abstract)
51. Thio CL, Locarnini S (2007) Treatment of HIV/HBV coinfection: clinical and virologic issues. AIDS Rev 9:40–53
52. Liang X, Bi S, Yang W, Wang L, Cui G, Cui F et al (2009) Epidemiological serosurvey of hepatitis B in China – declining HBV prevalence due to hepatitis B vaccination. Vaccine 27:6550–6557
53. He N, Chen L, Lin H, Zhang M, Wei J, Yang J et al (2011) Multiple viral coinfections among HIV/AIDS patients in China. Biosci Trends 5:1–9
54. Guo Y, Liu J, Meng L, Meina H, Du Y (2010) Survey of HBsAg-positive pregnant women and their infants regarding measures to prevent maternal-infantile transmission. BMC Infect Dis 10:26. http://www.biomedcentral.com/1471-2334/10/26. Accessed 31 Jan 2012
55. Zhang S, Li R, Wang Y, Liu Q, Zhou Y, Hu Y (2010) Seroprevalence of hepatitis B surface antigen among pregnant women in Jiangsu, China, 17 years after introduction of hepatitis B vaccine. Int J Gynaecol Obstet 109:194–197
56. Hill JB, Sheffield JS, Kim MJ, Alexander JM, Sercely B, Wendel GD (2002) Risk of hepatitis B transmission in breast-fed infants of chronic hepatitis B carriers. Obstet Gynecol 99:1049–1052
57. Wang J, Zhu Q, Wang X (2003) Breastfeeding does not pose any additional risk of immunoprophylaxis failure on infants of HBV carrier mothers. Int J Clin Pract 57:100–102
58. Beasley RP, Stevens CE, Shiao IS, Meng HC (1975) Evidence against breast-feeding as a mechanism for vertical transmission of hepatitis B. Lancet 2:740–741
59. Linnemann CC Jr, Goldberg S (1974) Letter: HBAg in breast milk. Lancet 2:155
60. Shi Z, Yang Y, Wang H, Ma L, Schreiber A, Li X et al (2011) Breastfeeding of newborns by mothers carrying hepatitis B virus: a meta-analysis and systematic review. Arch Pediatr Adolesc Med 2 May. http://archpedi.ama-assn.org/cgi/content/abstract/archpediatrics.2011.72v1. Accessed 27 Jan 2012
61. Mathers BM, Degenhardt L, Phillips B, Wiessing L, Hickman M, Strathdee SA et al (2008) Global epidemiology of injecting drug use and HIV among people who inject drugs: a systematic review. Lancet 372:1733–1745

62. Needle RH, Zhao L (2010) HIV prevention among injection drug users: strengthening US support for core interventions. http://csis.org/files/publication/100408_Needle_HIVPrevention_web.pdf. Accessed 3 July 2011
63. Choi S, Cheung Y, Chen K (2006) Gender and HIV risk behavior among intravenous drug users in Sichuan Province, China. Soc Sci Med 62:1672–1684
64. Gu J, Wang R, Chen H, Lau J, Zhang L, Hu X et al (2009) Prevalence of needle sharing, commercial sex behaviors and associated factors in Chinese male and female injecting drug user populations. AIDS Care 21:31–41
65. Liu H, Grusky O, Li X, Ma E (2006) Drug users: a potentially important bridge population in the transmission of sexually transmitted diseases, including AIDS, in China. Sex Transm Dis 33:111–117
66. Lau J, Zhang J, Zhang L, Wang N, Cheng F, Zhang Y et al (2007) Comparing prevalence of condom use among 15,379 female sex workers injecting or not injecting drugs in China. Sex Transm Dis 34:908–916
67. Lau J, Tsui H, Zhang Y, Cheng F, Zhang L, Zhang J et al (2008) Comparing HIV-related syringe-sharing behaviors among female IDU engaging versus not engaging in commercial sex. Drug Alcohol Depend 97:54–63
68. Gu J, Lau J, Chen H, Tsui H, Ling W (2011) Prevalence and factors related to syringe sharing behaviours among female injecting drug users who are also sex workers in China. Int J Drug Policy 22:26–33
69. van den Hoek A, Fu Y, Dukers NH, Chen Z, Feng J, Zhang L et al (2001) High prevalence of syphilis and other sexually transmitted diseases among sex workers in China: potential for fast spread of HIV. AIDS 15:753–759
70. El-Mohandes A, Herman AA, Nabil El-Khorazaty M, Katta PS, White D, Grylack L (2003) Prenatal care reduces the impact of illicit drug use on perinatal outcomes. J Perinatol 23:354–360
71. Liu Y, Wu Z, Mao Y, Rou K, Wang L, Zhang F (2010) Quantitatively monitoring AIDS policy implementation in China. Int J Epidemiol 39(Suppl 2):ii90–ii96
72. Jansson LM, Choo R, Velez ML, Harrow C, Schroeder JR, Shakleya DM et al (2008) Methadone maintenance and breastfeeding in the neonatal period. Pediatrics 121:106–114
73. Sullivan SG, Wu Z (2007) Rapid scale up of harm reduction in China. Int J Drug Policy 18:118–128
74. Zhang F, Dou Z, Ma Y, Zhang Y, Zhao Y, Zhao D et al (2011) Effect of earlier initiation of antiretroviral treatment and increased treatment coverage on HIV-related mortality in China: a national observational cohort study. Lancet Infect Dis 11:516–524

Chapter 18
The Role of the President's Emergency Plan for AIDS Relief in Infant and Young Child Feeding Guideline Development and Program Implementation

Michelle R. Adler, Margaret Brewinski, Amie N. Heap, and Omotayo Bolu

The basic science and clinical research investigating the relationship between HIV and breastfeeding has provided much of the evidence base for the development of public health policies and practice guidelines aimed at preventing mother-to-child transmission (PMTCT) of HIV. Programmatically, however, translating the evidence into practice has been challenging. In 2003, President George W. Bush established the President's Emergency Plan for AIDS Relief (PEPFAR) to "turn the tide against AIDS in the most afflicted nations of Africa and the Caribbean." Through multiple US agencies,[1] PEPFAR will have provided $63 billion between 2004 and 2013 in direct financial support and technical assistance for the implementation of HIV prevention, care, and treatment programs throughout the world. Focusing on PEPFAR's role in infant feeding guideline modification and implementation, this chapter reviews the history of infant feeding guideline revisions based on evolving research and evaluation, highlights the successes and challenges of translating this rapidly changing evidence into practice, and concludes with a discussion of potential strategies for the adoption and implementation of 2010 WHO PMTCT and infant feeding guidelines.

The Early Approach to HIV and Infant Feeding

It was discovered early in the epidemic that HIV could be transmitted via breastmilk [1]. In places where infant formula was available, clean water was accessible, and bottle-feeding was acceptable, it was clear that HIV-infected women should be advised not to breastfeed [2]. For low and middle income countries, however, multiple environmental, social, and economic factors complicated the decision

---

[1] US government agencies that receive funds for international HIV/AIDS programs: US Agency for International Development, Department of Health and Human Services (Centers for Disease Control and Prevention, National Institutes of Health, Health Resources and Services Administration, Food and Drug Administration, and Substance Abuse and Mental Health Services Administration, Department of State, Peace Corps, Department of Defense, Department of Commerce, and Department of Labor (http://www.pepfar.gov/agencies/index.htm).

M.R. Adler, M.D., M.P.H. (✉) • O. Bolu, M.B.B.S., M.Sc.
Division of Global HIV/AIDS, Maternal and Child Health Branch, Prevention of Mother to Child HIV Transmission (PMTCT), Centers for Disease Control and Prevention,
1600 Clifton Road NE, Mail Stop E-04, Atlanta, GA 30333, USA
e-mail: madler@cdc.gov

M. Brewinski, M.D., M.P.H. • A.N. Heap, M.P.H., R.D.
United States Agency for International Development, Office of HIV/AIDS, Washington, DC, USA

about the infant feeding guidance that should be given to HIV-infected women [3]. Formula feeding had been shown to increase the risk of infant mortality in the general population [4]. As a global public health response, many countries engaged communities in promoting the benefits of breastfeeding and endorsed the *International Code of Marketing of Breast-Milk Substitutes* regulating infant formula marketing [5]. While minimizing mother-to-child transmission (MTCT) of HIV was a high priority, care had to be taken not to compromise the breastfeeding promotion efforts that had led to progress in curbing infant and child mortality. Ultimately, this discourse transformed into an international public health framework for infant feeding that endorsed an individualized approach to infant feeding in the context of HIV [6]. Whereas countries were encouraged to intensify efforts to protect, promote, and support breastfeeding in the general population, HIV was seen as an "exceptionally difficult circumstance" within which women should be supported in making informed choices about infant feeding methods, balancing the risk of HIV transmission against the risk of infant mortality [7].

The 2003 *HIV and Infant Feeding Guidelines* accompanying this framework provided health care workers with criteria to help HIV-infected women decide whether or not to breastfeed. "When replacement feeding is acceptable, feasible, affordable, sustainable, and safe, avoidance of all breastfeeding by HIV-infected mothers is recommended" [8]. These criteria became known as "AFASS." Based on an early study suggesting that the provision of breastmilk alone may have a lower transmission rate than a combination of breastmilk and other foods [9], the guidelines also recommended that women who did not meet the AFASS criteria should be encouraged to exclusively breastfeed for 6 months followed by rapid weaning. However recognizing that exclusive replacement feeding and breastfeeding with early weaning might have serious repercussions, it was further recommended that all HIV-infected mothers should be provided with "…specific guidance and support for at least the first two years of the child's life" [8].

PEPFAR Mobilizes Countries to Implement 2003 WHO Infant and Young Child Feeding Guidelines

PEPFAR's support to scale-up PMTCT and pediatric care and treatment activities globally began in the context of this international framework emphasizing an individualized approach to infant feeding for HIV-infected mothers. With its significant financial and human resources, PEPFAR supported countries to develop national infant feeding policies, adapt counseling tools, and train health care workers to assist HIV-infected women in deciding how to feed their infants [10]. By the end of 2008, for example, the initial 15 PEPFAR-funded countries (Table 18.1) had national policies reflecting the international guidelines, and at PEPFAR-funded sites in over half of those countries, implementation was reported at greater than 80% [11]. With UNICEF and WHO as the lead international agencies on this issue, PEPFAR also provided considerable impetus in other countries, such as Malawi, Swaziland, Zimbabwe, Cambodia, China, and Thailand, to adapt national guidelines and move infant feeding discussions into action (Bulterys M, Personal communication. Atlanta, 30 July 2011).

The infant feeding policies adopted by some of these countries attempted to make formula available for HIV-infected mothers free of charge with the idea that, by removing the issue of affordability of replacement feeds, mothers could make an informed decision independent of their economic status. In a few countries, formula was available for all HIV-infected mothers. In others, formula provision was piloted only at specific sites or within specific programs. While PEPFAR did not routinely fund formula provision, PEPFAR partners and implementers worked in many sites where formula was available and played an instrumental role in evaluating both the individualized approach to counseling and the provision of free formula.

In Rwanda, a pilot program [12] reported success in decreasing transmission and improving infant survival through intensive community midwife outreach. However in Botswana [13] and Uganda [14], evaluations of population subsets reported increased morbidity and mortality in formula-fed infants. Although formula was given free to HIV-exposed infants for at least 6 months in both Botswana

Table 18.1 Initial 15 PEPFAR funded countries

Initial 15 PEPFAR-funded countries		
Botswana	Kenya	South Africa
Cote D'Ivoire	Mozambique	Tanzania
Ethiopia	Namibia	Uganda
Guyana	Nigeria	Vietnam
Haiti	Rwanda	Zambia

and in some sites in Nigeria [15], patients reported insufficient formula supply as a significant problem. Unhygienic practices for bottle cleaning and formula storage were common in Botswana, although were not reported as an issue in Rwanda where community health nurses routinely visited clients who were receiving formula.

Implementation Challenges Contribute to 2007 Infant Feeding Guideline Update

At the same time that PEPFAR programs were demonstrating implementation challenges, studies confirmed concerns about increased morbidity and mortality related to replacement feeding and early weaning [16]. Further, no HIV-free survival benefit was found with abrupt weaning as compared to gradual weaning in 4-month-old Zambian infants; to the contrary, early weaning correlated with growth faltering and serious diarrhea [17]. Finally, two published reports demonstrated reduced HIV transmission associated with exclusive as opposed to mixed feeding [18, 19]. Based on this developing body of evidence, the WHO HIV and infant feeding guidelines were reframed in 2007 [20]. While an individualized informed consent process remained the guiding principle, the primary recommendation shifted from replacement feeding to breastfeeding. While the 2003 guidelines had *recommended replacement feeding* unless mothers did not meet AFASS, the 2006 update *recommended breastfeeding* unless mothers could meet the AFASS criteria. The specific recommendations included:

- Exclusive breastfeeding for 6 months (same as 2003 recommendation).
- Complementary feeding if AFASS not met at 6 months until "a nutritionally adequate and safe diet can be provided."
- Gradual weaning over 2–3 days to 2–3 weeks to prevent maternal breast problems such as mastitis, and growth faltering and serious diarrheal illness in infants.

Despite Guideline Updates, PEPFAR-Supported Programs Demonstrated that Challenges Continued

Further studies and public health evaluations of PEPFAR-supported programs confirmed the initial studies and observations of increased mortality in populations of infants who stopped breastfeeding early. In rural Uganda, replacement feeding from birth and early weaning was associated with a sixfold increase in infant deaths [21]. In Botswana, a country where the Ministry of Health adopted a policy to provide formula for the infants of all HIV-infected mothers, a serious widespread diarrhea outbreak occurred in 2006. A Centers for Disease Control and Prevention (CDC) evaluation of 153 children hospitalized with diarrhea during the outbreak revealed that 97% were <2 years and 88% were not breastfeeding. Further, of the 33 children who died, only one was breastfeeding (Box 18.1) [22]. Finally, Cote d'Ivoire discontinued their free infant formula program in 2007 due to poor infant outcomes and high cost. These evaluations demonstrated that rather than improve HIV-free survival, early weaning significantly *increased* morbidity and mortality.

> **Box 18.1 Botswana: The Reality of Replacement Feeding**
>
> Botswana has been a leader in confronting the HIV epidemic in the developing world. As early as 1999 the government of Botswana, in an effort to drastically curb MTCT of HIV, committed to providing free formula for HIV-exposed infants. Despite decreasing HIV transmission, little impact has been made in reducing infant mortality in the last 20 years [23]. A clinical trial conducted between 2001 and 2003 found that among 1,193 HIV-exposed infants, those fed formula since birth had significantly higher mortality than those who breastfed for 6 months. Additionally, there was no difference in HIV-free survival at 18-months [16]. A CDC investigation of a pediatric diarrhea outbreak in Botswana in 2006 found that of 153 hospitalized infants with diarrhea, 88% were not breastfeeding and over half had preexisting malnutrition [22]. Many concerning practices were found among these formula-fed infants:
>
> - Inadequate supply of formula—often due to central stock-out or staff rationing.
> - Incorrect formula preparation.
> - Inadequate labeling of formula tins.
> - Lack of hygienic practices in the home (such as hand washing).
>
> However, none of these factors was individually found to be statistically associated with the outbreak. The only risk factor of significance was replacement feeding, demonstrating the complex array of survival benefits gained through breastfeeding, and that provision of free formula cannot replace [13].
>
> Botswana continues to offer infant formula today as part of its PMTCT program. Simultaneously, they have adapted the WHO Infant and Young Child Feeding (IYCF) curriculum to train health care workers to help HIV-infected women make informed feeding choices. A recent survey found that over 90% of these women still choose to formula feed, which may reflect both a bias in the counseling given by health care workers, and entrenched perceptions of women in the community regarding their viable options to safely feed their babies while still preventing MTCT of HIV (Lu L, unpublished). In order to implement policy recommending breastfeeding as the standard of care, a strong messaging campaign would be needed as a first step to obtaining buy-in from health care workers and the community.

Programs in which formula was provided with the intention of making replacement feeding affordable and accessible were often found to result in an increase in mixed feeding. In Nigeria, a survey investigating the social determinants of mixed feeding within a pilot formula-provision program demonstrated that family pressure to breastfeed played a significant role [15]. Fear of stigma proved to be a concern in South Africa, where women did not want to be seen carrying formula-feeding supplies from the clinic [24]. Interestingly in Rwanda, however, a Partners in Health program demonstrated that stigma was not a barrier to formula provision; in fact, women would often claim to be HIV-infected children in order to receive the additional assistance [25].

A third factor was health care worker bias against breastfeeding for HIV-infected mothers, often due to incorrect knowledge [26]. Most health workers significantly overestimated the risk of HIV transmission via breastfeeding, and only 8% of 340 nurses and counselors in a four-country study knew the actual transmission risk [27]. Finally, the lack of health care workers adequately trained and sensitized to provide high quality counseling could not be underestimated [26]. These PEPFAR programmatic experiences, in addition to other non-PEPFAR examples [28, 29], clearly reflected some of the sociocultural and human resource challenges in utilizing the individualized AFASS approach to infant feeding choice.

In addition to concerning outcomes for HIV-exposed infants, the messaging surrounding infant feeding guidelines also adversely affected HIV-negative populations. It was specifically noted in Kenya that special messaging around infant feeding for HIV-infected women derailed strides that had been made in promoting the Baby Friendly Hospital Initiative, a hospital-based program designed to promote exclusive breastfeeding during the first 6 months of life for all babies. In Botswana, feeding practices by HIV-uninfected mothers were adversely affected by PMTCT messages regarding infant feeding at clinics [30].

These multiple unintended consequences illustrated a need to again reevaluate infant feeding programs. The Opening Symposium at the 2007 CROI (Conference on Retroviruses and Opportunistic Infections) meeting, which featured presentations on PMTCT and the dangers of early weaning and/or replacement feeding, turned out to be the watershed event that was needed to push for a fundamental guideline revision [31–33].

Programmatic Evidence Highlight the Flaws of the Individualized AFASS Approach to Infant Feeding

Given the observations that HIV-free survival did not improve with replacement feeding interventions, and the problems many programs encountered with insufficient formula, family acceptance, fear of stigma, health care worker bias, and limited human resources, a number of PEPFAR-supported partners began to look at alternative options. Concerned that the individual approach to infant feeding decision-making was not practical for some countries with poor human resource capacity and limited funding, the International Center for AIDS Care and Treatment Programs (ICAP) developed a facility AFASS assessment score-card to enable facilities to consider local conditions that would impact a woman's ability to safely replacement feed. Facilities scoring high could continue to offer replacement feeding, while it was recommended that those scoring low should consider not offering the replacement feeding option [34]. This approach also took some of the decision making out of the hands of the potentially biased health care worker by emphasizing a facility-based policy.

In 2009, the PEPFAR Technical Working Groups (TWGs) for PMTCT/Pediatric Care and Treatment and Food and Nutrition reviewed their respective program areas, drawing upon field reports and program evaluations. They highlighted the challenges, unintended consequences, and lessons learned as a result of the rapid implementation of an infant feeding policy that focused on an individualized as opposed to a population-based approach. Specifically, they concluded that in resource-limited settings, very few situations could meet AFASS criteria because replacement feeding was:

- Not acceptable: in cultures where breastfeeding is the norm, replacement feeding was not suitable to mothers or families. Further, mothers feared being stigmatized if they bottle fed.
- Not feasible: the bottles, nipples, and refrigerated storage needed for replacement feeding were not available.
- Not affordable: families did not have the income needed to purchase formula consistently, including during clinic stock-outs, and after the age when subsidized formula provision ceased.
- Not sustainable: poor infrastructure and forecasting often contributed to formula stock-outs at health care facilities.
- Not safe: lack of clean water, poor hygiene practices, and inability to clean and sterilize bottles contributed to increases in diarrheal diseases.

Ultimately, both TWGs concluded that the individualized approach to infant feeding choice was impractical and put HIV-exposed infants at increased risk of morbidity and mortality. They recommended prioritizing Infant and Young Child Feeding (IYCF) within PMTCT programs, and revisiting the implementation strategy with the dual goals of reducing MTCT and improving infant health and survival.

Landmark Clinical Trials Pave the Way for Fundamental Guideline Revision

At the same time that PEPFAR was demonstrating implementation challenges with AFASS, several landmark investigations established that the use of antiretroviral drugs (ARVs) during the breastfeeding period significantly reduced the risk of HIV transmission through breastfeeding [35–41]. CDC provided substantial technical and financial support for a number of these studies, including the KiBS study in Kenya [41] and the BAN study in Malawi [38]. Additionally, the PEPFAR-CDC country office co-funded the ZEBS study in Zambia [17] in order to allow close follow-up for 24 months postpartum and implementation of antiretroviral therapy (ART) for mothers and children. This support from PEPFAR and CDC allowed for multiple trials to be conducted simultaneously resulting in a body of evidence upon which clear and confident guidance could be based. There was no longer a need to have mothers individually weigh the risks of HIV transmission and infant mortality. Breastfeeding could now be strongly recommended to all mothers with the simultaneous provision of ARV prophylaxis to minimize HIV transmission risk.

The 2010 WHO PMTCT and Infant Feeding Guidelines

The proven effectiveness of extended ARV prophylaxis during breastfeeding paved the way for international guidance to endorse a unified public health approach clearly focused on HIV-free survival. The WHO 2010 PMTCT guidelines reflect the important new addition of ARV prophylaxis during breastfeeding as a requisite component of a PMTCT strategy [42]. As detailed in Table 18.2, countries choose either Option A or Option B for maternal and infant prophylaxis based on national context and resource availability.

The *2010 HIV and Infant Feeding Guidelines* also highlight the provision of ARV prophylaxis for the duration of breastfeeding. In light of the availability of prophylaxis, the recommended duration of breastfeeding was extended to at least 12 months, with ongoing breastfeeding encouraged until a nutritionally adequate and safe diet without breast milk can be provided [43]. In order to simplify and standardize infant feeding counseling and messages, the guidelines suggest a population-based, national endorsement of a single infant feeding strategy (i.e., breastfeeding with ARV prophylaxis or exclusive replacement feeding) based on the overall country context and prevailing circumstances for the majority of women. Provisions were included for special circumstances within the national context. Box 18.2 provides a summary of major modifications and points of emphasis between the 2007

Table 18.2 2010 WHO PMTCT guideline options for ARV prophylaxis

Option A: maternal AZT during pregnancy, followed by infant prophylaxis during breastfeeding	Option B: maternal triple ARV regimen throughout pregnancy and breastfeeding, with prophylaxis for the infant in first 6 weeks of life only
For the mother: • Antepartum daily AZT • sd-NVP at onset of labor • AZT+3TC during labor and delivery • AZT+3TC for 7 days postpartum	For the mother, one of the following regimens: • AZT+3TC+LPV/r or • AZT+3TC+ABC or • AZT+3TC+EFV or • TDF+3TC (or FTC)+EFV
For the *breastfeeding infant*: daily NVP from birth until 1 week after breastfeeding has ended	For the *breastfeeding infant*: daily NVP from birth until 6 weeks of age
For the *non-breastfeeding infant*: daily AZT or NVP from birth until 6 weeks of age	For the non-breastfeeding infant: daily AZT or NVP from birth until 6 weeks of age

3TC lamivudine, *ABC* abacavir, *AZT* zidovudine, *EFV* efavirenz, *FTC* emtricitabine, *LPV/r* lopinavir/ritonavir, *NVP* nevirapine, *TDF* tenofovir disoproxil fumarate

Adapted from: Antiretroviral drugs for treating pregnant women and preventing HIV infection in infants: recommendations for a public health approach—2010 version [42]

> **Box 18.2** Summary of 2010 Infant Feeding Guidelines, Key Areas of Emphasis and Changes from Previous Guidelines
>
> - **Breastfeeding covered by ARVs is a strategy for maximizing infant HIV-free survival.** The new guidelines balance the risks of MTCT with the risks of increased morbidity and mortality associated with non-breastfeeding. *(Previous guidelines did not emphasize HIV-free survival.)*
> - **Treatment eligibility must be determined for all HIV+ pregnant women, with those eligible initiated on life-long ART immediately, regardless of gestational age.** Women eligible for ART represent about 40% of HIV+ pregnant women and account for >75% of overall MTCT, >80% of postpartum MTCT, and 85% of maternal deaths within 2 years of delivery. Initiating all eligible pregnant women on ART will be essential for realizing PMTCT goals of maximizing maternal health, reducing MTCT transmission to <5% and achieving infant HIV free survival. *(Previous guidelines recognized but did not emphasize the importance of initiating eligible mothers on antiretroviral treatment.)*
> - **ARVs should be continued during the breastfeeding period to reduce transmission.** Women started on ART for their own health should continue their ARVs for life, including during breastfeeding. For those not yet eligible for lifelong treatment, a highly effective prophylactic regimen for PMTCT should be provided from as early as 14 weeks gestation until 1 week after cessation of all breastfeeding. *(Previous guidelines did not emphasize the link between breastfeeding and ART or provide regimens for prophylaxis during breastfeeding.)*
> - **Recommendations for appropriate infant feeding in the context of HIV should be made at the national or subnational level.** Health authorities should decide whether mothers should be principally counseled and supported to *either* breastfeed and receive ARVs *or* avoid all breastfeeding and replacement feed from birth. *(Previous guidelines emphasized individual, informed choice by mothers based on counseling regarding the balance of risks between MTCT and infant mortality associated with early weaning.)*
> - **Where breastfeeding is recommended, exclusive breastfeeding should be practiced for the first 6 months of life, and breastfeeding continued thereafter with the addition of complementary foods, until at least 12 months of age. Breastfeeding should continue beyond 1 year until a safe and adequate replacement diet can be assured.** Breastfeeding should stop gradually over 1 month, and ARV prophylaxis for mother or infant should be continued for 1 week after breastfeeding is fully stopped. *(Previous guidelines promoted weaning at 6 months, or as soon as possible thereafter, when replacement feeding was acceptable, feasible, affordable, safe & sustainable—also known as "AFASS".)*
> - **The difference between mixed feeding and complementary feeding should be emphasized in messaging to both health care workers and the general population.** Mixed feeding refers to providing the infant with other milks, foods, or liquids in addition to breastmilk during the first 6 months of life (a practice that increases the risk of HIV transmission and infant morbidity). Complementary feeding refers to the introduction of additional sources of nutrition for infants older than 6 months that are essential for adequate growth and development. Mixed feeding should be discouraged, while complementary feeding promoted. *(Previous guidelines recommended this approach, but did not clearly describe the difference between mixed and complementary feeding.)*
> - **Mothers of infants and young children known to be HIV-infected should be strongly encouraged to breastfeed exclusively for the first 6 months and to continue breastfeeding, with adequate complementary feeding, up to 2 years or beyond.** Breastfeeding is crucial to maintaining the nutritional status of and providing passive immunity to HIV-infected children. *(Previous guidelines also emphasized this recommendation.)*
>
> Adapted from *Technical Update to the Field: 2010 Guidelines for IYCF in the Context of HIV* [44]

and 2010 *WHO HIV and Infant Feeding Guidelines*, demonstrating the shift from an approach aimed at minimizing HIV transmission toward one aimed at maximizing maternal and infant health and HIV-free survival.

The Road to Implementing the 2010 Guidelines

With the fundamental changes in the PMTCT and IYCF guidelines, PEPFAR will renew funding and technical assistance toward facilitating implementation. Lessons learned from operationalizing changes made in 2003 and 2006 have, and will continue to help shape PEPFAR's guidance to partners and countries, as well as guide program evaluations on the effectiveness of the new guidelines.

Providing Technical Guidance

In early 2010, for example, PEPFAR collaborated with WHO, AFRO, UNICEF, and other international partners to convene countries in order to disseminate the new WHO recommendations on ART, PMTCT, and IYCF, and provide technical guidance on their simultaneous adaptation. Linkages between different guidelines and programs were highlighted in an effort to assist countries in a coordinated rollout effort. Technical guidance from different program areas was also disseminated. PEPFAR played a lead role in developing the infant feeding guidance that included the following priority interventions:

- Contextualize infant feeding policies and counseling messages around HIV-free survival.
- Develop a unified national infant feeding policy. For most resource-limited countries, this means endorsing exclusive breastfeeding for 6 months, followed by complementary feeding until at least 12 months. Gradual weaning over a month or more is encouraged.
- In countries with diverse populations and geography, the infant feeding policy may need to be regionally specific rather than nationally dictated. In Kenya, for example, most women would breastfeed, but in Nairobi, a substantial percentage of women can safely replacement feed. Conversely, in China and Vietnam, while most women can safely replacement feed, breastfeeding should be recommended to some highland ethnic minority tribes who lack access to financial resources and safe water.
- Broaden communication strategies to inform health workers, other professional and management staff, and civil society regarding the reasons for the changes in national approaches and the opportunities to improve HIV-free survival.
- Emphasize the rapid scale up of ART and more efficacious PMTCT regimens for pregnant and lactating women.
- Invest in pre- and in-service training on HIV and nutrition for different cadres of health care workers, and design messaging aimed at correcting misconceptions surrounding the risk of HIV transmission through breast milk.
- Link infant feeding counseling to early infant diagnosis so that both HIV-uninfected and -infected infants are not prematurely weaned.
- Establish health and nutrition surveillance, referral and tracking systems linking clinics and communities to allow early identification and intervention to address maternal and infant malnutrition and health problems.

- Promote antenatal and postnatal family planning counseling, including the lactational amenorrhea method, to link exclusive breastfeeding with modern contraceptive methods.
- Document and spread early learning regarding how to implement and support new guidelines.

PEPFAR's technical assistance visits to countries as a follow-up to this meeting have and will continue to focus around these priority interventions.

Developing Indicators and Systems to Monitor and Evaluate PMTCT and IYCF Outcomes and Impact

In conjunction with the release of the revised guidelines, the Monitoring and Evaluation (M&E) Working Group of the Inter-agency Task Team (IATT) on the prevention of HIV infection among pregnant women, mothers, and their children identified a need to develop a standard set of PMTCT indicators that reflected the global goals of infant feeding outcomes and of HIV-free survival. IATT members, including representatives from PEPFAR and the Global Fund to Fight AIDS, Tuberculosis, and Malaria, agreed to harmonize indicators across programs in an effort to minimize burdensome and redundant reporting, and compare outcomes over time and between countries. The proposed indicators include three-key infant feeding outcome indicators (Table 18.3). Additionally, indicators to measure HIV-free survival as well as percentage of breastfed infants who are covered by PMTCT prophylaxis during breastfeeding are in the process of being developed. In order to assist countries in collecting the data needed to accurately report on these indicators, the M&E guidance document also offered methods and tools for measurement (including model registers) and encouraged national programs to use their own data for monitoring progress in PMTCT at national, regional, and facility levels.

These proposed standard national program indicators complement a set of more detailed infant feeding indicators (for the general population, though including HIV-exposed) that cover the continuum of feeding practices from 0 to 23 months:

1. Early initiation of breastfeeding.
2. Exclusive breastfeeding under 6 months.
3. Continued breastfeeding at 1 year.
4. Introduction of solid, semisolid, or soft foods.
5. Minimum dietary diversity.
6. Minimum meal frequency.
7. Minimum acceptable diet.
8. Consumption of iron-rich or iron-fortified foods.

Developed at a Global Consensus meeting in Washington, DC sponsored by WHO in November 2007, these more detailed indicators were designed to be used in general population-based surveys to

Table 18.3 Recommended National Indicators for Monitoring and Evaluating Infant Feeding Interventions

Recommended national indicators for monitoring and evaluating infant feeding interventions	
Indicator level	Name and definition
Outcome	Infant nutritional status: the number and percentage of HIV-exposed infants who have a weight for height Z-score less than −2 SD at 12 months of age
	Infant feeding: this indicator measures three specific areas of infant feeding: (1) percentage of HIV-exposed infants who are exclusively breastfeeding at 3 months of age, (2) percentage of HIV-exposed infants who are replacement feeding at 3 months of age, and (3) percentage of HIV-exposed infants who are mixed feeding at 3 months of age

Adapted from: WHO Draft Monitoring and Evaluating the Prevention of Mother to Child Transmission of HIV: a guide for national programmes [46]

assess individual components of the infant feeding process and identify drop-off points that can be used to target program interventions [45].

Leading by example, PEPFAR is considering adopting these IATT infant feeding indicators into PEPFAR reporting and is encouraging countries to incorporate them into national M&E programs. Further, PEPFAR is supporting the incorporation of the more detailed standardized infant feeding indicators into Demographic Health and AIDS-Indicator Surveys in many countries.

Establishing Continuous Quality Improvement Initiatives

In addition to M&E systems, continuous quality improvement (CQI) initiatives play a key role in infant feeding programming. Facility and community level CQI projects can serve to identify bottlenecks and address barriers early in national program implementation. These "early learning" sites can serve as effective models to facilitate and inform scale-up across the larger system.

Along with scale-up, ensuring that infant feeding is incorporated into a national CQI strategy is a critical step toward maintaining successful integration of infant feeding into the PMTCT continuum and into the overarching Maternal Child Health program at local, district, and national levels. With PEPFAR support, Elizabeth Glaser Pediatric AIDS Foundation, University Research Corporation, HEALTHQUAL International, and Institute for Healthcare Improvement, among other organizations, are already working on CQI initiatives in a number of countries. For example, PEPFAR is investing in a Nutrition Assessment, Counseling, and Support collaborative pilot program targeting the two-year postpartum period. It will assess the operational aspects of a common implementation framework using QI interventions with the ultimate aim of determining impact on HIV-free survival. Application of the CQI process will facilitate target setting, accountability, and efficiency allowing opportunities for task shifting, enhanced supervision, and early recognition of problems. Ultimately, PEPFAR anticipates these activities will provide a structure for ongoing improvement and ultimately maintenance of infant HIV-free survival.

Lessons Learned Pave the Way Forward

Guidance regarding infant feeding in the context of HIV has come full circle over the course of the past decade (Table 18.4). Prior to the HIV epidemic, a public health effort had been made to minimize the impact of formula marketing recognizing the impact it had on infant mortality. However, the HIV epidemic overshadowed that emphasis on infant survival with a sense of obligation to do everything possible to prevent infants from contracting HIV. Fortunately, science has enabled a return to a public health approach with an intervention that prevents transmission while maintaining the survival benefits of breastfeeding. A summary of key changes in the WHO guidelines on Infant Feeding in the Context of HIV from 1997 to 2010 can be found in Table 18.5.

Table 18.4 Summary of shifts in approaches to infant feeding

Fundamental shifts in approaches to infant feeding during the last decade		
Decrease HIV transmission while attempting to preserve child survival	→	Prioritize child survival while scaling up interventions to decrease HIV transmission
Replacement feed unless AFASS is not met	→	Breastfeed unless AFASS is met
Individualized approach taking into account cultural and epidemiologic context	→	Population approach allowing for individual choice

Table 18.5 Key changes to the WHO Guidelines on Infant Feeding in the Context of HIV: 1997–2010

1997/1998	2003	2006	2010
• Recognition of BF as mode of transmission and need for "appropriate alternatives to BF…available and affordable in adequate amounts" • BF identified as "the ideal way to feed the majority of infants," particularly for women of HIV negative or unknown status • Highlights need for increased availability of HIV testing and informed infant feeding choice for women living with HIV • Emphasis on prevention of HIV transmission to the baby, individual, informed infant feeding choice and cessation of BF by HIV-infected mothers as soon as they can provide adequate replacement feeds	• Affirms UN 4 prong PMTCT strategy • Introduces and defines AFASS concept as preferable if it can be achieved • Otherwise, exclusive BF is recommended during the first months of life with weaning as soon as AFASS conditions could be met • Ongoing emphasis on individual, informed infant feeding choice for women living with HIV • Calls on all governments to develop, implement, monitor, and evaluate a comprehensive Infant and Young Child Feeding (IYCF) policy • Aligned with the WHO *Global strategy for infant and young child feeding* and UN *HIV and infant feeding [47]: framework for priority action*, both also released in 2003	• Primary shift is from recommending replacement feeding unless AFASS cannot be met to recommending exclusive breastfeeding for 6 months unless AFASS could be met for replacement feeding • Gradual weaning over 2–3 weeks rather than abrupt weaning preferred • Use of antiretrovirals during the prenatal and intrapartum periods is greatly reducing MTCT, but the breastfeeding period remains a major challenge, particularly with mixed feeding commonly taking place due to many factors, including stigma, inadequate counseling and the practical difficulties of maintaining strict EBF or AFASS replacement feeding • Recognition that emphasis on replacement feeding for women living with HIV in some settings has negative effect on overall promotion of BF for all women	• Emphasis moves from reducing MTCT to maximizing maternal health and HIV-free infant health and survival • Growing concerns about increased infant morbidity and mortality with early weaning, particularly with diarrheal disease, and recognition that AFASS replacement feeding is extremely difficult for health workers and parents to assess, implement and assure • Accompanying PMTCT guidelines stress determination of ART eligibility and immediate initiation of treatment for eligible pregnant women, as they are at greatest risk for maternal morbidity and mortality and MTCT • For nontreatment eligible pregnant women, a combination antiretroviral regimen for PMTCT is initiated earlier and, for the first time, recommended to continue throughout the BF period specifically to reduce BF transmission • In light of the significant influence of BF on survival, the recommended duration of BF is extended to at least 12 months (in combination with ARV prophylaxis). Ongoing BF encouraged until a nutritionally adequate and safe diet without breast milk can be provided. The recommended duration of gradual weaning is extended to at least 4 weeks • Shift from individual maternal infant feeding choice to population-based endorsement of a single infant feeding strategy based on the overall country context and prevailing circumstances for the majority of women, with exceptions made for individual situations

Along the way, science and programmatic experience have highlighted lessons which should be heeded as the infant feeding and HIV debate continues.

- An individual approach does not work as a public health intervention. Although respecting autonomy is important, public messaging and community mobilization are lost. Additionally, the time constraint placed on health care workers does not allow for a thorough assessment of each client's situation, nor is the human resource capacity sufficient to maintain this approach with a level of quality. Finally, the individual approach impedes changes because new messages cannot be distributed rapidly.
- The speed with which guideline changes should occur must balance a need to provide new information to the field while recognizing that frequent change can undermine confidence and create confusion. These challenges can be seen at community, client, and health care worker levels.
- Cultural norms must be accounted for in-guideline development and program planning. Evidence has shown that promoting replacement feeding in a breastfeeding culture can lead to detrimental outcomes, such as mixed feeding, due to social and familial pressures.

Important studies and programmatic lessons learned have allowed for prioritization of both breastfeeding with ARV prophylaxis and HIV-free survival in the 2010 guidelines. The PEPFAR and UNIAIDS cochaired *Global Plan Towards the Elimination of New HIV Infections Among Children and Keeping Their Mothers Alive*, launched in June 2011, states these as explicit 2015 goals. As the 2010 guidelines are implemented, challenges remain. New messages around prophylaxis during breastfeeding will need to be disseminated hand in hand with health care worker training and ensuring access to ARV prophylaxis. The new evidence and new focus, implemented strategically, has the potential to greatly decrease new pediatric HIV infections while simultaneously contributing to reductions in under-5 mortality. While thoughtful implementation is necessary, it must be acknowledged that evidence is rapidly moving toward the potential of treating all HIV-infected individuals for life. With lessons learned over the last decade, preparations can already begin for the next guideline change.

Disclaimers The findings and conclusions in this chapter are those of the author(s) and do not necessarily represent the views of the Centers for Disease Control and Prevention/Agency for Toxic Substances and Disease Registry.

The authors' views expressed in this chapter do not necessarily reflect the views of the United States Agency for International Development or the United States Government.

References

1. Ziegler JB, Cooper DA, Johnson RO, Gold J (1985) Postnatal transmission of AIDS-associated retrovirus from mother to infant. Lancet 1(8434):896–898
2. Centers for Disease Control and Prevention (1985) Recommendations for assisting in the prevention of perinatal transmission of human T lymphotropic virus type III/lymphadenopathy-associated virus and acquired immunodeficiency syndrome. MMWR 34(48):721–726, 731–732
3. WHO Special Programme on AIDS and the Division of Family Health (1987) Statement from the consultation on breast-feeding/breast milk and human immunodeficiency virus (HIV) infection
4. WHO Collaborative Study Team on the Role of Breastfeeding on the Prevention of Infant Mortality (2000) Effect of breastfeeding on infant and child mortality due to infectious diseases in less developed countries: a pooled analysis. WHO Collaborative Study Team on the Role of Breastfeeding on the Prevention of Infant Mortality. Lancet 355(9202):1104
5. 39th World Health Assembly (1986) International code of marketing of breast-milk substitutes. WHA39/1986/REC/1
6. WHO Technical Consultation on Behalf of the UNFPA/UNICEF/WHO/UNAIDS Inter-Agency Task Team on Mother-to-Child Transmission of HIV (2000) New data on the prevention of mother-to-child transmission of HIV and their policy implications: conclusions and recommendations

7. WHO, UNICEF, UNFPA, UNAIDS, World Bank, UNHCR, WFP, FAO, IAEA (2003) HIV and infant feeding: framework for priority action
8. UNICEF, UNAIDS, WHO, UNFPA (2003) HIV and infant feeding: guidelines for decision-makers. World Health Organization, Geneva
9. Coutsoudi A (1999) Influence of infant feeding patterns on early mother-to-child transmission of HIV-1 in Durban, South Africa: a prospective cohort study. Lancet 354:471–476
10. WHO (2006) Infant and young child feeding counselling: an integrated course
11. PEPFAR. FY 2009 State of the Program Area (SOPA) Report: PMTCT (2010) https://www.pepfar.net/C8/Care%20and%20Treament%20-%20PMTCT%20%20Ped/default.aspx
12. Stulac SN, Franke MF, Rugira IH et al (2007) Successful implementation of a formula feeding program for mothers and infants lacking clean water access. Abstract 251. HIV/AIDS implementers' meeting abstract book. Kigali, Rwanda
13. Lu LS, Baek C, Smith M et al (2007) Problems in use of infant formula by HIV-infected women in Botswana's National PMTCT Program. Abstract 1435. HIV/AIDS implementers' meeting abstract book. Kigali, Rwanda
14. Kagaayi J, Gray RH, Wabwire FM et al (2008) Survival, by feeding modality, of infants born to HIV-positive mothers in Makai. HIV/AIDS implementers' meeting abstract book. Abstract 1898. Kampala, Uganda
15. Charurat ME, Datong P, Matawal B (2008) Implementation of a replacement feeding program for mothers and infants in North Central Nigeria: results and lessons learned. HIV/AIDS implementers' meeting abstract book. Abstract 834. Kampala, Uganda
16. Thior I, Lockman S, Smeaton LM et al (2006) Breastfeeding plus infant zidovudine prophylaxis for 6 months vs formula feeding plus infant zidovudine for 1 month to reduce mother-to-child HIV transmission in Botswana: a randomized trial: the Mashi Study. JAMA 296:794–805
17. Kuhn L, Aldrovandi GM, Sinkala M et al (2008) Effects of early, abrupt weaning on HIV-free survival of children in Zambia. N Engl J Med Vol 359(2):130–141
18. Coovadia HM, Rollins NC, Bland RM et al (2007) Mother-to-child transmission of HIV-1 infection during exclusive breastfeeding: the first six months of life. Lancet 369:1107–1116
19. Iliff PJ, Piwoz EG, Tavengwa NV et al (2005) Early exclusive breastfeeding reduces the risk of postnatal HIV-1 transmission, and increases HIV-free survival. AIDS 19:699–708.21
20. WHO, UNICEF, UNAIDS, UNFPA (2006) HIV and infant feeding update based on the technical consultation held on behalf of the interagency task team on prevention of HIV infection in pregnant women, mothers, and their infants, October 2006. Geneva, Switzerland
21. Homsy J, Moore D, Barasa A et al (2010) Breastfeeding, mother-to-child HIV transmission, and mortality among infants born to HIV-infected women on highly active antiretroviral therapy in rural Uganda. JAIDS 53(1):28–35
22. Creek TL, Kim A, Lu L et al (2010) Hospitalization and mortality among primarily nonbreastfed children during a large outbreak of diarrhea and malnutrition in Botswana, 2006. JAIDS 53(1):14–19
23. UNICEF (2011) The State of the World's Children 2011: Adolescence—An age of opportunity. UNICEF, New York, NY
24. Baxen PO, Molefe LM, Ntombela N, Tshiula J (2007) Feeding of infants and young children in Kwa-Zulu Natal South Africa. Rwanda HIV/AIDS implementers' meeting abstract book. Abstract 1179. Kigali, Rwanda
25. Farmer P (2010) Partner to the poor: a Paul Farmer reader. University of California Press, Berkeley, CA, p 145
26. Chatterjee A, Mebrahtu S, Henderson P, Chopra M (2008) Multi-country assessment of infant feeding support to HIV-positive women accessing PMTCT services. HIV/AIDS implementers' meeting abstract book. Abstract 1776. Kampala, Uganda
27. Chopra M, Rollins N (2008) Infant feeding in the time of HIV: assessment of infant feeding policy and programmes in four African countries scaling up prevention of mother to child transmission programmes. Arch Dis Child 93(4):288–291
28. Woldesenbet S, Jackson D (2009) The impact of quality of antenatal HIV counselling on HIV-free survival. Abstract WEPED226. In: 5th IAS conference on HIV pathogenesis, treatment and prevention, Cape Town, South Africa, 2009
29. de Paoli MM, Mkwanazi NB, Richter LM, Rollins N (2008) Early cessation of breastfeeding to prevent postnatal transmission of HIV: a recommendation in need of guidance. Acta Paediatr 97(12):1663–1668
30. PMTCT Advisory Group and Infant Feeding Study Group et al (2002) Evaluation of a pilot program and a follow-up study of infant feeding practices during the scaled-up program in Botswana. Evaluation and Program Planning 25:421–431
31. Creek T, Arvelo W, Kim A et al (2007) A large outbreak of diarrhea with high mortality among non-breastfed children in Botswana: implications for HIV prevention strategies and child health. In: Plenary Session 7, Conference on retrovirus and opportunistic infections, Los Angeles, CA
32. Bulterys M (2007) PMTCT of HIV in resource-poor settings—why are we doing so badly? In: Plenary Session 7, Conference on retrovirus and opportunistic infections, Los Angeles, CA
33. Coovadia HM, Coutsoudis A, Rollins N et al (2007) Prevention of HIV transmission from breastfeeding. In: Plenary Session 7, Conference on retrovirus and opportunistic infections, Los Angeles, CA

34. Oliveira Tsiouris F, Riese S, Fayorsey R, Abrams E (2008) Developing a public health approach to infant feeding in the context of HIV: a facility assessment of the AFASS criteria. Abstract 1628. HIV/AIDS implementers' meeting abstract book. Kampala, Uganda
35. Kumwenda NI, Hoover DR, Mofenson LM et al (2008) Extended antiretroviral prophylaxis to reduce breast-milk HIV-1 transmission. N Engl J Med 359(2):119–129
36. Taha TE, Hoover DR, Kumwenda NI et al (2007) Late postnatal transmission of HIV-1 and associated factors. J Infect Dis 196(1):10–14
37. Kesho Bora Study Group (2011) Triple antiretroviral compared with zidovudine and single-dose nevirapine prophylaxis during pregnancy and breastfeeding for prevention of mother-to-child transmission of HIV-1 (Kesho Bora study): a randomised controlled trial. Lancet Infect Dis 11(3):171–180
38. Chasela CS, Hudgens MG, Jamieson DJ et al (2010) Maternal or infant antiretroviral drugs to reduce HIV-1 transmission. N Engl J Med 362:2271–2281
39. Six Week Extended-Dose Nevirapine (SWEN) Study Team (2008) Extended-dose nevirapine to 6 weeks of age for infants to prevent HIV transmission via breastfeeding in Ethiopia, India, and Uganda: an analysis of three randomised controlled trials. Lancet 372:300–313
40. Shapiro RL, Hughes MD, Ogwu A et al (2010) Extended-dose nevirapine to 6 weeks of age for infants to prevent HIV transmission via breastfeeding in Ethiopia, India, and Uganda: an analysis of three randomised controlled trials. N Engl J Med 362:2282–2294
41. Thomas TK, Masaba R, Borkowf CB et al (2011) Triple-antiretroviral prophylaxis to prevent mother-to-child HIV transmission through breastfeeding – the Kisumu Breastfeeding Study, Kenya: a clinical trial. PLoS Med 8(3):e1001015
42. WHO (2010) Antiretroviral drugs for treating pregnant women and preventing HIV infection in infants: recommendations for a public health approach – 2010 version
43. WHO, UNAIDS, UNFPA, UNICEF (2010) Guidelines on HIV and infant feeding: principles and recommendations for infant feeding and a summary of evidence
44. Brewinski M, Heap A, Henneberg C, Phelps BR, Quick T (2010) Technical update to the field: 2010 guidelines for infant and young child feeding in the context of HIV. USAID. http://www.k4health.org/toolkits/pmtct/technical-update-field-2010-who-guidelines-infant-and-young-child-feeding-context-hiv
45. WHO (2008) Indicators for assessing infant and young child feeding practices. Part 1 definitions. Conclusions of a consensus meeting held 6–8 November 2007. Washington DC, USA. http://www.who.int/nutrition/publications/infantfeeding/9789241596664/en/index.html
46. WHO, UNICEF (2010) Draft monitoring and evaluating the prevention of mother to child transmission of HIV: a guide for national programmes. Unpublished, but available after October 2011 at http://www.who.int/hiv/pub/me/en/index.html
47. WHO (2003) Global strategy for infant and young child feeding. Geneva, Switzerland

Chapter 19
HIV-1 and Breastfeeding in the United States*

Kristen M. Little, Dale J. Hu, and Ken L. Dominguez

While breastfeeding remains a significant source of mother-to-child HIV transmission (MTCT) globally, it is the recommended infant feeding option for HIV-infected women in resource-limited settings [1]. However, HIV-infected women in the USA—where breast milk alternatives are acceptable, feasible, affordable, sustainable, and safe—have been counseled to avoid all breastfeeding since 1985 [2]. A number of studies have found that despite such recommendations against breastfeeding by HIV-infected women, a very small proportion of HIV-infected women in the USA continue to breastfeed their infants [3–5] for various reasons. Many of these women received late or no prenatal care, inadequate antiretroviral (ARV) prophylaxis, or were not diagnosed with HIV until at or after labor and delivery. While breastfeeding has never been a major source of perinatal HIV infections in the USA, studies have identified the practice as a risk factor for MTCT in the USA [4]. Complete avoidance remains the only sure way to prevent late postnatal HIV transmission through breastfeeding.

This chapter explores current breastfeeding practices and infant feeding recommendations in the context of maternal HIV infection among women in the USA. Rates of HIV-infection among women of childbearing age are highlighted, as well as existing opportunities for the prevention of mother-to-child transmission (PMTCT), including advising HIV-infected pregnant women to avoid breastfeeding. This chapter also examines the benefits and potential health risks associated with the avoidance of breastfeeding for mothers and infants, as well as the legal ramifications for HIV-infected women who, despite knowing their own positive HIV status, opt to breastfeed or provide their pumped breast milk to their infants.

---

*The findings and conclusions in this article are those of the authors and do not necessarily represent the official position of the Centers for Disease Control and Prevention.

K.M. Little, M.P.H.
Rollins School of Public Health, Atlanta, GA, USA

D.J. Hu, M.D., M.P.H.
Division of Viral Hepatitis (DVH), National Center for HIV/AIDS, Viral Hepatitis,
Sexually-Transmitted Disease and Tuberculosis Prevention (NCHHSTP),
Centers for Disease Control and Prevention, Atlanta, GA, USA

K.L. Dominguez, M.D., M.P.H. (✉)
Division of HIV/AIDS Prevention (DHAP), NCHHSTP, Centers for Disease Control and Prevention,
MS E-45. 1600 Clifton Road, Atlanta, GA 30333, USA
e-mail:KLD0@cdc.gov

Women and HIV in the USA

In 2006, more than 25% of all incident cases of HIV in the USA were among women over the age of 13 [6], and approximately 278,000 adult or adolescent women were living with a diagnosis of HIV infection [7]. As many as a quarter of these women do not know that they are HIV-infected, putting them at an increased risk of MTCT [8]. In addition, HIV/AIDS disproportionately impacts black and Hispanic/Latino women. Compared to white female adults and adolescents, black women had nearly 18 times as high a rate of HIV diagnoses in 2007. Hispanic/Latino women had a rate of diagnoses nearly four times as high as that of white women [7].

Perinatal HIV transmission also disproportionately affects black and Hispanic/Latino infants. While the rate of perinatal HIV infections continues to fall both overall and among racial/ethnic groups, significant racial disparities still exist among perinatally HIV-infected children, and rates of perinatal HIV infection remain several-fold higher among black or Hispanic/Latino infants compared to their white counterparts. In fact, between 2004 and 2007 more than 85% of all reported children diagnosed with perinatal HIV infection were black or Hispanic/Latino [9].

Racial/ethnic disparities in perinatal HIV are related to factors critical to PMTCT, including early prenatal care, HIV testing during pregnancy, and adequate antiretroviral prophylaxis [9]. These interventions have resulted in an overall reduction of MTCT rates in the USA from approximately 25% to less than 2% [10]. PMTCT interventions, however, have not benefited all racial/ethnic groups equally. Based on a study of Medicaid data from four states, black and Hispanic women compared to white women were found to be significantly less likely to receive early and adequate prenatal care, or to begin Medicaid-covered prenatal care at least 5 months before delivery [11]. Another study found that black and Hispanic women were less likely to have initiated ARV therapy before their current pregnancy than white women [12]. Women who do not know their HIV status before labor and delivery are more likely to transmit HIV to their infants [13, 14]. They are also more likely to initiate breastfeeding than women who know their status before or during pregnancy [5].

Epidemiology of Breastfeeding and HIV

The first case of HIV transmission through breastfeeding was reported in 1985 after an Australian woman was infected with HIV during a postnatal blood transfusion, and subsequently passed the virus on to her breastfeeding infant [15]. In the same year, researchers isolated HIV in cell-free breast milk for the first time [16]. Additional case reports of probable transmission of HIV from a mother to her child during breastfeeding began emerging in subsequent years [17]. Prospective epidemiologic studies conducted in the late 1980s and published in the early 1990s clearly implicated breastfeeding in late postnatal HIV transmission from mother to child [18, 19].

Globally, one third to one half of all perinatal HIV infections can be attributed to breastfeeding, which may have led to approximately 200,000 new mother-to-child HIV infections worldwide in 2007 [20, 21]. In the absence of PMTCT interventions, mother-to-child transmission rates in developed countries range from 14 to 23% among women who avoid breastfeeding and 25–42% among breastfeeding populations in developing countries [22–24]. Without PMTCT interventions, breastfeeding results in an estimated additional 14% risk of perinatal HIV transmission [25].

Maternal risk factors for MTCT of HIV through breastfeeding include younger maternal age, higher parity, low CD4+ cell counts (as an indication of advanced HIV disease progression), increased viral load, and breast abnormalities, such as mastitis, abscesses, and cracked nipples. Infant risk factors include oral candidiasis before the age of 6 months, duration of breastfeeding, and stomatitis [26–28].

HIV and Infant Feeding: National Recommendations

Soon after the first documented cases of late postnatal HIV transmission from mother to child, CDC released its first recommendation on breastfeeding and HIV, stating that "HTLV-III/LAV-infected women should be advised against breastfeeding to avoid postnatal transmission to a child who may not yet be infected" [2].

While infant feeding guidelines for HIV-infected women in resource-limited settings have evolved considerably since these initial recommendations, guidelines in the USA have changed very little. Due to the benefits of breastfeeding in resource-limited countries with high infectious disease rates and poor sanitation, there was a need to assess and compare the risk associated with HIV transmission from breastfeeding with the mortality risk from not breastfeeding [29]. While analyses did support existing recommendations in the USA and other developed countries that HIV-infected women should not breastfeed their children, there was a strong impetus to clarify the situation in resource-limited settings.

Studies from sub-Saharan Africa on extended maternal and/or infant ARV prophylaxis have shown promising results in preventing late postnatal HIV transmission through breastfeeding [26]. Early results from the studies, such as the Breastfeeding, Antiretrovirals, and Nutrition (BAN) trial indicate that extended maternal or infant ARV prophylaxis significantly reduces the risk of vertical HIV transmission through breastfeeding [30]. These findings have prompted the World Health Organization to further refine their stance on breastfeeding and perinatal HIV prevention in resource-limited settings where breastfeeding alternatives are not feasible, affordable, acceptable, or safe [31]. The 2010 guidelines are the first to include recommendations for the provision of ARV drugs to infants or mothers in order to reduce the risk of HIV transmission through breastfeeding [31].

Additional research has evaluated the risks of sexual and perinatal transmission in the context of highly active antiretroviral therapy (HAART). Early findings indicate that individuals receiving treatment with HAART have significantly reduced risk of sexual HIV transmission [32, 33]. Officials and researchers in Switzerland have gone so far as to conclude that "an HIV-infected person on antiretroviral therapy (ART) with completely suppressed viraemia (effective ART) is not sexually infectious" if the individual is not concurrently infected with another sexually transmitted infection [34]. Despite such conclusions, further studies have indicated that persons with undetectable HIV viral loads may still transmit HIV to their sexual partner horizontally through sexual relations, or vertically through MTCT [35–38]. Therefore, current recommendations in the USA are for HIV-infected persons to use condoms when having sexual relations, and for HIV-infected pregnant women to receive ARV prophylaxis and to avoid breastfeeding, regardless of whether one's viral load is undetectable [35].

Nevertheless, there have been new and ongoing discussions on HIV and infant feeding practices in the USA after a number of studies have demonstrated the impact of maternal HAART on MTCT (see Chap. 12). The decreased risk associated with effective maternal HAART therapy have prompted speculation that this may allow for changes in current breastfeeding recommendations for HIV-infected women in the USA and other developed countries. The Panel on Treatment of HIV-Infected Pregnant Women and Prevention of Perinatal Transmission did not, however, alter their position on HIV and infant feeding in their most recent recommendations. In their 2010 "recommendations for use of antiretroviral drugs in pregnant HIV-1-infected women for maternal health and interventions to reduce perinatal HIV-1 transmission in the United States," the Panel concluded that acceptable, feasible, affordable, sustainable, and safe breastfeeding alternatives are available for HIV-infected women in the USA. The panel recommended that HIV-infected women, including those on combination or triple ARV therapies, avoid breastfeeding [35].

The Panel highlighted that, while studies had shown breastfeeding transmission of HIV to be significantly reduced by extended therapy with ARV drugs, the risk could not be completely eliminated. The Panel was also concerned about the possibility of developing drug-resistant HIV strains in infants infected during breastfeeding despite ARV prophylaxis, which has occurred in several studies [39, 40].

There is also growing concern over the low risk of ARV-associated potential adverse health outcomes among perinatally ARV-exposed infants, including premature birth, mitochondrial toxicities, and childhood cancers [41–44]. Because the risk for HIV transmission outweighs the smaller risk for such adverse health outcomes, however, perinatal HIV prophylaxis continues to be the standard of care for HIV-infected pregnant women [35].

Recent recommendations also highlight the risk of new HIV infections in late pregnancy or among women currently breastfeeding. The high viremia typical of acute HIV infection puts women at an increased risk of transmitting the virus to their infants [45]. Current testing guidelines recommend that pregnant women in high-risk areas be retested for HIV during the third trimester [46]. For women who are newly infected with HIV while breastfeeding, PMTCT interventions are less clear, though retesting high-risk women during the breastfeeding period would likely be an effective means of PMTCT. As HAART coverage of pregnant and delivering women in the USA increases and rates of MTCT remain low, the proportion of all mother-to-child HIV infections due to newly acquired HIV-infections in breastfeeding women is also likely to increase [47].

The Panel on Treatment of HIV-Infected Pregnant Women and Prevention of Perinatal Transmission highlighted these challenges in their 2010 update, writing:

> Although new perinatal HIV infections are becoming rare in resource-rich countries, infections continue to occur, and the birth of an infected infant is a sentinel event representing missed opportunities and barriers to prevention. Important obstacles to eradication of perinatal transmission in the United States include the continued increase of HIV infection among women of childbearing age; absent or delayed prenatal care, particularly in women using illicit drugs; acute (primary) infection in late pregnancy and in women who are breastfeeding; poor adherence to prescribed antiretroviral regimens; and lack of full implementation of routine, universal prenatal HIV counseling and testing. [35]

The prevention of primary HIV infections among women of childbearing age, HIV testing during pregnancy, receipt of prenatal care and other PMTCT interventions, and the avoidance of breastfeeding by HIV-infected women has greatly reduced MTCT in the USA. Elimination of perinatal HIV in the USA depends upon increasing the coverage of these interventions, including counseling HIV-infected pregnant women to utilize breast milk alternatives for their infants.

Breastfeeding Among HIV-Infected Women in the USA

Despite recommendations, a small proportion of HIV-infected women in the USA still breastfeed their infants. Several studies have found that maternal HIV diagnosis at or after labor and delivery was significantly associated with breastfeeding [5, 48].

An analysis of data from 15 US jurisdictions in the Enhanced Perinatal Surveillance System (EPS) examined risk factors for MTCT as well as missed opportunities for PMTCT [4]. From 2005 to 2008, 179 infants in EPS were diagnosed with perinatal HIV infection, for an overall mother-to-child transmission rate of 2.2%. Eighty of the 7,654 HIV-infected women (1.0%) included in EPS with data on infant feeding practices reported breastfeeding their infants. After controlling for other covariates, late HIV testing, no ARV use during pregnancy, and breastfeeding were all significantly associated with perinatal HIV transmission. Women who breastfed had 4.4 times greater odds of having an HIV-infected infant (95% CI: 2.1–9.4). In addition, 10% of all infant HIV infections in EPS during this period were attributable to breastfeeding [4]. While this route of transmission has not been recognized as a substantial driver of MTCT in the USA, data from EPS indicates that breastfeeding continues to contribute to domestic perinatal HIV transmission [4].

An additional study of EPS data from 1999 through 2008 examined risk factors for breastfeeding among HIV-infected women [48]. This analysis included data on 18,044 HIV-infected pregnant women in 26 participating areas. Among HIV-infected women delivering live-born infants for whom infant feeding practices were known, 303 women (1.7%) reported breastfeeding their infants.

Compared to black/African American women, white women were 1.6 times more likely to breastfeed (95% CI: 1.2–2.2). The study did not find a statistically significant difference in breastfeeding between white and Hispanic/Latina women [48].

Notably, this study also found that the timing of maternal HIV diagnosis was significantly associated with breastfeeding. Women who were not diagnosed with HIV until after labor and delivery were over 30 times more likely to breastfeed than women who learned of their status during pregnancy, or were diagnosed at labor and delivery (OR: 30.1, 95% CI: 21.9–41.6) [48]. Similarly, women who were not diagnosed until labor and delivery were almost twice as likely to breastfeed as women who knew their HIV status before pregnancy (OR: 1.9, 95% CI: 1.4–2.6). In an adjusted logistic regression model, not receiving ARVs during pregnancy (aOR: 2.8, 95% CI: 1.2–6.8), not receiving ARVs in labor and delivery (aOR: 6.2, 95% CI: 2.7–14.1), not being diagnosed with HIV before or during pregnancy (aOR: 12.2, 95% CI: 5.4–27.9), and being married (aOR: 2.3, 95% CI: 1.3–4.1) were all significantly associated with breastfeeding. Interestingly, infant HIV infection, race/ethnicity, and receipt of prenatal care were not significantly associated with breastfeeding [48].

A study of MTCT in New York State from 1988 to 2008 also identified a small proportion of HIV-infected women who breastfed their infants [3]. One hundred and sixty one of 7,269 HIV-infected women (1.8%) reported breastfeeding. Women who breastfed were over four times more likely to transmit HIV to their infants than women who did not breastfeed (OR: 4.42, 95% CI: 2.68–7.29). While the authors did not examine risk factors for breastfeeding, they found that perinatal HIV transmission was significantly associated with maternal HIV diagnosis at or after labor and delivery, a lack of or inadequate prenatal care, and receipt of less than three components (prenatal, intrapartum, and neonatal) of ARV prophylaxis. Other studies have linked these same risk factors to breastfeeding among HIV-infected mothers [3, 5].

Earlier research on the rates of breastfeeding among HIV-infected mothers in the USA obtained similar results. A study of HIV-infected mothers in Los Angeles and Massachusetts from 1988 to 1993 found that 79 of the 1,193 women included in the analysis reported breastfeeding [5]. Using the 2,667 HIV-infected childbearing women identified from the Serosurvey of Childbearing Women (SCBW) in Los Angeles and Massachusetts as the denominator, the authors calculated a minimum breastfeeding proportion of 3% for the study period. As with previous studies, the timing of maternal HIV diagnosis was significantly associated with breastfeeding. Women who did not know their HIV status before delivery were more than eight times more likely to breastfeed than women who were known to be HIV-infected before or during pregnancy (RR=8.2, 95% CI: 4.7–14.2) [5]. Timely HIV testing during pregnancy and/or labor and delivery is vital in preventing MTC of HIV transmission at every phase of pregnancy, including the late postnatal period.

Similar findings have been observed in studies from other developed countries. In an analysis of data from the French Perinatal Cohort, researchers found that approximately 0.6% of HIV-infected women reported breastfeeding their infants. Women who were not treated for HIV during their pregnancy were eight times more likely to breastfeed their infants when compared to women who received treatment (3.8 vs. 0.5%; $p<0.01$) [49].

While HIV-infected women in the USA have long been advised to avoid breastfeeding, preventing late postnatal HIV transmission depends on identifying HIV infections in women of childbearing age, as well as those who are pregnant and breastfeeding. Routine, opt-out HIV-testing of all pregnant women and repeat testing at labor and delivery of women in high risk areas have greatly reduced the number of missed prevention opportunities in the USA by identifying maternal HIV infections before labor and delivery [13]. However, maternal seroconversion in late pregnancy or during the postnatal period remains a difficult issue. Though rare, breastfeeding-associated HIV transmission has been reported among previously seronegative women unknowingly infected with HIV during the postnatal period [50]. Late pregnancy or postnatal HIV-seroconversion among breastfeeding women is of particular concern, given the high viremia associated with an acute HIV infection, and the elevated viral loads found in colostrum and early milk [51].

Breastfeeding and Infant Feeding Alternatives

Breastfeeding has a variety of beneficial health effects for both infants and mothers, including reduced infant morbidity and mortality from diarrhea, respiratory tract infections, otitis media, urinary tract infections, sudden infant death syndrome (SIDS), and overweight and obesity [52]. Women who breastfeed have decreased postpartum bleeding, more rapid uterine contraction, faster return to pre-pregnancy weight, and a reduced risk of cancers of the breasts and ovaries [52].

Despite the benefits associated with breastfeeding, formula feeding remains a common infant feeding practice in the USA and many parts of the world. Though recommendations from organizations, such as the American Academy of Pediatrics (AAP), the American College of Obstetricians and Gynecologists (ACOG), and the American Academy of Family Physicians (AAFP), advise women to exclusively breastfeed for at least 6 months, breastfeeding rates at 6 months in the USA remain significantly lower than the goals set by the Healthy People 2010 goals [53]. In 2001, 70% of women in the USA initiated breastfeeding, but only 46% of women initiated exclusive breastfeeding. Significantly, fewer women continued breastfeeding. In fact, only 33% of women were still breastfeeding at 6 months, and only 17% of women were exclusively breastfeeding [54]. By 2010, 75% of infants were ever breastfed, a rate that met the Healthy People 2010 goals. However, the proportion of women initiating exclusive breastfeeding remained the same, and the percentage of women exclusively breastfeeding at 6 months fell from 17% in 2001 to 13.3% in 2010 [55].

Formula feeding is a culturally acceptable, safe, and affordable feeding option in the USA, and is the recommended breastfeeding alternative for HIV-infected women. Because of the known benefits of breastfeeding for both mothers and infants, research has explored various approaches to treating the breast milk of HIV-infected women. Potential strategies include boiling, pasteurization, freezing, and chemical treatment of expressed breast milk [56]. None of these approaches, however, are currently recommended for HIV-infected women in the USA, as they may not completely eliminate the risk for breast milk-associated HIV transmission [35].

Human donor milk banks may offer another breastfeeding alternative for HIV-infected women in the USA who would prefer to feed their infants human breast milk. However, the provision of donated human breast milk has had a contentious history in the USA. Human milk banks were more common in the USA prior to the start of the HIV/AIDS epidemic in the late 1970s. Recognition of the risk of HIV transmission through various bodily fluids, including breast milk, led to a sharp decrease in the use of human donor milk banks [57]. Traditionally, such milk banks have provided human milk prescribed to premature and other critically ill infants born to mothers who are unable to breastfeed or provide their own breast milk to their infants.

Not-for-profit donor milk banks in the USA are not currently regulated or endorsed by any federal agency, including the Food and Drug Administration (FDA), and operate instead under the aegis of the Human Milk Banking Association of North America's (HMBANA) *Guidelines for the Establishment and Operation of a Human Milk Bank*. These guidelines provide direction for the collection, storage, and dissemination of donor human milk by milk banks in the USA. In order to prevent the spread of infectious diseases, the guidelines state that:

> All human milk donors should be screened according to the American Association of Blood Banks' standards for screening blood donors. All milk accepted for donation should be pasteurized unless the recipient's condition requires fresh-frozen milk, in which case the milk bank director should consult with the medical director and advisory board to approve the dispensing of microbiologically screened, fresh-frozen milk from suitable donors. [58]

Currently, only nine HMBANA human milk banks operate in the USA, with six others in various stages of development [50]. One for-profit company, Prolacta Bioscience, also collects donated human milk and is the only company using donated human milk to produce nutritional products for infants [59].

The US government recommendations regarding the transmission of infectious agents through human breast milk were first published in the early 1990s [60]. The Centers for Disease Control and Prevention published *Guidelines for Preventing Transmission of Human Immunodeficiency Virus Through Transplantation of Human Tissue and Organs* in 1994 and an update in 2001. In it, CDC recommended that breast milk donors, like other human tissue donors, should sign consent forms indicating that they understand the information provided by the milk bank and that they will decline to donate if they are at potential risk of spreading HIV. The recommendations also advised milk banks to test donors for HIV as a part of the screening process [60].

A 2010 meeting of the FDA's Pediatric Advisory Committee (PAC) provided additional, updated information to the public about human milk banking [61]. It concluded that the infectious disease risk associated with donated human breast milk is minimal when taken from HMBANA member banks or Prolacta Bioscience Inc., due to their standardized recommendations for donor screening and post-collection storage and treatment processes for donated breast milk. However, serious concerns were raised regarding the practice of internet or person-to-person exchange of human milk, which has grown in popularity in recent years [62]. The committee continues to highly discourage internet exchange of human breast milk [61].

Despite such scrutiny, research on the benefits of donated human breast milk for infants and financial support for human breast milk banking remains limited. A number of government programs have withdrawn their support for the practice, among them the United States Department of Agriculture's (USDA) Women, Infants, and Children (WIC) program, which "no longer authorizes banked human breast milk as a WIC-eligible formula" [58]. Currently, no government policies support the use of donated human breast milk, including the provision of donated breast milk as an alternative feeding option for infants of HIV-infected mothers. Utilization of banked human breast milk requires a physician's prescription, is typically not covered by health insurance plans, and is no longer provided under Medicaid, rendering the option unaffordable for many women [63]. However, the use of banked milk as an alternative feeding strategy for HIV-infected women in the USA has not been thoroughly explored.

Current Issues

The decision to breastfeed may carry both health and penal risks for HIV-infected women in some parts of the USA. Currently, 34 states have laws pertaining to the criminalization of HIV transmission [64]. At least five states, including California, Idaho, Maryland, North Carolina, and South Carolina, have HIV transmission laws that explicitly mention the provision of human breast milk by HIV-infected women as a punishable act. For example, the "California Health and Safety Code § 1621.5" states:

> It is a felony punishable by imprisonment in the state prison for two, four, or six years, for any person who knows that he or she has HIV/AIDS to donate blood, body organs or other tissue, semen to any medical center or semen bank that receives semen for purposes of artificial insemination, or breast milk to any medical center or breast milk bank that receives breast milk for purposes of distribution, whether he or she is a paid or a volunteer donor. [65]

While many of these laws are only rarely enforced, there are documented instances where Child Protective Services has intervened to remove a child from the care of an HIV-infected woman choosing to breastfeed against the recommendation of her physician [66]. HIV-infected women are also vulnerable to prosecution for neglect or "intent to harm" if a child is infected with HIV because the mother refused to avoid breastfeeding. Given emerging research on HAART and HIV-transmission through breastfeeding, new questions have begun to arise about breastfeeding in the context of well-controlled maternal HIV infections in the USA. Given the small but lingering risk of MTCT associated with breastfeeding—even among HIV-infected women with undetectable viral loads—national recommendations against breastfeeding have not been altered, and HIV-infected women are still advised to utilize alternative feeding options. HIV-infected women who choose to breastfeed despite recommendations remain vulnerable to prosecution in many parts of the USA.

Conclusion

The rapid reduction in the rate of perinatal HIV transmission in the USA has been one of the great public health success stories of recent decades. Interventions, such as HIV testing before and during pregnancy, ARV prophylaxis, elective cesarean deliveries, and avoidance of breastfeeding have reduced MTCT rates from approximately 25% to less than 2% in the USA. However, missed prevention opportunities remain an obstacle to the elimination of perinatal HIV. Breastfeeding among HIV-infected women is still a small—but often significant—risk factor contributing to MTCT in the USA. Women with a late HIV diagnosis and inadequate ARV prophylaxis are significantly more likely to breastfeed than women diagnosed before labor and delivery, or women who receive ARV drugs. A timely diagnosis is the key to preventing perinatal HIV transmission in utero, during the peripartum period, and postnatally through breastfeeding.

Despite advances in ART, HIV-infected women in the USA continue to be advised to avoid breastfeeding due to the lingering risk of transmission and the availability of safe infant feeding alternatives. The prevention of late postnatal HIV transmission through breastfeeding, however, still depends on the early identification of HIV-infection in women of childbearing age. This remains true not only for pregnant or delivering women, but also those who may become newly HIV-infected during the breastfeeding period. Continued efforts at preventing, identifying, and treating HIV-infections in women and girls will prevent future cases of perinatal HIV infection at any period during and after pregnancy.

References

1. WHO (2010) Guidelines on HIV and infant feeding 2010. 1–58. http://www.who.int/child_adolescent_health/documents/9789241599535/en/index.html. Accessed 1 2010
2. CDC (1985) Recommendations for assisting in the prevention of perinatal transmission of human T-lymphotropic virus type III/lymphadenopathy-associated virus and acquired immunodeficiency syndrome. MMWR 34(48):721–726, 731–732
3. Birkhead GS, Pulver WP, Warren BL et al (2010) Progress in prevention of mother-to-child transmission of HIV in New York State: 1988–2008. J Public Health Manag Pract 16(6):481–491
4. Whitmore SK, Taylor AW, Espinoza L, Shouse RL, Lampe MA, Nesheim S (2012) Correlates of mother-to-child transmission of HIV in the United States and Puerto Rico. Pediatrics 129(1):e74–e82. http://pediatrics.aappublications.org/content/early/2011/11/30/peds.2010-3691
5. Bertolli JM, Hsu H, Frederick T et al (1996) Breastfeeding among HIV-infected women, Los Angeles and Massachusetts, 1988–1993. In: XI international conference on AIDS. Vancouver, Canada, Abstract #158
6. Prejean J, Song R, An Q, Hall HI (2008) Subpopulation estimates from the HIV incidence surveillance system – United States, 2006. MMWR 57(36):985–989
7. CDC (2008) HIV prevalence estimates – United States, 2006. MMWR 57(39):1073–1076
8. Marks G, Crepaz N, Janssen RS (2006) Estimating sexual transmission of HIV from persons aware and unaware that they are infected with the virus in the USA. AIDS 20(10):1447–1450
9. Lampe MA, Nesheim S, Shouse RL et al (2010) Racial/ethnic disparities among children with diagnoses of perinatal HIV infection – 34 States, 2004–2007. MMWR 59(4):97–101
10. Taylor A, Zhang X, Whitmore SK, Rhodes P, Blair J (2009) Estimated number of perinatal HIV infections in the United States. In: 2009 National HIV prevention conference, Atlanta, GA, 2009, p 351
11. Gavin NI, Adams EK, Hartmann KE, Benedict MB, Chireau M (2004) Racial and ethnic disparities in the use of pregnancy-related health care among medicaid pregnant women. Matern Child Health J 8(3):113–126

12. Cunningham CK, Balasubramanian R, Delke I et al (2004) The impact of race/ethnicity on mother-to-child HIV transmission in the United States in pediatric AIDS clinical trials group protocol 316. J Acquir Immune Defic Syndr 36:800–807
13. Peters V, Liu KL, Dominguez K (2003) Missed opportunities for perinatal HIV prevention among HIV-exposed infants born 1996–2000, pediatric spectrum of HIV disease cohort. Pediatrics 111:1186–1191
14. Bulterys M, Jamieson DJ, O'Sullivan MJ et al (2004) Rapid HIV-1 testing during labor: a multicenter study. JAMA 292(2):219–223
15. Ziegler JB, Cooper DA, Johnson RO, Gold J (1985) Postnatal transmission of AIDS-associated retrovirus from mother to infant. Lancet 1:896–898
16. Thiry L, Sprecher-Goldberger S, Jonckheer T et al (1985) Isolation of AIDS virus from cell-free breast milk of three healthy virus carriers. Lancet 2(8460):891–892
17. Lepage P, Van de Perre P, Carael M et al (1987) Postnatal transmission of HIV from mother to child. Lancet 2(8555):400
18. Hira SK, Mangrola UG, Mwale C et al (1990) Apparent vertical transmission of human immunodeficiency virus type 1 by breast-feeding in Zambia. J Pediatr 117(3):421–424
19. Van de Perre P, Simonon A, Msellati P et al (1991) Postnatal transmission of human immunodeficiency virus type 1 from mother to infant. A prospective cohort study in Kigali, Rwanda. N Eng J Med 325(9):593–598
20. World Health Organization (2007) HIV transmission through breastfeeding: a review of available evidence. World Health Organization, Geneva, Switzerland
21. Joint United Nations Program on HIV/AIDS, AIDS (2007) Epidemic update. United Nations, Geneva, Switzerland
22. Sturt AS, Dokubo EK, Sint TT (2010) Antiretroviral therapy (ART) for treating HIV infection in ART-eligible pregnant women (review). Cochrane Database Syst Rev 3:1–100
23. De Cock KM, Fowler MG, Mercier E et al (2000) Prevention of mother-to-child HIV transmission in resource-poor countries: translating research into policy and practice. JAMA 283(9):1175–1182
24. WHO (2010) PMTCT strategic vision. World Health Organization, Geneva, Switzerland, pp 1–40
25. Dunn DT, Newell ML, Ades AE, Peckham CS (1992) Risk of human immunodeficiency virus type 1 transmission through breastfeeding. Lancet 340(8819):585–588
26. Kourtis A, Jamieson I, deVincenzi A et al (2007) Prevention of human immunodeficiency virus-1 transmission to the infant through breastfeeding: new developments. Am J Obstet Gynecol 197(3 Suppl):S113–S122
27. Rollins NC, Filteau SM, Coutsoudis A, Tomkins AM (2001) Feeding mode, intestinal permeability, and neopterin excretion: a longitudinal study in infants of HIV-infected South African Women. J Acquir Immune Defic Syndr 28(2):132–139
28. John-Stewart G, Mbori-Ngacha D, Ekpini R et al (2004) Breast-feeding and transmission of HIV-1. J Acquir Immune Defic Syndr 35(2):196–202
29. Hu DJ, Heyward WL, Byers RH Jr et al (1992) HIV infection and breast-feeding: policy implications through a decision analysis model. AIDS 6(12):1505–1513
30. Chasela CS, Hudgens MG, Jamieson DJ et al (2010) Maternal or infant antiretroviral drugs to reduce HIV-1 transmission. N Engl J Med 362(24):2271–2281
31. WHO (2010) Guidelines on HIV and infant feeding 2010: principles and recommendations for infant feeding in the context of HIV and a summary of evidence. http://www.who.int/child_adolescent_health/documents/9789241599535/en/index.html. Accessed 15 Dec 2010
32. Fisher M, Pao D, Brown AE et al (2010) Determinants of HIV-1 transmission in men who have sex with men: a combined clinical, epidemiological and phylogenetic approach. AIDS 24(11):1739–1747
33. Granich R, Crowley S, Vitoria M et al (2010) Highly active antiretroviral treatment for the prevention of HIV transmission. J Int AIDS Soc 13(1):1–8
34. Vernazza P, Hirschel B, Bernasconi E (2008) HIV-infected persons on effective antiretroviral therapy (and free of other STDs) are sexually non-infectious. Bull Med Swiss 89(5):1–10
35. Panel on Treatment of HIV-Infected Pregnant Women and Prevention of Perinatal Transmission (2010) Recommendations for use of antiretroviral drugs in pregnant HIV-1-infected women for maternal health and interventions to reduce perinatal HIV transmission in the United States. May 24, 2010. http://aidsinfo.nih.gov/ContentFiles/PerinatalGL.pdf. Accessed 21 Jan 2011
36. Sturmer M, Doerr HW, Berger A, Gute P (2008) Is transmission of HIV-1 in non-viraemic serodiscordant couples possible? Antivir Ther 13(5):729–732
37. Ibanez A, Puig T, Elias J et al (1999) Quantification of integrated and total HIV-1 DNA after long-term highly active antiretroviral therapy in HIV-1-infected patients. AIDS 13(9):1045–1049
38. Furtado MR, Callaway DS, Phair JP et al (1999) Persistence of HIV-1 transcription in peripheral-blood mononuclear cells in patients receiving potent antiretroviral therapy. N Engl J Med 340(21):1614–1622
39. Hudelson SE, McConnell MS, Bagenda D et al (2010) Emergence and persistence of nevirapine resistance in breast milk after single-dose nevirapine administration. AIDS 24(4):557–561

40. Moorthy A, Gupta A, Bhosale R et al (2009) Nevirapine resistance and breast-milk HIV transmission: effects of single-dose and extended-dose nevirapine prophylaxis in subtype C HIV-infected infants. PLoS One 4(1):e4096
41. Brogly SB, Ylitalo N, Mofenson LM et al (2007) In utero nucleoside reverse transcriptase inhibitor exposure and signs of possible mitochondrial dysfunction in HIV-uninfected children. AIDS 21(8):929–938
42. Blanche S, Tardieu M, Rustin P et al (1999) Persistent mitochondrial dysfunction and perinatal exposure to antiretroviral nucleoside analogues. Lancet 354(9184):1084–1089
43. Benhammou V, Warszawski J, Bellec S et al (2008) Incidence of cancer in children perinatally exposed to nucleoside reverse transcriptase inhibitors. AIDS 22(16):165–177
44. Thorne C, Patel D, Newell ML (2004) Increased risk of adverse pregnancy outcomes in HIV-infected women treated with highly active antiretroviral therapy in Europe. AIDS 18(17):2337–2339
45. Garcia PM, Kalish LA, Pitt J et al and Woman and Infants Transmission Study Group (1999) Maternal levels of plasma human immunodeficiency virus type 1 RNA and the risk of perinatal transmission. N Engl J Med 341:394–402
46. Branson BM, Handsfield HH, Lampe MA et al (2006) Revised recommendations for HIV testing of adults, adolescents, and pregnant women in health-care settings. MMWR 55(RR-14):1–17
47. Bulterys M (2011) Centers for Disease Control and Prevention, Division of Global Health, Atlanta, GA, Personal Communication 15 May 2011
48. Whitmore SK, Nesheim S, Lampe MA, Taylor AW (2009) Characteristics of breastfeeding HIV-infected women delivering live infants, enhanced perinatal surveillance, 26 areas, US 1999–2008. In: National HIV prevention conference. Atlanta, GA, 2009, Poster #065M
49. Mayaux MJ, Teglas JP, Blanche S (2003) Characteristics of HIV-infected women who do not receive preventative antiretroviral therapy in the French perinatal cohort. J Acquir Immune Defic Syndr 34(3):338–343
50. Nicholson O, Michalik DE, Patel S, LaRussa P, Neu N (2007) Acute human immunodeficiency virus infection in a breast-fed infant in New York City. Pediatr Infect Dis J 26(7):653–655
51. Rousseau CM, Nduati RW, Richardson BA et al (2003) Longitudinal analysis of human immunodeficiency virus type 1 RNA in breast milk and of its relationship to infant infection and maternal disease. J Infect Dis 187:741–747
52. Bartick M, Reinhold A (2010) The burden of suboptimal breastfeeding in the United States: a pediatric cost analysis. Pediatrics 125(5):e1048–e1056
53. Centers for Disease Control and Prevention (2010) Breastfeeding report card – United States 2010. http://www.cdc.gov/breastfeeding/data/reportcard.htm. Accessed 15 Dec 2010
54. American Academy of Pediatrics (2005) Breastfeeding and the use of human milk. Pediatrics 115(2):496–506
55. Centers for Disease Control and Prevention (2010) Breastfeeding report card – United States 2010. In:, P.A. Division of Nutrition, and Obesity (ed) Breastfeeding report card. Department of Health and Human Services. Atlanta, GA, p 1–4. http://www.cdc.gov/breastfeeding/pdf/BreastfeedingReportCard2010.pdf. Accessed 30 Jan 2012
56. Hartmann SU, Berlin CM, Howett MK (2006) Alternative modified infant-feeding practices to prevent postnatal transmission of human immunodeficiency virus type 1 through breast milk: past, present, and future. J Hum Lact 22(1):75–88, 89–93
57. Arnold LDW (2000) Becoming a donor to a human milk bank. Leaven 36(2):19–23
58. Arnold LDW (2008) U.S. health policy and access to banked donor human milk. Breastfeed Med 3(4):221–229
59. Prolacta Bioscience Inc. (2011) Prolacta bioscience: advancing the science of human milk. http://www.prolacta.com/index.php. Accessed 21 Jan 2011
60. Rogers MF, Simonds RJ, Lawton KE, Moseley RR, Jones WK (1994) Guidelines for preventing transmission of human immunodeficiency virus through transplantation of human tissue and organs. MMWR 43(RR-8):1–17
61. FDA Pediatric Advisory Committee (2011) FDA pediatric advisory committee flash minutes. http://www.fda.gov/downloads/AdvisoryCommittees/CommitteesMeetingMaterials/PediatricAdvisoryCommittee/UCM238627.pdf. Accessed 21 Jan 2011
62. Shute N (2011) Moms who can't nurse find milk donors online. National public radio. http://www.npr.org/2011/01/24/133110199/moms-who-cant-nurse-find-milk-donors-online. Accessed 24 Jan 2011
63. Arnold LDW (1998) How to order banked donor milk in the United States: what the health care provider needs to know. J Hum Lact 14(1):65–67
64. Positive Justice Project (2010) Ending and defending against HIV criminalization: a manual for advocates. In: State and federal laws and prosecutions. HIV Law and Policy, New York. http://www.hivlawandpolicy.org/resources/view/564. Accessed 30 Jan 2012
65. Bennett-Carlson R, Faria D, Hanssens C (2010) State and federal laws and prosecutions. In: P.J. Project (ed) Ending and defending against HIV criminalization: a manual for advocates. The Center for HIV Law and Policy, New York, p 1–221
66. Mead A (1999) Sophie's choice. Salon.com 1999. http://www.salon.com/life/feature/1999/12/08/brassard/print.html. Accessed 7 Dec 1999

Part VI
DEBATE: Should Women With HIV-1 Infection Breastfeed Their Infants? Balancing the Scientific Evidence, Ethical Issues and Cost-Policy Considerations

Chapter 20
Pendulum Swings in HIV-1 and Infant Feeding Policies: Now Halfway Back

Louise Kuhn and Grace Aldrovandi

Introduction: The Pre-HIV Era

As one of the defining characteristics of mammalian reproduction, it should come as no surprise that breastfeeding is the norm, the healthiest practice for both mothers and infants regardless of where they live [1]. Benefits of breastfeeding have been noticed by health practitioners since the middle ages with poignant records of the outcomes of foundlings given human milk compared to those fed with artificial feeds [2]. By the mid-twentieth century, the industry producing and selling infant formula was so confident that their product was equivalent to mother nature's "product" that a vast population-level experiment was conducted with tragic results. Infant formula began to be actively promoted in sub-Saharan Africa leading to the well-publicized increases in infant death [3].

Years of Denial

The first case of HIV transmission through breastfeeding was described in 1985, not long after the disease that would become a global pandemic was first noticed among gay men. Almost immediately, guidelines for HIV-infected women in the USA recommended artificial feeding [4]. However, the magnitude of the HIV epidemic among women in sub-Saharan Africa was not fully appreciated and the complexities of balancing the risks and benefits of replacement feeding among HIV-infected women in developing countries are so complex that initial international recommendations were vague [5]. Furthermore, at this time the epidemiology of breast milk HIV transmission largely relied on a methodologically flawed meta-analysis [6] that would, nevertheless, turn out to be remarkably accurate in its estimates of the risks of HIV transmission. This resulted in some uncertainty as to whether the magnitude

---

L. Kuhn, Ph.D. (✉)
Gertrude H. Sergievsky Center, College of Physicians and Surgeons, 630 W 168th Street,
New York, NY 10032, USA

Department of Epidemiology, Mailman School of Public Health, Columbia University,
New York, NY, USA
e-mail: lk24@columbia.edu

G. Aldrovandi, M.D.
Department of Pediatrics, Children's Hospital Los Angeles, University of Southern California,
Los Angeles, CA, USA

of postnatal transmission was substantial enough to require changes in policy [5]. Through at least 1992, WHO recommended that HIV-infected women in the *developing* world continue to breastfeed [7].

Focus in the HIV research community turned to quantifying the risk of HIV transmission through breastfeeding. This quantification was most elegantly accomplished in a randomized trial conducted in Nairobi, Kenya [8]. Comparing women randomized to no breastfeeding vs. those randomized to some breastfeeding (median duration 17 months), the risk of postnatal HIV transmission through breastfeeding was reported to be 16% (95% CI 7–26%) consistent with the meta-analysis estimate: 14% (95% CI 7–21) [6, 8]. This all-or-nothing approach to breastfeeding dominated thinking in the field for several years but is problematic for a number of reasons. Primarily, it is unhelpful as it polarizes decisions around two suboptimal alternatives. It also glosses over the duration of breastfeeding and the quality of breastfeeding (exclusive vs. nonexclusive)—parameters with considerable influence on the absolute magnitude of postnatal HIV transmission.

Recognizing the competing risks involved in avoiding breastfeeding, the Nairobi study group selected HIV-free survival as their primary study endpoint. This approach combines two adverse endpoints, namely HIV infection and death of uninfected children. This approach tempers enthusiasm for support of artificial feeding by reminding us that shifts away from breastfeeding carry a cost in terms of the lives of HIV-uninfected infants. However, only an HIV-uninfected child death is considered sufficiently severe to be counted as equivalent to an HIV infection. Thereby, the spectrum of other benefits of breastfeeding for maternal and child health is discounted. Using this endpoint, the Nairobi study reported a net benefit of formula feeding over breastfeeding [8] providing the impetus for a major pendulum swing in international infant feeding policy.

The Pendulum Swings Away from Breastfeeding

Trapped between a rock and a hard place, it was now incumbent on the WHO to provide guidance in the public health minefield of infant feeding policies for HIV-infected women. WHO policy shifted towards support of formula feeding with a big IF. Formula feeding was supported for HIV-infected women if it was Affordable, Feasible, Acceptable, Sustainable and Safe (AFASS) [9, 10]. This cumbersome acronym allowed some policy makers and implementers to be deluded into thinking that there were no real dangers of artificial feeding that could not simply be overcome with AFASS-enhancing programs.

In part, the growing disregard of the dangers of shifts away from breastfeeding was supported by the results of the Nairobi study [8]. It remains unclear why this study reported so few adverse outcomes of formula feeding. No study to date has been able to replicate such good outcomes with formula. All studies in subsequent years, many of them considerably larger and some of them randomized and with at least equivalent methodological rigor, have reported, at best, no benefit with shifts away from breastfeeding [11, 12] or, in program settings, worse HIV-free survival [13, 14] (Table 20.1). We speculate that strict selection of study participants from atypically good socioeconomic circumstances and extensive monitoring and support during the trial limited its generalizability. Lack of any type of antiretroviral intervention and low rates of exclusive breastfeeding undoubtedly also led to the high rates of HIV transmission observed. The small sample size was vulnerable to chance fluctuations and may have contributed to the lack of balance between the groups in important confounders that some have pointed out [15].

AFASS Takes Hold

With AFASS policies in place, the population experiment of the effects of withholding breast milk on child survival could once again be repeated. Because of, or in spite of, information being given to HIV-infected women, the past 10 years have seen major changes in how women in sub-Saharan Africa

Table 20.1 Studies reporting the effects on HIV-free survival when breastfeeding is curtailed

	Study design	Comparisons	HIV-free survival
Nairobi, Kenya [8]	Randomized trial ($n=401$)	Formula from birth vs. breastfeeding (median 17 months)	Net benefit of formula
Botswana (MASHI) [11]	Randomized trial ($n=1,200$)	Formula from birth vs. breastfeeding for 6 months	Equivalent outcomes
Lusaka, Zambia (ZEBS) [29, 87, 88]	Randomized trial ($n=958$)	Early weaning at 4 months vs. breastfeeding (median 16 months)	Equivalent outcomes (in intent to treat analysis)
South Africa [13]	Program evaluation	Formula from birth vs. some breastfeeding	Worse outcomes with formula if poor socioeconomic status
Cote d'Ivoire [12]	Epidemiologic study of self-selected feeding choices ($n=557$)	Formula from birth vs. breastfeeding to 4 months	Equivalent outcomes
Rakai, Uganda [14]	Program evaluation ($n=182$)	Formula from birth vs. breastfeeding	Worse outcomes with formula
Rwanda [18]	Epidemiologic study ($n=532$)	Formula from birth vs. breastfeeding for 6 months	Equivalent outcomes
Western Kenya [19]	Program evaluation ($n=2,477$ but high drop-out)	Formula from birth vs. breastfeeding to 4 months	Equivalent outcomes

feed their infants and little support for exclusive breastfeeding. In some settings, there has been almost complete avoidance of breastfeeding and in others, much shorter durations of breastfeeding than usual in these communities. In circumstances where adequate data before and after these changes were collected at least the adverse consequences of these changes could be scrutinized. One group who initially theorized that shifts away from breastfeeding simply to avoid HIV would *not* result in adverse health outcomes [16], observed in their own program, substantial elevations in mortality among women who elected not to breastfeed [14]. Several other programs too reported that even after the benefits of HIV prevention were taken into account worse or, at best, no benefit of artificial feeding were observed [14, 17–20].

Two study teams in Malawi, one in Kenya and one in Uganda recommended early weaning (complete stoppage of breastfeeding) to women participating in their trials and subsequently found marked elevations in diarrheal morbidity and mortality [21–24], compared, in some cases, to earlier cohorts when no specific recommendations were given to shorten breastfeeding duration. All these studies included close monitoring and follow-up, as well as education and counseling which theoretically should minimize risks of weaning. Two of the studies were interrupted by their Data Safety and Monitoring Boards concerned about the elevations in morbidity after weaning. Although it could be argued that historical comparisons are a weak study design, these observations are biologically plausible and increased rates of diarrheal morbidity among nonbreastfed children are consistently reported even in settings such as the UK and the USA without barriers to clean water and healthcare [25–27]. These increases in child morbidity and mortality were all the more palpable as access to antiretrovirals as well as other child-related services improved over time. Epidemiologic analyses of mortality among breastfed and nonbreastfed infants and young children between birth and 24 months in two trials in Malawi revealed that breastfeeding was associated with a 2.9-fold lower risk of mortality among exposed-uninfected infants after adjustment for confounders [28]. For a selection of some of the adverse effects, see Table 20.2.

Using HIV-free survival as the primary outcome also serves to neglect and overlook mortality among HIV-infected children. It is now well-established that HIV-infected children who are formula-fed

Table 20.2 Effects on morbidity and mortality of uninfected infants born to HIV-infected mothers when breastfeeding was curtailed

	Study design	Comparisons	Morbidity and mortality in uninfected children
Randomized trials			
Nairobi, Kenya [8]	Randomized trial ($n=401$)	Formula vs. breastfeeding	Trend towards higher 2-year mortality (24%) in formula (24%) vs. breastfeeding (20%) group
Botswana (MASHI) [11]	Randomized trial ($n=1,200$)	Formula vs. short breastfeeding	Higher mortality at 7 months in formula (9.3%) vs. breastfeeding (4.9%) groups
Lusaka, Zambia (ZEBS) [29, 87, 88]	Randomized trial ($n=958$)	Early weaning vs. long breastfeeding	Two to fourfold increase in uninfected child mortality due to weaning through 18 months
Historical controls			
Kampala, Uganda [23]	Observations during a trial vs. previous study ($n=1,307$)	Early weaning vs. longer breastfeeding	Higher diarrhea-related and all cause mortality in cohort encouraged to wean earlier
Malawi [21]	Comparison to prior trial with longer breastfeeding ($n=3,845$)	Early weaning vs. long breastfeeding	Higher diarrhea-related morbidity and mortality and all cause mortality in cohort encouraged to wean early
Kisumu, Kenya [24, 36]	Comparison to prior study with longer breastfeeding ($n=491$)	Early weaning vs. longer breastfeeding	Higher diarrhea-related morbidity Water safety intervention ineffective
Epidemiologic studies			
South Africa [13]	Program evaluation	Formula vs. breastfeeding	Formula had higher adverse outcomes (HIV and uninfected death combined) if poor socioeconomic status
Cote d'Ivoire [12]	Self-selected feeding choice ($n=557$)	Formula vs. short breastfeeding	Equivalent HIV-free survival
Malawi [28]	Combined studies ($n=2,000$)	Multivariate analysis of actual feeding practices	Significant reduction (hazard ratio=0.44) in mortality if breastfed (both infected and uninfected children)
Rakai, Uganda [14]	Program evaluation ($n=182$)	Formula vs. breastfeeding	Sixfold increase in mortality if formula-fed
Rwanda [18]	Self-selected feeding choice ($n=532$)	Formula vs. short breastfeeding	Nonsignificant trend towards higher mortality in formula (5.6%) vs. breast-fed (3.3%)
Pune, India [20]	Program evaluation ($n=148$)	Formula vs. breastfed	Significant higher risks of hospitalization if formula-fed
Rural Uganda [17]	Self-selected feeding practices ($n=109$)	Early weaning vs. longer breastfeeding	Sixfold increase in death if wean before 6 months
Western Kenya [19]	Program evaluation ($n=2,477$ but high drop-out)	Formula vs. short breastfeeding	Equivalent HIV-free survival
Botswana [32, 34]	Public Health outbreak investigation	Actual practices	25-fold increase in diarrhea deaths if not breastfed

or who are weaned off breast milk early are at high risk of dying prematurely [11, 28, 29]. It is not practical to make infant feeding recommendations based on the child's HIV status. Decisions about infant feeding are usually made during pregnancy and, if avoidance of breastfeeding or early weaning is selected, may be difficult to reverse. Even with the availability of early infant diagnosis, the child's status is not known to the mother for weeks. Moreover, incorrect information about theoretical risks of "super-infection" and drug resistance and toxicity has been used to discourage women with newly identified infected children from breastfeeding. To its credit, the WHO has clearly stated that known HIV-infected children should breastfeed without any equivocation about duration or AFASS. As HIV-infected children can progress rapidly [30], breastfeeding is essential for these infants to survive long enough to access pediatric HIV care and treatment programs to benefit from antiretroviral therapy.

The Myth of AFASS: Can Formula Ever Be Safe?

Implicit in the concept of AFASS is that formula feeding is safe under certain circumstances. To attempt to unpack this myth in more detail, we consider the potential mechanisms whereby breastfeeding protects infants' health. For heuristic purposes, we separate the biological basis for the harm of nonbreastfeeding into three overarching mechanisms: (1) contamination, i.e., artificial feeding places the infant at risk through introducing environmental contaminants and creating a less hygienic feeding method; (2) poor nutrition, i.e., abstinence from breastfeeding could compromise an infant's nutritional status if formula is not mixed correctly or not given in appropriate quantities; and (3) the absence of immune protection.

Contamination: One of the most commonly stated reasons for why breastfeeding needs to be protected among HIV-infected women in sub-Saharan Africa is lack of clean water. It is certainly true that lack of clean water and inadequate sanitation facilities exaggerate the dangers of artificial feeding [31]. The dramatic epidemic of diarrhea-related deaths that occurred in Botswana among formula-fed infants after a period of severe flooding is a clear example of the dangers of contaminated water even in settings that usually have safe water supply [32–34]. Provision of a sustained supply of adequate clean water at the point of use in the household is clearly a major priority for public health [35]. But if safe water is available is formula feeding safe?

An exemplary demonstration of the multifactorial nature of breastfeeding's benefits came from an interesting confluence of circumstances in a clinical trial in rural Kenya. During a study to evaluate, the effects of antiretroviral therapy during lactation on the prevention of postnatal HIV transmission, HIV-infected women were encouraged to stop all breastfeeding by 6 months which was the time when antiretroviral therapy was also stopped. The study was temporarily suspended when elevated rates of diarrhea morbidity were noticed around the time of weaning [24]. A state-of-the-art home water quality improvement program was introduced. Despite the known benefits of this intervention in other settings, for weaning-related morbidity it was ineffective. The intervention reduced diarrhea while infants were being breastfed but not once they had stopped [36]. These results clearly demonstrate that while water contamination plays a role in exacerbating the risks of artificial feeding [31], clean water is insufficient to fully mitigate artificial feeding's risks.

Training around "safe" preparation of formula feeds has featured prominently in infant feeding programs in sub-Saharan Africa. The assumption is that if women can simply be sufficiently motivated to boil all water, wash their hands and follow all hygiene rules, infant formula can be given safely. In practice, following these guidelines in most homes in low resource settings, often without indoor water sources or electricity, is extremely difficult [37, 38]. In a study conducted in KwaZulu-Natal, South Africa, about 80% of formula samples mothers prepared at home after instructions from the counselors were contaminated with fecal bacteria [39]. About 20% of the samples that the counselors prepared at the clinic while showing the mothers how to do everything correctly were also contaminated [39].

Taken together, what these results demonstrate is that infant formula is not the only source of exposure to pathogens among infants and young children, especially in contaminated environments. Much to their parents' chagrin, young children explore their world with their hands and mouths. Breast milk has evolved to protect children from these pathogens [40]. Some breastfeeding, even if it is nonexclusive, is more protective against diarrhea morbidity and mortality than no breastfeeding [41], even if the quantity of breast milk consumed is relatively small [42]. It is noteworthy that breastfeeding reduces the risk of respiratory illness and pneumonia, outcomes where contaminated water plays little or no role [25, 43–45]. Breastfeeding also protects against severe infectious disease in settings with a predominantly safe water supply [25–27].

Poor nutrition: The cost of infant formula places it beyond the financial reach of all but small elites in most sub-Saharan African countries. Lack of access to formula or limited access resulting in over-dilution to stretch the available formula as far as possible is often invoked as the explanation as to why HIV-infected women need to breastfeed. This concern gives rise to the inference that simply providing adequate quantities of infant formula would solve the challenges of infant feeding for HIV-infected women since infant formula is specifically developed to mirror the nutritional composition of breast milk as closely as possible. Theoretically, health service provision of formula should be able to address the affordability challenge but given the reality of weak health service infrastructures, ensuring a sustained supply has been a challenge in many programs. Audits of the South African national formula program have described stock-outs and rationing in both urban and rural sites [46]. Population mobility introduces further complications for sustained access.

Other than these gross limitations in terms of access to formula, even from a nutritional point of view the product falls short in other respects. Breast milk is physiologically regulated such that the content varies from the beginning to the end of the feed so that a child can be most quickly satiated even with a short feed but can continue to feed for comfort and not become overfed on longer feeds [47]. The composition of human milk also varies even between feeds based on the amount the child consumes and over time being regulated to adapt to the unique needs of a specific child [48]. This individualization cannot be achieved with formula. Obesity and metabolic syndrome in children and young adults are now increasingly recognized as being linked to formula feeding [49–52] indicating that even with adequate access and sufficient quantities, infant formula remains nutritionally substandard. There are burgeoning efforts in the infant formula industry to make formula more like human milk by adding immunologically active components, such as long chain fatty acids and probiotics [53–55]. Breast milk is more than food and attempting to achieve only nutritional parity will not bring formula to the level of protection of infant health that breast milk can provide.

Absence of immune protection: Since neither clean water nor adequate supplies can explain the benefits of breastfeeding we are left with the clear inference that it is not so much what breastfeeding *keeps out* that is important, but what breastfeeding *puts in*. Breast milk contains a vast spectrum of immunologically active components that include antigen-specific antibodies and cellular immune components as well as almost every soluble factor known to have immunologic activity to protect against disease [56–58]. Passive transfer of maternal antibodies across the placenta is now well known to be an important means by which the infant, whose immune system is not fully mature at the time of birth, is protected immunologically. This process continues during breastfeeding with passive transfer of immunologic components as well as immunomodulatory effects of breast milk components on the infant gut and developing immune system [56, 57]. A substantial component of this activity is by dampening the immune response creating "tolerance" in the infant and preventing activation after exposure to pathogens [59]. Since HIV has so many dysregulatory effects on the human immune system, theoretically the immunologic quality of breast milk from HIV-infected women might be compromised. Although this topic needs considerable more study, one study from Botswana showed that HIV-infected and uninfected women had similar quantities of the immunologic components that they measured [60] suggesting that despite HIV infection, breast milk immunologic quality is not compromised.

Paying Close Attention to the Numbers

Despite the decades of research into the adverse effects of artificial feeding and the considerable body of basic science, clinical and epidemiologic research that do not support the safety of formula feeding, the argument by analogy is often invoked: namely, women in the USA formula feed their infants all the time and those babies are doing just fine. Or, the n-of-one argument: I (or my child) was formula-fed and I (he/she) am (is) doing just fine.

To appreciate the reasons for why some of those babies in the USA might (or might not [61, 62]) be doing just fine, it is important to make the distinction between an *absolute* risk and a *relative* risk. An absolute risk is the frequency with which an event occurs in the population, e.g., the infant mortality rate might be 10 deaths per 1,000 live-births. A relative risk requires a comparison. For example, we might say the infant mortality rate is 10/1,000 live-births if women breastfeed, but 20/1,000 live-births if women avoid all breastfeeding, i.e., a twofold increased risk. The ratio of rates in the two groups is referred to as the relative risk. The *relative risk* associated with artificial feeding is elevated in all populations, but what makes the north different from the south is that the absolute rates of morbidity and mortality are generally low. Moreover, breastfeeding may protect against morbidity, but since most morbidity in these settings is not fatal, arguably the benefits can be ignored. For low resource settings, women face a double whammy: the absolute background rates of mortality are several fold higher, so even small elevations translate into large numbers of infant deaths, and the relative risks are higher too because environmental deprivation and barriers to health care exacerbate the biological inferiority of formula.

Benefits of breastfeeding are multifactorial. Although a strong public health program may be able to minimize risks of environmental contamination and poor nutrition, programs can do nothing to mitigate the risks conferred by the absence of the immunologically active components of breast milk. The fact that breastfeeding confers benefits to infant health even in wealthy countries [25, 26] suggests that there is a biological threshold below which it is not possible to go even with the strongest programs.

Clinical trial data are important in formulating policy but caution is required when extrapolating results on the risks of artificial feeding from clinical trials. In most clinical studies, participants are highly motivated, receive the best possible educational interventions and are provided with close monitoring and a health service safety set. In a rigorous and well-monitored clinical trial in urban Botswana, a country with some of the best economic indicators in sub-Saharan Africa, HIV-exposed, uninfected infants randomized to infant formula from birth had a twofold increase in mortality compared to those randomized to breastfeeding [11]. In contrast, in two separate programs in Uganda, infant mortality was increased more than sixfold among women who considered formula feeding an AFASS choice for themselves [14, 17]. Under the best-case scenario, when infant formula is provided under carefully monitored conditions, with adequate access to medical care and sufficient education and support and with optimal selection of women considered to have adequate personal resources to safely formula feed, there is still about a twofold increased risk of mortality. In programmatic settings, the risks of death are several fold higher.

The Subtext of Poverty and Human Rights

The subtext of poverty and human rights makes discourse around infant feeding and HIV complex. It is obviously, morally, and ethically unacceptable that some babies are born into poverty, into settings where there is no clean water, insufficient food at home and where there are background risks of fatal infections that make use of infant formula a foolhardy choice at best. The concept of AFASS attempted to be a gentle reminder to guard against the inappropriate promotion and use of formula. AFASS was impractical and failed in the field because it required that HIV-infected women themselves make the determination for whether or not infant formula was AFASS for their circumstances. An interesting study in South Africa stratified participants based on objective socioeconomic criteria. Women who chose above their station, i.e., chose formula when their socioeconomic station would have precluded it had the worst outcomes [13]. Counseling to explain that breastfeeding should only be undertaken if you are too poor to afford to safely formula feed is likely to be a challenge for even the most tactful of counselors.

In addition to the insensitivity of such counseling, provision of free or subsidized formula poses complex ethical challenges [63, 64]. In situations of scarcity, infant formula is perceived as a valuable and precious commodity. When it is further endorsed by the health service, it is also perceived as a safe and superior method of feeding. Qualitative research has highlighted the coercive dynamics of free formula and there are many examples of confusion and misinformation [38, 46, 65–68]. AFAS – S (minus the last S) appears to make an ethical demand that formula be made available by programs, in an affordable, acceptable, and sustainable way, in all settings (especially those where they are *not* affordable to the population at large). Availability is set up as the limitation rather than the intrinsic lack of safety. The impulse to address the gross economic inequalities between the developed and the developing world by simply implementing the same "standard of care" (provision of formula) is an understandable one. However, in the field of HIV and breastfeeding this one-size-fits-all approach has now done considerable harm.

Misunderstandings About Exclusive Breastfeeding

In a field of controversy and confusion, no single finding has generated as much of its own than the observation that exclusive breastfeeding is associated with reduced risk of HIV transmission. Initially, when first reported from Durban, South Africa that mothers who gave their infants only breast milk through 3 months of age were less likely to transmit HIV than mothers who breastfed and gave other solids or liquids before this age [69, 70], the average response was disbelief. How can *more* breastfeeding lead to *less* HIV transmission? A biological puzzle indeed, but when nature gives us clues as to some of the complexities of HIV pathogenesis it is worth paying attention [71]. But only few gave thought to the likely mechanisms involved [72], and the primary response was confusion as to why women in this study were counseled to breastfeed exclusively at all. It is hot in Durban, don't the babies get thirsty? So it took some time for infectious disease specialists to get up to speed with a standard midwifery syllabus that includes exclusive breastfeeding as one of the primary principles of lactation support, embedded in the baby-friendly principles endorsed by all international health agencies [73]. Since breast milk alone can support all of an infant's nutrition and fluid requirements through at least 6 months of age [74], supplements are unnecessary and potentially dangerous. Exclusive breastfeeding also establishes a regularity of breast milk supply and demand reducing mastitis and other breast problems [75, 76] which represent additional risk factors in HIV transmission. The finding that the *quality* of breastfeeding, ascertained by the extent of exclusive breastfeeding, is related to the risk of HIV transmission has now been confirmed in at least three additional large studies [70, 77–79]. Thus, estimates of postnatal transmission gathered from settings where support of exclusive breastfeeding is lacking or in communities with poor uptake of recommendations to breastfeed exclusively are likely to be higher than those collected in settings more favorable to and supportive of exclusive breastfeeding.

Exclusive breastfeeding also reminded us that although breastfeeding is a biological process, it is also a cultural practice [80]. What is healthiest and what is normative do not necessarily coincide. Cultural practices that displace breastfeeding, such as giving herbal supplements to infants, can be detrimental to both mother and infant [81]. The lament was then raised that exclusive breastfeeding was too difficult and cultural barriers too entrenched to propose this as a viable strategy to reduce HIV transmission. Despite the fact that programs to support exclusive breastfeeding were being successfully developed and evaluated in Asia and Latin America during this time [27, 82–85], these new innovations were not incorporated into HIV programs. To our knowledge, no randomized study of how to support exclusive breastfeeding among HIV-infected women has yet been undertaken—a serious missed opportunity for operational prevention research.

An even more serious confusion that arose out of discussions of the observations that exclusive breastfeeding reduces HIV transmission was the conclusion that all breastfeeding should stop when breastfeeding is no longer exclusive. WHO in an attempt to be responsive to new scientific findings issued a specific recommendation about the importance of supporting exclusive breastfeeding for HIV-infected women who elected to breastfeed [86]. For reasons that remain unclear, almost everyone understood these recommendations to mean that exclusive breastfeeding should be supported for *no longer than* 6 months and then women should be encouraged to stop all breastfeeding abruptly. We stand accused as one of the teams of investigators who thought that it was so plausible that a short period of exclusive breastfeeding followed by abrupt cessation of breastfeeding was the optimal strategy to preserve HIV-free survival that we designed a randomized trial to test just that. Unfortunately, we were dead wrong. As we published our results showing that early weaning neither improved HIV-free survival nor was safe in terms of protecting survival of exposed-uninfected and infected infants [29], we found ourselves up against a community who had already begun implementing this approach thinking it was the recommendation of the WHO!

As part of our trial, 958 HIV-infected women in Lusaka, Zambia were randomized to either stop breastfeeding abruptly at 4 months or to continue breastfeeding for their own preferred duration. Infant formula and a specially developed, fortified weaning cereal was provided for infants in the intervention group. Since the cereal required cooking, contamination of water sources would, theoretically, be less of a concern. Infants in either study arm were weighed regularly and were provided with food supplements if there was any evidence of failure to thrive. Routine childhood interventions, including vaccines, vitamin A, and prophylactic cotrimoxazole was provided to children in both groups. The context of the trial also provided a health services safety net and intensive counseling and education, including about safe water and hygiene [29]. Despite our best educational efforts, early weaning was not well-accepted by the study population. Thus, it became essential, in the interpretation of our results, to analyze the data based on actual feeding behaviors. What we observed was that infants born to women who adhered to their assignment and weaned early as instructed, had worse outcomes than those whose mothers ignored their random assignment and continued breastfeeding; as did infants born to women who refused to adhere to their assignment to the control group and weaned early [87]. Benefits of breastfeeding on infant and young child survival persisted into the second year of life to around 18 months [88]. Benefits of continued breastfeeding were also observed for child growth [89] and for diarrheal morbidity and mortality [90].

HIV transmission persists throughout the duration of breastfeeding but early weaning is a late starter as much of the transmission has already occurred by the time breastfeeding ends. The older the child when breastfeeding ends, the less there is to gain in terms of avoiding HIV infection. As a result, risks of early weaning take on greater weight. With only small benefits of HIV prevented, small increases in mortality easily offset this benefit. We have reported no benefit of cessation of breastfeeding at 4 months for the combined outcome of HIV infection or death (HIV-free survival) compared to standard practice of breastfeeding ad lib in the primary intent to treat analyses [29]. With further analysis of the actual practices, we have found that the magnitude of benefit associated with early weaning (i.e., the amount of HIV prevented) was almost identical to the magnitude of the harm caused by early weaning (i.e., the numbers of uninfected child deaths caused) [87]. Women who stopped breastfeeding by 5 months had an additional 1.1% transmission rate after 4 months but a 17.4% uninfected child mortality rate through 24 months. In contrast, women who continued to breastfeed for 18 months had an additional 11.2% transmission rate but an uninfected child mortality rate of 9.7% [87].

Fig. 20.1 Results from the Zambia exclusive breastfeeding study (ZEBS) which showed no net benefit for HIV-free survival of early weaning because reductions in HIV transmission were counter-balanced by increases in uninfected child mortality [93]

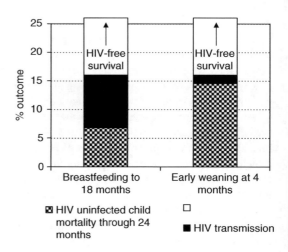

Thus, the total number of adverse events was almost identical between early weaning and prolonged breastfeeding but the composition was different with HIV making up a larger proportion of the adverse events in the breastfeeding women (Fig. 20.1).

Our data from Zambia were collected in the absence of either maternal antiretroviral treatment or extended antiretroviral prophylactic regimens that continue during breastfeeding. It is now clearly established that these interventions reduce postnatal HIV transmission considerably [91–93]. Thus, when antiretrovirals are given, the magnitude of mortality caused by artificial feeding is even larger than the magnitude of HIV transmission prevented. In our trial, among women who were not yet at an advanced enough disease stage to require antiretroviral therapy for their own health, stopping breastfeeding at 4 months led to a threefold *increase* in the combined outcome of child HIV infection or death occurring between 4 and 24 months [87].

A Parallel Antiretroviral Research Agenda

And so antiretrovirals stepped up to save the day. Parallel with the research on infant feeding and HIV was a systematic research agenda evaluating the use of antiretrovirals to prevent mother-to-child transmission [93]. Starting with the first proof of principle in 1996 [94], later studies investigated how to shorten the duration of prophylaxis as much as possible to make the interventions affordable and feasible in low resources settings. When the shortest possible intervention, single-dose nevirapine, was found to be almost 50% effective in reducing transmission peripartum (largely infections occurring in labor or during delivery) [95], a new age of implementation began which made the prevention of pediatric HIV infection a real possibility in low resource settings. Over almost 15 years, breastfeeding was a thorn in the side of antiretroviral prevention programs, since the short-course regimens were designed to attack transmission occurring peripartum with little to no coverage over the breastfeeding period. Breastfeeding simply added new infections after the antiretrovirals had ceased, eroding the benefits [95]. Eventually, the research agenda turned to evaluating longer courses of antiretrovirals continuing through the breastfeeding period. Prophylaxis of the infant with nevirapine proved to be an effective approach [91, 96, 97]; in contrast to prophylaxis with zidovudine which was not [11]. The first study evaluated 6 weeks of nevirapine, the second 14 weeks, and the third 24 weeks of daily nevirapine, all showing reductions in transmission when prophylaxis was given but resumption of transmission when prophylaxis was discontinued [91, 96, 97]. It was finally up to WHO to bravely pull the plug and declare that there was now sufficient evidence that use of nevirapine as prophylaxis was effective whenever it was used during breastfeeding and endorsing its use over the duration of breastfeeding [98].

At the same time, as the agenda on how to use antiretrovirals to prevent mother-to-child transmission was unfolding, a parallel agenda to make antiretroviral therapy available to adults in low resource settings was gathering momentum. Rationing initially made drugs available only to the sickest individuals, with guidelines recommending treatment only for persons with CD4 counts below 200 cells/mm^3. Advocates were quick to motivate that pregnant women should be prioritized. As therapy needs to continue lifelong, its initiation during pregnancy would theoretically reduce postnatal HIV transmission via breastfeeding in addition to its established benefits for reducing peripartum transmission and protecting maternal health. Studies providing antiretroviral drugs to pregnant women who met treatment criteria reported low rates of transmission even when women continued to breastfeed [99, 100]. Studies then turned to evaluating antiretroviral drugs given to women with higher CD4 counts, or to all women regardless of CD4 count, with the goal less to protect maternal health but to prevent transmission through all routes [18, 92, 99, 101–105]. Prophylaxis using a usually therapeutic antiretroviral regimen would then be stopped when breastfeeding ended. To our knowledge, no study has yet to provide antiretrovirals over a normal duration of breastfeeding, i.e., 18 months or longer as most of these studies were designed in the era when early weaning was in place as the de facto recommendation. Nevertheless, one study with 6 months of breastfeeding has reported <1% postnatal transmission rate with antiretroviral therapy started during pregnancy indicating that maternal therapy can almost eliminate transmission occurring via breastfeeding [92].

There is a difference of opinion as to whether maternal therapy to all women regardless of CD4 count is a better public health approach to prevent mother-to-child HIV transmission than maternal therapy only to those who meet criteria with infant prophylaxis to the rest [106]. Since guidelines for who to treat have expanded to include all with CD4 counts <350 cells/mm^3 [107] the argument with respect to postnatal HIV is somewhat moot. More than 80% of postnatal transmission occurs among women with CD4 counts below 350 cells/mm^3 [108] so whatever interventions are given to the remaining women has only minor impact on the overall risk of transmission. The biggest "bang for the buck" is treating pregnant women with CD4 count <350 cells/mm^3. The challenge remains to meet the unmet needs for antiretroviral therapy among pregnant women since there is a limited time window and fragmentation of health services create further barriers for this priority population.

The Pendulum Swings Halfway Back

New WHO guidelines now support breastfeeding to 12 months for HIV-infected women albeit with some equivocation related to setting [98]. The success of antiretrovirals in reducing postnatal transmission to very low levels facilitated this partial swing back to arguably a reasonable middle-ground position. The now growing reports of the adverse consequences of shifts away from breastfeeding among HIV-infected women are grim reminders of why antiretrovirals during breastfeeding are so essential. The challenge now is implementation: both in terms of providing access to antiretrovirals and in terms of supporting breastfeeding. Effective use of antiretroviral drugs can now reduce HIV transmission to such low levels that there are few circumstances where replacement feeding can be justified.

Acknowledgments We would like to acknowledge support from the National Institutes of Child Health and Human Development (HD 57161, HD 39611, and HD 40777).

The authors have nothing to disclose.

References

1. Ip S, Chung M, Raman G, Trikalinos TA, Lau J (2009) A summary of the Agency for Healthcare Research and Quality's evidence report on breastfeeding in developed countries. Breastfeed Med 4(Suppl 1):S17–S30
2. Mathews-Grieco SF. Breastfeeding, wet nursing and infant mortality in Europe (1400–1800). Historical perspectives on breastfeeding. United Nations Children's Fund: UNICEF 1991, pp 15–60
3. Jelliffe DB, Jelliffe EF (1978) Human milk in the modern world. Oxford University Press, New York
4. Centers for Disease Control and Prevention (1985) Recommendations for assisting in the prevention of perinatal transmission of human T-lymphotropic virus type III/lymphadenopathy-associated virus and acquired immunodeficiency syndrome. MMWR 34:721–726, 731–732
5. World Health Organization (1990) Health factors which may interfer with breastfeeding. Infant feeding: the physiological basis. WHO, Geneva, pp 41–54
6. Dunn DT, Newell ML, Ades AE, Peckham CS (1992) Risk of human immunodeficiency virus type 1 transmission through breastfeeding. Lancet 340:585–588
7. World Health Organization Global Programme on AIDS. Consensus statement from the WHO/UNICEF consultation on HIV transmission and breastfeeding. Report No WHO/GAPA/INF/92 1 1992, WHO, Geneva–Switzerland
8. Nduati R, John G, Mbori-Ngacha D et al (2000) Effect of breastfeeding and formula feeding on transmission of HIV-1: a randomized clinical trial. JAMA 283:1167–1174
9. World Health Organization (2000) New data on the prevention of mother-to-child transmission of HIV and their policy implications. Conclusions and recommendations. WHO technical consultation on behalf of the UNFPA/UNICEF/WHO/UNAIDS inter-agency task team on mother-to-child transmission of HIV. 11–13 Oct 2000. www.who.int/reproductive-health/publications/new_data_prevention_mtct_hiv/index.html
10. World Health Organization (2003) HIV and infant feeding. Guidelines for decision-makers. Geneva. http://www.who.int/child-adolescent-health/New_Publications/NUTRITION/HIV_IF_DM.pdf
11. Thior I, Lockman S, Smeaton LM et al (2006) Breastfeeding plus infant zidovudine prophylaxis for 6 months vs formula feeding plus infant zidovudine for 1 month to reduce mother-to-child HIV transmission in Botswana: a randomized trial: the Mashi study. JAMA 296:794–805
12. Becquet R, Bequet L, Ekouevi DK et al (2007) Two-year morbidity-mortality and alternatives to prolonged breastfeeding among children born to HIV-infected mothers in Cote d'Ivoire. PLoS Med 4:e17
13. Doherty T, Chopra M, Jackson D, Goga A, Colvin M, Persson LA (2007) Effectiveness of the WHO/UNICEF guidelines on infant feeding for HIV-positive women: results from a prospective cohort study in South Africa. AIDS 21:1791–1797
14. Kagaayi J, Gray RH, Brahmbhatt H et al (2008) Survival of infants born to HIV-positive mothers by feeding modality in Rakai, Uganda. PLoS One 3:e3877. doi:10.1371/journal.pone.0003877
15. Bulterys M (2000) Breastfeeding in women with HIV. JAMA 284:956–957
16. Brahmbhatt H, Gray RH (2003) Child mortality associated with reasons for non-breastfeeding and weaning: is breastfeeding best for HIV-positive mothers? AIDS 17:879–885
17. Homsy J, Moore D, Barasa A et al (2010) Breastfeeding, mother-to-child HIV transmission, and mortality among infants born to HIV-infected women on highly active antiretroviral therapy in rural Uganda. J Acquir Immune Defic Syndr 53:28–35
18. Peltier CA, Ndayisaba GF, Lepage P et al (2009) Breastfeeding with maternal antiretroviral therapy or formula feeding to prevent HIV postnatal mother-to-child transmission in Rwanda. AIDS 23:2415–2423
19. Nyandiko WM, Otieno-Nyunya B, Musick B et al (2010) Outcomes of HIV-exposed children in western Kenya: efficacy of prevention of mother to child transmission in a resource-constrained setting. J Acquir Immune Defic Syndr 54:42–50
20. Phadke MA, Gadgil B, Bharucha KE et al (2003) Replacement-fed infants born to HIV-infected mothers in India have a high early postpartum rate of hospitalization. J Nutr 133:3153–3157
21. Kafulafula G, Hoover DR, Taha TE et al (2010) Frequency of gastroenteritis and gastroenteritis-associated mortality with early weaning in HIV-1-uninfected children born to HIV-infected women in Malawi. J Acquir Immune Defic Syndr 53:6–13
22. Kourtis AP, Fitzgerald D, Hyde L et al (2007) Diarrhea in uninfected infants of HIV-infected mothers who stop breastfeeding at 6 months: the BAN study experience. Los Angeles, CA, 25–28 Feb 2007
23. Onyango-Makumbi C, Bagenda D, Mwatha A et al (2010) Early weaning of HIV-exposed uninfected infants and risk of serious gastroenteritis: findings from two perinatal HIV prevention trials in Kampala, Uganda. J Acquir Immune Defic Syndr 53:20–27
24. Thomas T, Masaba R, van Eijk A et al (2007) Rates of diarrhea associated with early weaning among infants in Kisumu, Kenya. Los Angeles, CA, 25–28 Feb 2007
25. Chantry CJ, Howard CR, Auinger P (2006) Full breastfeeding duration and associated decrease in respiratory tract infection in US children. Pediatrics 117:425–432

26. Quigley MA, Kelly YJ, Sacker A (2007) Breastfeeding and hospitalization for diarrheal and respiratory infection in the United Kingdom Millennium Cohort Study. Pediatrics 119:e837–e842
27. Kramer MS, Chalmers B, Hodnett E et al (2001) Promotion of breastfeeding intervention trial (PROBIT): a randomized trial in the Republic of Belarus. JAMA 285:413–420
28. Taha TE, Kumwenda NI, Hoover DR et al (2006) The impact of breastfeeding on the health of HIV-positive mothers and their children in sub-Saharan Africa. Bull WHO 84:546–554
29. Kuhn L, Aldrovandi GM, Sinkala M et al (2008) Effects of early, abrupt cessation of breastfeeding on HIV-free survival of children in Zambia. N Engl J Med 359:130–141
30. Violari A, Cotton MF, Gibb DM et al (2008) Early antiretroviral therapy and mortality among HIV-infected infants. N Engl J Med 359:2233–2244
31. Habicht JP, DaVanzo J, Butz WP (1988) Mother's milk and sewage: their interactive effects on infant mortality. Pediatrics 81:456–461
32. Creek T, Arvelo W, Kim A et al Role of infant feeding and HIV in a severe outbreak of diarrhea and malnutrition among young children, Botswana, 2006. Los Angeles, CA, 25–28 Feb 2007
33. Mach O, Lu L, Creek T et al (2009) Population-based study of a widespread outbreak of diarrhea associated with increased mortality and malnutrition in Botswana, January–March, 2006. Am J Trop Med Hyg 80:812–818
34. Creek TL, Kim A, Lu L et al (2010) Hospitalization and mortality among primarily nonbreastfed children during a large outbreak of diarrhea and malnutrition in Botswana, 2006. J Acquir Immune Defic Syndr 53:14–19
35. Marino DD (2007) Water and food safety in the developing world: global implications for health and nutrition of infants and young children. J Am Diet Assoc 107:1930–1934
36. Harris JR, Greene SK, Thomas TK et al (2009) Effect of a point-of-use water treatment and safe water storage intervention on diarrhea in infants of HIV-infected mothers. J Infect Dis 200:1186–1193
37. Dunne EF, Angoran-Benie H, Kamelan-Tano A et al (2001) Is drinking water in Abidjan, Cote d'Ivoire, safe for infant formula? J Acquir Immune Defic Syndr 28:393–398
38. Doherty T, Chopra M, Nkonki L, Jackson D, Greiner T (2006) Effect of the HIV epidemic on infant feeding in South Africa: "when they see me coming with the tins they laugh at me". Bull WHO 84:90–96
39. Andresen E, Rollins NC, Sturm AW, Conana N, Griener T (2007) Bacterial contamination and over-dilution of commercial infant formula prepared by HIV-infected mothers in a prevention of mother-to-child transmission (PMTCT) programme in South Africa. J Trop Pediatr 53:410–414
40. McClellan HL, Miller SJ, Hartmann PE (2008) Evolution of lactation: nutrition v. protection with special reference to five mammalian species. Nutr Res Rev 21:97–116
41. Victora CG, Smith PG, Vaughan JP et al (1987) Evidence for protection by breast-feeding against infant deaths from infectious diseases in Brazil. Lancet 2:319–322
42. Brown KH, Black RE, Lopez de Romana G, Creed de Kanashiro H (1989) Infant-feeding practices and their relationship with diarrheal and other diseases in Huascar (Lima) Peru. Pediatrics 83:31–40
43. Cesar JA, Victora CG, Barros FC, Santos IS, Flores JA (1999) Impact of breast feeding on admission for pneumonia during the postnatal period in Brazil: nested case–control study. BMJ 318:1316–1320
44. Bahl R, Frost C, Kirkwood BR et al (2005) Infant feeding patterns and risks of death and hospitalization in the first half of infancy: multicentre cohort study. Bull WHO 83:418–426
45. Arifeen S, Black RE, Antelman G, Baqui A, Caulfield L, Becker S (2001) Exclusive breastfeeding reduces acute respiratory infection and diarrhea deaths among infants in Dhaka slums. Pediatrics 108:E67
46. Chopra M, Rollins N (2008) Infant feeding in the time of HIV: rapid assessment of infant feeding policy and programmes in four African countries scaling up prevention of mother to child transmission programmes. Arch Dis Child 93:288–291
47. Neville MC (1995) Determinants of milk volume and composition. In: Jensen RG (ed) Handbook of milk composition. Academic, San Diego, pp 87–114
48. Neville MC, Keller RP, Seacat J, Casey CE, Allen JC, Archer P (1984) Studies on human lactation. I. Within-feed and between-breast variation in selected components of human milk. Am J Clin Nutr 40:635–646
49. Hummel S, Pfluger M, Kreichauf S, Hummel M, Ziegler AG (2009) Predictors of overweight during childhood in offspring of parents with type 1 diabetes. Diabetes Care 32:921–925
50. Koletzko B, von Kries R, Monasterolo RC et al (2009) Can infant feeding choices modulate later obesity risk? Am J Clin Nutr 89:1502S–1508S
51. Owen CG, Martin RM, Whincup PH, Davey SG, Gillman MW, Cook DG (2005) The effect of breastfeeding on mean body mass index throughout life: a quantitative review of published and unpublished observational evidence. Am J Clin Nutr 82:1298–1307
52. Owen CG, Whincup PH, Kaye SJ et al (2008) Does initial breastfeeding lead to lower blood cholesterol in adult life? A quantitative review of the evidence. Am J Clin Nutr 88:305–314
53. Lonnerdal B (2008) Personalizing nutrient intakes of formula-fed infants: breast milk as a model. Nestle Nutr Workshop Ser Pediatr Program 62:189–198, discussion 198–203
54. Heird WC (2007) Progress in promoting breast-feeding, combating malnutrition, and composition and use of infant formula, 1981–2006. J Nutr 137:499S–502S

55. Koletzko B, Baker S, Cleghorn G et al (2005) Global standard for the composition of infant formula: recommendations of an ESPGHAN coordinated international expert group. J Pediatr Gastroenterol Nutr 41:584–599
56. Labbok MH, Clark D, Goldman AS (2004) Breastfeeding: maintaining an irreplaceable immunological resource. Nat Rev Immunol 4:565–572
57. Goldman AS (1993) The immune system of human milk: antimicrobial, antiinflammatory and immunomodulating properties. Pediatr Infect Dis J 12:664–671
58. Morrow AL, Rangel JM (2004) Human milk protection against infectious diarrhea: implications for prevention and clinical care. Semin Pediatr Infect Dis 15:221–228
59. Hanson LA (2007) Session 1: feeding and infant development: breastfeeding and immune function. Proc Nutr Soc 66:384–396
60. Shapiro RL, Lockman S, Kim S et al (2007) Infant morbidity, mortality, and breast milk immunologic profiles among breast-feeding HIV-infected and HIV-uninfected women in Botswana. J Infect Dis 196:562–565
61. Chen A, Rogan WJ (2004) Breastfeeding and the risk of postneonatal death in the United States. Pediatrics 113:e435
62. Bartick M, Reinhold A (2010) The burden of suboptimal breastfeeding in the United States: a pediatric cost analysis. Pediatrics 125:e1048–e1056
63. Coutsoudis A, Goga AE, Rollins N, Coovadia HM (2002) Free formula milk for infants of HIV-infected women: blessing or curse? Health Policy Plan 17:154–160
64. Coutsoudis A, Coovadia HM, Wilfert CM (2008) HIV, infant feeding and more perils for poor people: new WHO guidelines encourage review of formula milk policies. Bull WHO 86:210–214
65. Sibeko L, Coutsoudis A, Nzuza S, Gray-Donald K (2009) Mothers' infant feeding experiences: constraints and supports for optimal feeding in an HIV-impacted urban community in South Africa. Public Health Nutr 10:1–8
66. Bland RM, Rollins NC, Coovadia HM, Coutsoudis A, Newell ML (2007) Infant feeding counselling for HIV-infected and uninfected women: appropriateness of choice and practice. Bull WHO 85:289–296
67. Doherty T, Chopra M, Nkonki L, Jackson D, Persson LA (2006) A longitudinal qualitative study of infant-feeding decision making and practices among HIV-positive women in South Africa. J Nutr 136:2421–2426
68. Desclaux A, Alfieri C (2009) Counseling and choosing between infant-feeding options: overall limits and local interpretations by health care providers and women living with HIV in resource-poor countries (Burkina Faso, Cambodia, Cameroon). Soc Sci Med 69:821–829
69. Coutsoudis A, Pillay K, Spooner E, Kuhn L, Coovadia HM (1999) Influence of infant feeding patterns on early mother-to-child transmission of HIV-1 in Durban, South Africa. Lancet 354:471–476
70. Coutsoudis A, Pillay K, Kuhn L, Spooner E, Tsai WY, Coovadia HM (2001) Method of feeding and transmission of HIV-1 from mothers to children by 15 months of age: prospective cohort study from Durban, South Africa. AIDS 15:379–387
71. Smith MM, Kuhn L (2000) Exclusive breast-feeding: does it have the potential to reduce breast-feeding transmission of HIV-1? Nutr Rev 58:333–340
72. Lunney KM, Iliff P, Mutasa K et al (2010) Associations between breast milk viral load, mastitis, exclusive breastfeeding, and postnatal transmission of HIV. Clin Infect Dis 50:762–769
73. Baby Friendly Hospital Initiative (BFHI). http://www.unicef.org/programme/breastfeeding/baby.htm. Accessed 2007
74. World Health Organization (2006) Planning guide for national implementation of the global strategy for infant and young child feeding. http://www.who.int/nutrition/publications/Planning_guide.pdf
75. Flores M, Filteau S (2002) Effect of lactation counselling on subclinical mastitis among Bangladeshi women. Ann Trop Paediatr 22:85–88
76. Georgeson JC, Filteau SM (2000) Physiology, immunology, and disease transmission in human breast milk. AIDS Patient Care STDS 14:533–539
77. Coovadia HM, Rollins NC, Bland RM et al (2007) Mother-to-child transmission of HIV-1 infection during exclusive breastfeeding in the first 6 months of life: an intervention cohort study. Lancet 369:1107–1116
78. Iliff P, Piwoz E, Tavengwa N et al (2005) Early exclusive breastfeeding reduces the risk of postnatal HIV-1 transmission and increases HIV-free survival. AIDS 19:699–708
79. Kuhn L, Sinkala M, Kankasa C et al (2007) High uptake of exclusive breastfeeding and reduced early post-natal HIV transmission. PLoS One 2(12):e1363. doi:10.1371/journal.pone.0001363
80. Pak-Gorstein S, Haq A, Graham EA (2009) Cultural influences on infant feeding practices. Pediatr Rev 30:e11–e21
81. Fjeld E, Siziya S, Katepa-Bwalya M, Kankasa C, Moland KM, Tylleskar T (2008) 'No sister, the breast alone is not enough for my baby' a qualitative assessment of potentials and barriers in the promotion of exclusive breastfeeding in southern Zambia. Int Breastfeed J 3:26
82. Bhandari N, Bahl R, Mazumdar S et al (2003) Effect of community-based promotion of exclusive breastfeeding on diarrhoeal illness and growth: a cluster randomised controlled trial. Lancet 361:1418–1423
83. Haider R, Ashworth A, Kabir I, Huttly SR (2000) Effect of community-based peer counsellors on exclusive breastfeeding practices in Dhaka, Bangladesh: a randomised controlled trial. Lancet 356:1643–1647

84. Coutinho SB, Cabral de Lira PI, Lima MC, Ashworth A (2005) Comparison of the effect of two systems for the promotion of exclusive breastfeeding. Lancet 366:1094–1100
85. Su LL, Chong YS, Chan YH et al (2007) Antenatal education and postnatal support strategies for improving rates of exclusive breastfeeding: randomised controlled trial. BMJ 335:596, Epub 2007 Aug 1
86. World Health Organization (2006) Consensus statement. WHO HIV and infant feeding technical consultation held on behalf of the inter-agency task team (IATT) on prevention of HIV infections in pregnant women, mothers and their infants. 25–27 Oct 2006. http://www.who.int/child-adolescent-health/publications/NUTRITION/consensus_statement.htm
87. Kuhn L, Aldrovandi GM, Sinkala M et al (2009) Differential effects of early weaning for HIV-free survival of children born to HIV-infected mothers by severity of maternal disease. PLoS One 4:e6059. doi:10.1371/journal.pone.0006059
88. Kuhn L, Sinkala M, Semrau K et al (2010) Elevations in mortality due to weaning persist into the second year of life among uninfected children born to HIV-infected mothers. Clin Infect Dis 54:437–444
89. Arpadi SM, Fawzy A, Aldrovandi GM et al (2009) Growth faltering due to breastfeeding cessation among uninfected children born to HIV-infected mothers in Zambia. Am J Clin Nutr 90:344–350
90. Fawzy A, Arpadi S, Kankasa C et al (2011) Early weaning increases diarrhea morbidity and mortality among uninfected children born to HIV-infected mothers in Zambia. J Infect Dis 203:1222–1230
91. Chasela CS, Hudgens MG, Jamieson DJ et al (2010) Maternal or infant antiretroviral drugs to reduce HIV-1 transmission. N Engl J Med 362:2271–2281
92. Shapiro RL, Hughes MD, Ogwu A et al (2010) Antiretroviral regimens in pregnancy and breast-feeding in Botswana. N Engl J Med 362:2282–2294
93. Mofenson LM (2010) Antiretroviral drugs to prevent breastfeeding HIV transmission. Antivir Ther 15:537–553
94. Connor EM, Sperling RS, Gelber R et al (1994) Reduction of maternal-infant transmission of human immunodeficiency virus type 1 with zidovudine treatment. N Engl J Med 331:1173–1180
95. Jackson JB, Musoke P, Fleming T et al (2003) Intrapartum and neonatal single-dose nevirapine compared with zidovudine for prevention of mother-to-child transmission of HIV-1 in Kampala, Uganda: 18-month follow-up of the HIVNET 012 randomised trial. Lancet 362:859–868
96. Six Week Extended-Dose Nevirapine (SWEN) Study Team (2008) Extended-dose nevirapine to 6 weeks of age for infants to prevent HIV transmission via breastfeeding in Ethiopia, India, and Uganda: an analysis of three randomised controlled trials. Lancet 372:300–313
97. Kumwenda NI, Hoover DR, Mofenson LM et al (2008) Extended antiretroviral prophylaxis to reduce breast-milk HIV-1 transmission. N Engl J Med 359:119–129
98. World Health Organization (2009) HIV and infant feeding: revised principles and recommendations: rapid advice. http://whqlibdocwhoint/publications/2009/9789241598873_eng.pdf. Accessed 1 Apr 2010
99. Tonwe-Gold B, Ekouevi DK, Viho I et al (2007) Antiretroviral treatment and prevention of peripartum and postnatal HIV transmission in West Africa: evaluation of a two-tiered approach. PLoS Med 4:e257
100. Taha TE, Kumwenda J, Cole SR et al (2009) Postnatal HIV-1 transmission after cessation of infant extended antiretroviral prophylaxis and effect of maternal highly active antiretroviral therapy. J Infect Dis 200:1490–1497
101. Thomas T, Masaba R, Ndivo R et al (2008) Prevention of Mother-to-child transmission of HIV-1 among breastfeeding mothers using HAART: the kisumu breastfeeding study, Kisumu, Kenya, 2003–2007. 15th conference of retrovirus and opportunistic infections, Boston, USA, Abstract 45aLB
102. Palombi L, Marazzi MC, Voetberg A, Magid MA (2007) Treatment acceleration program and the experience of the DREAM program in prevention of mother-to-child transmission of HIV. AIDS 21(Suppl 4):S65–S71
103. Kilewo C, Karlsson K, Ngarina M et al (2009) Prevention of mother-to-child transmission of HIV-1 through breastfeeding by treating mothers with triple antiretroviral therapy in Dar es Salaam, Tanzania: the Mitra Plus study. J Acquir Immune Defic Syndr 52:406–416
104. Marazzi MC, Nielsen-Saines K, Buonomo E et al (2009) Increased infant human immunodeficiency virus-type one free survival at one year of age in sub-saharan Africa with maternal use of highly active antiretroviral therapy during breast-feeding. Pediatr Infect Dis J 28:483–487
105. Kesho Bora Study Group (2011) Triple antiretroviral compared with zidovudine and single-dose nevirapine prophylaxis during pregnancy and breastfeeding for prevention of mother-to-child transmission of HIV-1 (Kesho Bora study): a randomised controlled trial. Lancet Infect Dis. doi:10.1016/S1473-3099(10)70288-7
106. Becquet R, Ekouevi DK, Arrive E et al (2009) Universal antiretroviral therapy for pregnant and breast-feeding HIV-1-infected women: towards the elimination of mother-to-child transmission of HIV-1 in resource-limited settings. Clin Infect Dis 49:1936–1945
107. World Health Organization (2009) Antiretroviral therapy for HIV infection in adults and adolescents. Geneva. http://www.who.int/hiv/pub/arv/rapid_advice_art.pdf
108. Kuhn L, Aldrovandi GM, Sinkala M, Kankasa C, Mwiya M, Thea DM (2010) Potential impact of new World Health Organization criteria for antiretroviral treatment for prevention of mother-to-child HIV transmission. AIDS 24:1374–1377

Chapter 21
Should Women with HIV-1 Infection Breastfeed Their Infants? It Depends on the Setting

Grace John-Stewart and Ruth Nduati

Introduction

Breastfeeding is the ideal infant food—it provides both optimal nutrition and numerous factors that contribute to infant immunity, growth, cognition, and health. It also enhances maternal–infant bonding and child-spacing and may provide long-term benefits to mothers. An estimated 7.7 million children under 5 years of age die annually, with >30% dying of infectious diseases [1, 2]. *Breastfeeding has been identified as the most effective intervention to prevent under-5 mortality* [3]. It was therefore a huge public health and policy challenge to discern the best infant feeding strategy when it was discovered that HIV-1 could be transmitted through breastfeeding. While nonbreastfeeding could entirely prevent transmission of a rapidly fatal infection, implementation of artificial feeding could be associated with increased infant mortality and morbidity. Over the past two decades, mothers, clinicians, and policy makers have wrestled with balancing infant risk of HIV-1 acquisition against risk of infant mortality in the context of concurrently changing interventions that decrease transmission of HIV-1.

During the past decade, an estimated 350,000–500,000 children acquired HIV-1 infection each year, over 90% in sub-Saharan Africa [4]. Without interventions, mother-to-child transmission (MTCT) of HIV-1 risk ranges from 25 to 40% and occurs in utero, at delivery and via breastfeeding [5]. Breastmilk HIV-1 transmission contributes 30–50% of infant HIV-1 infections [5]. In the last decade, there has been enormous progress in identifying effective interventions and implementing programs to decrease mother-to-child HIV-1 transmission, including breastmilk HIV-1 transmission. Women who receive antiretroviral therapy early in pregnancy and throughout breastfeeding can have very low transmission risk (as low as ~1% has been achieved in one clinical trial) [6]. As breastfeeding has become safer with antiretrovirals, the relative benefit of nonbreastfeeding has declined.

In resource-limited regions that have high HIV-1 prevalence, the benefits of breastfeeding with antiretrovirals outweigh the risks of nonbreastfeeding. In this chapter, we review the evidence and ethical and cost-policy considerations pertinent to breastmilk HIV-1 transmission. For interested readers, WHO, UNAIDS, UNICEF, and UNFPA have summarized evidence and guidelines in an

G. John-Stewart, M.D., Ph.D. (✉)
Departments of Global Health, Medicine, Epidemiology and Pediatrics,
University of Washington, Seattle, WA, USA
e-mail: gjohn@uw.edu

R. Nduati, M.B.Ch.B., M.Med., M.P.H.
Department of Pediatrics, University of Nairobi, Nairobi, Kenya

excellent document "Guidelines on HIV and Infant Feeding." [7] Other chapters of this book provide detailed information on breastmilk and research related to breastmilk transmission of HIV-1, thus, we focus on evidence related to policy, practice, and decision-making.

Scientific Evidence

Risk of Breastmilk HIV-1 Transmission

Evidence for breastmilk HIV-1 transmission is diverse and strongly conclusive. Early studies noted detection of HIV-1 in breastmilk and higher transmission risk in breastfeeding than formula feeding women in observational cohorts and in comparisons of regions in which HIV-1-infected women breastfed versus those in which they did not [8]. An early study that was able to directly implicate breastfeeding HIV-1 transmission and exclude in utero and intrapartum HIV-1 transmission evaluated breastfeeding women who were HIV-1 negative at delivery, but who acquired HIV-1 after delivery [9]. These women had very high (29%) risk of transmitting virus to their breastfeeding infants, typically within 3 months of their own seroconversion [9]. In acute HIV-1 infection, viral loads are initially very high, which leads to much higher risk of breastmilk HIV-1 transmission in women with acute HIV-1 than in women with chronic HIV-1. A randomized clinical trial of breast and formula feeding among chronically HIV-1-infected women in Nairobi, Kenya, observed 20% transmission in those randomized to formula versus 36% transmission in those randomized to breastfeed [10]. This randomized trial estimate of breastmilk transmission risk (16%) was similar to a meta-analysis estimate from previous observational studies (14%) [8]. Finally, in several studies, levels of breastmilk HIV-1 RNA and HIV-1 DNA were highly associated with transmission risk of HIV-1 to infants [11–13]. *Several studies have accrued data to support the association between breastfeeding and HIV-1 transmission (see Table 21.1 summarizing Bradford Hill criteria). The evidence for breastmilk HIV-1 transmission is of high quality and coherent.*

Table 21.1 Causal inference regarding breastmilk HIV-1 transmission (Bradford Hill Criteria)

Strength of association	Highly significant risk of breastmilk HIV-1 transmission ~16% risk of transmission
Consistency	Studies from many countries and continents noting increased HIV-1 transmission in breastfeeding women
Specificity of association	Infants with HIV-1 almost always acquired their HIV-1 from HIV-1-infected mothers
Temporality	Late postnatal HIV-1 transmission occurs in infants who were previously uninfected who breastfeed
Biological gradient	Higher breastmilk HIV-1 RNA and HIV-1 DNA levels in mothers who transmit HIV-1 to their infants
Plausibility	HIV-1 detected in breastmilk
Coherence	Transmission related to levels of breastmilk HIV-1 and decreased by antiretrovirals or nonbreastfeeding
Experiment	Randomized clinical trials demonstrating • Lower HIV-1 infection risk in infants who do not breastfeed versus who do breastfeed • Less HIV-1 in breastfeeding infants who receive antiretrovirals (ARV) compared to those who do not • Less HIV-1 in breastfed infants of women receiving maximally suppressive ARV therapy
Analogy	Similar to Human T-cell Leukemia Virus-1 (HTLV-1)

Timing of Breastmilk HIV-1 Transmission

Breastmilk HIV-1 transmission occurs throughout lactation. There is some evidence of increased risk early postpartum. The majority of risk difference (75%) in the Nairobi randomized clinical trial of breast versus formula feeding occurred within the first 6 months [10]. Without a randomized clinical trial design, it is not possible to accurately estimate *early* breastmilk HIV-1 transmission because infections occurring between 2 days and 1 month or life may include late in utero, intrapartum, and early breastfeeding transmissions of HIV-1. Thus, in studies that did not randomize on feeding modality, estimates of timing of breastmilk HIV-1 transmission are restricted to infections occurring after 1 month of life ("late postnatal transmission"). In one such study from Malawi, transmission risk was estimated to be 0.7% per month during months 1–5, 0.6% per month during months 6–11, and 0.3% per month in months 12–17 [14]. In contrast to this study, a large meta-analysis estimated a relatively constant risk of transmission after 1 month at 8.9 transmissions/100 person-years of breastfeeding [15]. In summary, as long as an HIV-1-infected woman breastfeeds there is some risk of transmitting HIV-1 to her infant. Timing of HIV-1 transmission has influenced recommendations regarding optimal duration of breastfeeding.

Risk Factors for Breastmilk HIV-1 Transmission

Factors associated with increased breastmilk transmission of HIV-1 include high plasma and breastmilk viral loads [11–13, 16]. Breast infections which potentially increase local recruitment of HIV-1-infected cells or decrease the systemic/mucosal compartmental barriers are associated with increased breastmilk HIV-1 transmission. As noted earlier, women with acute HIV-1 infection have an increased risk of HIV-1 transmission to the infant (29% vs. 14%) [8, 9, 17]. In addition, early introduction of other feeds ("mixed feeding"), particularly other milks or protein, has been associated with increased transmission of HIV-1 [18–21]. *By far the most important intervention to prevent mother-to-child transmission of HIV-1 is the use of antiretroviral drugs as treatment or prophylaxis during the antenatal period, labor and delivery, and through breastfeeding, as discussed below.*

Antiretrovirals to Prevent Breastmilk HIV-1 Transmission

The first randomized clinical trial to demonstrate impact of antiretrovirals for prevention of mother-to-child transmission (PMTCT) of HIV-1 (ACTG 076) compared zidovudine to placebo and showed a 67% reduction in the risk of MTCT of HIV-1 [22]. Following this study which was restricted to nonbreastfeeding women, studies in breastfeeding populations confirmed that antiretrovirals decreased infant HIV-1 infection. Short-course zidovudine or nevirapine in breastfeeding populations decreased HIV-1 among those who received it [23]. Single-dose nevirapine results in large decreases in breastmilk HIV-1 RNA sustained over 2–3 weeks [24]. PMTCT regimens that combined treatment of ART-eligible women with short-course antiretrovirals for those not requiring ART for their own health resulted in lower transmission risk (~5.7%) [25]. Recently, several studies have noted very high efficacy of ARV in pregnancy and during breastfeeding resulting in transmission risk of <3% in programs or clinical trials (Table 21.2 and Fig. 21.1). The Mma Bana study in Botswana noted ~1% transmission among breastfeeding women receiving ARV [6]. Decreased risk of infant HIV-1 infection in these studies is certainly a result of the ARV regimen used, but it is not often appreciated that treatment of ART-eligible women (who have significantly higher transmission risk) and shortened breastfeeding (6 months) also contributed substantively to the decreased infant HIV-1 infections in these studies.

Table 21.2 Studies with randomized feeding modality for HIV-1-infected mothers

	Randomization of feeding option	18 month or 2-year cumulative mortality or HIV-1 (combined adverse outcomes)
Nairobi 2000 (no antiretrovirals) [10]	Breast versus formula	Replacement feeders had better infant outcome (2 years); $p=0.02$ Adverse outcomes—infant death or HIV-1 • 30% replacement feeders • 42% breastfeeders
MASHI Botswana 2005 (ART to immunosuppressed; short-course zidovudine/and sometimes NVP and zidovudine during BF) [30]	Breast versus formula	Replacement feeders and breastfeeders had comparable outcomes (18 months), $p=0.80$ Adverse outcomes—infant death or HIV-1 • 13.9% replacement feeders • 15.1% breastfeeders
ZEBS Zambia 2009 SD NVP [31]	Early versus deferred cessation of breastfeeding	Early cessation and continued breastfeeders had comparable outcomes (24 months); $p=0.13$ Adverse outcomes—infant death or HIV-1 • 31.6% early cessation at 4 months • 36% indefinite breastfeeding

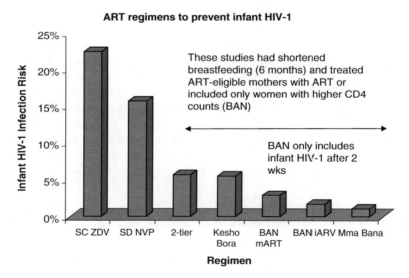

Fig. 21.1 Comparison of transmission risk with different antiretroviral regimens. *SC-ZDV* short-course zidovudine, *SD-NVP* HIVNET 012 regimen, single-dose NVP to mother/single-dose NVP to infant, *2-tier*: HAART if CD4<200; combined SC-ZDV/SD-NVP prophylaxis if CD4>200, *Kesho Bora* maternal triple drug prophylaxis prepartum and postpartum, *BAN mART* maternal ARV regimen postpartum during breastfeeding, *BAN iARV* infant daily NVP during breastfeeding, *Mma Bana* maternal HAART prenatal and postnatal

While maternal ART decreases maternal viral load and may provide infant prophylaxis due to infant ingestion of antiretrovirals in breastmilk, antiretrovirals to infants alone also provide effective prophylaxis against transmission. In infant prophylaxis studies from Malawi, infants who received nevirapine during breastfeeding had significantly lower transmission risk [26]. In the BAN study, rates of transmission were nonsignificantly lower among infants receiving antiretroviral prophylaxis with nevirapine than among infants whose mothers received ARV [27] In this study, mothers started

ARV postpartum which may explain the lower efficacy of maternal ARV compared to the Mma Bana study, because ARV may take ~4–6 weeks to decrease plasma and breastmilk HIV-1 RNA to undetectable levels. Although clinical trials studied women who stopped breastfeeding at 6 months, it is assumed that ART will continue to exert benefits throughout the breastfeeding period. Potential drawbacks to long-term ART or infant ARV during breastfeeding include drug toxicities and the need for adherence throughout breastfeeding.

Ethical Principles in Delineating Breastfeeding and HIV Messages and Policy

Ethical principles to consider in individual and policy-level decision making include provision of accurate information for autonomous decision-making by mothers, presenting the risks and benefits, establishing a beneficent policy with the highest possible good and lowest possible risk (nonmaleficence). In the case of infant feeding, the highest attainable standard of care needs to be considered.

Beneficence/Nonmaleficence

In settings with safe water and sanitation such as the USA or EU, even though the risk of perinatal transmission of HIV-1 is low (1% or less) with antiretroviral prophylaxis, the HIV-1 transmission risk is still higher than zero. Because there is negligible risk of mortality for replacement feeding in these settings and potential risk of transmission of a fatal infection to infants despite use of ART, breastfeeding is not recommended. In the future, it is possible that other potential long-term benefits of breastfeeding (such as benefits in cognition seen in a large randomized feeding trial in Belarus) may spur consideration of breastfeeding with ART even in these settings [28].

The risk/benefit balance for breastfeeding by HIV-1-infected mothers in settings with high prevalence of infectious diseases has been determined by combining the two main potential adverse outcomes (infant HIV-1 and infant death) into one measure [29]. It is clearly unacceptable for infants to either die or to get HIV-1; thus, a metric that considers these two outcomes equivalently as "HIV-free survival" or conversely stated combined risk of death or infant HIV-1 measures the likelihood that children *neither die nor acquire HIV-1*. The other less overtly discussed consideration is mother's well being.

Table 21.2 summarizes the combined risk of infant death and HIV-1 in all randomized studies of feeding modality in HIV-1. Since the first randomized comparison of breast and formula by HIV-1-infected mothers, this combined outcome measure has been used to compare interventions for impact on infant outcomes. The randomized trial in Nairobi noted superiority of formula feeding to breastfeeding, despite shared water sources and slum housing [10]. In the Nairobi trial, formula-fed infants had a significantly lower combined risk of infant death and HIV-1 [10]. This trial was conducted before antiretroviral regimens were used in Africa for PMTCT. In a subsequent randomized trial of breast versus formula feeding (MASHI) from Botswana, in which immunosuppressed women received antiretroviral therapy and all women received short-course antiretrovirals, replacement feeding conferred no benefit in terms of combined risk of infant HIV-1 and mortality [30]. The ZEBS study noted no significant difference in combined risk of HIV-1 or infant death between women who stopped breastfeeding early versus those who breastfed up to 2 years. [31]. Although studies sometimes included subset analyses (risk in HIV-1-uninfected infants or earlier time-point comparisons), the final 18 month or 2 year combined outcome best summarizes the net risk/benefit for infants by feeding randomization.

The other important consideration is that feeding decisions may have an impact on two individuals—mother and infant. While an early study suggested that breastfeeding may have adverse impact on maternal survival, subsequent studies have not shown deleterious effects on maternal health due to breastfeeding [32–36]. Nonbreastfeeding may compromise confidentiality regarding maternal HIV-1 diagnosis and have unanticipated social harm due to revealing maternal HIV-1 status to the community.

Standard of Care

The standard of care regarding recommendations for breastfeeding is contextual—it is based on the risk/benefit balance in different scenarios. Some implementers have suggested that a universal standard of care would involve no breastfeeding by HIV-1-infected mothers because unsafe water and poor sanitation are surmountable and that standard of care in all settings should be to not breastfeed. For example, a pilot study in Rwanda elected to recommend replacement feeding and to include fuel and formula to increase the safety of nonbreastfeeding [37]. However, this approach may result in infant mortality if there is not continued attention to provision of a tight safety net during expansion. Although formula feeding in Botswana may have contributed to prevention of infant HIV-1, widespread implementation of formula in Botswana also resulted in an episode of increased infant mortality in formula feeders due to bacterial contamination of water [38]. In general, programs have a less robust safety net to monitor replacement feeding and several studies have noted increased diarrhea and hospitalizations among infants who do not breastfeed [39–44].

Autonomy

Mothers must make decisions based on accurate information regarding potential benefits and risks and these decisions have impact for themselves and for their infants. The complexity of decision-making at all levels (individual, caregiver, clinician, policy-maker) regarding breastmilk HIV-1 transmission cannot be overstated. Breastfeeding is a primal essential pillar of infant health and HIV-1 infection is a stigmatizing incurable disease. At the individual level, a variety of potential maternal and infant perspectives could be anticipated:

- *If I don't breastfeed, my child may get diarrhea but I can manage diarrhea. If I do breastfeed, my child may get HIV which is a terminal disease.*
- *If I don't breastfeed, my mother-in-law will suspect I have HIV and my husband will abandon me.*
- *I breastfed and my last baby got HIV, I would like to not breastfeed.*
- *I formula fed and my last baby got HIV, I would like to breastfeed.*
- *I formula fed and my baby died. I would like to breastfeed.*
- *I breastfed and my last baby was uninfected. I would like to breastfeed.*
- *My mother breastfed and I got HIV. I wish she had formula fed.*
- *My mother formula fed and I still got HIV and my father left us. I wish she had breastfed.*

For individuals, the context of risk is framed by peers and providers and changing media messages. In high HIV-1 prevalence settings, HIV-1 and prevention messaging around HIV-1 has extended from urban to rural settings. Maternal decision-making is influenced by previous pregnancies, options available at the local clinics, and recent community messages. Partners may also play a role in maternal decision-making regarding infant feeding. *For the care provider and policy-maker the challenge has been how to distill and communicate the evidence base and recommendations accurately to allow an informed autonomous choice that optimizes beneficence and non-maleficence for mother and child. Rapid changes in messaging results in a mismatch between caregivers, mothers, and policy-makers— where on the ground the messages of yesterday have finally taken hold at the same time as changed messages are developed that need to be delivered* [39]. Changes in messages, unclear messages, or conveying some uncertainty about the evidence-base as new studies provide data may lead to confusion or mistrust of information and recommendations in general. In 2010, WHO recommended to change the decision-making regarding infant feeding from individual to national level [7].

Ethics at the Policy Level

At global and national level, countries have deliberated on WHO recommendations to tailor-make their national policy and during the past decade there have been several changes in recommendations as data accrued [7]. In the highest HIV-1 prevalence settings in Africa, countries have differed in their feeding recommendations. Botswana was notable for early adoption of a replacement feeding policy while other settings opted to promote breastfeeding. Urban and rural settings may differ in feeding recommendations based on differences in potential replacement feeding morbidity and mortality. Prevalence of regional infectious diseases influences national decision-making regarding whether replacement feeding should be an available option in public sector programs. In addition to the individual mother–child decision-making, there has been concern that provision or recommendations on replacement feeding would undermine breastfeeding in the uninfected population. However, the potential stigma of replacement feeding in the context of HIV-1 may have attenuated this risk. Throughout the deliberations on breastfeeding and HIV-1 policy, regional infant mortality and infectious diseases has influenced regional decision-making because of its integral relationship to the mortality risk of nonbreastfeeding. Over the past decade, the changes in wording of WHO/UNAIDS guidelines for HIV-infected mothers reflect the rapidly changing evidence base and the continued desire to optimize infant survival while minimizing risk of HIV-1. Antiretrovirals have provided an effective way to preserve breastfeeding for infants born to HIV-1-infected mothers.

Summary

In summary, breastfeeding transmission contributes a substantial proportion of HIV-1 transmissions from mother to infant, but can be effectively prevented by maternal antiretrovirals or infant antiretroviral prophylaxis. Globally, it is important to continue to promote breastfeeding because of its substantial benefits in preventing infant mortality and morbidity. For some women in selected settings with safe water and sanitation, nonbreastfeeding is recommended. However, for most HIV-1-infected mothers worldwide, breastfeeding with maternal ART or infant antiretrovirals will be the best option.

References

1. Rajaratnam JK, Marcus JR, Flaxman AD et al (2010) Neonatal, postneonatal, childhood, and under-5 mortality for 187 countries, 1970–2010: a systematic analysis of progress towards Millennium Development Goal 4. Lancet 375:1988–2008
2. UN (2010) The millennium development goals report. p 27
3. Labbok MH, Clark D, Goldman AS (2004) Breastfeeding: maintaining an irreplaceable immunological resource. Nat Rev Immunol 4:565–572

4. UNAIDS (2010) Report on the Global HIV/AIDS epidemic
5. De Cock KM, Fowler MG, Mercier E et al (2000) Prevention of mother-to-child HIV transmission in resource-poor countries: translating research into policy and practice. JAMA 283:1175–1182
6. Shapiro RL, Hughes MD, Ogwu A et al (2010) Antiretroviral regimens in pregnancy and breast-feeding in Botswana. N Engl J Med 362:2282–2294
7. WHO (2010) Guidelines on HIV and infant feeding: principles and recommendations for infant feeding in the context of HIV and a summary of evidence
8. Dunn DT, Newell ML, Ades AE, Peckham CS (1992) Risk of human immunodeficiency virus type 1 transmission through breastfeeding. Lancet 340:585–588
9. Van de Perre P, Simonon A, Msellati P et al (1991) Postnatal transmission of human immunodeficiency virus type 1 from mother to infant. A prospective cohort study in Kigali, Rwanda. N Engl J Med 325:593–598
10. Nduati R, John G, Mbori-Ngacha D et al (2000) Effect of breastfeeding and formula feeding on transmission of HIV-1: a randomized clinical trial. JAMA 283:1167–1174
11. Rousseau CM, Nduati RW, Richardson BA et al (2004) Association of levels of HIV-1-infected breast milk cells and risk of mother-to-child transmission. J Infect Dis 190:1880–1888
12. Rousseau CM, Nduati RW, Richardson BA et al (2003) Longitudinal analysis of human immunodeficiency virus type 1 RNA in breast milk and of its relationship to infant infection and maternal disease. J Infect Dis 187:741–747
13. Semba RD, Kumwenda N, Hoover DR et al (1999) Human immunodeficiency virus load in breast milk, mastitis, and mother-to-child transmission of human immunodeficiency virus type 1. J Infect Dis 180:93–98
14. Miotti PG, Taha TE, Kumwenda NI et al (1999) HIV transmission through breastfeeding: a study in Malawi. JAMA 282:744–749
15. Coutsoudis A, Dabis F, Fawzi W et al (2004) Late postnatal transmission of HIV-1 in breast-fed children: an individual patient data meta-analysis. J Infect Dis 189:2154–2166
16. John GC, Nduati RW, Mbori-Ngacha DA et al (2001) Correlates of mother-to-child human immunodeficiency virus type 1 (HIV-1) transmission: association with maternal plasma HIV-1 RNA load, genital HIV-1 DNA shedding, and breast infections. J Infect Dis 183:206–212
17. Humphrey JH, Marinda E, Mutasa K et al (2010) Mother to child transmission of HIV among Zimbabwean women who seroconverted postnatally: prospective cohort study. BMJ 341:c6580
18. Coutsoudis A, Pillay K, Spooner E, Kuhn L, Coovadia HM (1999) Influence of infant-feeding patterns on early mother-to-child transmission of HIV-1 in Durban, South Africa: a prospective cohort study. South African Vitamin A Study Group. Lancet 354:471–476
19. Coovadia HM, Rollins NC, Bland RM et al (2007) Mother-to-child transmission of HIV-1 infection during exclusive breastfeeding in the first 6 months of life: an intervention cohort study. Lancet 369:1107–1116
20. Iliff PJ, Piwoz EG, Tavengwa NV et al (2005) Early exclusive breastfeeding reduces the risk of postnatal HIV-1 transmission and increases HIV-free survival. AIDS 19:699–708
21. Kuhn L, Sinkala M, Kankasa C et al (2007) High uptake of exclusive breastfeeding and reduced early post-natal HIV transmission. PLoS One 2:e1363
22. Connor EM, Sperling RS, Gelber R et al (1994) Reduction of maternal-infant transmission of human immunodeficiency virus type 1 with zidovudine treatment. Pediatric AIDS Clinical Trials Group Protocol 076 Study Group. N Engl J Med 331:1173–1180
23. Wiktor SZ, Ekpini E, Karon JM et al (1999) Short-course oral zidovudine for prevention of mother-to-child transmission of HIV-1 in Abidjan, Cote d'Ivoire: a randomised trial. Lancet 353:781–785
24. Chung MH, Kiarie JN, Richardson BA, Lehman DA, Overbaugh J, John-Stewart GC (2005) Breast milk HIV-1 suppression and decreased transmission: a randomized trial comparing HIVNET 012 nevirapine versus short-course zidovudine. AIDS 19:1415–1422
25. Tonwe-Gold B, Ekouevi DK, Viho I et al (2007) Antiretroviral treatment and prevention of peripartum and postnatal HIV transmission in West Africa: evaluation of a two-tiered approach. PLoS Med 4:e257
26. Kumwenda NI, Hoover DR, Mofenson LM et al (2008) Extended antiretroviral prophylaxis to reduce breast-milk HIV-1 transmission. N Engl J Med 359:119–129
27. Chasela CS, Hudgens MG, Jamieson DJ et al (2010) Maternal or infant antiretroviral drugs to reduce HIV-1 transmission. N Engl J Med 362:2271–2281
28. Kramer MS, Aboud F, Mironova E et al (2008) Breastfeeding and child cognitive development: new evidence from a large randomized trial. Arch Gen Psychiatry 65:578–584
29. Alioum A, Dabis F, Dequae-Merchadou L et al (2001) Estimating the efficacy of interventions to prevent mother-to-child transmission of HIV in breast-feeding populations: development of a consensus methodology. Stat Med 20:3539–3556
30. Thior I, Lockman S, Smeaton LM et al (2006) Breastfeeding plus infant zidovudine prophylaxis for 6 months vs formula feeding plus infant zidovudine for 1 month to reduce mother-to-child HIV transmission in Botswana: a randomized trial: the Mashi Study. JAMA 296:794–805

31. Kuhn L, Aldrovandi GM, Sinkala M et al (2008) Effects of early, abrupt weaning on HIV-free survival of children in Zambia. N Engl J Med 359:130–141
32. Nduati R, Richardson BA, John G et al (2001) Effect of breastfeeding on mortality among HIV-1 infected women: a randomised trial. Lancet 357:1651–1655
33. Coutsoudis A, Coovadia H, Pillay K, Kuhn L (2001) Are HIV-infected women who breastfeed at increased risk of mortality? AIDS 15:653–655
34. Coutsoudis A, England K, Rollins N, Coovadia H, Newell ML, Bland R (2010) Women's morbidity and mortality in the first 2 years after delivery according to HIV status. AIDS 24:2859–2866
35. Kuhn L, Kasonde P, Sinkala M et al (2005) Prolonged breast-feeding and mortality up to two years post-partum among HIV-positive women in Zambia. AIDS 19:1677–1681
36. Otieno PA, Brown ER, Mbori-Ngacha DA et al (2007) HIV-1 disease progression in breast-feeding and formula-feeding mothers: a prospective 2-year comparison of T cell subsets, HIV-1 RNA levels, and mortality. J Infect Dis 195:220–229
37. Farmer PE, Nizeye B, Stulac S, Keshavjee S (2006) Structural violence and clinical medicine. PLoS Med 3:e449
38. Creek TL, Kim A, Lu L et al (2006) Hospitalization and mortality among primarily nonbreastfed children during a large outbreak of diarrhea and malnutrition in Botswana. J Acquir Immune Defic Syndr 53:14–19
39. Chopra M, Doherty T, Mehatru S, Tomlinson M (2009) Rapid assessment of infant feeding support to HIV-positive women accessing prevention of mother-to-child transmission services in Kenya, Malawi and Zambia. Public Health Nutr 12:2323–2328
40. Homsy J, Moore D, Barasa A et al (2010) Breastfeeding, mother-to-child HIV transmission, and mortality among infants born to HIV-Infected women on highly active antiretroviral therapy in rural Uganda. J Acquir Immune Defic Syndr 53:28–35
41. Kafulafula G, Hoover DR, Taha TE et al (2010) Frequency of gastroenteritis and gastroenteritis-associated mortality with early weaning in HIV-1-uninfected children born to HIV-infected women in Malawi. J Acquir Immune Defic Syndr 53:6–13
42. Kuhn L, Aldrovandi GM (2009) Clean water helps but is not enough: challenges for safe replacement feeding of infants exposed to HIV. J Infect Dis 200:1183–1185
43. Phadke MA, Gadgil B, Bharucha KE et al (2003) Replacement-fed infants born to HIV-infected mothers in India have a high early postpartum rate of hospitalization. J Nutr 133:3153–3157
44. Kagaayi J, Gray RH, Brahmbhatt H et al (2008) Survival of infants born to HIV-positive mothers, by feeding modality, in Rakai, Uganda. PLoS One 3:e3877

Part VII
Epilogue

Chapter 22
The Future of Breastfeeding in the Face of HIV-1 Infection: Science and Policy

Marc Bulterys and Athena P. Kourtis

This book presents a comprehensive and detailed overview of transmission of HIV-1 to the infant via the infected mother's breast milk. There is no question that breast milk is the ideal food for the infant—with numerous nutritional, immunologic, cognitive, and psychological benefits for the mother/infant dyad, it is a food evolutionarily designed and uniquely tailored to the needs of the developing infant. Promotion of breastfeeding is now recognized throughout the world—and especially in resource-limited settings—as one of the most critical interventions to prevent infant and young child mortality [1–4]. Derrick Jelliffe, one of the founding fathers of developing world pediatrics and child health/nutrition, aptly described the result of promotional practices of the formula industry in the 1960s and 1970s as "commerciogenic malnutrition" [5–7].

Unfortunately, some infectious agents can also be transmitted via breast milk. It was discovered early in the HIV/AIDS epidemic that the HIV-1 virus could be readily transmitted through breastfeeding [8–10]. Chapter 1 of this book outlines the epidemiology and global magnitude of HIV-1 transmission through breastfeeding. Indeed, breastfeeding accounts for more than 40% of all mother-to-child HIV-1 transmission worldwide [11], or, according to UNAIDS' latest estimates, up to 160,000 new infant infections annually. In addition to HIV-1, other retroviruses, some herpesviridae, and even other microorganisms occasionally are transmitted through breast milk (see Chapter 2). The resulting infections may have serious consequences for the health of infants depending on the agent, age, stage of infant development, and whether specific prophylactic measures can be used, as in the case of hepatitis B.

Even though this book is primarily focused on the consequences of HIV-1 transmission for the infant, the health outcomes of breastfeeding for the nursing HIV-infected mother must also receive due attention. Whether breastfeeding contributes to nutritional depletion in the HIV-infected mother in resource-poor and often famine-ridden settings is still debated in the literature; in Chapter 3 this question is discussed, along with the social/psychological consequences of practicing breastfeeding or not.

M. Bulterys, M.D., Ph.D. (✉)
Division of Global HIV/AIDS (DGHA), Center for Global Health, Centers for Disease Control and Prevention (CDC), 1600 Clifton Road, NE, Atlanta, GA 30333, USA

CDC Global AIDS Program, Beijing, China

Adjunct Professor of Epidemiology, UCLA School of Public Health, Los Angeles, CA, USA
e-mail: zbe2@cdc.gov

A.P. Kourtis, M.D., Ph.D., M.P.H.
Division of Reproductive Health, NCCDPHP, Centers for Disease Control and Prevention,
4770 Buford Highway, NE, MSK34, Atlanta, GA 30341, USA

The biology of transmission of HIV-1 through breastfeeding is complex and multifactorial, influenced by viral factors such as viral load and characteristics and host factors including genetics, immunity, and behavior. The main factors related to risk of transmission are viral load in the mother's plasma and in the breast milk. Many other questions, such as whether the virus in breast milk is compartmentalized—which seems in most recent studies not to be the case despite earlier suggestions—whether the transmitted virus is cell-free or cell-associated and whether it has special characteristics that determine transmissibility—remain largely unanswered. These questions are succinctly reviewed in Chapter 5.

The issue of antiretroviral drug resistance in Chapter 6 will undoubtedly receive increased attention as antiretroviral prevention of mother-to-child transmission (PMTCT) prophylaxis and treatment programs in many countries increase to reach national coverage. Drug resistance could compromise both future prevention of HIV-1 transmission efforts and treatment of the few infants who still become infected despite intensive prophylaxis strategies. Animal models of HIV transmission through breast milk and their potential applications to human biology are discussed in Chapter 7; such models have been very helpful in the development of human prophylaxis applications. Penetration of antiretroviral agents into the breast milk and into the infant's plasma in Chapter 8 is an emerging field of investigation: still the available information for many drugs is limited. Complementary to virologic aspects, the issues of early infant HIV-1 diagnosis, given the complicating factors of the transfer of maternal antibodies and continued exposure to the virus through breastfeeding, are discussed in Chapter 4.

Many aspects of the immune system of breast milk are only now being discovered, from its innate immunity and its anti-inflammatory and antimicrobial properties, to both humoral and cellular specific adaptive immunity; this is an increasingly complex and fascinating field with many future applications for the development of immuno-prophylaxis and, the holy grail of HIV immunology, a vaccine. It is covered in detail in Chapters 9–11 and in Chapter 13.

In recent years, much progress has been made in preventing transmission of HIV-1 through breastfeeding, primarily through the use of antiretroviral agents administered to the mother or to the infant during the time of breastfeeding, as highlighted in Chapter 12. As a result of the evidence gleaned from recent clinical trials, the World Health Organization (WHO) now recommends maternal prophylaxis started during pregnancy and continued during delivery and postpartum for the duration of breastfeeding or infant prophylaxis during breastfeeding [12]. Other approaches aimed at neutralizing HIV in the breast milk through heating or chemical methods are still being explored and may have a future role to play (Chapter 14). The use of supplemental vitamins and micronutrients as an adjunct to preventing breastfeeding transmission is discussed in Chapter 15; results of several randomized trials have not shown a clear beneficial effect.

Exclusive breastfeeding and early rapid weaning, two strategies that were recommended as methods for curtailing HIV-1 transmission to infants, are also presented. Whereas the former is still strongly recommended as the preferred method of feeding for the first 6 months of life for infants of HIV-infected mothers in resource-limited settings, the latter recommendation has been abandoned as several studies convincingly demonstrated an increase in morbidity and mortality from diarrheal illnesses in infants who weaned early (Chapter 14). In addition, rapid weaning was associated with increased viral load in the milk. Maintaining exclusive breastfeeding in the first half year remains a practical challenge in many settings where early introduction of other foods or liquids is a centuries-old cultural norm; its role in preventing HIV-1 transmission in the face of concomitant antiretroviral prophylaxis may be less critical.

Chapter 16 gives an account of the extensive research performed over more than two decades at a single site in sub-Saharan Africa (Malawi), research that has yielded a wealth of new knowledge and has advanced the field considerably. An account of the circumstances and progress made in one large Asian country (China in Chapter 17), very different in terms of the dynamics and chronology of the HIV/AIDS epidemic and also along socioeconomic parameters, gives the reader two different perspectives and the unique challenges and opportunities they offer. The work of the Centers for Disease

Control and Prevention (CDC) and the United States Agency for International Development (USAID) through the President's Emergency Plan for AIDS Relief (PEPFAR) has led to very significant gains in reaching goals for access to antiretroviral treatment among children and their families and in PMTCT of HIV-1 in many countries around the world. This work has been highlighted in Chapter 18.

Two groups of authors agreed to defend the two sides of a debate on whether HIV-infected women in resource-limited settings should indeed breastfeed their infants (Chapters 20 and 21). Their discussion is both informative and enjoyable, and helps the reader follow the evolution of thinking reflected in the successive WHO recommendations on this complex topic. Finally, the question of whether HIV-infected women in the USA and other resource-rich settings should consider breastfeeding their infants in light of highly effective prophylaxis regimens is discussed in all its dimensions in Chapter 19.

The remarkable evolution and gains in knowledge about the pathogenesis, mechanisms, and factors affecting breastfeeding transmission of HIV-1, the great progress effected by clinical trials that focused on its prevention, and the drastic policy changes these new findings have generated are all extensively and expertly covered. So where are we now? Has the problem of HIV-1 transmission through breastfeeding, a primary challenge in preventing mother-to-child transmission of HIV, been solved? For the first time, the possibility of daring to envision the goal of eliminating mother-to-child transmission of HIV-1 worldwide is arising—a dream for many of the dedicated researchers in this area. Ambassador Eric Goosby, US Global AIDS Coordinator, addressed this eloquently: "Ensuring that all babies are born HIV-free must be a global priority and not left to a lottery of geography. Children everywhere deserve a healthy start in life and they deserve a mother not just to help bring them into the world but to raise them [13]."

With elimination of perinatal and pediatric HIV infections as the goal, new challenges and new areas in urgent need of investigation have emerged. The use of antiretroviral regimens is not yet 100% effective in prevention of HIV-1 transmission and is not free of toxicities for either the mother or the infant. Indeed, the use of postnatal maternal antiretrovirals is associated with multiclass HIV-1 drug-resistant virus in those infants who do become infected and may make future treatment of such infants a very difficult challenge [14, 15]. Lack of correct implementation of such prophylaxis or incomplete drug regimen adherence could lead to rampant resistance that could further complicate future mother-to-child and sexual HIV-1 prevention efforts.

Antiretroviral treatment of pregnant and breastfeeding HIV-infected women who need it for their own health is the strategy component shown to have the highest impact in preventing infant infections and in ensuring maternal survival [16–18]. Universal access to treatment for such women, and to prophylaxis for women with higher CD4+ T cell counts, should be an urgent implementation priority for programs worldwide. Limited access to antiretrovirals in many resource-limited settings due to costs and lack of resources or political will, poor health infrastructure, lack of integration of HIV prevention and treatment with maternal and reproductive services, and social reasons such as stigma is still a major obstacle in the path and especially affects those communities with the least resources and a general lack of hope [19, 20]. Tackling these issues is not easy, but swift and coordinated action needs to be undertaken on the policy front in order for the goal of eliminating mother-to-child transmission of HIV-1 to be reached worldwide. On the science front, new antiretroviral agents that attack the virus at different points, including before cell entry, or that make drug resistance less likely to emerge need to be developed and tested in mothers and infants. Developing an effective HIV vaccine should remain a research priority. Addressing other maternal and infant co-infections should not be forgotten, as they contribute to the morbidity and mortality burden and increase the risk of HIV infection. It is only with a comprehensive preventive strategy offering a very high degree of safety, efficacy, and effectiveness that breast milk, the optimal infant food, can be embraced universally as the preferred method of infant feeding among HIV-infected women.

Acknowledgment The opinions expressed in this chapter are those of the authors and do not necessarily represent the official views of the Centers for Disease Control and Prevention.

References

1. WHO Collaborative Study Team on the Role of Breastfeeding on the Prevention of Infant Mortality (2000) Effect of breastfeeding on infant and child mortality due to infectious diseases in less developed countries: a pooled analysis. Lancet 355:451–455
2. Labbok MH, Clark D, Goldman AS (2004) Breastfeeding: maintaining an irreplaceable immunological resource. Nat Rev 4:565–572
3. Read JS (2003) Human milk, breastfeeding, and transmission of human immunodeficiency virus type 1 in the United States. Pediatrics 112:1196–1205
4. Jelliffe DB, Jelliffe EF (1978) Human milk in the modern world. Oxford University Press, New York
5. Jelliffe DB, Stanfield JP (1978) Diseases of children in the subtropics and tropics, 3rd edn. Edward Arnold Publishers Ltd., London
6. Editorial (1992) In memoriam – Derrick Jelliffe. J Trop Pediatr 38:145
7. Victora CG, Smith PG, Vaughan JP et al (1987) Evidence for protection by breast-feeding against infant deaths from infectious diseases in Brazil. Lancet 2:319–322
8. Ziegler JB, Cooper DA, Johnson RO, Gold J (1985) Postnatal transmission of AIDS-associated retrovirus from mother to infant. Lancet i:896–897
9. Van de Perre P, Simonon A, Msellati P et al (1991) Postnatal transmission of human immunodeficiency virus type 1 from mother to infant: a prospective cohort study in Kigali, Rwanda. N Engl J Med 325:593–8
10. Bulterys M, Fowler MG, Van Rompay KK, Kourtis AP (2004) Prevention of mother-to-child transmission of HIV-1 through breast-feeding: past, present, and future. J Infect Dis 189:2149–2153
11. UNAIDS (2010) Report on the global HIV/AIDS epidemic. UNAIDS, Geneva
12. World Health Organization (2010) Guidelines on HIV and infant feeding 2010. http://wholibdoc.who.int/publications/2010/9789241599535_eng.pdf. Accessed 20 Jun 2011
13. Goosby E (2011) Remarks by Ambassador Eric Goosby at launch of global plan to work toward elimination of new HIV infections among children by 2015. New York, 9 June 2011
14. Mofenson LM (2010) Protecting the next generation – eliminating perinatal HIV-1 infection. N Engl J Med 362:2316–2318
15. Fogel J, Li Q, Taha TE et al (2011) Initiation of antiretroviral treatment in women after delivery can induce multiclass drug resistance in breastfeeding HIV-infected infants. Clin Infect Dis 52:1069–1076
16. Bulterys M, Wilfert CM (2009) HAART during pregnancy and during breastfeeding among HIV-infected women in the developing world: has the time come? AIDS 23:2473–2477
17. Kuhn L, Reitz C, Abrams EJ (2009) Breastfeeding and AIDS in the developing world. Curr Opin Pediatr 21:83–93
18. Kuhn L, Aldrovandi G, Sinkala M et al (2010) Potential impact of new World Health Organization criteria for antiretroviral treatment for prevention of mother to child HIV transmission. AIDS 24:1374–1377
19. Barnett T, Weston M (2008) Wealth, health, HIV and the economics of hope. AIDS 22(Suppl 2):S27–S34
20. Schwartländer B, Stover J, Hallett T et al (2011) Towards an improved investment approach for an effective response to HIV/AIDS. Lancet 377:2031–2041

Index

A

Abbott m2000 assay, of HIV-1 infection, 54, 54t
Acquired immunodeficiency syndrome (AIDS), 4f, 5f, 8f. *See also* HIV-1 infection
Active immunization, in animal models, 97–99, 190
Acute otitis media, breastfeeding as protective against, 133
Adhesion molecules, in breast milk lymphocytes, 163
Affordable, Feasible, Acceptable, Sustainable, and Safe (AFASS), formula feeding guidelines, 274, 279–280
Africa. *See also* specific countries
    AIDS statistics in, 4f
    cytomegalovirus infection in, 186–187
    endemic HIV-1 infection in, pregnancy in, 5
    family-based infant feeding decisions in, 46
    HIV-1 assays in, 53, 54t, 56t
    HIV-2 in breast milk in, 28
    infant morbidity and mortality in, breastfeeding duration and, 275t, 276t
    maternal morbidity and mortality in, breastfeeding and, 41–43, 42t
    mixed feeding rates in, 45
    new pediatric infections in, 55f
    PEPFAR in, 248–249, 249t
    prolonged breastfeeding in, epidemiology of, 7, 161
        HIV-1 transmission risk and, 140
    prophylactic antiretrovirals in, in labor and delivery, 17
        percentage during pregnancy of, 8f
        studies of, 17, 175
    SIV animal infection in, 90
    Sub-Saharan (*see* specific countries; Sub-Saharan Africa)
African Green monkeys, natural SIV infection in, 149–150
AIDS. *See also* HIV-1 infection
    regional epidemiology of, 4f, 5f, 8f
Alkyl sulfates, expressed breast milk treated with, 200
ALVAC/AIDSVAX vaccine trial, 98
ALVAC-SIV vaccine study, 99
ALVAC vaccine trials, 152, 190
AMATA study, of prophylactic antiretrovirals, 19, 176
American Academy of Pediatrics, HIV-1 testing recommended by, 55
Animal models
    of breastfeeding-associated/pediatric HIV transmission, 89–102, 91f, 93f, 96f
    drug therapy in, 99–101
    future directions in, 101–102
    immunization in, 97–99
    infant inoculation in, 95
    maternal, 92–95, 93f
    overview of, 89–90, 101–102
    pathogenesis and, 95–96
    primate, 90, 91f
    SIV, 90–92, 96f, 148–150
    clinical studies based on, 100, 101
    of HIV-1 infection, maternal immunization against, 165
    prevention and treatment of, 97–101
    immunoglobulin A responses in, 145–146
Antenatal HIV-1 counseling, in China, 5, 7
Antibiotics, HIV-1 transmission and, 220t, 221f, 223f, 223t, 226–227
Antibodies, to HIV-1
    animal models of, 148–150
    in breastfed infant stool, 152
    in breast milk, 139–140, 143–147
    in infants, 11
    maternal transfer of, 147
    qualitative detection of, 144
    in vitro evaluation of, 147–148
    in vivo significance of, 148
    polyreactive, 143–144
Antibody-dependent cellular cytotoxicity, virus neutralization in, 148
Anti-idiotypic antibodies, in breast milk, 129
Anti-inflammatory effects, of breast milk, 130–131
Antimicrobial factors, in breast milk, 11–12, 141
Antioxidants, in breast milk, 131
Antiproteases, in breast milk, 131

Antiretroviral prophylaxis, in animal models, 99–101
  in breast milk, 109–116, 112t
    assays for, 111
    concentrations of, 71–72, 110, 112t
    differential exposures in, 116
    single time points *versus* extensive sampling in, 116
    specific exposures to, 112–116
    studies of, 111
  clinical trials of, 17–19, 13t–16t
  during breastfeeding, 39, 173–181, 178t–179t, 181f
    in Chinese program, 240–241
    clinical trials of, 177, 180–181, 226–230, 252
    effectiveness of, 174, 282–283, 291–293
    evidence base for, 552
    in infants, 175–177
    in mothers, 174–175
    in United States, 263–264
    WHO guidelines and, 177, 179, 240
  effectiveness of, 282–283, 292f
  future considerations in, 181
  infant plasma concentrations of, 112t, 113–116
  maternal plasma concentrations in, breast milk and, 112t, 113–116
  pregnancy and postpartum, 109–110
  maternal risks of, 176–177
  regimens of, 18–20, 252t
Antiretroviral therapy. *See also* Highly active antiretroviral therapy; specific classes and agents
  in breastfeeding HIV-infected women, morbidity and mortality and, 41, 43
  in HIV-exposed infants, breastfeeding and, 39
    WHO guidelines for, 19–20
  HIV RNA and DNA effect of, 71–72
    on drug resistance, 84
  mother-to-child HIV-1 transmission reduced by, 3
  prophylactic (*see* Antiretroviral prophylaxis)
  resistance to, 81–86
    in breast milk, 75
    breast milk-blood relationship in, 82–84
    detection of, 81–82
    genetic basis of, 81
    in genital tract, 75
    policy implications in, 85–86
    predisposing factors in, 83–84
    reports of, 84–85
  vaccine development to supplant, 101–102
Antiseptic birth canal cleansing, WASH study of, 220t, 221f, 224
Arachidonic acid, in breast milk, 128, 131
Asia. *See also* specific countries
  HIV epidemiology in, 4–5, 4f–5f
  PEPFAR in, 248, 249t
Atazanavir, breast milk exposure to, 112t, 114–115
Autonomy, in breastfeeding messages and policy, 294–295
AZT. *See* Zidovudine

B
Baby Friendly Hospital Initiative, in China, 238
  PEPFAR guidelines and, 251
Bacille Calmette-Guérin vaccination, infant T-cell response to, 187
Bangladesh, malnutrition in, breastfeeding and survival and, 132
B-cell-derived humoral defenses, in breastfeeding-associated HIV-1 transmission, 139–153, 146f
B cells, breast milk, 121–122, 142, 163–164
  in developing immune system, 187
Belarus, antimicrobial breastfeeding benefits in, 133
Beneficence, in breastfeeding messages and policy, 293
Bifidobacteria, in breast-fed infants, 126
bioMerieux NucliSense EasyQ assay, 54, 54t
Blood transfusion, HIV-1-infected postpartum, 7–8, 69, 240, 262
Botswana, adverse effects of formula feeding in, 250b, 251, 277, 279, 292t, 294
  breastfeeding study in, 480
  prophylactic antiretroviral studies in, 19, 84, 175, 291–293
Brazil, exclusive breastfeeding *versus* mixed feeding in, 132, 198
Breast abscess, HIV-1 transmission risk and, 8
Breastfeeding
  antiretroviral prophylaxis during, 9, 53, 173–181, 178t–179t, 181f
    pharmacology of, 109–116, 112t
  avoidance of, in Chinese HIV-1 prevention program, 5, 237, 240
    in early HIV epidemic, 273
    in early PEPFAR guidelines, 247–251
    feasibility of, 199
    in HTLV-1 infection, 30
    in resource-rich countries, 83
  benefits of, 39, 173, 237, 273, 289
    anti-infective, 11–12, 142, 186–187
    HIV-1 transmission risk *versus*, 3, 173–174, 237
    immune, 165–166
    WHO position on, 39, 252, 252t, 254
  in China, 237–242
  duration and patterns of, HIV-1 transmission risk and, 9, 139–140, 161, 199–200
  exclusive (*see* Exclusive breastfeeding)
  in HIV-1 infection, arguments against in some settings, 289–295, 290t, 292t
    arguments in support of, 273–283, 275t, 276t, 282f
  complexity of, 153
  debate on, 289–295
  exclusive *versus* mixed feeding in, 197–199
  maternal guilt about, 47
  maternal morbidity and mortality and, 39–41, 43
  PEPFAR guidelines for, 249, 252, 255, 255t
  protective effects of, 12, 161
  rationale for, 83

Index

social issues and, 44–45
timing in HIV-1 transmission and, 10, 291
in the United States, 261–268
WHO guidelines on, 177, 180, 252, 252t, 254
HIV-1 transmission through (*see* Breastfeeding-associated HIV-1 transmission)
immunologic breast milk factors in, 9, 122–126, 142–144
maternal nutritional requirements during, 39, 40t
Breastfeeding, Antiretrovirals, and Nutrition (BAN) study, 18, 84, 176–177, 180, 252, 263, 292–293
Breastfeeding and HIV Transmission Study Group, 43
Breastfeeding-associated HIV-1 transmission, animal models of, 89–102, 91f, 93f, 96f
antiretrovirals and, clinical trials of, 13t–16t, 17–19
drug resistance and, 81–86
effectiveness of, 282–283, 292f
observational studies of, 111
pharmacology of, 109–116, 112t
prophylactic, 17–19, 174–181
breast milk factors in, 12
breast milk micronutrients and, 205–210
challenges related to, 20–21
in China, 237–242
declining incidence of, 5, 175
diagnosis, early infant, 51–59, 52t, 54t, 56t, 57t
tests for, 51–59, 52t, 54t, 56t, 57t
early risk quantification studies of, 274
epidemiology of, 3–21, 4f, 5f, 8f, 51, 89, 139, 173
ethical principles and, 293–295
exclusive breastfeeding and, 280–281
future research directions in, 20–21
history of, 52, 69
immunology of, antimicrobial, 121–133, 123t, 124f, 127t
cellular, 161–166
humoral defenses in, 139–153, 141t, 146f
low efficiency of, 139–140, 187
maternal health outcomes and, 39–47, 40t, 42t
mechanisms of, 11–12, 111, 140–142, 141t
PEPFAR in prevention of, 247–258, 249t, 250b, 252t
prevention of, antiretroviral, 173–181, 178t–179t, 291–293
Chinese national program for, 237–242
early weaning in, 199–200
exclusive breastfeeding in, 197–199, 202
heat treatment of breast milk in, 200–202
immune approaches for, 185–192
microbicide breast milk treatment in, 200–201
research on, in Malawi, 217–230, 219f–221f, 220t, 223f
risk factors in, 7–10, 290–291, 290t
infant, 141, 141t
maternal, 9, 12, 140, 141t
virologic, 7–9
timing of, 10, 291
virologic determinants of, 69–75

Breastfeeding-associated transmission, of cytomegalovirus, 30–32
of hepatitis viruses, 32–34, 241
of HIV-1. *See* Breastfeeding-associated HIV-1 transmission
of HIV-2, 28
of HTLV-1, 29
Breast milk. *See also* Breastfeeding
amount ingested, antiretroviral concentration and, 110
HIV-1 transmission risk and, 140, 198–199
antiretroviral pharmacology in, 109–116, 112t
assays measuring concentration of, 110–111
breast milk exposure in, 111–116
drug excretion into, 110
maternal plasma concentration and, 109–110
milk ingestion and, 110
components of, 122–123, 142–144
with possible infant immune effects, 127t
cytomegalovirus in, 30–31
HIV-1 concentrations in, 11
HIV-1 infectiousness of, 69–70, 139–140
HIV-1 molecular forms in, 141
HIV-1 origins in, 11
HIV-1 transmission through (*see* Breastfeeding-associated HIV-1 transmission)
immune factors in, 121–133, 124f, 127t, 139–153, 141t, 146f, 161–166
(*see also* Immune system)
infective and anti-infective factor balance in, 142, 186–188
micronutrients in, 205–210
plasma HIV RNA correlated with, 74
resistance to infection of, 11, 69–70, 187
Burkina Faso, family in infant feeding decisions in, 46
mixed feeding and formula feeding in, 45–46
prophylactic antiretrovirals in, 17, 19

C

Caesarian section, prophylactic, 5
Candidiasis, as infant host factor, 10
CAPRISA 004 trial, of tenofovir, 100
Carbohydrate recognition domain, in breast milk, 143
ß-Carotene, in breast milk, 206–207
Cavidi ExaVir assay, in early HIV-1 infection diagnosis, 52
CCR5 chemokine receptor, in breast milk, 73, 143–144
CD3 cells, in breast milk, 122, 142
CD4 cells, breastfeeding and, maternal morbidity and mortality in, 41
in breast milk, 163
maternal antiretroviral prophylaxis guided by, 283
mother-to-child HIV-1 transmission risk and, 9
CD8 cells, in breast milk, 163, 164
CD14 soluble form, in breast milk, 126
Cell-free and cell-associated HIV, in viral transmission, 69–72, 141, 161

Cellular immunity, in breast milk, 161–166
  components of, 162–164
  future directions in, 166
  T-cell function in, 164–166
Centers for Disease Control and Prevention, infant feeding in HIV recommendations of, 263
  PEPFAR studies supported by, 252
Central Asia, HIV-1 infection trends in, 5
Chemokine receptors, in breast milk, 143, 164
Children. *See also* Infants
  HTLV-1 infection in, 29
Chimpanzees, SIVcpz infection in, 90
China
  breastfeeding-associated HIV-1 transmission in, 237–242
    breastfeeding risk and benefits and, 237
    future considerations in, 242
    hepatitis B infection and, 241
    injection drug use and, 241–242
    national prevention program for, 238–239
    postpartum maternal HIV infection and, 240
  breastfeeding in, among HIV-infected women, 240–241
    prevalence and patterns of, 238, 239f
  CARES program in, 239
  mother-to-child HIV-1 transmission in, declining, 5
  postpartum HIV-1-infected blood transfusion in, 7–8
Coinfections, in HIV-1-infected infants, 187
  maternal hepatitis B as, 241
Colostrum, HIV-specific humoral response in, 144, 148
  HIV viral loads in, 70–71
  immunoglobulins in, 123, 144
  mother-to-child HIV-1 transmission through, 10
Community, in infant feeding decision-making, 45–46
Compartmentalization, in breast milk HIV-1 transmission, 74–75
  drug resistance and, 82–83
Complement, in breast milk, 126
Contamination, of infant formula, 251, 277
Costs, of antiretroviral therapy, 100
  of early nonantiretroviral interventions, 224
  of infant formula, 251, 278
Côte d'Ivoire, zidovudine trial in, 17
Cultural issues, in breastfeeding, 44–47
Cytokines, breast milk, 127–131, 127t
  in inflammatory response, 130
Cytolytic activity, of CD8 cells, 165–166
Cytomegalovirus, 30–32
  breastfeeding-associated transmission of, 27, 30–32
  congenital infection with, 30–31
  HIV-1 coinfection with, 186–187
  long-term outcomes of infection with, 32
  prevention of infection with, 32
Cytotoxic T lymphocytes, breast milk, 122

D
DC-SIGN molecule, in breast milk, 143
Dendritic cells, as infant host factor, 10
DermaVir vaccine trials, 190
Developing countries. *See* Resource-limited countries

Diarrhea, breastfeeding as protective against, 132, 249, 250b, 277
Docosahexaenoic acid, in breast milk, 128, 132
Dried blood spot samples, in HIV-1 assays, 53–54, 56t
Drug abuse, in China, 241–242
Drug resistance, 81–86. *See also* Antiretroviral therapy
  in breast milk, 75
  in genital tract, 75

E
Early weaning, benefits of, 199–200
  HIV disease severity and, 200
  risks of, 9, 249, 275–277, 275t, 281
Efavirenz, breast milk exposure to, 112t, 114,
Endemic HIV-1 infection, in pregnant African women, 5–6
Enhanced Perinatal Surveillance System, HIV transmission risk factors in, 264–265
Entero-mammary axis hypothesis, 116
Envelope sequence, in breast milk HIV-1 transmission, 73–74
Epithelial cells, HIV-1-infected cells and, 11, 73
Epstein-Barr virus, in breast milk, 27
Ethical principles, in breastfeeding messages and policy, 293–295
Europe, HIV-1 infection in, breastfeeding avoidance in, 3, 83
  epidemiology of, 4, 4f
Exclusive breastfeeding, benefits of, 280
  in China, 238
  misunderstandings about, 280–282
  mixed feeding *versus*, 197–199
  in WHO guidelines, 253b

F
Family decision-making, in infant feeding methods, 44–46
Fatty acids, in breast milk, 128, 129, 131–132, 169
Fetal catheterization model, of HIV transmission, 92–93
Flash heating, of expressed breast milk, 201–202
Formula feeding. *See* Infant formula
France, breast milk HIV-2 in, 28
Free radicals, breast milk effects on, 131
French Guyana, mother-to-child transmission of HTLV-1 in, 29

G
Gambia, breast milk HIV-2 in, 28
Gastrointestinal infant mucosa, as HIV-1 transmission site, 73, 140, 151, 198–199
  as infant HIV-1 host factor, 10
Genetic mutations, in animal models of HIV transmission, 91–92
  heritability of drug resistance, 75
  in HIV drug resistance, 81–82, 85
Genital tract, drug resistance in, 75
Genotyping assays, of drug resistance, 81–82

Index

GenProbe HIV-1 assays, 54, 54t
Ghana, pediatric mortality in, breastfeeding and, 132
Global Programme for Vaccines and Immunization, tuberculosis recommendations of, 43
Gut mucosal activation markers, infant, 12

H

Haptocorrin, in breast milk, 125
Heat treatment, of expressed breast milk, 200–202
Hepatitis B virus, breastfeeding and, 32–33, 241
    immunization against, 33
Hepatitis C virus, breastfeeding and, 33–34
Herbal infant feedings, cultural issues in, 45, 280–281
Herpes simplex virus, in breast milk, 27
Herpesviruses, in breast milk, 27, 34
Highly active antiretroviral therapy.
    See also Antiretroviral therapy
    during breastfeeding, 263–264
    breast milk concentrations of, 72
    breast milk HIV-1 concentrations and, 11
    drug resistance to, 84–85
    studies of, 229–230
    WHO guidelines for, 85–86
High-performance liquid chromatography, in antiretroviral assay, 111
HIV-1
    assays of, 51–59, 52t, 54t, 56t, 57t
        (see also specific tests)
        in Chinese national program, 240
        rapid, 57–58
    chimeric constructs and, 91
    high particle ingestion of, 139–140
    SIV versus, 92
HIV-2, 28
HIV-1 DNA, assays of, 52–54, 54t, 56t, 57t
    in viral transmission, 70–72
HIV-1-free survival, breastfeeding duration and, 275t
    limitations of, 275–276, 278
HIV immune globulin, clinical trial of, 190
HIV-1 infection, breastfeeding in resistance to, 11
    coinfections with, 186, 241
    declining trend in, 5
    drug-resistant, in infants, 84–85
    epidemiology of, in China, 238–239
        populations most affected by, 4–5
        by region, 4f, 5f, 8f
    formula feeding perceived as evidence of, 45–47
    highly active retroviral therapy in,
        DNA and RNA and, 11
    impaired mucosal humoral immunity in, 145–147, 146f
    in infants, disease progression in, 186
        early diagnosis of, 51–59, 52t, 54t, 56t, 57t
        isolation from breast milk of, 70–71
    maternal, treatment of, 174
        untreated, 109

nutritional requirements in, 39–40
T and B lymphocytes in, 164
transmission of (see HIV-1 transmission)
HIVNET 012 trial, 17, 226
    experimental precursor to, 100
HIV-1 RNA, assays of, 52, 52t, 54–55, 54t, 56t, 57t
    in viral transmission, 70–72, 74
HIV-1 transmission. See also Breastfeeding-associated HIV-1 transmission
    advances in prevention of, 3
    cell-free versus cell-associated, in breast milk, 69–72
    mother-to-child (see Mother-to-child HIV-1 transmission)
    perinatal versus breast milk in, 72–73
    viral variants and subtypes in, 73–75
HIV-1 vaccines
    development of, 102
        adaptive antibodies and, 144
        concerns regarding, 191
        for maternal immunization, 165
        mucosal vaccines in, 189
        nonclinical human studies in, 188–189
        pediatric, 101–102
        SIV models in, 150, 165–166, 188–190
    pediatric clinical trials of, 98, 152, 191
HLA antigens, mother-to-child HIV-1 transmission risk and, 9
Holder pasteurization, of expressed breast milk, 201
Hormones, breast milk, 128
HPTN 024 study, of antibiotics, 226, 221f, 223f, 220t, 223t
HPTN 046 study, of prophylactic antiretrovirals, 18, 178t
HTLV-1 and HTLV-1, 28–30
Human herpesviruses, in breast milk, 27
Human immunodeficiency virus type 1. See HIV-1
Human immunodeficiency virus type 2. See HIV-2
Human milk banking, 266–267
Human rights, infant feeding in HIV infection and, 279–280
Human T-cell lymphotrophic virus, areas endemic for, 28
    breastfeeding-associated, 28–30
    sequelae of infection with, 29
Humoral defenses
    in breastfeeding-associated HIV-1 infection,
        in HIV-exposed children, 150–153
        maternal breast milk immunity, 142–150
    in breastfeeding-associated HIV-1 transmission, 139–153, 146f

I

Immunomodulatory factors, in breast milk, micronutrients as, 205–210
Immune approaches. See also HIV-1 vaccines
    to prevention of mother-to-child HIV-1 transmission, 185–192
Immune cells, in breast milk, 127, 142
Immune globulin, clinical trial of HIV, 190

Immune system, breast milk
  antibody transfer in, 139–140, 143–147, 278
    anti-inflammatory properties in, 121, 130–132
    antimicrobial properties in, 121, 132–133, 142
    B-cell-derived humoral defenses in, 139–153, 146f
    cellular immunity in, 161–166
    components of, 122–126, 142–144
    cytokines in, 127–128, 127t, 130–131
    developmental role of, 126–128
    HIV-specific humoral response in, 144–150
    mechanisms of mucosal protection in, 124f, 126
    priming of, 129
    tolerance in, 122, 128–129
  neonatal, 121–122, 124f, 127t, 151–152
    development of, 186
Immunization. See also HIV-1 vaccines
  in animal models, 97–101
Immunoglobulin A antibodies, in breast milk, 145–147
Immunoglobulin G antibodies, HIV neutralized by, 139
  HIV-specific, in breast milk, 144–150
  in early infection diagnosis, 52
Immunoglobulins
  breast milk, 123–124, 142–147
    HIV-specific humoral responses in, 145–147
    HIV-1 transmission risk and, 9, 144
  in pasteurized breast milk, 201
Immunomodulatory factors, in breast milk, 11, 121–133, 124f, 127t, 278
  micronutrients as, 205, 210
IMPAACT PROMISE 1077 trial, of antiretroviral prophylaxis, 20, 179t
India, epidemiology of HIV in, 4
  infant mortality in, breastfeeding and, 132
  prophylactic antiretrovirals in, 18
Indinavir, breast milk exposure to, 115
Infant feeding. See also Breastfeeding; Infant formula
  evolution of, 256, 256t
  PEPFAR guidelines for, 249, 252, 255t, 255–256
  studies of, 292t
  WHO guidelines for, 252, 252t, 254, 257t
Infant formula, in Chinese national program, 237–242
  coercive dynamics of free, 280
  inappropriate counseling on, 280
  melamine-tainted, 238
  mortality risk with, 248, 249, 250b, 275, 279
  in Nairobi study, 274
  in PEPFAR guidelines, 247–251
  risks of, 251, 277–279
  social stigma associated with, 44–47
  WHO guidelines for use of, 9
Infants, breastfeeding benefits in, 3, 39–40.
    See also Breastfeeding
  cytomegalovirus infection in, 30
  defenses against HIV-1 infection in, immune, 151–152
    nonimmune, 150–151
    preimmune, 151
  extended antiretroviral prophylaxis in, 18
  gut-protective effects of breast milk in, 130
  HIV-1 infection in, animal models of, 95–101, 96f
    drug-resistant, 84–85
    early diagnosis of, 51–59, 52t, 54t, 56t, 57t

    epidemiology of, 51
    mother-to-child transmission of
      (see Mother-to-child HIV-1 transmission)
      rapid assays of, 57–58
      timing of diagnostic tests for, 53
      virulence of, 96
  HIV-2 infection in, 28
  host factors in, in HIV-1 transmission risk, 9–10
  HTLV-1 infection in, 29
  immune system of, 121–122, 124f, 151–152
  morbidity and mortality in, breastfeeding duration and, 275t, 276t
    breastfeeding in prevention of, 289
  prophylactic antiretrovirals in
    (see also Antiretroviral prophylaxis)
    trials of, 18
  reactions of vaccines in, 191
Infection, breast milk in protection against, 132–133
Infectiousness, of breast milk-borne HIV-1, 69–71, 139–140
Inflammatory response, modulation by breast milk of, 130–132
Injection drug use, in HIV-infected Chinese women, 241–242
Innovative technologies, in HIV-1 assays, 57t, 58–59
INSTI test, for HIV-1 infection, 58
Interferon-γ, in infant immune response, 151, 165
Interleukins, in breast milk, 129, 130
International Collaborative AIDS Research (ICAR) study, 220t, 221–222, 223f, 223t
Intestinal microflora, of breast-fed infants, 126
In utero HIV-1 infection, mother-to-child transmission of, 10

J
Japan, HTLV-1 in, 29, 30
Johns Hopkins Research Project (Malawi), overview of, 218–221, 219f
  studies in, early observational, 221–222
    of nonantiretroviral intervention, 222–225
    of postexposure prophylaxis, 226–227
Joint United Nations Program on HIV-1/AIDS, 4, 185

K
Kenya, family in infant feeding decisions in, 46
  maternal morbidity and mortality in, infant feeding methods and, 41, 42t
  mother-to-child HIV-1 transmission in, timing of, 10–11
  virologic subtypes in, 8–9
  studies in, of breastfeeding risk quantification, 274
    of prophylactic antiretrovirals, 19, 84, 175–176, 252
    of vaccine, 191
    of vitamin/mineral-HIV-1 transmission correlation, 207, 209
Kesho Bora trial, of prophylactic antiretrovirals, 19, 174, 176–177, 180–181
Kisumu Breastfeeding Study (KiBS), 19, 176, 252

Index

L

Labor and delivery
    antiretroviral therapy in, 3–4
        clinical trials of, 17
        mother-to-child HIV-1 transmission in, 139
        preventive interventions in China during, 5
Laboratory testing, for early HIV-1 infection diagnosis, 51–59, 52t, 54t, 56t, 57t
Lactadherin, in breast milk, 125
Lactation. *See* Breastfeeding
Lactobacilli, in breast-fed infants, 126
Lactoferrin, in breast milk, 125, 131, 132
Lactoperoxidase, in breast milk, 126
Lamivudine, breast milk concentrations of, 111, 112t, 113
Latin America, pediatric mortality in, breastfeeding and, 132
Legal issues, in breastfeeding in HIV infection, 267–268
Liat HIV-1 assay, 58
Long-chain polyunsaturated fatty acids, in breast milk, 128, 129, 131–132
Lopinavir, breast milk exposure to, 115, 112t
Low birth weight infants, cytomegalovirus infection in, 31
Lymphocytes, in breast milk, 122, 142, 162–166
Lysozyme, in breast milk, 125

M

Macaque monkeys, experimental SIV infection in, 90–92, 96f, 148–150, 188–190
Macrophages, in breast milk, 123, 162–163
Malawi, breastfeeding-associated HIV-1 transmission risk in, 140
    epidemiology of HIV-1 infection in, 218–221, 219f, 221f
    HIV-1 testing in, 58
    maternal morbidity and mortality in, breastfeeding and, 42t
    studies in, Johns Hopkins Research Project, 217–230, 219f–221f, 220t, 223f, 223t
        of mother-to-child HIV-1 transmission, 217–230, 219f–221f, 220t, 223f, 223t
        of prophylactic antiretrovirals, 18, 19, 175, 252, 292
        of vitamin A-HIV-1 transmission correlation, 206–207
Malnutrition, breastfeeding and survival in, 132
Mass spectometry, as breast milk antiretroviral concentration assay, 111
Mastitis, HIV-1 transmission risk in, 9, 198
    mechanisms of, 12
    virology of, 70–72
Maternal breast factors, in HIV-1 transmission risk, 12
Maternal guilt, about breastfeeding in HIV infection, 47
Maternal health outcomes, in breastfeeding HIV-infected women, 39–47, 40t, 42t, 294
    morbidity in, 39–43, 42t
    mortality and, 39–43, 42t
    social issues and, 44–46
    tuberculosis and, 43
    vitamin supplements and, 44
Maternal host factors, in HIV-1 transmission risk, 9–10
Maternal immunoglobulin G antibodies, in early HIV-1 infection diagnosis, 52
Maternal nutritional requirements, during breastfeeding, 39, 40t
Maternal postnatal HIV-1 infection, in HIV-1 transmission through breastfeeding, 7–8
MCH study, of vitamin A supplementation, 220t, 221f, 224–225
Melamine infant formula scandal, in China, 238
Memory B cells, in breast milk, 142
Methadone maintenance treatment, in China, 242
Microbicide, in expressed breast milk, 200
Micronutrients, breast milk, mother-to-child HIV-1 transmission and, 205–210
    deficiency of, in breastfeeding women, 44
    HIV-1 transmission risk and, 9
    immunomodulating effects of, 205, 210
    requirements in breastfeeding of, 39, 40t
Mitra Plus study, antiretroviral, 176, 180
Mitra study, antiretroviral, 18, 84, 175, 180
Mixed feeding, exclusive breastfeeding *versus*, 12, 45, 73, 197–199
    rates in Africa of, 45
Mma Bana trial, of prophylactic antiretrovirals, 19, 174, 177, 180, 291–293
Modified vaccinia virus Ankara (MVA) vaccine, 191
Monkeys, SIV subtype infections in, 90–95, 96f
Mortality, breastfeeding-associated, 39–43, 42t
    formula feeding-associated risk of, 248, 249, 250b, 275
Mother-to-child HIV-1 transmission. *See also* Breastfeeding-associated HIV-1 transmission
    antiretroviral prophylaxis of
        clinical trials of, 17–19, 226–230
        drug resistance and, 81, 83–86
        experimental, 92–95, 93f
        extended, 18–19
        pharmacology of, 109–116
        in resource-limited areas, 83
    breast milk micronutrients in, 205–210
    CD8 cells in, 165–166
    early diagnosis in infants of, 51–59, 52t, 54t, 56t, 57t
    epidemiology of, 139, 186
    prevention of, advances in, 3
        PEPFAR in, 247–258, 249t, 250b, 252t, 255t
        WHO guidelines in, 177, 180
Mother-to-child transmission. *See also* Breastfeeding-associated transmission
    HIV-1. *See* Mother-to-child HIV-1 transmission
    non-HIV-1, 27–34
Mozambique, drug studies in, 85, 175
Mucosa-associated lymphoid tissue, breast milk immunoglobulins and, 142
Mucosal vaccines, in SIV and SHIV models, 189–190
Multivitamins, study of, 44
MVA-SIV vaccine study, 98–99

N

Nairobi study, of breastfeeding risk quantification, 274, 293
Natural antibodies, polyreactive, 143–144
Natural killer cells, in resistance to HIV-1, 11
Nelfinavir, breast milk exposure to, 112t, 115
Neutralization of virus, in animal models, 149–150
  protective effect of maternal, 148
Neutrophils, breast milk and, 130
Nevirapine, breast milk exposure to, 112t, 114
  prophylactic, in animal model, 100
    in breastfed infants, 39, 81–82, 291
      clinical trials of, 175–176, 221f, 223f, 220t, 223t, 227–230
      resistance to, 82–86
Non-nucleoside reverse transcriptase inhibitors, breast milk exposure to, 114, 112t
  genetics of resistance to, 85
North America, epidemiology of HIV-1 infection in, 5, 4f–5f
Nucleic acid techniques, in early HIV-1 infection diagnosis, 52, 55, 58–59, 52t, 54t, 56t, 57t
Nucleoside reverse transcriptase inhibitors, breast milk exposure to, 112–114, 112t
  genetics of resistance to, 85
Nucleotides, breast milk, 128
Nutrient deficiencies, from formula feeding, 278
  mother-to-child HIV-1 transmission and, 9
Nutritional requirements, during pregnancy and breastfeeding, 39, 40t
NVAZ studies, of prophylactic antiretrovirals, 220t, 221f, 223f, 223t, 226–227

O

Obstetrical interventions, in China, 5
Oligosaccharides, breast milk, 125
Omega-3 and omega-6 fatty acids, in breast milk, 129
Oral inoculation, in animal HIV-1 transmission models, 95–96
OraQuick HIV-1 test, 57

P

Panel on Treatment of HIV-Infected Pregnant Women, recommendations of, 263–264
p24 antigen assay, in early HIV-1 infection diagnosis, 52, 52t, 55–58, 57t
Passive immunization, in animal models, 97, 190
Pasteurization, of breast milk
  in CMV prevention, 32
  expressed, 200–202
Pediatric AIDS Clinical Trial Group (PACTG) studies, 17, 191
  experimental precursor to, 100
Pediatric HIV-1 infection, declining incidence of, 5, 7, 51, 175
  early diagnosis of, 51–59, 52t, 54t, 56t, 57t
Peripartum antiretroviral therapy
  mother-to-child HIV-1 transmission reduced by, 3–4
  trials of, 17

Peru, mother-to-child transmission of HTLV-1 in, 29
PETRA trial, 17
Phylogenetics, of breast milk HIV-1 concentrations, drug resistance and, 82–83
Plasma, HIV-1 concentrations in, 11
Point-of-care technologies, in HIV-1 assays, 57t, 58–59
Policy making for infant feeding, ethics in, 295
Polymerase chain reaction, in early HIV-1 infection diagnosis, 52
Polyreactive antibodies, in breast milk, 143
  in HIV-1 vaccine development, 143–144
Population sequencing, in drug resistance detection, 81–82
Post-Exposure Prophylaxis of Infants (PEPI) study, 18, 84, 174, 176, 220t, 221f, 228–230
Postnatal maternal HIV-1 infection, through breastfeeding, 7–8
Postpartum HIV-1 infant infection, 10.
  See also Mother-to-child HIV-1 transmission
  breastfeeding duration and, 139–140
Postpartum HIV-1-infected blood transfusion, in Australia, 69
  in China, 7
Postpartum period, maternal plasma antiretroviral concentrations in, 110
Poverty, infant feeding in HIV infection and, 279–280
Pregnancy, antiretrovirals in, 8f
  endemic HIV-1 infection in Africa and, 5
  maternal plasma antiretroviral concentrations in, 110
  mother-to-child HIV-1 transmission in, 139
    timing of, 10, 291
  nutritional requirements during, 39, 40t
  prophylactic antiretroviral therapy in, 3
  rapid HIV-1 assays in, 57–58
Prenatal screening, for HIV-1, 186
  for HTVL-1, 30
Preparation for AIDS Vaccine Evaluation (PAVE) study, 220t, 221–222, 223f, 223t, 225
President's Emergency Plan for AIDS Relief (PEPFAR), 247–258
  in Botswana, 250b
  challenges in, 248–251
  countries supported by, 248–249, 249t
  early HIV/infant feeding approach in, 247
  guideline revisions in, 251, 252, 252t
  implementation of, 249t, 253–258, 255t
  international partnerships with, 253–254
  monitoring and evaluation in, 255, 255t
Preterm infants, cytomegalovirus infection in, 31
Pretoria pasteurization, of expressed breast milk, 201
Primates, SIV in nonhuman, 89–92, 96f
Proficiency testing panel, for HIV-1 assays, 58–59
Prolonged breastfeeding, HIV-1 transmission risk in, 7, 139–140, 161, 197–199
PROMISE-PEP trial, 176, 178t–179t
PROMISE trial, of antiretroviral prophylaxis, 20
Prophylactic antiretroviral therapy.
  See Antiretroviral prophylaxis
Prostaglandin E/d2/D inhibitors, 132
Protease inhibitors, breast milk exposure to, 112t, 114–116
Protein binding, antiretroviral concentration predicted by, 110

Index

Q
Quality assurance, in HIV-1 assays, 58–59

R
Racial and ethnic disparities, in United States HIV-1 infections, 262
Real-time viral load assays, of HIV-1 infection, 54, 54t
Recommended dietary allowances, in pregnancy and lactation, 40t, 41
Relative risk *versus* absolute risk, of formula feeding, 279
Replacement feeding. *See* Infant formula
Replicative capacity, in breast milk HIV-1 transmission, 74
Resource-limited countries.
　　See also specific countries
　antimicrobial breastfeeding benefits in, 132
　antiretroviral prophylaxis in, 283
　breastfeeding recommendations in, 173–174, 177, 180, 205, 247–248, 252, 253b
　in early HIV epidemic, 273–274
　challenges in, HIV-1 screening and, 185–186
　　HIV-1 testing and, 58
　ethics of recommendations for, 293
　HIV-1-infected breastfed infants in, early diagnosis in, 51–59, 52t
　infection protection through breastfeeding in, 132–133
　low-cost prophylactic interventions in, 223
　risks of infant formula in, 277–279
Resource-rich countries. *See also* specific countries
　antimicrobial breastfeeding benefits in, 132–133
　breastfeeding recommendations in HIV infection in, 247–248
　HIV-1 nucleic acid assays in, 52t, 54t, 55, 56t, 57t
Respiratory infections, breastfeeding protective against, 132–133
Rhesus macaques, experimental SIV infection in, 90–92, 91f, 148–150, 189–190
Ritonavir, breast milk exposure to, 112t, 115–116
Roche HIV-1 assays, 52t, 54
Rubella vaccine virus, in breast milk, 27–28
Rwanda, mother-to-child HIV-1 transmission in, breast milk immunoglobulins and, 148
　timing and, 10–11
　pediatric mortality in, breastfeeding and, 132
　prophylactic antiretroviral studies in, 18, 19, 175

S
Safety, of infant formula in resource-limited countries, 251, 277–278
Saliva, hostility to HIV-1 of, 11
　infant immune responses in, 152
Secretory component, in breast milk, 124
Secretory immunoglobulin A, in breast milk, 121, 123–124, 142, 145
Secretory leukocyte protease inhibitor, in breast milk, 125
Selenium, in breast milk, 209
　deficiency of, HIV-1 transmission risk and, 9

Serological assays, in early diagnosis of infant HIV-1 infection, 52, 52t, 55,
SHIV, 197–198
SHIV89.6P vaccine study, in animal model, 98
SIMBA trial, of antiretroviral prophylaxis, 18, 175–176
Simian-human immunodeficiency virus (SHIV), in vaccine development, 189–190
Simian immunodeficiency virus (SIV), 150, 165, 188–189
　in HIV transmission models, 90–92, 96f, 139
　in vaccine models, 150, 165, 188–189
Six Week Extended Dose Nevirapine (SWEN) study, of prophylactic antiretrovirals, 18, 83–85, 175–176
Smoking, in China, 238
Social issues, in breastfeeding HIV-infected women, 44–47
　in early HIV epidemic, 222
　in infant formula use, 251
Sodium dodecyl sulfate, expressed breast milk treated with, 200
Solar-powered pasteurization, of expressed breast milk, 201
South Africa
　breastfeeding in, maternal morbidity and mortality and, 40, 42t
　　mixed feeding *versus*, 198
　family and community in infant feeding decisions in, 46
　formula feeding stigmatized in, 44–46
　heat treatment of breast milk in, 201–202
　prophylactic antiretrovirals in, 19
　rapid p24 antigen assay in, 58
　studies in, of exclusive breastfeeding, 280
　　of formula feeding, 277–279
　　of topical microbicide, 100
　　of vitamin A-HIV-1 transmission correlation, 206
Standard of care, in breastfeeding messages and policy, 294
STEP HIV vaccine study, 98
Stigma, formula feeding associated with, 44–46
Sub-Saharan Africa. *See also* specific countries
　coinfections in, 188
　HIV-1 infection in, epidemiology of, 4–5, 4f, 5f, 7, 8f, 51, 289
　　mother-to-child transmission of, 139, 289
　infant formula safety issues in, 227–228
　unavailability of infant formula in, 44
Subtypes, HIV-1, 74

T
Tanzania, maternal morbidity and mortality in, breastfeeding and, 42t
　mother-to-child HIV-1 subtype transmission in, 8
　studies in, of prophylactic antiretrovirals, 18
　　of vitamin/mineral-HIV-1 transmission correlation, 206–209

T cells
  breast milk, 121–122, 142, 163–166
    antigen-specific, 163
  in developing immune system, 186–188
Tenofovir, in animal models, 164–166
  as topical microbicide, 100
Thailand, HIV vaccine trial in, 98, 188–189
  prophylactic zidovudine trial in, 17
Timing
  of mother-to-child HIV-1 transmission, 10–11, 291
    virologic determinants of, 73
Tolerance, in neonatal immune system, 121
Toll-like receptors, in breast milk, 1
Total nucleic acid HIV-1 assays, 52t, 58
Transforming growth factor-ß, in breast milk, 128–131
Transmission, HIV-1. See HIV-1 transmission
Trial of Vitamins Study, 43
Tropism, in breast milk HIV-1 transmission, 73–75
Tuberculosis, in breastfeeding HIV-infected women, 43

U
Uganda, formula-associated infant mortality in, 279
  studies in, of HIV immune globulin, 190
    of prophylactic antiretrovirals, 17, 175–176
UMA trial, of prophylactic antiretrovirals, 178t–179t
UNAIDS, demographic data of, 4
  global report on mother-to-child HIV-1 transmission of, 185
United Arab Emirates, breastfeeding-associated HCV infection in, 34
United Kingdom, antimicrobial breastfeeding benefits in, 133
United States
  HIV-1 infection in, 261–268
    antiretroviral prophylaxis in, 263–264
    breastfeeding avoidance in, 261, 263–264
    breastfeeding benefits and, 266
    breastfeeding recommended in, 263
    breastfeeding reported in, 264–265
    early identification of, 265, 268
    epidemiology of, 262
    formula feeding in, 266
    human donor milk banks in, 266–267
    infant feeding recommendations in, 263–264
    legal issues in breastfeeding in, 267–268
    new maternal, 264
    racial and ethnic disparities in, 262
    risks factors for mother-to-child transmission of, 263
  postneonatal mortality in, breastfeeding and, 133
  prevention of mother-to-child HIV-1 transmission in, breastfeeding avoidance in, 83
  epidemiology of, 4
  vitamin A-HIV-1 transmission correlation study in, 206
United States Department of Health and Human Services, HIV-1 testing recommendations of, 58
United States Food and Drug Administration, rapid HIV-1 assays approved by, 57

V
Vaccines. See also Immunization
  in animal models, 97–99, 101–102
  HIV-1 (see HIV-1 vaccines)safety issues in, 191
Varicella immunization, contraindicated in breastfeeding, 28
Vaxgen vaccine trial, 188
Vertical transmission. See Mother-to-child transmission
Viral factors, in HIV-1 transmission through breastfeeding, 7–9
Viral load, assays of, 54–55, 54t
  in HIV-1-infected infants, 54, 186
  microbicides and heat in reduction of, 200–202
Viral variants, in postnatal HIV-1 mother-to-child transmission, 8–9
Viremia, in postnatal primary HIV-1 infection, 7
Virological assays, in early diagnosis of infant HIV-1 infection, 52, 52t, 54t, 56t, 57t
Virologic determinants, of breast milk HIV-1 transmission, 69–75
  cell-free versus cell-associated, 69–72
  mastitis and, 71–72
  overview of, 69, 75
  site of, 73
  timing of, 72–73
  tropism in, 73
Virulence, of experimental virus strains, 95–96
Viruses, in breast milk
  cytomegalovirus, 27–28, 30–32
    hepatitis B, 32–33
    hepatitis C, 33–34
    HTLV-1, 28–30
    non-HIV-1, 27–34
  HIV-1. See HIV-1
Vitamin A, in breast milk, 206–207
  deficiency of, 9
  supplements of, 206–207, 220t, 221f, 424–425,
Vitamin D, in breast milk, 208
  deficiency of, 9, 208
Vitamins B, C, and E, in breast milk, 207–208
  deficiencies of, 207
  supplements of, 207–208
Vitamin supplements, breastfeeding-associated HIV-1 transmission and, 44, 206–210
VSV-SIVgpe vaccine study, 99

W
WASH study, 220t, 221f, 224
Weaning, early
  benefits of, 119–200
    risks of, 9, 275–278, 281
  HIV-1 transmission risk in, 12
  maternal mortality and, 43
  PEPFAR guidelines for, 249
Weight loss, breastfeeding and postpartum maternal, 40–41
West Africa, breast milk HIV-2 in, 28
  cytomegalovirus infection in, 187

Index

World Health Organization (WHO), antiretroviral
  guidelines of, in pregnancy, 19, 176
  breastfeeding guidelines of, 9, 173, 252, 252t, 253b, 254
    antiretrovirals in, 53, 86, 240
    for HIV-1 exposed infants, 19–20, 173–174
  on early weaning, 200
  HIV-1 testing recommendations of, 58–59
  infant feeding guidelines of, 252, 252t, 253b, 254
    PEPFAR implementation of, 254–256
    shifts in, 257t, 274

Z

Zambia, breastfeeding in
  early weaning and, 198–200, 281–282
  maternal morbidity and mortality and, 40, 42t

Zambia Exclusive Breastfeeding
  Study (ZEBS), 174, 180, 199–200,
  252, 281–282, 282f, 293
Zidovudine, breast milk concentrations
  of, 84, 112–113, 112t
  prophylactic, in animal models, 100–101
  clinical trials of, 17–20, 112–113, 174–176,
    220t, 221f, 221t, 223f, 226–230
Zimbabwe, heat treatment of breast milk in, 202
  mother-to-child HIV-1 transmission in, 7
  mixed feeding *versus* breastfeeding in, 12, 198
  studies in, of drug resistance, 85
  of vitamin A-HIV-1 transmission correlation, 207
Zinc, in breast milk, 208–209
ZVITAMBO trial, of breastfeeding-related
  HIV-1 transmission, 12, 41, 198